T0249915

CRC Handbook of the Zoology of Amphistomes

Author

Otto Sey, Ph.D.
Department of Zoology
University of Agricultural Science
Keszthely, Hungary

CRC Press
Taylor & Francis Group
Boca Raton London New York

CRC Press is an imprint of the
Taylor & Francis Group, an **informa** business

PREFACE

The group of amphistomes is one of a well integrated taxa of the digenetic trematodes and together with other groups, Turbellaria and Cestoda, is placed to the phylum, Platyelminthes. They are parasitic in each grade of vertebrates and are distributed over every continent of the globe. Due to the scientific interest in them and to their practical importance, a vast amount of information has been accumulated since the activities of the pioneering workers — Diesing, Rudolphi, Fischoeder — of the amphistomology.

This handbook describes species of amphistomes presently known and attempts to summarize the literary data available on their systematics, ecology, and zoogeography.

Due to the close similarity of many species of amphistomes to one another, as well as to the correspondences in their strongly developed body musculature, the usage and reorganization of cryptic characters are indispensable. Data of the ontogenetic cycle of the species studied have been utilized to enlarge the spectra of the holomorphological character states. To eliminate the difficulties of the amphistome, diagnosis keys are constructed as simple as possible, for higher and lower taxa alike. Wherever possible every species is provided with a short diagnosis together with an illustration of the adult worm. A new classification has been reconstructed while keeping the phylogenetic relationships in the foreground. Within the zoogeographical analysis, the host-parasite co-evolution, distributional patterns, and characterization of faunas of the zoogeographical realms have been presented. On the basis of the holomorphological character states a possible evolutionary scenerio has been formulated.

Studying world literature, a great number of monographs, journals and papers have been assessed: the *Helminthological Abstracts*, the *Biological Abstracts*, and the *Referativnyĭ Zhurnal* have proved to be particularly useful. Obtaining literature data was terminated on December 31, 1988 with the hope that our list is reasonably complete. The names include 254 species, 101 genera, 10 families, 3 superfamilies, and 1 order. The bibliography contains 1100 references and there are 499 illustrations including 488 figures.

Among the many who directly helped me by sending wet specimens, museum samples, reprints on his or her papers, my particular thanks are due to J. C. Boray in Australia; J. A. Dinnik, S. Prudhoe, A. Jones, D. I. Gibson, R. A. Bray, L. F. Khalil in Great Britain; E. Kritscher in Austria; J. Pačenovský, B. Kotrlá in Czechoslovakia; N. K. Gupta, H. D. Srivastava, S. B. Srivastava, D. T. Soota, M. Hafeezullah in India; Nguyen Thi Lee in Vietnam; R. I. Sayed, M. S. Ahdel-Rahman in Egypt; A. Eslami in Iran; A. Kohn in Brazil; R. A. Lamothe in Mexico; F. Puylaert in Belgie; J.-L., Albaret, M. Graber, S. Deblock in France; E. Arru in Italy; Yu. Visnyakov, P. Samnaliev in Bulgaria; S. L. Eduardo in Philippines; Cl. Vaucher, Switzerland; D. R. Brooks, W. Threlfall in Canada; R. Oleröd in Sweden; I. V. Velitchko, V. P. Sharpilo in the Soviet Union; J. R. Lichtenfels, H. W. Stunkard in the United States. Appreciation is due, indirectly, to the contributors whose long list of names is printed at the end of the book.

Special thanks are due to Mrs. Edit Boskovits for the preparation of the histological sections, to Mr. László Veres for the artistic illustrations, and to Mrs. Dr. Zsuzsa E. Sipos for the conscientious typing of the manuscript.

Editor's Note: The author's use of the French word *follicule* (follicle) has been retained throughout this volume for the sake of uniformity.

THE AUTHOR

Otto Sey, Ph.D., was born in 1936 in Hungary. His university studies were pursued at the Faculty of Biological Sciences, József Attila University, Szeged where he obtained the degree of B.Sc. in 1963. In the same year he commenced teaching at the Zoological Department, Janus Pannonius University, Pécs. Since 1989 he has worked, as the head and Professor, at the Zoology Department, University of Agricultural Sciences, Keszthely, Hungary.

Dr. Sey has specialized in animal and veterinary parasitology. His post-graduate training was pursued at the Parasitology Department, University of Veterinary Sciences, Budapest. His research interest has been concentrated on parasitological problems, at first on helminth parasites and parasitic diseases transmitted by marine food fish and muskrats and later on helminth parasites of birds and mammals having fishing and hunting importance as well as helminthiases of domestic animals.

As a result of these studies, Dr. Sey obtained the degree of M.Sc. in 1964. In his thesis the helminth parasites of the rodent *(Ondatra zibethica)*, acclimatized in Hungary, were examined from the taxonomic and ecological point of view.

Between 1964 and 1985 he paid several visits to European parasitological institutes (e.g., Great Britain, France, Germany) and spent some time in helminthological institutes outside Europe: Egypt (1973), six months; India (1978), three months; Vietnam (1985), two months. These professional visits gave him a comprehensive experience of the veterinary problems of domestic animals living in tropical climates and allowed him insights into the control of diseases caused by these helminths. As the result of further post-graduate studies, Dr. Sey obtained the degree of Ph.D. in 1974. The subject of this thesis was the various aspects (causative agents and their life-cycle, intermediate snail hosts, epidemiology, pathology and treatment) of paramphistomiasis of domestic ruminants.

Since 1970, extensive experimental work has been done on various aspects (diagnosis, life-cycle, pathology etc.) of trematode diseases (mainly paramphistomiasis, fasciolopsiasis etc.). These examinations have been carried out in a broad collaboration with veterinary experts at home and abroad.

The thesis for fulfillment of the degree of D.Sc. has been submitted to the Scientific Qualifying Board of the Hungarian Academy of Sciences. In this thesis the phylogenetic systematics, developmental biology, and geographical distribution of all known amphistomes of vertebrates have been analyzed and critically summarized. In this respect it is a unique work on amphistomes. The results of this research work have been compiled in more than 90 papers, published in both Hungarian and foreign periodicals.

Dr. Sey is a member of the Hungarian Society of Parasitologists, the Hungarian Sociaty of Zoologists, and the Zoological Committee of the Hungarian Academy of Sciences. In addition to his mother tongue he writes and speaks English well and reads French, German, and Russian.

TABLE OF CONTENTS

1. MORPHOLOGY OF THE AMPHISTOMES

1.1. DESIGNATION AND SCOPE OF THE AMPHISTOMES

The taxon Amphistomida, in the present work, comprises a natural group of digenetic trematodes which has the rank of order in the modern classification of higher taxa of trematodes. Morphologically, its generalized in-group feature is the posteroterminally spaced ventral sucker. Biologically, it is one of the few groups of flukes whose representatives are parasitic in each higher taxon of vetebrate hosts. Biogeographically, it is broadly distributed in countries with tropical and temperate climates.

The scope of amphistomes has varied considerably since the 19th Century when the name *Amphistoma* was established by Rudolphi.[795] It referred to the erroneous observation that they have a mouth opening at both ends of the body. Rudolphi's category, however, also embraced those flukes which, in the present interpretation, belong to another natural taxon, Strigeida. The generic name, *Amphistomum* was given by Nitzsch[657] (cit. Skrjabin),[890] and out of Rudolphi's species those were allocated to this genus which were characterized by the presence of an oral opening at the anterior end of the body and by a ventral sucker at the posteroterminal one.

By the turn of the century, it became evident that the species assigned to the genus *Amphistomum* was a heterogeneous assembling of flukes like these. In an endeavor to establish the homogeneous structure of this taxon, the genus *Amphistomum* was divided into various genera (*Paramphistomum, Cladorchis,* etc.) by Fischoeder,[309,311] and they were transferred to the family Paramphistomidae, a category created by the same author. Subsequent investigators have accepted Fischoeder's nomenclature, and superfamiliar categories were constituted from this name (e.g., Paramphistomoidea Stiles et Goldberger, 1910; Paramphistomata Szidat, 1936; Paramphistomiformes Travassos, Freitas et Kohn, 1963; Brooks, O'Grady et Glen, 1985).

Recently, Odening,[668] who intended to designate the natural higher taxa of trematodes, proposed the usage of the name Amphistomida (Lühe, 1909) for the order, which embraces amphistomes of all kinds. Odening[664,668] pointed out that it is reasonable to retain the name of the superfamiliar category even if the rank of the original group has been altered by subsequent taxonomic operations. Moreover, because the taxon Amphistomida has nowadays become like a natural taxon, as it was in Lühe's time, and because the first superfamiliar category was nominated by him and, finally, because there is no nomenclatural regulation to apply one of the valid generic names to nominate higher taxon, Odening's notion is fully acceptable to the priority of the taxonomic name, Amphistomida (Lühe, 1909) Odening, 1974. ("Amphistomida" was formed of Lühe's Amphistomata with the aid of the "-ida" added to the root of the word.)

Since the beginning of this century, the scope of the amphistomes has been considerably enlarged not only by the discovery of newer forms, but also by the allocation of groups of flukes to amphistomes, the taxonomic position having been debated. The family Microscaphidiidae was assigned to amphistomes by Odhner,[672] Opistholebitidae and Opisthoporidae by Fukui,[334] Gyliauchenidae by Ozaki,[686] Cephaloporidae by Travassos,[986] Mesometridae and Notocotylidae by Szidat,[941] Metacetabulidae by Skrjabin,[890] Heronimidae by Crandall,[188] and Zonocotylidae by Padilha.[691]

Odhner[672] was of the opinion that the monostome microscaphidiid trematodes are amphistomes which lost their ventral sucker, and the organological resemblances (pharyngeal appendages, lymphatic system, etc.) make them similar to amphistomes. Groschaft et al.[377] clarified that the pharyngeal appendages are glandular in nature in microscaphidiids, and, thus, they are not homologous with that of amphistomes. Since that time, it has been pointed out that the lymphatic system is present not only in the taxa in question, but also in several others phylogenetically removed.[839]

Cable,[153] studying the development of *Opistholebes diodontis* in the final host, came to the conclusion that the amphistome plan of the taxa Opistholebitidae and probably of the Gyliauchenidae and Cephaloporidae should be considered to be secondary in both ontogeny and phylogeny.

Stunkard[926] reexamined MacCallum's species, *Paramphistomum aspidonectes*, for which Fukui[334] set up the genus *Opisthoporus*, and the family Opisthoporidae, respectively, and he stated that it should be transferred to the family of Pronocephalidae.

Odening[664,668] emphasized that the taxa Notocotylidae and Mesometridae assigned by Szidat[941] to amphistomes is not acceptable due to their monostome structure in adult and developmental stages (Notocotylidae).

On the Metacetabulidae taxon, it was recently elucidated by Bilquees[95] that the supposed rudimental ventral sucker, which was the basis of putting amphistomes in this taxon,[890] is not a sucker at all, but a secondary, excretory bladder, and, thus, these flukes have no amphistome structure.

Besides the above-mentioned families, the species *Khalilloossia ali-ibrahimi* Hilmy, 1949 (Cephaloporidae); *Choanomyzus tasmaniae* Manter et Crowcroft, 1950 (Opistholebetidae); and *Podocnemitrema papillosus* Alho et Vicente, 1964 (Microscaphidiidae) have also been allocated to amphistomes. Now these species belong to the families indicated in parentheses whose taxonomic position and nonamphistomal nature were discussed above.

At the same time, it was convincingly demonstrated by Cable[153] that the monostome taxon Heronimidae (shared traits with amphistomes are the pattern of epidermal cells of the miracidia and the position of the ventral sucker of the cercariae) and Zonocotylidae demonstrated by Padilha[691] (shared feature with amphistomes is the pattern of epithelial cells of miracidium) should be considered as amphistomes.

From the controversial views presented above on the scope of amphistomes and from the modern demand that the outline of a group of flukes be based on the totality of characteristics, it follows that the size of the group can be formularized by their holomorphological features at hand. In this context, amphistomes are characterized by the following traits.

Adults, of amphistome type or monostome evolving from the amphistome plan. Usually conical, strongly muscular of body. Hermaphroditic and with a well-developed posteroterminal acetabulum. Oral opening surrounded with muscular pharynx; esophagus muscular with sphincter (muscular thickening or bulb) or without it. Ceca usually straight, rarely undulating. Lymphatic system present; excretory system stenostome; excretory bladder pear-shaped; two main descending trunks opening to bladder. Parasitic in each group of vertebrates.

Adolescariae usually encyst freely; cercariae large, amphistome type with unforked tail. Body with or without dense pigmentation or light pigmentation. A pair of eyespots and many cystogenous cells present (except for *Heronimus*). Protonephridial system well developed, excretory bladder nonepithelial; newly born cercariae poorly developed. Rediae and sporocysts rarely absent. Miracidia amphistomum type with 6:8:4:2 and 6:6:4:2 pattern of epidermal cells, without eyespot (except for *Heronimus*). Eggs: large in size, egg shell thin (freshwater developed) or thick (marine developed); viviparity, ovoviviparity, ovoparity present. Life cycle dixenous type; intermediate host usually freshwater pulmunate snails.

1.2. TECHNIQUES FOR STUDY OF ADULT AMPHISTOMES

The specific diagnosis of amphistomes requires the realization of more and more character states or, ideally, the full scope of these which are manifested by adult specimens. These character complexes may be gross morphological (shape and size of body as well as internal organs), histomorphological (structure of the muscular organs), and cytogenetic (karyotype architecture). Whatever features have been taken into consideration, it is important to follow the proper methods which result in preparations that are equipped with the traits required. The methods described below concentrate on four types of preparations: (1) whole mounts, (2) scanning electron microscopy (SEM), (3) histological sections, and (4) cytological preparations. Collection, relaxation, if any, and fixation are common in each case (except for karyology).

Collection

Amphistomes are obtainable either in the lower part of the alimentary tract or in the stomach, mainly in the rumen. They can be kept until the time of relaxation in physiological saline. The prefixative watery treatment (which has often been used) is not recommended because such soaked specimens showed considerable changes in surface topography, shape of body, and arrangement of the internal organs. It is advisable to put the specimens in fixative solution as soon as possible after removal from the digestive tube.

Relaxation

When researchers have the opportunity to relax specimens before fixation, it is advisable to do so. Improved relaxation is the best way to preserve the shape of the body and topography of the internal organs. The proper relaxation can be obtained by killing the flukes in hot water.

The specimens are placed in a 50-ml glass tube in physiological saline (enough solution to overlap all the flukes) and are shaken several times vigorously. They are poured into a 500-ml beaker, and the water is heated to 70 to 80°C. Then, they are gradually poured into the beaker until it is full. After some minutes the parasites die in a completely relaxed state. Fixation should be done shortly after this.

Fixation

When the flukes have settled to the bottom, the supernatant is discarded and 70% alcohol is added. The quantity of the fluid should be several times higher than that of the flukes. Lack of relaxation of worms can be put immediately in fixative solution. Although formalin has been widely used as fixative solution, its use is not advisable in the case of amphistomes. The samples fixed in formalin become not only harder, but residue is also formed around the eggshells, which prevents the penetration of solutions to be applied later. Fixed specimens can be stored in 70% alcohol for a long time. Well-fixed test material is the basic prerequisite for every kind of later preparational operation.

Processing Specimens for Whole Mounts

The fixed specimens are washed in distilled water for 1 to 2 h and are stained in alcoholic borax carmin. The flukes are first overstained, then differentiated in acid alcohol, which allows the internal organs to retain stain but the body wall to lose most of its stain.

By this method, the structure of the digestive tract and of the reproductive organs is clearly visible. Such organ systems as excretory and lymphatic systems can be examined by Tandon's[950,952] methods.

The main steps for study of the excretory system are as follows: pressed specimens are fixed in a concentrated solution in acidic corrosive sublimate for about 12 h. Then, they are washed, first in tap water, later in distilled water, for a period from 1 h to several hours until the corrosive sublimate is properly removed from the external surface. These specimens then are treated with 1% potassium hydroxide for 1 to 24 h and are washed again all through. Excretory ducts are visible as dark tubes and are proper for study of the components in temporary glycerine preparations.

To study the lymphatic system, the pressed specimens are fixed in a concentrated solution of acidic corrosive sublimate for 10 to 12 h. Then they are washed, first in distilled, than in tap water for 1 h to several hours. After washing, they are treated with 0.5% solution of sodium thiosulfate and are washed again all through. The lymphatic system appears to be yellow, and specimens are studied in temporary glycerine preparations.

After staining and differentiation, dehydration follows to make these preparations permanent. Thorough dehydration is essential and takes place in ascending grades of alcohol (80, 90, 100%) from 1 h to several hours depending on the size of worms. After this, specimens are put in cleaning solution (clove oil, beach wood oil, xylene) for several hours.

In the case of relatively flattened amphistomes, specimens are closed under cover glass in balsam. When the specimens are conical and fleshy (the majority of amphistomes belongs to this group), another method should be followed. Such specimens are mounted on slides with ventral, dorsal, or lateral sides upward, and they are covered with cover glass supported by two glass slabs of appropriate thickness on the glass. The space between the glass and cover glass is filled with balsam.

Processing Specimens for Median Sagittal Section

Since Näsmark,[647] the taxonomic value of the structure of the muscular organs of amphistomes (pharynx, terminal genitalium, acetabulum) has been recognized, and a preparation of histological sections has become one of the broadly applied methods used in amphistome diagnosis.

The specimens fixed in 70% alcohol are dehydrated and embedded in wax. Flukes are placed on the block with ventral side upward. The ventral side of the embedded worm faces the edge of the microtome knife to avoid destruction of the terminal genitalium. During sectioning it is necessary to control the right plane of cutting. The right plane of cutting is achieved when it is median sagittal because this plane shows the traits that have taxonomic importance. The sections are cut 8 to 12 μm thick using a sliding microtome, and they are arranged on glass slides in the proper sequence, stained with hematoxylin and eosin, dehydrated, and mounted in balsam.

Processing Specimens for Scanning Electron Microscopy

Specimens which are seemingly in good condition are only chosen for electron microscopy. They are carefully examined under stereo microscopy and then embedded, sectioned, stained, and identified. Only those that show identity with these specimens are chosen for further examination from the lots of the test material. They are washed several times in distilled water, then dehydrated in ascending grades of alcohol, dried to the critical point, mounted on metal stubs, coated with gold-palladium, and examined with a scanning electron microscope.

Processing Preparations for Karyological Studies

Cells for chromosomal examinations can be obtained from squashes and sections of the testes or from larval (rediae, cercariae) embryos. Freshly dissected testes are squashed in aceto-orcein (2%) after which they are dehydrated and mounted. For sectioning, the materials are fixed in Carnoy's, Flemming's, or Bouin's solution. The sections are stained with Heidenhain's hemotoxylin or Feulgen's stain.

First or second generation of rediae are obtained from the snails 8 to 10 weeks after infestation and fixed in Carnoy's solution. Temporary aceto-orcein squash preparations are made within 1 week after fixation. Meiotic stages of chromosomes are recorded in photographs.

1.3. GENERAL APPEARANCE OF THE AMPHISTOMES

1.3.1. BODY SHAPE AND SIZE

Amphistomes are one of the groups of digenetic trematodes which have, in major part, a conical shape of body *(Calicophoron, Paramphistomum)*. Several other species have more or less flattened forms with fleshy bodies *(Gastrodiscus, Watsonius)*, and there are a few with a strongly flattened appearance *(Platyamphistoma, Platycladorchis)*. Body shape can strongly be affected by the method used for fixation. When fixation is, however, satisfactory, the shape of body may be characteristic of certain genera (e.g., *Explanatum*) or families (e.g., Gastrodiscidae), but caution is reasonable. Body size is also affected by several factors such as age of specimens, type of fixation used, crowding effect, type of definitive hosts, etc. Dinnik and Dinnik[229] observed that specimens of *Paramphistomum* (= *Calicophoron*) *microbothrium* obtained from younger infestations were smaller than those from older ones. Dinnik and Dinnik[229] also showed that specimens of *C. microbothrium* recovered from experimentally infected goats were smaller than those from cattle. Horak[422] demonstrated that *C. microbothrium* develops larger in cattle than in sheep or goats. A crowding effect was demonstrated by Willey[1059] at *Zygocotyle lunata* and by Tandon[954] at *Fischoederius elongatus* and *Gastrothylax crumenifer*. However, the body size, even if it is of a varying nature, is one of the characteristic states.

1.3.2. BODY SURFACE

The body of an amphistome is covered with smooth aspinose tegument. In the living state, it has a glassy luster; in preserved specimens there are transverse wrinkles on the surface as a result of the fixation. The structure of the tegument of amphistomes was studied by Bogitsh,[103] who found that in *Megalodiscus temperatus* it is syncytial in nature and is essentially the same as that described for other trematodes. It consists of two regions: (1) at the surface, a lager of cytoplasm lines on a basal lamina and (2) beneath the basal lamina and muscle layers there are nucleated cell bodies connected with the surface layer by narrow, tubular extensions.

Some species have a transverse ridge anteriorly *(Kalitrema)* and caudal appendages *(Brumptia)* or posterolateral projections *(Alassostomoides)*. Two types of papillae are found along the body surface. One of them, the so-called tegumental papilla, is common and as found in every species examined (full details later). The other type of papilla is longer, mammiform, and found on the ventral surface of the species *Gastrodiscus, Homalogaster, Gastrodiscoides*, and *Choerocotyloides, Watsonius*. Papillae of the first type are retractile and contain muscle fibers and lymph canals with minute ducts open to the surface. Papillae of *Homalogaster* are not retractile, contain lymph canal and fine muscle fibers, and are without ducts. In the rest of the species, ventral papillae are not so prominent; rather, they have papillar ridges the whole length of *(Choerocotyloides)* or on certain parts of the ventral surface *(Gastrodiscoides, Watsonius)*.[453]

Along the body surface there are several apertures of different organs. The oral opening is usually terminal or subterminal, surrounded with papillae of various type and size. The genital pore is ventral, usually in the anterior third of body, situated on a muscular elevation or shallow depression of the body surface, surrounded with papillae of various types. In certain species (Gastrothylacidae), it opens into the spacious ventral pouch. The aperture of the voluminous ventral sucker (acetabulum) may be ventral, ventroterminal, or terminal. The size of acetabulum may be characteristic in some genera *(Gigantoctyle, Explanatum)*, and due to its strongly muscular nature, it can be regarded as a relatively stable morphological characteristic. The much more valuable taxonomic characteristics are, however, exhibited by the structure of the muscular series of the acetabulum which is discussed below. Along the dorsal surface in the median line, there are two minute pores: the openings of the excretory

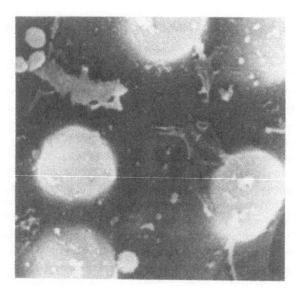

FIGURE 1.   Dome to conical, nonciliated papilla *(Dadaytrema oxy-cephalum)*. (Magnification × 1290.)

system and Laurer's canal. The position of the openings relative to each other has taxonomic value. In amphistomes, except for the subfamily Paramphistominae, Laurer's canal opens before the pore of the excretory system; hence, the crossing of these two ducts are not seen on a median sagittal section. In species of Paramphistominae, the excretory pore is situated in front of Laurer's canal; thus, the crossing of these vessels exists.

1.3.3. TEGUMENTAL PAPILLAE

These papillae, when they are present, are distributed in five regions of the body surface: around the oral, genital, and acetabular openings as well as around the orifice of the ventral pouch. They are usually arranged either in several rows or in smaller or bigger groups, loosely or very closely packed. Of the papillae, those around the oral opening and the ones on the forebody appear to be the most variable.

The presence of these papillae has been observed by several previous investigators of amphistomology. The limitations of the methodology of examination prevented the investigators from revealing the exact nature of these formations. Hence, their bearing has never been seriously considered. Of the amphistome trematodes, several species have been examined by SEM by Nollen and Nadakanukaren,[658] Tandon and Maitra,[955-958] Eduardo,[266-268,271-276,278] and Sey.[851] The papillae have been characterized and typified by Eduardo[271] and Sey[851]

Types of tegumental papillae and their occurrence among species examined:

1.   Dome to conical, nonciliated papilla (Figure 1)
      Basidiodiscus ectorchis, Carmyerius cruciformis, C. gregarius, C. mancupatus, Calicophoron calicophorum, C. clavula, C. daubneyi, C. microbothrium, C. microbothrioides, C. papillosum, C. phillerouxi, C. raja, C. sukari, C. sukumum, Cotylophorom cotylophorum, C. panamense, Dadaytrema oxycephalum, Diplodiscus mehrai, Gastrothylax crumenifer, F. elongatus, Gastrodiscoides hominis, Gigantocotyle symmeri, Homalogaster paloniae, Megalodiscus temperatus (in part), *Orthocoelium dawesi, O. dinniki, O. giganthopharynx, O. indonesiense, O. scoliocoelium, Paramphistomum epiclitum, P. gotoi, P. hiberniae, P. ichikawai, P. leydeni, P. liorchis, Stephanopharynx compactus, Stunkardia dilymphosa*

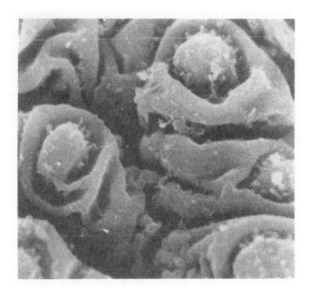

FIGURE 2. Dome to conical, ciliated papilla *(Carmyerius schoute-deni)* (Magnification × 1063.)

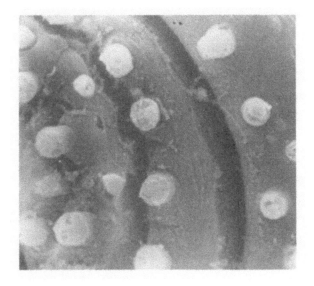

FIGURE 3. Short and stumpy papilla sitting on a tegumental elevation *(Hawkesius hawkesi).* (Magnification × 80.)

2. Dome to conical, ciliated papilla (Figure 2)
   Bilatorchis papillogenitalis, Carmyerius schoutedeni, Cotylophoron barreilliense, Glyptamphistoma paradoxum, Leiperocotyle gretillati, Orthocoelium tamilense (= O. dicranocoelium), Megalodiscus temperatus (in part)
3. Short and stumpy papilla covered with hairlike formations
   Leiperocotyle okapi, Orthocoelium dicranocoelium, Paramphistomum gracile, Stunkardia dylimphosa
4. Short and stumpy papilla sitting on a tegumental elevation (Figure 3)
   Hawkesius hawkesi, Platyamphistoma polycladiforme

FIGURE 4.   Long papilla with bulblike branches *(Homalogaster pa-loniae)*. (Magnification × 225.)

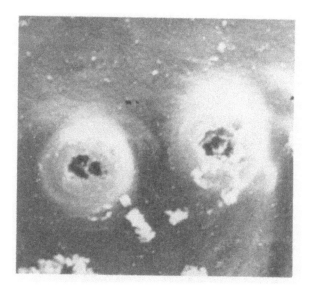

FIGURE 5.   Crater-form papilla *(Dadaytrema oxycephalum)*. (Magnification × 1400.)

5.   Long nonbranching and nonciliated papilla
   Carmyerius gregarius, C. mancupatus, Fischoederius cobboldi, Hawkesius hawkesi
6.   Long papilla with simple uniciliated branches
   Balanorchis anastrophus
7.   Long papilla with bulblike branches (Figure 4)
   Homalogaster paloniae, Nilocotyle sp., *Platyamphistoma polycladiforme, Sellsitrema sellsi*
8.   Crater-form papilla (Figure 5)
   Dadaytrema oxycephalum

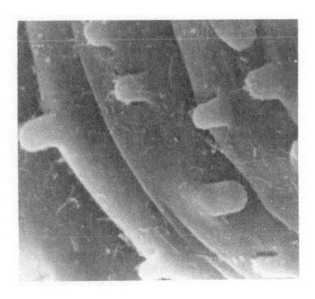

FIGURE 6.   Columnlike papilla *(Neocladorchis multilobularis)*.
(Magnification × 712.)

9.   Columnlike papilla (Figure 6)
      Neocladorchis multilobularis
10.   Papilla probably absent
      Cotylophoron macrosphinctris, Gigantocotyle duplicitestorum, G. gigantocotyle, Paramphistomum cervi, P. cephalophi, Watsonius noci

Before an appraisement of the taxonomic value of the type of tegumental papillae, the sources of errors should be pointed out which can impair such examinations: (1) misleading identification can easily occur, especially among species closely related, provided that the gross morphology, definitive host(s), etc. do not predestine the specific identity because the very specimens subjected to SEM examinations are not fit for further identification; (2) as the tegumental papillae are usually minute elements, they are easily destroyed due to the prefixative treatment, condition of storage, age of samples, etc. Raising this question becomes reasonable when tegumental papillae are not found: is such a state a primary or a secondary condition? Examinations made in this field suggest that the presence of these papillae on the body surface can be said to be a general phenomenon. The tegumental papillae, however, can be regarded as another set of the morphological characteristics and appear to have some taxonomic value combined with other traits.

1.4. ORGAN SYSTEMS

The internal organ systems of amphistomes are composed of digestive, reproductive, excretory, lymphatic, and nervous systems. Basically, they are similar to those of other digeneans with some pecularities characteristic of amphistomes. Besides these organ systems, there is a special organ of the species of the family Gastrothylacidae, called a ventral pouch. It is a large hole along the ventral side of these species. It begins with a transverse orifice close behind the oral opening, and it may be shorter *(Velasquezotrema)* or longer (Gastrothylacidae). In the latter case it terminates close in front of the testes. In cross section, it shows various forms which are more or less constant and are characteristic of species when fixed by suitable methods. The cross section may be either triangular, with the apex dorsal

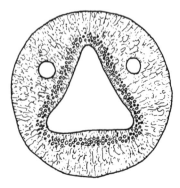 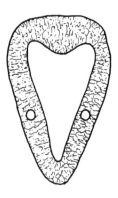 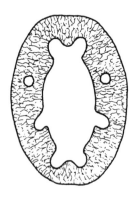

FIGURE 7. Cross section of ventral pouch *(Gastrothylax crumenifer).*

FIGURE 8. Cross section of ventral pouch *(Fischoederius elongatus).*

FIGURE 9. Cross section of ventral pouch *(Carmyerius exoporus).*

(Figure 7) or ventral (Figure 8), or hexangular (Figure 9), etc. The internal surface of the ventral pouch is light brown in color, showing resemblance to hemoglobin.[519] The actual function of it, however, is not yet known.

1.4.1. DIGESTIVE SYSTEM

It shows uniformity and relative simplicity throughout the group. Three different main parts are always distinguished: (1) pharynx with primary pharyngeal sacs (Figure 10) or with pharyngeal bulb and secondary pharyngeal sacs (Figure 11) or without appendages, (2) esophagus with or without muscular thickening, and (3) blind ceca. The muscular apparatus situated around the oral opening is a pharynx and not a sucker in amphistomes. It is homologous with the pharynx and not with the oral sucker, as was convincingly demonstrated by Näsmark.[647] It comprises two, morphologically well-defined parts: the body of the pharynx itself and the various appendages characteristic of all amphistomes except for the species of Gastrothylacidae and Paramphistomidae. The structure of the pharyngeal appendages is similar to that of the pharynx, but in some species *(Helostomatis)* the primary pharyngeal sacs can divide into an anterior muscular part and a less muscular, posterior one. The body of the pharynx has a varied structure of its own muscular elements which will be discussed in full in Section 1.5.1.

The esophagus is usually a shorter or a longer tube connecting the pharynx with the gut ceca. Typically, it consists of two layers, an outer longitudinal muscle and an inner circular muscle, and its inner surface is covered with integument similar to that of the pharynx (Figure 12). The longitudinal muscle fibers are usually weak and few. The circular muscle layer, especially at the posterior part of the esophagus, is more or less thickened by concentration of fibers in several rows. Depending on the development of this layer, it can be divided into three groups: sphincter, (Figure 13), muscular thickening (Figure 14), and bulb (Figure 15). The esophagus can be regarded functionally as part of the pharyngoesophageal complex capable of realizing the complicated mechanism of taking on and forwarding the nutritive materials. This seems to be supported by the correlations which are recognizable, for the most part, among the development of the musculature of the pharynx, the structure and size of appendages, and the development of the esophageal musculature. The tegumental layer usually lines the esophagus in its full length, but in some species *(Fischoederius cobboldi,*[334] *Calicophoron bothriophoron,*[368] *C. sukari,*[271] etc.) only the anterior part is covered with integument. The ceca are usually situated in the lateral fields (in the majority of species) or dorsal (e.g., *Fischoederius*), straight on, or wavy with characteristic positions of the end parts (e.g., *Calicophoron microbothrium, C. daubneyi*). Their length is also variable: it may be short, terminating at about the middle part of body (e.g., *Colocladorchis,*

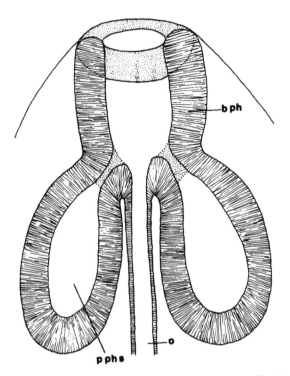

FIGURE 10. Pharynx with primary pharyngeal sacs. (By permission of the *Parasitol. Hung.*, 1987, Budapest.)

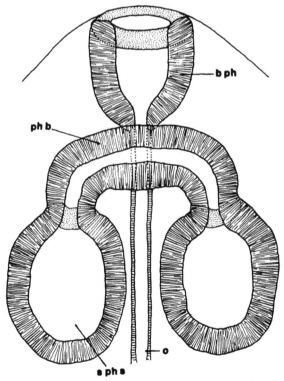

FIGURE 11. Pharynx with pharyngeal bulb and secondary pharyngeal sacs. (By permission of the *Parasitol. Hung.*, 1987, Budapest.)

FIGURE 12.   Esophagus
with typical structure
(outer longitudinal and inner circular muscle lay-
ers).

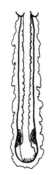

FIGURE 13.   Esophagus
with sphincter.

FIGURE 14.   Esophagus
with muscular thickening.

FIGURE 15.   Esophagus
with muscular bulb.

Catadiscus) or long, extending to the acetabular zone (in the majority of species). Rarely, can the ceca be divided into an anterior swollen and a posterior slender part (e.g., *Pfenderius heterocaeca*). Cross section shows a composition of ciliated epithelium, circular and longitudinal muscle fibers.

1.4.2. REPRODUCTIVE SYSTEM
1.4.2.1. Male Genital Organs
They are composed of testes, vasa efferentia, vas deferens, and ductus ejaculatorius.

Testes — The number of testes is usually two, except for the species in Diplodiscidae, where a single testis is present. The testes are voluminous, compact, globular bodies, lobed or branched. They are situated mostly intercecally occasionally extracecally, in part (*Sandonia, Basidiodiscus*) or completely (*Kalitrema, Osteochilotrema*), in the anterior, middle, or posterior third of the body. The shape and size are variable but more or less constant and can be used for diagnosis. The arrangement of testes is also characteristic of species, and it may be tandem, oblique tandem, diagonal, horizontal, and dorsoventraloblique. The position of the testes relative to the ovary is often used as a diagnostic feature. The ovary is either post- or intertesticular, except for *Balanorchis* where it is pretesticular.

Vasa efferentia — These are usually two slender tubes of varying lengths and are lined by a ciliated epithelium similar to that of the testes. Daday[197] reported its absence in *Pseudodiplodiscus cornu*, and Looss[557] found in *Diplodiscus subclavatus* and Ruiz[797] in *Catadiscus freitaslenti* that there are two vasa efferentia in spite of the single testis.

Vas deferens — After junction of the two vasa efferentia, the single tube is called vas deferens. It is usually long and varying in width, composed of vesicula seminalis, pars

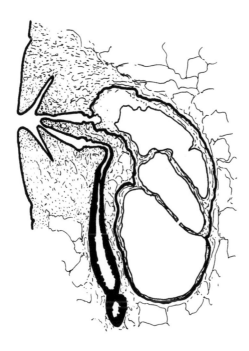

FIGURE 16. Terminal genitalium with cirrus sac.

musculosa, pars prostatica, and ductus ejaculatorius. Where the cirrus sac is present, there
are vesicula seminalis externa and vesicula seminalis interna, the latter is enclosed in the
cirrus sac (Figure 16). The latter organ is a muscular sac which surrounds the distal part of
the deferens. It is characteristic of many species of amphistomes, parasitic mainly in lower
vertebrates.

In some of the species, the distal parts of both the male and the female genital organs
are enclosed in a joint sac, which was termed a hermaphroditic pouch by Fukui[334] (Figure
17).

In *Gastrodiscus,* the cirrus sac is not a true one because it encloses pars musculosa and
the dilated chamber (Figure 18). The dilated chamber is a part of the vas deferens between
the pars musculosa and ejaculatory duct in certain amphistomes (Figure 19). It corresponds
to the ciliated chamber[1068] and sperm canal dilation of Prudhoe et al.[758] Where it was studied,
it proved to be ciliated in some species *(Gastrodiscoides hominis, Homalogaster paloniae,
Choerocotyloides onotragi, Watsonius deschiensi)* but not ciliated in others *(Gastrodiscus
aegyptiacus, Pseudodiscus collinsi).*

In species having neither a cirrus sac nor hermaphroditic pouch, the vesicula seminalis
is usually a very much coiled, thin-walled tube full of spermatozoa (Figure 20). Pars musculosa is a shorter *(Paramphistomum)* or a longer *(Calicophoron)* part of the vas deferens.
It is characterized by the well developed circular musculature which, however, varies with
species and age (Figure 20). Outerly, it is covered with longitudinal muscle fibers. The pars
musculosa is continued anteriorly into the pars prostatica. It varies in length and is usually
wavy, composed of circular and longitudinal muscle layers, around which there is a closely
packed layer of prostatic cells. The ductus ejaculatorius is usually a short tube which opens
into the terminal genitalium.

According to the above description there are five types of vas deferens:

1.   Vas deferens composed of vesicula seminalis externa, vesicula seminalis interna, and
pars prostatica. The latter two parts are enclosed in a cirrus sac (characteristic of many
species parasitic in lower vertebrates (Figure 16).

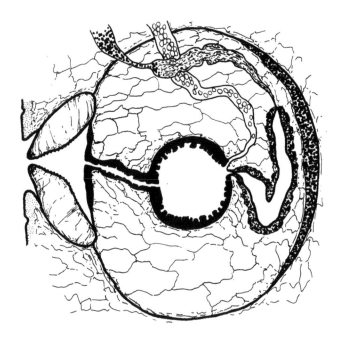

FIGURE 17.   Terminal genitalium with hermaphroditic pouch.

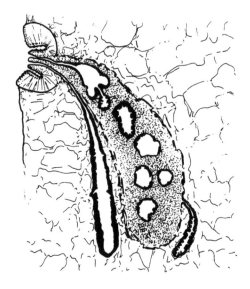

FIGURE 18.   Terminal genitalium with modified cirrus sac.

2.   Vas deferens consisting of vesicula seminalis interna and pars prostatica. These two
       parts are enclosed in a large pouch together with the metraterm *(Balanorchis, Brumptia,
       Choerocotyle, Hawkesius)* (Figure 17).
3.   In *Gastrodiscus* the cirrus sac is composed of pars prostatica and the dilated chamber
       (Figure 18).
4.   Vas deferens composed of vesicula seminalis, pars musculosa, dilated chamber, and
       ejaculatory duct *(Gastrodiscus, Gastrodiscoides, Choerocotyloides, Homalogaster,
       Pseudodiscus, Watsonius)* (Figure 19).
5.   Vas deferens composed of vesicula seminalis, pars musculosa, and pars prostatica
       without a cirrus sac (characteristic of many species parasitic in mammals) (Figure 20).

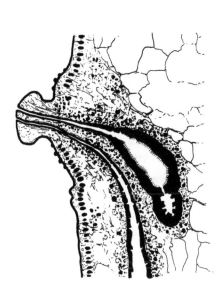 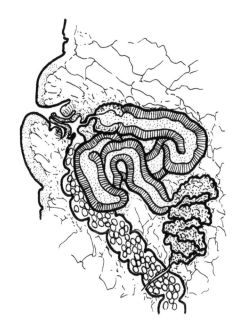

FIGURE 19.   Terminal genitalium with dilated chamber.

FIGURE 20.   Terminal genitalium without cirrus sac.

1.4.2.2. Female Genital Organs

This organ system is composed of the ovary, the oviduct, the Laurer's canal, the vitelline follicules and their ducts, Mehlis' gland, the ootype, the uterus, and the metraterm.

The ovary is a compact, spherical, occasionally lobed organ that is usually much smaller than the testis. In some species, the ovary is either equal with the testes in size *(Brevicaecum, Diplodiscus nigromaculati)* or bigger than that of the testes *(Cotylophoron barreilliense)*. Its position may be posttesticular (in the majority of species), intertesticular *(Colocladorchis, Macropotrema,* etc.), or pretesticular *(Balanorchis)*. It usually lies intercecally in the median line, in various parts of body.

The oviduct is a short tube. At the junction of the ovary and the oviduct, there is a muscular chamber (ovicapt) which probably controls the release of ova.

Laurer's canals leads from the oviduct to the dorsal surface, and its pore is before or behind the excretory opening.

The vitellaria is various in appearance and in its development. It is either follicular (greatest number of species) or compact *(Zonocotyle, Catadiscus dolichocotyle,* etc.). Vitelline follicules may be variable in both number and size. Typically, they are situated along the lateral sides of the body or along the ceca occupying longer or shorter distances of them. Rarely, they can be confluent posteriorly *(Dermatemytrema, Australotrema)* or anteriorly *(Balanorchis, Paramphistomum ichikawai)*, or both *(Cotylophoron panamense)*. A short part of the oviduct is surrounded by a group of glandular cells and is collectively known as Mehlis' gland. Its probable function is to activate the spermatozoa before fertilization.

The ootype is a modified part of the oviduct where fertilization and encapsulation of the ova occur. In the amphistome, there is no receptaculum seminis. Spermatozoa may accumulate in the proximal part of the uterus.

The position and the running of the uterus are very variable. The uterine coils are usually intercecal; rarely, partially extracecal (Travassosiniinae, *Sandonia,* etc.); posterior to the testes *(Dadaytrema, Schizamphistomum)*; anterior to the testes *(Nematophila grandis, Fischoederius cobboldi)*; or situated dorsal to the testes *(Dadayius marenzelleri, Paraibatrema*

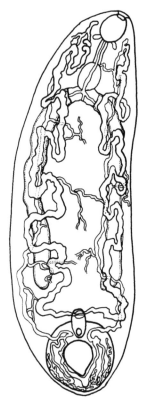

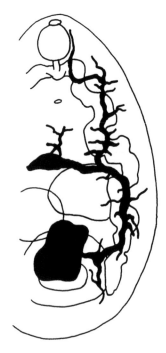

FIGURE 21.   Excretory system of *Stunkardia stunkardi*. (After R. S. Tandon, 1971.)

FIGURE 22.   Excretory system with two excretory bladders. (After T. Fukui, 1929.)

inesperatum). At the level of the genital pore, it turns to the ventral surface *(Calicophoron, Paramphistomum).* The metraterm is the muscular end part of the female genital organ. The metraterm and the ejaculatory duct often form a joint vessel called ductus hermaphroditicus which opens into the terminal genitalium.

The eggs of amphistomes are relatively large, operculated, and with a knob on the antioperculated end where they were studied.

1.4.3. EXCRETORY SYSTEM

This system is a stenostome in type, consisting of solenocytes and capillaries, main canals, connecting vesicle, connecting canal, excretory bladder, excretory canal, and excretory pore. The solenocytes are cellular units from which long cilia project into the blind termination of the system. They open to capillaries which continue to the main canals. There are two main canals, and they run on either side of the body near the alimentary tracts (Figure 21). Fukui[334] distinguished three types of main canals:

1.   They run along the ceca and empty into the vesicle. Between the main canals there are several transverse canals at various points. The connecting vesicle is present, situated on the dorsal side at the level of the anterior edge of the anterior testis. From either side the connecting canals proceed to the main canals *(Gastrothylax, Explanatum,* etc.) (Figure 22).

2.   They are similar to the preceding ones but without a connecting vesicle or connecting canal *(Homalogaster, Protocladorchis pangasii,* etc.) (Figure 23).

3.   They are wavy along the lateral sides and around the ceca they form a fairly complicated pattern *(Pfenderius, Schizamphistomum,* etc) (Figure 24).

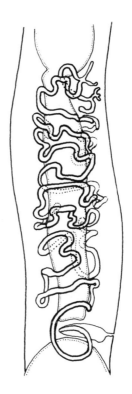

FIGURE 23. Excretory system with one excretory bladder. (After T. Fukui, 1929.)

FIGURE 24. Excretory system of *Schizamphistomum* spp. with strongly convoluted lateral tubes. (Courtesy of D. Blair.)

The excretory bladder is nonepithelial, various in form and size, situated in the posterior part of the body, usually among the posterior testis, acetabulum, and body surface proper (in most species), or, rarely, is intertesticular *(Solenorchis)*. The excretory canal is a fine tube which leads from the bladder to the dorsal surface. It is lined by integument continuous with that of the body. The excretory pore is usually at the level of the testes or, rarely, in the vicinity of the pharynx. The relative position of the excretory pore and that of Laurer's canal is of higher taxonomic value. These pores open separately (in most species), rarely with a joint orifice *(Microrchis megacotyle, Carmyerius wenyoni)*.

1.4.4. LYMPHATIC SYSTEM

In the amphistomes, besides the excretory system, there is another tubular system termed the lymphatic system.[560] It is embedded in the body paranchyma, and the whole body is covered with a network of its fine branches (Figure 25). Ultrastructural[919] and histochemical[564] investigations or both[251,876] indicate that its most probable function is to transport nutritive materials to the tissues.

Taking the number of the main canals into account, there seem to be four types in this system:

1. The main canal is two in number *(Cleptodiscus, Dermatemytrema, Paramphistomum,* etc.).
2. The main canal is four in number *(Nemathophila, Stunkardia, Solenorchis,* etc.).
3. The main canal is six in number *(Neocladorchis, Taxorchis,* etc.).
4. The main canal is variable (2 to 6) in number *(Zygocotyle, Orientodiscus)*.

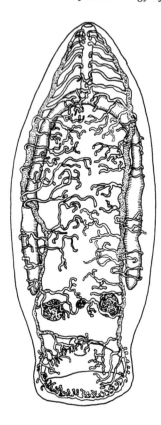

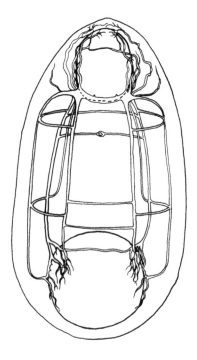

FIGURE 25. Lymphatic system of *Gastrothylax crumenifer*. (After R. S. Tandon, 1960.)

FIGURE 26. Nervous system of *Calicophoron calicophorum*. (Courtesy of S. K. Lee-Poon.)

When the presence of the lymphatic system was discovered in amphistomes by Looss,[560] he attributed an important role to it in the systematics of these flukes. Later, this system was recovered in several other digenean groups (Cyclocoelidae, Apocreadiidae, etc.) and in the monogenetic family Sphyranuridae. Until now, the structure of the lymphatic system has been described in more than 20 species, and there is no correlation between the phylogenetic relationships and the number of the main canals. It seems to be reasonable to regard the features of the lymphatic system as one of the characteristic states of the species and/or genera.

1.4.5. NERVOUS SYSTEM

Apart from the early accounts by Looss,[558] Otto,[682] and Brandes,[115] the structure of the nervous system has been studied in a limited number of species: *Opisthodiscus diplodiscoides;*[1] *Hawkesius hawkesi;*[334] *Fischoederius elongatus, Calicophoron calicophorum, Parorientodiscus magnus* (= *Stunkardia dilymphosa*);[526] *Fischoederius cobboldi;*[390] *Ceylonocotyle* (= *Orthocoelium*) *scoliocoelium;*[793] *Fischoederius cobboldi* and *Olveria indica.*[615,616]

These examinations show close similarity in the structure of this organ system. It consists of one pair of cerebral ganglia which are connected by a dorsal transverse commissure. This nerve mass, which constitutes the brain, lies dorsal to the esophagus. From the cerebral ganglia four pairs of longitudinal nerves proceed anteriorly and three pairs posteriorly, and both of them are connected by some commissures. The innervation of the internal organs occurs by branches of the main nerves. All of the posterior main nerves contribute to establish the network found on the ventral and dorsal surfaces (Figure 26).

1.5. HISTOMORPHOLOGY OF THE MUSCULAR ORGANS

The majority of amphistomes are characterized as having strongly developed body musculature and by the presence of a muscular apparatus around the oral opening (pharynx), muscular elements around the genital pore (terminal genitalium), and a muscular ventral sucker (acetabulum). These three organs have usually been called muscular organs. Due to their strongly muscular nature, their structure can be examined on histological section. Specific and diagnostic value, involving structural elements of these organs, were recognized in the early stages of amphistomal studies[311,334,562,915] but Näsmark[647] was the first to elaborate a new classification based on histological structure of pharynx, genital atrium (now properly named terminal genitalium), and the observable section of the acetabulum prepared in the median sagittal plane. By revelation of features in structures of the muscular organs, the scope of specific traits has further been enlarged, and the characteristics that are relatively constant have proved more valuable than the rather variable gross morphological ones.

Detailed examinations were carried out by Näsmark[647] on the species of the subfamily Paramphistominae which mainly includes amphistomes of mammalian hosts. He established a special terminology for designation of types of various muscular organs, characteristic of amphistomes only. The framework outlined by him proved to be a useful means in diagnosing amphistomes from whatever species of the definitive hosts. Subsequent authors followed Näsmark's diagnostic principles and revealed further new types of these muscular organs. Nowadays, we have more information about 148 of the total number of 254 amphistome species. Such a survey renders it possible to rearrange all of the types previously described on the basis of a general guiding principle.

1.5.1. PHARYNX AND KEYS TO ITS TYPES

The muscular apparatus situated around the oral opening is a pharynx, including two morphologically clearly defined parts: the body of the pharynx itself and the appendages which are known in certain groups of amphistomes.

The body of the pharynx is usually the true muscular organ, consisting of different circular, longitudinal, and radial muscle elements. The appendages are blindly ended sacs with poorly developed musculature, and they are located dorsoventrally. The esophagus may have musculature or not. In the former case it can be divided into three categories according to its development: sphincter, muscular thickening, and esophageal bulb. Schematic representation of a general type of pharynx shows all the traits, with the relevant terminology, necessary for diagnosis (Figure 27).

KEY TO ABBREVIATIONS

Pharynx (Figures 10, 11, and 27)

as	— anterior sphincter
bph	— body of pharynx
bs	— basic circular
ec	— exterior circular
el	— exterior longitudinal
ic	— interior circular
il	— interior longitudinal
ls	— lip sphincter
mc	— middle circular
o	— esophagus
phb	— pharyngeal bulb

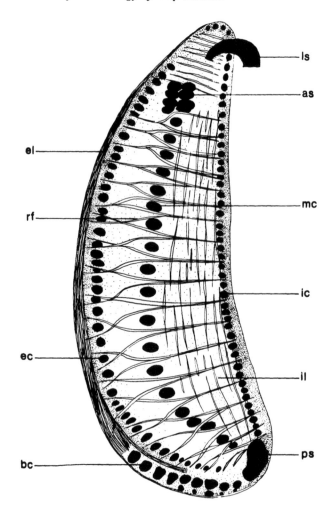

FIGURE 27.   Schematic representation of a general type of pharynx
(median sagittal section). (After K. E. Näsmark, 1937.)

pphs — primary pharyngeal sac
ps   — posterior sphincter
rf   — radial fibers
sphs — secondary pharyngeal sac

On the basis of the structure of the pharynx in a broader sense, five main groups and
further subgroups have been differentiated with several types within each group.

1.5.1.1. Types of Pharynx with Two Intramural Primary Pharyngeal Sacs
1. Nematophila type Sey, 1973a (Figure 28)*

Anterior and posterior sphincters present. Posterior bigger, well developed. Middle
circular units in anterior third of pharynx. Interior circular units well developed along its
length. Interior longitudinal fibers poorly developed, consisting of some weak muscle fibers.
Exterior longitudinal fibers well developed. Radial fibers form small bundles and ramify in
more peripheral parts of pharynx. Lip sphincter and exterior circular units absent.

* Types of pharynx (median sagittal section) are shown in Figures 28 to 69.

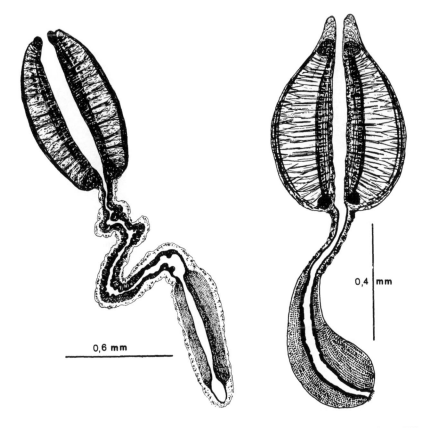

FIGURE 28. *Nematophyla* type Sey, 1973.   FIGURE 29. *Pfenderius* type Sey, 1987.

2. Pfenderius type Sey, 1987 (Figure 29)

Anterior and posterior sphincters present. Anterior one somewhat bigger. Interior circular units poorly developed. Middle circular units moderately developed. Lip sphincter, basal and circular units absent. Interior longitudinal units moderately developed, exterior longitudinal ones well developed. Radial musculature evenly distributed, and fibers ramify in periphery of pharynx.

3. Pseudocladorchis type Näsmark, 1937 (Figure 30)

Anterior and posterior sphincters present. Posterior sphincter strongly developed. Interior circular units moderately developed. Interior longitudinal fibers fine and considerably wavy along its running. Radial fibers form coarse bands and ramify in peripheral parts of pharynx.

4. Pseudodiplodiscus type Sey, 1983a (Figure 31)

Anterior sphincter present, more peripherally located. Middle circular and interior circular units moderately developed. Interior longitudinal fibers consist of a few weakly developed ones. Radial musculature consists of scattered bands of muscle fibers which show ramification in periphery. Lip and posterior sphincter, basal, exterior circular units absent.

5. Scleroporum type Näsmark, 1937 (Figure 32)

Anterior and posterior sphincters present. Anterior one has a particular shape consisting of an anterior and a posterior part. Posterior sphincter enormously developed. Strong muscle fibers present in presphincteral space. Middle circular layer strongly developed, consisting of uniform size of fibers in anterior and middle parts, and decreases in size at posterior part.

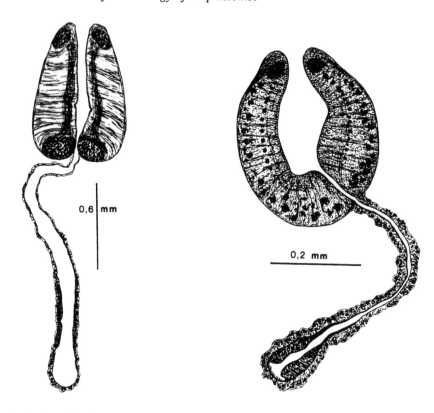

FIGURE 30. *Pseudocladorchis* type Näsmark,   FIGURE 31.   Pseudodiplodiscus type Sey, 1983.
1937.

Basal and interior longitudinal fibers well developed along the middle part and diminishing in breadth toward both ends. Internal longitudinal fibers well developed; coarse, radial fibers dominate its structure. Radial musculature well developed with straight fibers. Lip sphincter and exterior circular units absent.

6. Solenorchis type Sey, 1980a (Figure 33)

Anterior and posterior sphincters present. Posterior one bigger and prominent. Presphincteral space without circular units. Middle circular units well and evenly developed. Basal circular units well developed, interior circular units poorly developed. Interior longitudinal fibers few, weakly developed; exterior longitudinal fibers well developed; equally thick along pharynx. Radial fibers evenly developed and without any sign of ramification. Lip sphincter and exterior circular units absent.

7. Spinolosum type Sey, 1987a (Figure 34)

Anterior sphincter present at anterior third of pharynx. Middle circular units weakly developed, interior circular units consisting of some muscle elements anteriorly. Interior and exterior longitudinal fibers poorly developed. Radial fibers loosely packed, forming small bundles and showing ramification in terminal part. Lip and posterior sphincters, basal and exterior circular units absent.

8. Stichorchis type Näsmark, 1937 (Figure 35)

Anterior and posterior sphincters present. Anterior sphincter bigger and consists of an anterior and a posterior part. Middle circular layer normally developed. Interior circular and basal circular units slightly developed. Interior longitudinal fibers poorly developed. Radial

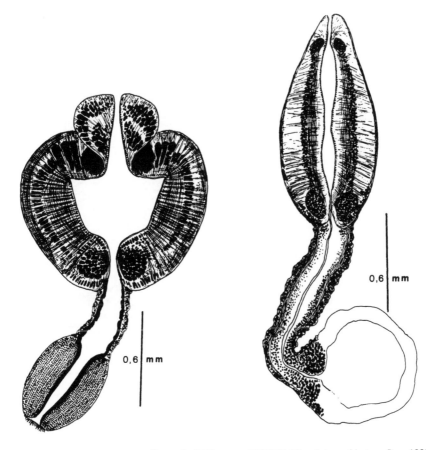

FIGURE 32. Scleroporum type Näsmark, 1937.   FIGURE 33. Solenorchis type Sey, 1980.

fibers poorly developed, forming a few, invariably running bundles which ramify in periphery. Lip sphincter and exterior circular units absent.

KEY TO TYPES OF PHARYNX

1. Anterior and posterior sphincters present... 2
    Anterior sphincter present.. 7
2. No muscle units in presphincteral space 3
    Muscle units present in presphincteral space .... Scleroporum
3. Middle circular units present... 4
    Middle circular units absent.................................... Pseudocladorchis
4. Anterior sphincter smaller than posterior... 5
    Anterior sphincter bigger than posterior.. 6
5. Middle circular units well developed..................................... Solenorchis
    Middle circular units poorly developed, extending in anterior quarter of pharynx......
    ....Nematophila
6. Anterior sphincter transverse oval, big in size; interior longitudinal fibers poorly
    developed.. Stichorchis
    Anterior sphincter spherical, small; interior longitudinal fibers moderately
    developed.. Pfenderius
7. Anterior sphincter in anterior third of pharynx, middle circular units weakly developed,
    extending in anterior half of pharynx .... Spinolosum

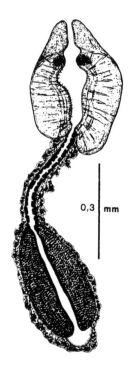

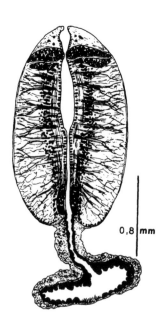

FIGURE 34.   Spinolosum type Sey, 1987. (Courtesy of D. Blair.)

FIGURE 35.   Stichorchis type Näsmark, 1937.

Anterior sphincter near terminal, middle circular units normally developed............
.... Pseudodiplodiscus

1.5.1.2. Types of Pharynx with Two Extramural Pharyngeal Sacs
1. Amurotrema type Sey, 1985a (Figure 36)

Circular musculature consists of only interior and middle circular units. Interior circular units small, difficult to see; middle circular layer well developed. Interior longitudinal fibers absent. Radial musculature moderately developed, fibers form bundles and concentrate in middle and posterior parts. No ramification. Lip, anterior, posterior sphincters, basal and exterior circular units absent.

2. Balanorchis type Eduardo, 1982a (Figure 37)

Interior circular units moderately developed, middle exterior, and basal circular musculature absent. Randomly situated longitudinal muscle fibers present, well developed, and coarse. Radial fibers well developed, forming no bundles.

3. Cladorchis type Näsmark, 1937 (Figure 38)

Anterior sphincter present, situated subterminally, arranged in transverse direction, and divided into two parts by radially running walls. Interior circular units very small. Interior longitudinal fibers poorly, unevenly developed. Radial musculature distributed, forming a few bundles of fibers. Musculature of enormously developed primary pharyngeal sacs consists of a loose network of fibers. Lip and posterior sphincters, middle, exterior, and basal circular units absent.

4. Ferrum-equinum type Sey, 1983a (Figure 39)

Anterior sphincter present, enormously developed. Circular muscle units present in presphincteral space, some of them extending laterally behind sphincter. Middle circular

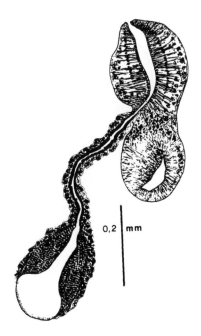

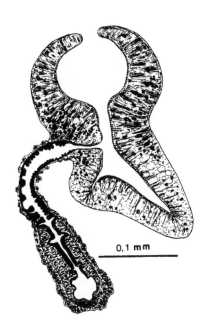

FIGURE 36. Amurotrema type Sey, 1985.

FIGURE 37. Balanorchis type Eduardo, 1982.

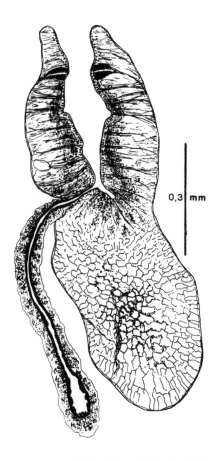

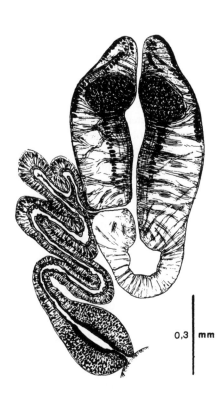

FIGURE 38. Cladorchis type Näsmark, 1937.

FIGURE 39. Ferrum-equinum type Sey, 1983.

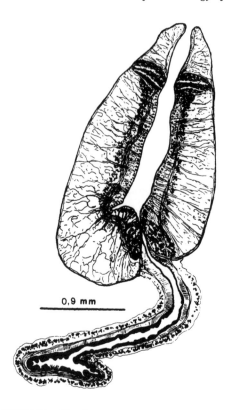

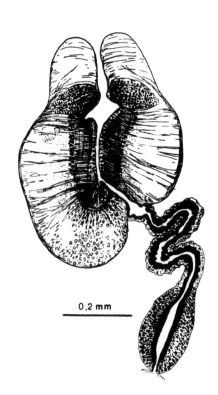

FIGURE 40.   Gigantheus type Sey, 1987. (By permission of the *Parasitol. Hung.*, 1987, Budapest.)

FIGURE 41.   Megacotyle type Sey, 1983. (By permission of the *Parasitol. Hung.*, 1983, Budapest.)

layer well developed, interior circular units poorly developed. Interior longitudinal one consists of some scattered fibers. Radial musculature poorly developed, fibers running irregularly in various directions, sometimes more or less form bundles. Lip and posterior sphincters, exterior and basal circular units absent.

5. Gigantheus type Sey, 1987 (Figure 40)

Anterior sphincter subterminal transverse oval in position. Posterior sphincter consists of loosely packed fibers. Interior circular units small, middle and basal circular well developed. Interior longitudinal fibers poorly developed. Radial musculature weakly developed, unevenly distributed throughout. Lip sphincter, exterior circular units absent.

6. Megacotyle type Sey, 1983a (Figure 41)

Anterior sphincter present, enormous in size. Middle circular units well developed, interior circular units small. Circular muscle units absent in presphincteral space. Interior longitudinal fibers poorly developed with some fibers. Radial muscle fibers form bundles, unevenly distributed. Lip, posterior sphincters, basal exterior circular units absent.

7. Megalodiscus type Sey, 1983a (Figure 42)

Anterior sphincter present, more or less compact. Middle circular units well developed, interior circular units very small. Interior longitudinal fibers weakly developed, radial fibers developed without showing a tendency to form bundles. Lip, posterior sphincters, basal and exterior circular units absent.

8. Microrchis type Näsmark, 1937 (Figure 43)

Anterior sphincter present, well collected and strongly developed. Exterior circular units

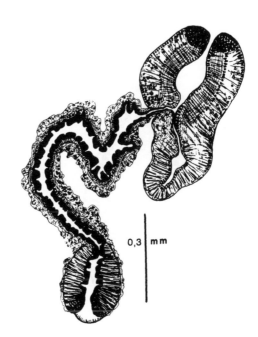

FIGURE 42.   Megalodiscus type Sey, 1983.   FIGURE 43.   Microrchis type Näsmark, 1937.

well developed in anterior half of pharynx. Interior circular units small, usually developed only in anterior two thirds of pharynx. Middle circular units moderately developed, its units bigger in anterior half of pharynx. Basal circular units well developed. Interior longitudinal fibers well developed, closely packed. Radial muscle fibers distributed evenly. Lip and posterior sphincters absent.

9. Onotragi type Jones, 1986 (Figure 44)

In anterior third of pharynx, loosely aggregated circular units present. Anterior sphincter voluminous, subterminal. Interior circular units poorly developed. Middle circular well developed, exterior circular units along anterior third of pharynx. Basal circular units present. Interior longitudinal fibers few, weakly developed. Exterior longitudinal fibers poorly developed. Radial fibers moderately developed. Lip, posterior sphincters absent.

10. Olveria type Sey, 1987 (Figure 45)

A unique feature of this type is the occurrence of muscular esophagus along its full length. Musculature of pharynx itself fairly poor. Circular units present, sometimes poorly developed. No other circular muscular units. Radial musculature moderately developed, consisting of separate fibers.

11. Pseudochiorchis type Tandon, 1970 (Figure 46)

Anterior and posterior sphincters present, anterior sphincter bigger than posterior one. External circular units present along margin of pharynx between anterior and posterior sphincters. Interior circular units very small. Middle circular units moderately developed. Radial musculature consists of unevenly distributed fibers. Lip sphincter and exterior circular units absent.

12. Stunkardia type Sey, 1987 (Figure 47)

Anterior and posterior sphincters present, anterior one smaller than posterior one. Middle

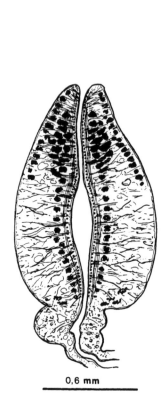

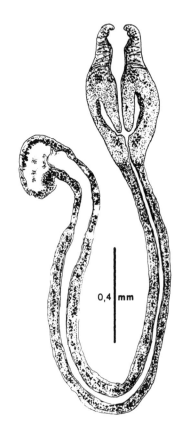

FIGURE 44.   Onotragi type Jones, 1986. (Courtesy of A. Jones.)

FIGURE 45.   Olveria type Sey, 1987. (By permission of the *Parasitol. Hung.*, 1987, Budapest.)

circular units moderately developed, interior circular ones poorly developed. Basal circular units present. Interior longitudinal fibers poorly developed. Radial muscle fibers scattered and coarse, form unevenly distributed bundles. Lip sphincter and exterior circular units absent.

13. Subclavatus type Sey, 1983a (Figure 48)
Anterior sphincter present, comprised of loosely packed fibers. Presphincteral space without circular muscle units. Interior circular units normally developed. Interior longitudinal fibers poorly developed, consisting of some fibers. Radial musculature moderately developed, fibers form evenly distributed bundles, peripheral parts ramify. Lip, posterior, basal, middle, and exterior circular units absent.

14. Taxorchis type Sey, 1987 (Figure 49)
Anterior sphincter present consisting of loosely packed fibers. Middle circular units very small, seen on cross section, muscle units situated in anterior half of pharynx. Middle circular units small. Internal longitudinal fibers consisting of some very small fibers. Radial fibers unevenly distributed forming some bundles with ramification in periphery. Lip, posterior sphincters, basal and exterior circular units absent.

15. Zygocotyle type Sey, 1987 (Figure 50)
Anterior sphincter consisting of loosely packed units, well developed. External circular units present, along anterior third margin of pharynx. Interior circular units moderately

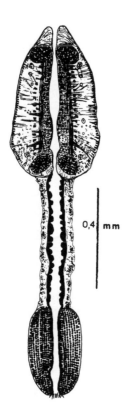
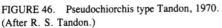
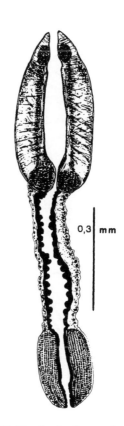

FIGURE 46. Pseudochiorchis type Tandon, 1970.
(After R. S. Tandon.)

FIGURE 47. Stunkardia type Sey, 1987.
(By permission of the *Parasitol. Hung.*, 1987,
Budapest.)

developed. Obliquely running fibers, seen on median sagittal plane, situated in a row, near pheriphery, similar to spaces of muscle bands. Of longitudinal muscle fibers, external longitudinal ones are visible, moderately developed. Lip; posterior sphincters; basal, middle, and external circular units absent.

KEY TO THE TYPES OF PHARYNX

1. Circular musculature weakly developed, interior circular units present only 2
   Circular musculature not weakly developed...................................... 3
2. Primary pharyngeal sacs well developed .... Balanorchis
   Primary pharyngeal sacs poorly developed................................. Olveria
3. Primary pharyngeal sacs longer than pharynx...................................... 4
   Primary pharyngeal sacs shorter than pharynx 5
4. Anterior circular muscle units loosely packed, middle circular units poorly
   developed .... Taxorchis
   Anterior circular muscle units form sphincter, middle circular units weakly
   developed .... Cladorchis
   Anterior circular units absent, middle circular series well developed .... Amurotrema
5. Anterior and posterior sphincters present.. 6
   Anterior sphincter present... 8
   Muscle units of anterior sphincter loosely packed, occupying smaller or bigger parts of
   pharynx 9
6. Pharynx moderately developed.. 7
   Pharynx enormous in size... Gigantheus

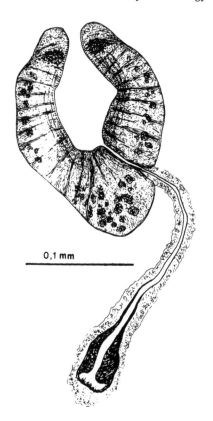

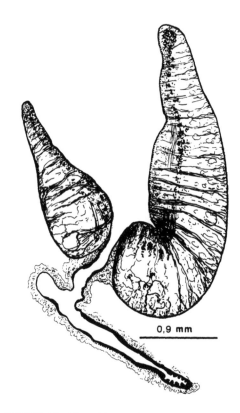

FIGURE 48.   Subclavatus type Sey, 1983.

FIGURE 49.   Taxorchis type Sey, 1987. (By permission of the *Parasitol. Hung.*, 1987, Budapest.)

7. Anterior sphincter bigger than posterior............................ Pseudochiorchis
   Anterior sphincter smaller than posterior.................................. Stunkardia
8. Anterior sphincter terminal.. 10
   Anterior sphincter subterminal 11
9. Middle circular series present... Onotragi
   Middle circular absent .... Zygocotyle
10. Anterior sphincter compact, interior and middle circular series present .... Microrchis
    Muscle units of anterior sphincter loosely packed, middle circular series packed ......
    ....Megalodiscus
11. Circular muscle units present in presphincteral space, exterior circular series extends to
    middle part of pharynx... Ferrum-equinum
    Presphincteral space without circular units, middle circular series present, exterior circular series absent .... Megacotyle
    Presphincteral space without circular muscle units, middle and exterior circular series
    absent... Subclavatus

1.5.1.3. Type of Pharynx with Unpaired Primary Pharyngeal Sac
1. Stephanopharynx type Näsmark, 1937 (Figure 51)

Inner circular units weakly developed. Middle circular units in anterior third of pharynx, moderately developed, near middle circular units. Inner longitudinal fibers few. External longitudinal fibers poorly developed. Radial fibers evenly distributed and form bundles consisting of some fibers. Lip, anterior and posterior sphincters; basal and external circular units absent.

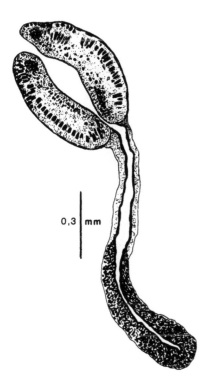

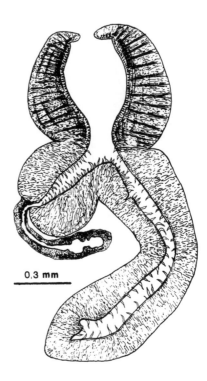

FIGURE 50. Zygocotyle type Sey, 1987. (By permission of the *Parasitol. Hung.*, 1987, Budapest.)

FIGURE 51. Stephanopharynx type Näsmark, 1937.

1.5.1.4. Types of Pharynx with Pharyngeal Bulb and Secondary Pharyngeal Sacs
1. Brumptia type Näsmark, 1937 (Figure 52)

Anterior sphincter present, slightly developed, its muscular units loosely packed. Middle circular moderately developed, interior, circular units small and equal in size. Internal longitudinal fibers embracing a quarter of pharynx. Muscle fibers form a weak bundle outside middle of circular units. Radial musculature constitutes coarse fiber bundles, ramifying peripherally. Exterior longitudinal normally developed. Posterior and lip sphincters; exterior circular and basal circular units absent.

2. Choerocotyle type Sey, 1987 (Figure 53)

Anterior sphincter present, subterminal, comprising loosely packed but strong muscle units. In anterior half of pharynx, muscle units situated as part of external circular units and occupy more or less irregular position. Middle circular units considerably smaller in posterior half of pharynx. These units continue backward into pharyngeal bulb. Middle circular units well developed. Interior longitudinal fibers moderately developed, muscle fibers present outside middle circular units. Radial musculature consists of fiber bundles, evenly distributed with ramification peripherally. External longitudinal fibers normally developed. Posterior and lip sphincters and basal circular layers absent.

3. Gastrodiscus type Näsmark, 1937 (Figure 54)

Anterior sphincter poorly developed, consisting of some muscle units. Interior circular units small and equal sized throughout. Middle circular units closely packed, anteriorly connect to anterior sphincter. Internal longitudinal fibers thin; some bundles outside middle circular units present. External longitudinal fibers well developed. Radial musculature constitutes thin bundles of fibers, no ramification in periphery. Posterior and lip sphincters, exterior and basal circular unit absent.

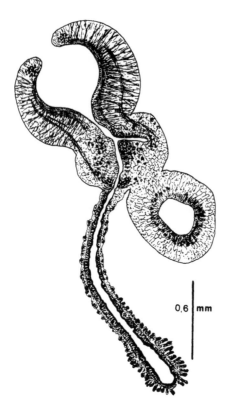

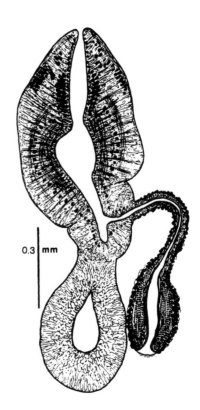

FIGURE 52. Brumptia type Näsmark, 1937. (From O. Sey and M. Graber, 1979.)

FIGURE 53. Choerocotyle type Sey, 1987. (By permission of the *Parasitol. Hung.*, 1979, Budapest.)

4. Macropotrema type Sey, 1987 (Figure 55)

Anterior sphincter well developed, compact and terminal. Inner circular units unevenly developed, bigger ones in posterior half of pharynx. Middle circular units well developed. Inner longitudinal fibers few. External longitudinal normally developed. Radial musculature comprising evenly distributed bundles of fibers, ramifying in periphery. Posterior and lip sphincters, external and basal circular units absent.

5. Pseudodiscus type Näsmark, 1937 (Figure 56)

Interior circular units poorly developed, continue posteriorly in pharyngeal bulb. Middle circular units extremely small, somewhat smaller anteriorly. Interior longitudinal fibers inside middle circular units, very fine. Some oblique fibers present outside middle circular units. External longitudinal fibers poorly developed. Radial musculature consisting of thick bundles of fibers with fairly considerable ramifications. Anterior, posterior, and lip sphincters, external and basal circular series absent.

6. Watsonius type Näsmark, 1937 (= Hawkesius Sey, 1985, = Homalogaster Sey, 1984) (Figure 57)

Interior circular units small, equal in size. Middle circular units comprise well-developed units. Size of units noticeably longer anteriorly. Interior longitudinal fibers consist of relatively coarse bundles. Radial musculature comprises sparse, irregularly running bundles of fibers, showing tendency to ramify peripherally. External longitudinal fibers evenly developed and normal in extension. Anterior, posterior, and lip sphincters, external and basal units absent.

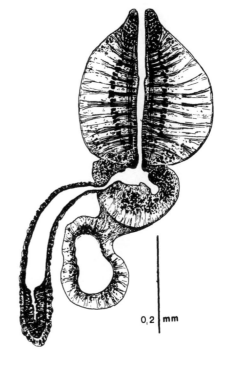

FIGURE 54. Gastrodiscus type Näsmark, 1937.

FIGURE 55. Macropotrema type Sey, 1987. (By permission of the *Parasitol. Hung.*, 1987, Budapest.)

KEY TO TYPES OF PHARYNX

1. Anterior sphincter present.. 2
   Anterior sphincter absent 5
2. Anterior circular units loosely packed... 3
   Anterior circular units form compact sphincter...................... Macropotrema
3. Exterior circular units present.. Choerocotyle
   Exterior circular units absent.. 4
4. Interior longitudinal fibers continue in pharyngeal bulb.................... Brumptia
   Interior longitudinal fibers absent in pharyngeal bulb .... Gastrodiscus
5. Pharyngeal bulb and secondary pharyngeal sacs longer than pharynx.................
   Middle circular series well developed.................................... Watsonius
   Middle circular series poorly developed .... Pseudodiscus

1.5.1.5. Types of Pharynx without Pharyngeal Bulb and Secondary Pharyngeal Sacs

1. Calicophoron type Dinnik, 1964 (= Calicophoron and Ijimai types Näsmark, 1937) (Figure 58)

Interior circular units small in size with possible variation. Middle circular units poorly developed but easily seen on transverse section. Exterior circular units more or less distinct. Basal circular units small, as a rule, lie in two rows. Interior longitudinal fibers with varying thickness. Exterior longitudinal fibers normally developed. Radial fibers well developed, without ramification. Anterior, posterior, and lip sphincters absent.

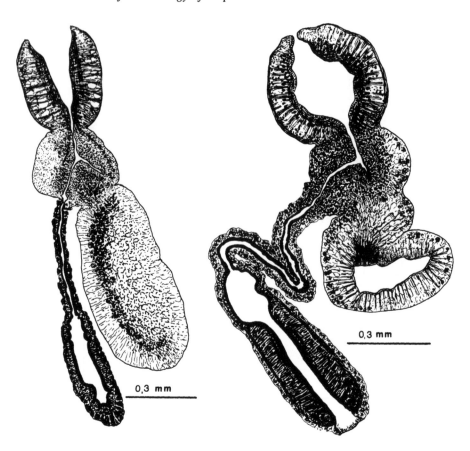

FIGURE 56.   Pseudodiscus type Näsmark, 1937.   FIGURE 57.   Watsonius type Näsmark, 1937.

2. Cephalophi type Eduardo, 1982a (Figure 59)

Anterior sphincter present with loosely packed muscle units. Interior muscle units very small and evenly distributed. External circular units confined only to anterior part of pharynx. Basal circular units fairly well developed. Interior longitudinal fibers moderately developed, one third of pharynx's breadth, exterior longitudinal fibers normally developed. Radial muscle fibers moderately developed. Posterior and lip sphincters, middle circular series absent.

3. Dicranocoelium type Näsmark, 1937 (Figure 60)

Lip sphincters present, located on lateral sides of anterior part of pharynx, forming horseshoe-shaped structures; interior circular units small; basal circular units usually consist of a single row. Interior longitudinal fibers fairly well developed; several exterior longitudinal fibers near exterior border. Radial fibers poorly developed. Anterior, postrior sphincters, middle and exterior circular units absent.

4. Explanatum type Näsmark, 1937 (Figure 61)

Anterior and posterior sphincters present, but anterior one divides into several elements. Interior circular units moderately developed; middle circular units strongly developed and distinct; exterior circular units moderately to strongly developed: basal circular units well developed. Interior longitudinal fibers occupy half the volume of pharynx; exterior longitudinal fibers normally developed. Radial fibers coarse and sparse, forming bundles with ramification periphery. Lip sphincter absent.

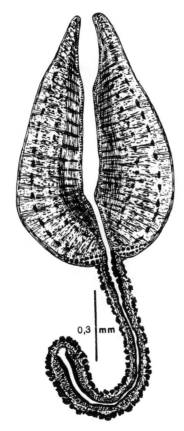

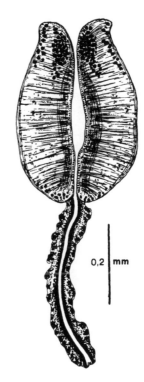

FIGURE 58.   Calicophoron type Dinnik, 1964.

FIGURE 59.   Cephalophi type Eduardo, 1982a. (By permission of Dr. W. Junk Publishers, 1982, Dordrecht.)

5. Gemellicotyle type Sey, 1987 (Figure 62)

Anterior sphincter present, with loosely packed muscle units; interior circular units very small; middle circular well developed and distinct. Interior longitudinal fibers poorly developed and few, exterior longitudinal fibers normally developed. Radial fibers moderately developed, forming no bundles. Posterior and lip sphincters, basal, exterior circular series absent.

6. Gigantopharynx type Eduardo, 1982a (Figure 63)

Interior circular units poorly developed; exterior circular units well developed and present in posterior third of pharynx; basal circular units well developed. Interior longitudinal fibers absent but some weak fibers found in anterior part of pharynx; exterior longitudinal fibers distinct and close to exterior wall. Radial fibers well developed forming bundles. Anterior, posterior, and lip sphincters, middle circular layer absent.

7. Gregarius type Sey, 1987 (Figure 64)

Interior circular units small but distinct; middle circular units present in anterior third of pharynx, well developed; exterior circular units close to exterior wall. Interior longitudinal fibers fine and moderately developed, exterior longitudinal fibers moderately developed. Radial fibers form evenly distributed bundles, without any tendency to ramify. Anterior posterior, and lip sphincters; basal, circular series absent.

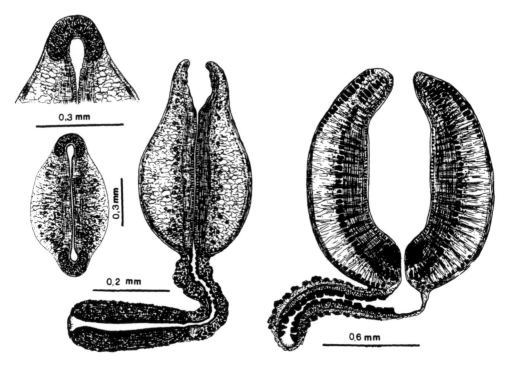

FIGURE 60.   Dicranocoelium type Näsmark, 1937.

FIGURE 61.   Explanatum type Näsmark, 1937.

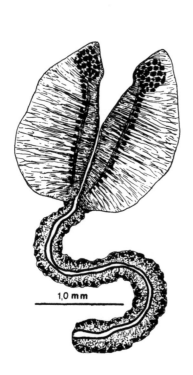

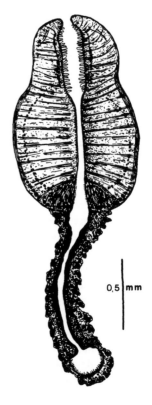

FIGURE 62.   Gemellicotyle type Sey, 1987. (By permission of the Editor, Dr. B. S. Chauhana Commemoration Volume, 1975, Bhubaneswar.)

FIGURE 63.   Gigantopharynx type Eduardo, 1982. (By permission of Dr. W. Junk Publishers, 1985, Dordrecht.)

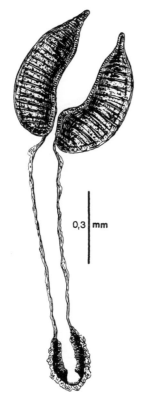

0,3 mm

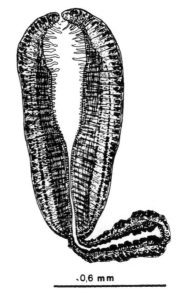

‹0,6 mm

FIGURE 64. Gregarius type Sey, 1987.
(By permission of the *Parasitol. Hung.*,
1987, Budapest.)

FIGURE 65. Liorchis type Näsmark, 1937.

8. Liorchis type Näsmark, 1937 (= Pseudoliorchis type Velitchko, 1966) (Figure 65)

Interior circular units very small; middle circular series well mashed, most clearly seen in posterior half of pharynx; exterior circular series clear and well mashed, strongly developed in posterior half of pharynx; basal circular units well developed arranged in one or two rows. Interior longitudinal fibers distinct, greatest breadth in posterior part of pharynx; exterior longitudinal fibers well and clearly developed. Radial fibers well developed. Internal wall of anterior part of pharynx covered by long papillae. Anterior, posterior, and lip sphincters absent.

9. Orthocoelium type Eduardo, 1985 a (Figure 66)

Anterior sphincter present, subterminal, consisting of closely packed muscle units; interior circular units moderately developed: middle and exterior circular units well developed, uniting in their anterior and posterior limits: basal circular units moderately developed. Interior longitudinal fibers moderately developed but distinct, exterior longitudinal fibers normally developed. Radial fibers well developed and closely packed. At posterior part of pharynx circular fibers present, lip sphincter absent.

10. Paramphistomum type Näsmark, 1937 (Figure 67)

Interior circular units normal in size, evenly distributed. Exterior circular units more or less distinct, near margin or rarely somewhat middle in position: basal circular units well developed in one or two rows. Interior longitudinal fibers with varying breadth, sometimes with indistinct exterior limit. Exterior longitudinal fibers varying in thickness, but easily seen. Radial fibers varying in appearance, coarse and sparse fibers forming bundles, without ramification. Anterior, posterior, and lip sphincters; middle circular series absent.

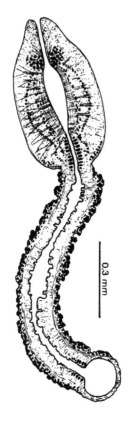

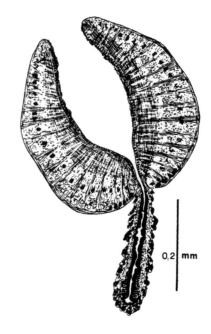

FIGURE 66.   Orthocoelium type Eduardo, 1982.

FIGURE 67.   Paramphistomum type Näsmark,
1937.

11. Pisum type Näsmark, 1937 (Figure 68)

Anterior and posterior sphincters present; interior circular units small but distinct; middle circular units strongly developed; exterior circular units small and insignificant; basal circular units well developed. Interior longitudinal fibers strongly but unevenly developed; exterior longitudinal fibers strongly but unevenly developed. Radial fibers forming evenly distributed bundles and ramify considerably. Lip sphincter absent.

12. Serpenticaecum type Eduardo and Peralta, 1987 (Figure 69)

Interior, middle exterior and basal circular units moderately developed. Interior and exterior longitudinal fibers few, weakly developed. Radial fibers moderately developed. Lip, anterior and posterior sphincters absent. Lumen lined by anterior long and finger-like and by short and dome-shaped papillae.

KEY TO TYPES OF PHARYNX

1. Anterior and posterior sphincters present.. 2
   Anterior sphincter present... 3
   Anterior, posterior, and lip sphincters absent 4
   Lip sphincter present... Dicranocoelium
2. Posterior sphincter bigger than anterior sphincter .... Pisum
   Anterior sphincter bigger than posterior sphincter, strong exterior circular units in anterior part of pharynx .... Explanatum
3. Anterior sphincter present, exterior and middle circular units well developed..........
   ....Orthocoelium

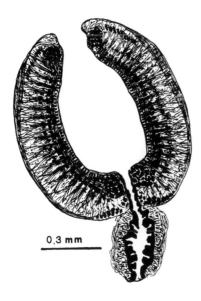

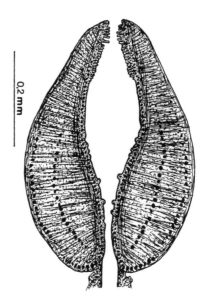

FIGURE 68. Pisum type Näsmark, 1937.

FIGURE 69. Serpenticaecum Eduardo and Peralta, 1987. (Courtesy of Eduardo and Peralta.)

Anterior sphincter consists of loosely packed units, exterior circulated units in anterior third of pharynx, middle circular units absent..............................Cephalophi
Anterior sphincter consists of loosely packed units and middle circular units present.. Gemellicotyle
4. Inner surface of pharynx covered with papillae 5
   Inner surface of pharynx without papillae... 7
5. Middle circular units present.. 6
   Middle circular units absent.. Gigantopharynx
6. Inner surface of pharynx covered with two kinds of papillae: fingerlike and dome shape ....Serpenticaecum
   Inner surface of pharynx covered with one type of papillae: fingerlike .... Liorchis
7. Exterior and middle circular units weakly developed ....Calicophoron
   Exterior circular units poorly developed, middle circular units present only in anterior third of pharynx..Gregarius
   Middle circular units absent.. Paramphistomum

1.5.2. TERMINAL GENITALIUM AND KEYS TO ITS TYPES

The genital pore of amphistomes is situated on a more or less developed but distinct elevation. It is surrounded by fold(s) in subsequent concentric row(s) which enclose shallow cavities. These folds and cavities, seen on median sagittal section (in this plane they are called papilla and atrium, respectively) constitute the fundamental elements of the structure of this area. The folds and the wall of this area are furnished with different types of muscular elements whose position, arrangement, and relative constancy have been proved to be of important diagnostic value. This area was disclosed for the first time, from a diagnostic point of view, by Näsmark,[647] designating and characterizing parts of this terminal structure. His findings provided the basis for further examinations. His description, however, was incomplete in certain details and was not free of confusion. These missing details were rightly solved by Eduardo,[271] naming all parts of this organ and establishing the name, terminal genitalium, for the whole terminal structure of the genital system. The terminal genitalium is used here in a wider sense, the cirrus sac, when it is present, being taken into

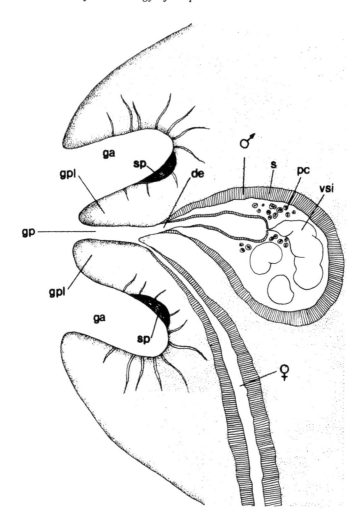

FIGURE 70.   Schematic representation of a general type of terminal genitalium (median sagittal section) with cirrus sac. (By permission of the *Parasitol. Hung.*, 1987, Budapest.)

consideration (Figure 70). Schematic representation of the terminal genitalium, without the cirrus sac, and the terminology were adopted from Näsmark[647] and Eduardo[271] (Figure 71).

Types of terminal genitalium can be divided into four groups on the basis of its structural elements.

KEY TO ABBREVIATIONS

Terminal Genitalium (Figures 70 and 71)

cs   — cirrus sac
de   — ductus ejaculatorius
ga   — genital atrium
gf   — genital fold
gp   — genital pore
gpl — genital papilla
gs   — genital sphincter

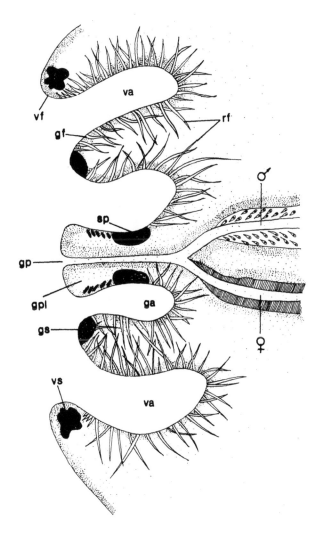

FIGURE 71. Schematic representation of a general type of terminal genitalium (median sagittal section) without cirrus sac. (After K. E. Näsmark, 1937.)

pc — pars prostatica
rf — radial fibers
sp — sphincter papillae
va — ventral atrium
vf — ventral fold
vs — ventral sphincter
vsi — vesicula seminalis interna

1.5.2.1. Types of Terminal Genitalium with True or Modified Cirrus Sac
1. Cladorchis type Sey, 1987 (Figure 72)*

Whole terminal genitalium characterized by forming well-developed genital sucker. Genital papilla usually slender; genital sphincter present, slightly or well developed. Cirrus sac equal with diameter of terminal genitalium with poorly developed musculature. Radial fibers evenly and strongly developed.

* Types of terminal genitalium (median sagittal section) are shown in Figures 72 to 131.

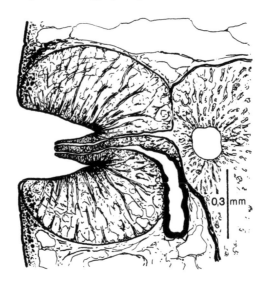

FIGURE 72.   Cladorchis type Sey, 1987. (By permission of
the *Parasitol. Hung.*, 1987, Budapest.)

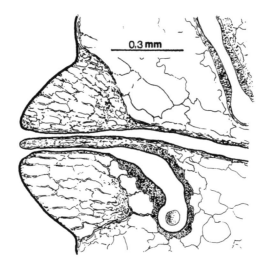

FIGURE 73.   Dadayius type Sey, 1987. (By permission of the
Parasitol. Hung., 1987, Budapest.)

2. Dadayius type Sey, 1987 (Figure 73)

Genital opening surrounded by sucker, well marked off from neighboring parenchyma.
Genital papilla clumsy, strongly developed circular musculature entirely absent. Cirrus sac
small, its musculature poorly developed. Radial fibers moderately developed.

3. Gastrodiscus type Sey, 1975 (Figure 74)

Genital papilla and genital fold present, well developed. At junction of these papillae
some circular muscle units present. Cirrus sac is modified in structure. Radial fibers form
some coarse bundles. Cirrus sac and musculature of its wall moderately developed.

4. Nematophila type Sey, 1973b (Figure 75)

Genital sphincter and sphincter papillae present, strongly developed. Genital atrium
small but distinct. Genital papilla long and stout, genital fold well developed. Cirrus sac

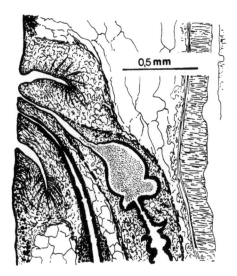

FIGURE 74. Gastrodiscus type Sey, 1975.

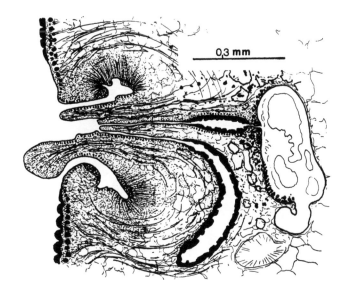

FIGURE 75. Nematophila type Sey, 1973.

very small, with poorly developed musculature. Genital-atrial-radial layer well developed. True ventral fold and ventral atrium absent.

5. Scleroporum type Sey, 1983a (Figure 76)

Genital fold poorly developed, true genital atrium absent but slight atrium formed with body wall. Cirrus sac two or three times longer than diameter of genital opening. Musculature of its wall moderately developed. Radial musculature poorly developed.

6. Spinolosum type Sey, 1983a (Figure 77)

Genital papilla moderately developed, short and stumpy. Genital papilla and body wall form a well-developed atrium. Circular musculature anteriorly absent. Radial fibers weakly developed. Length of cirrus sac agrees with diameter of genital opening. Musculature of its wall moderately developed. Radial fibers poorly developed.

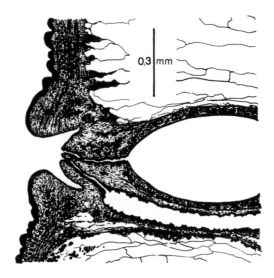

FIGURE 76. Scleroporum type Sey, 1983.

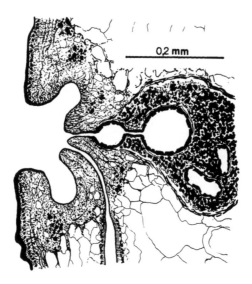

FIGURE 77. Spinolosum type Sey, 1983.

7. Stichorchis type Sey, 1975 (Figure 78)

Genital papilla stumpy with moderately developed genital sphincter. Genital fold and, thus, true genital atrium absent. Borderline of terminal genitalium clearly separated from surrounding parenchyma. Cirrus sac small with poorly developed musculature. Radial fibers form some strong bundles.

8. Waltheri type Sey, 1987 (Figure 79)

Strongly developed circular musculature present on inner rim of genital papilla and genital fold. Genital papilla short and broad. Cirrus sac small with poorly developed musculature. A row of circular muscle units present at junction of genital papilla and genital fold. Radial musculature consists of a coarse bundle of fibers.

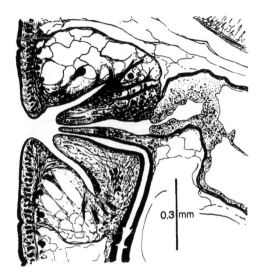

FIGURE 78. Stichorchis type Sey, 1975.

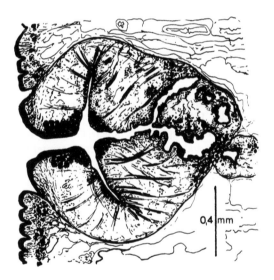

FIGURE 79. Waltheri type Sey, 1987. (By permission of the
Parasitol. Hung., 1987, Budapest.)

KEY TO TYPES OF TERMINAL GENITALIUM

1. Terminal genitalium with genital papilla 2
   Terminal genitalium with genital papilla and genital fold . . . . Nematophila
2. Genital papilla normally developed 3
   Genital papilla enormously developed . . . . Dadayius
3. Genital sucker absent 5
4. Genital sucker present 4
4. Circular muscle units in rim of sucker forming sphincter papilla and genital
   sphincter . . . . Waltheri
   Genital sphincter present . . . . Cladorchis
5. Genital papilla without sphincter 6
   Genital papilla with sphincter . . . . Stichorchis

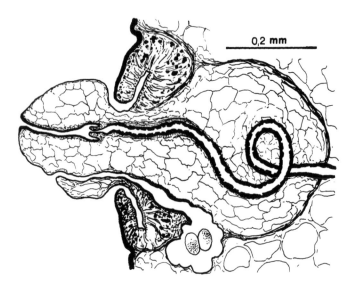

FIGURE 80.   Balanorchis type Eduardo, 1982.

6.   Genital atrium without radial musculature 7
      Genital atrium with radial muscle bundles ....Gastrodiscus
7.   Length of cirrus sac equal with diameter of terminal genitalium...........Spinolosum
      Cirrus sac 2 or 3 times longer than diameter of terminal genitalium......Scleroporum

1.5.2.2. Types of Terminal Genitalium with Hermaphroditic Pouch
1. Balanorchis type Eduardo, 1982a (Figure 80)

Characterized by presence of hermaphroditic pouch, enclosing metraterm in it. A unique feature of this type is that male and female ducts open separately, easily seen when this pouch is protruded. In retracted state there is a single genital pore and fold surrounding it which should be regarded as genital papilla and not a genital fold as indicated by Eduardo.[271]

2. Bicaudata type Sey et Graber, 1979 (Figure 81)

Characterized by presence of large hermaphroditic pouch enclosing vas deferens, vesicula seminalis interna, pars prostatica, and metraterm. Ductus hermaphroditicus opens into enlarged atrium surrounded by well-developed genital papilla. Musculature of wall of hermaphroditic pouch well developed. Some circular muscle units present at tip of genital papilla.

The structure of the terminal genitalium was described by MacCallum,[573] and additional information was published later by Travassos,[979] Maplestone,[591] and Stunkard.[925] Such a type of genital opening like this was proposed to be called "hermaphroditic pouch" by Fukui,[334] but it was not named.

3. Epuluensis type Sey, 1987 (Figure 82)

Genital papilla present, stout in appearance, circular muscular units absent. Metraterm and pars prostatica enclosed in hermaphroditic pouch. Radial muscle units form coarse bundles. True genital atrium absent but in strongly retracted position, slight atrium may form between genital papilla and body wall.

4. Hawkesi type Sey, (Figure 83)

Whole genital opening small, genital papilla present, *clumsy*. *Metraterm is barrel shaped*, pars prostatica enclosed in hermaphroditic pouch. Radial muscle units poorly developed. True genital atrium absent.

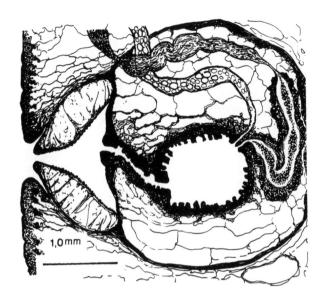

FIGURE 81. Bicaudata type Sey and Graber, 1979. (After G. A. McCallum, 1917.)

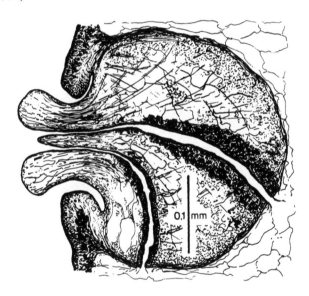

FIGURE 82. Epuluensis type Sey, 1987. (By permission of the *Parasitol. Hung.*, 1987, Budapest.)

KEY TO TYPES OF TERMINAL GENITALIUM

1. Terminal genitalium enormously developed 2
   Terminal genitalium small 3
2. Cirrus sac enclosing metraterm and vesicula seminalis and pars prostatica . . . Bicaudata
   Cirrus sac enclosing metraterm and pars prostatica . . . . Epuluensis
3. Male and female ducts open separately . . . . Balanorchis
   Hermaphroditic duct present . . . . Hawkesi

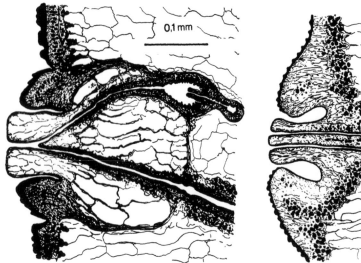

FIGURE 83.   Hawkesius type Sey, 1985.

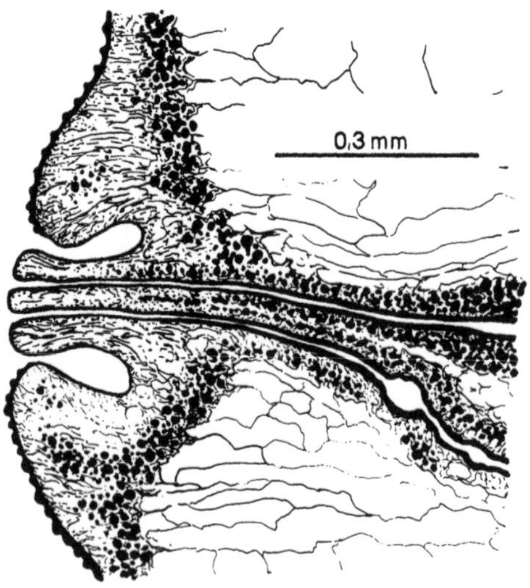

FIGURE 84.   Cruciformis type Sey, 1985.

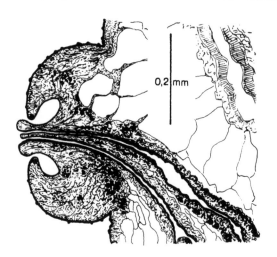

FIGURE 85.   Elongatus type Sey, 1983b.

1.5.2.3. Types of Terminal Genitalium without Cirrus Sac and Hermaphroditic Pouch and with Tegumental Papillae

1. Cruciformis type Sey, 1983b (Figure 84)

Genital papilla slender, without sphincter. Genital fold normally developed, with tegumental papillae along its outer surface. Genital atrium usually developed. Radial musculature poorly developed.

2. Elongatus type Sey, 1983b (Figure 85)

Genital papilla and well-developed genital fold present, latter with poorly developed genital sphincter. Papillae present along outer surface of genital papilla. Radial musculature poorly developed.

3. Endopapillatus type Sey, 1983b (Figure 86)

Genital papilla, genital fold, and ventral fold (but not always typically) present. Sphincter papillae and genital sphincter present, moderately developed; ventral sphincter absent. Papilla

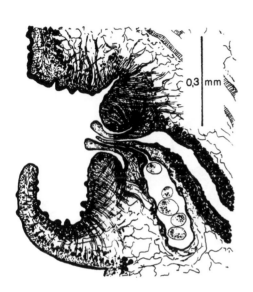

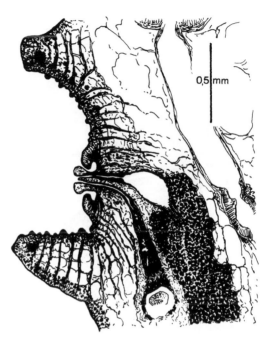

FIGURE 86. Endopapillatus type Sey, 1983.

FIGURE 87. Gregarius type Sey, 1983.

present along inner surface of ventral fold. Ventral atrium usually spacious. Genital-atrial-radial layer well developed.

4. Gregarius type Sey, 1983b (Figure 87)

Genital papilla, genital and ventral folds present. Sphincter papillae poorly, genital sphincter moderately, ventral sphincter well developed. Papillae present along inner surface of ventral atrium. Ventral atrium spacious. Radial muscle units well developed, more or less evenly distributed.

5. Homalogaster type Sey, 1984b (Figure 88)

Genital papilla protruded, voluminous and stout. Sphincter papillae absent. Tegumental papillae present on ventral papilla. Genital fold and genital atrium absent. Radial musculature poorly developed.

6. Noci type Sey, 1984b (Figure 89)

Genital papilla small, without sphincter. Genital fold enormous, without circular muscle units. Tegumental papillae arranged along slightly protruded genital fold. Radial musculature poorly developed.

7. Olveria type Sey, 1987 (Figure 90)

Poorly developed terminal genitalium. Genital papilla weakly developed with well-developed sphincter papillae. Genital atrium straight, lined with tegumental papillae. Genital fold thick and broad. Genital-atrial-radial fibers strongly developed.

8. Papillogenitalis type Eduardo, 1980d (Figure 91)

Genital papilla well developed with sphincter papillae. Genital fold well defined, genital sphincter well developed, distinct and compact. External surface of genital fold and area surrounding genital pore lined with tegumental papillae. True ventral atrium absent. Radial fibers well developed.

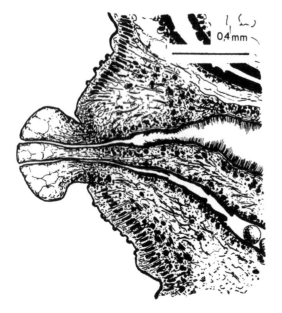

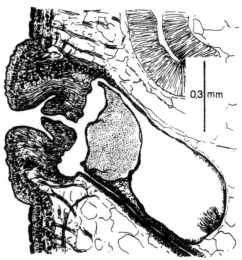

FIGURE 88.   Homalogaster type Sey, 1984.

FIGURE 89.   Noci type Sey, 1984.

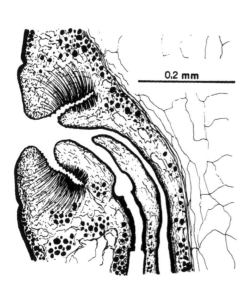

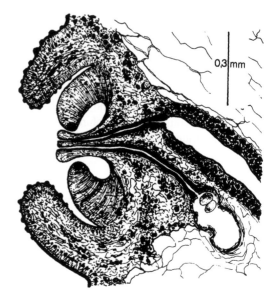

FIGURE 90.   *Olveria* type Sey, 1987. (By permission of the *Parasitol. Hung.*, 1987, Budapest.)

FIGURE 91.   Papillogenitalis type Eduardo, 1980.

9. Parvipapillatum type Eduardo, 1982a (Figure 92)

Genital papilla moderately developed, slender genital fold small. Circular muscle units entirely absent. Papillae present along genital fold and along base of genital papilla. Radial muscle fibers moderately developed. Two ventral atriae present but in strongly retracted form, a straight atrium visible.

10. Pseudodiscus type Sey, 1987 (Figure 93)

Genital papilla and genital fold present. Genital fold with moderately developed genital

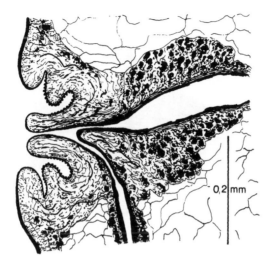

FIGURE 92.   Parvipapillatum type Eduardo, 1982.

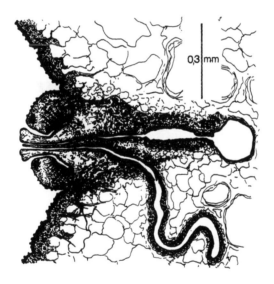

FIGURE 93.   Pseudodiscus type Sey, 1987. (By permission
of the *Parasitol. Hung.*, 1987, Budapest.)

sphincter. Papillae around genital pore and outer surface of genital fold. Genital-atrial-radial
layer well developed.

11. Raja type Eduardo, 1982a (Figure 94)
Genital papilla and genital fold present. Sphincter papillae and genital sphincter well
developed and compact. Papillae present around genital pore or in protruded state, along
ventral papilla.[647] Genital-atrial-radial-layer well developed.

12. Schoutedeni type Sey, 1983b (Figure 95)
Genital papilla well developed with a moderately developed papilla at its bottom. Genital
fold well developed genital sphincter absent. Papillae present along outer surface of genital
fold. Genital-atrial-radial layer well developed. Genital atrium shallow.

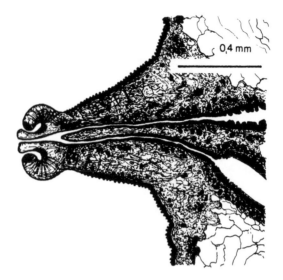

FIGURE 94.   Raja type Näsmark, 1937.

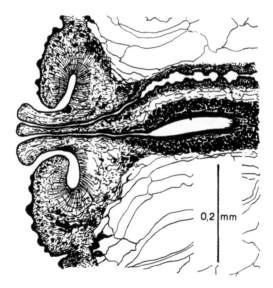

FIGURE 95.   Schoutedeni type Sey, 1983.

13. Serpenticaecum type Eduardo and Peralta, 1987 (Figure 96)

Genital papilla slender with well developed sphincter papillae. Genital fold moderately, ventral fold well developed. Internal surface of the latter are covered with tegumental papillae. Radial fibers moderately developed.

14. Stephanopharynx type Näsmark, 1937 (Figure 97)

Genital papilla well developed, with a well developed sphincter papillae. Genital fold well developed, thick and broad. Genital sphincter absent. Tegumental papillae present on surface of genital fold. Genital-atrial-radial layer well developed.

15. Synethes type Sey, 1983b (Figure 98)

Genital papilla, genital and ventral folds present. Sphincter papillae and genital sphincter

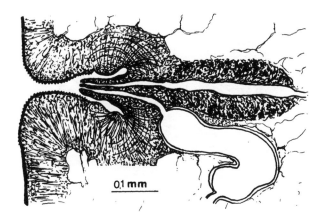

FIGURE 96. Serpenticaecum Eduardo and Peralta, 1987. (By permission of Dr. W. Junk Publishers, 1987, Dordrecht.)

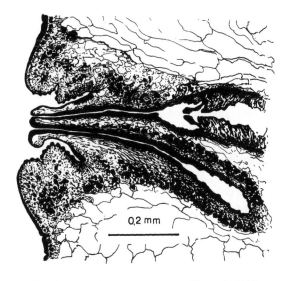

FIGURE 97. Stephanopharynx type Näsmark, 1937.

well developed and ventral sphincter strongly developed; papillae present along outer surface of ventral fold. Ventral atrium shallow. Genital-atrial-radial-layer well developed.

KEY TO TYPES OF TERMINAL GENITALIUM

1. Ventral and genital folds and genital papilla present 2
   Genital fold and genital papilla present 4
   Genital fold and genital papilla present with or without circular musculature 5
2. Ventral and genital sphincters and sphincter papillae present 3
   Genital sphincter and sphincter papillae present...................... Papillogenitalis
   Sphincter papillae present..Serpenticaecum
3. Tegumental papillae present along external surface of ventral fold.......... Synethes
   Tegumental papillae present along internal surface of ventral fold ....Gregarius
4. Tegumental papillae on surface of ventral papilla............................... Raja
   Tegumental papillae along internal surface of ventral fold ....Endopapillatus
5. Circular musculature present 6
   Circular musculature absent.. 10

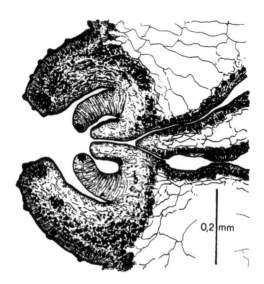

0,2 mm

FIGURE 98.   Synethes type Sey, 1983.

6. Sphincter papillae present... 7
   Genital sphincter present... 9
7. Sphincter papillae moderately developed... 8
   Sphincter papillae strongly developed...................................... Olveria
8. Sphincter papillae oval to longish.................................. Stephanopharynx
   Sphincter papillae on base of genital papilla, small and spherical........ Schoutedeni
9. Genital fold moderately developed, radial musculature strongly developed ....
   .... Pseudodiscus
   Genital fold enormously developed, radial fibers poorly developed..........Elongatus
10. Genital papilla voluminous, protruding ventral papilla covered with tegumental
    papillae .... Homalogaster
    Base of genital papilla and entire genital fold covered with papillae...Parvipapillatum
    Genital papilla and genital atrium normally developed.................... Cruciformis
    Genital papilla poorly developed, genital atrium shallow ....Noci

1.5.2.4. Types of Terminal Genitalium without Cirrus Sac, Hermaphroditic Pouch, and Tegumental Papillae

1. Bothriophoron type Näsmark, 1937 (Figure 99)

Ventral and genital folds and genital papilla present. Ventral sphincter loosely packed, genital sphincter and sphincter papillae well developed and compact. Ventral atrium large, genital atrium small. Radial fibers moderately developed.

2. Brevisaccus Eduardo and Javellana, 1987 (Figure 100)

Genital papilla small, with weakly developed genital sphincter. Genital fold clumsy with well-developed sphincter papillae. Radial musculature well developed. Terminal genitalium is almost surrounded by a thick rim turned inward.

3. Bubalis type Sey (Figure 101)

Genital papilla slender, without sphincter papillae. Genital fold well developed with sphincter at its tip. Genital atrium present. Genital-atrial-radial layer well developed.

4. Buxifrons type Näsmark, 1937 (Figure 102)

Genital papilla thick and clumsy farthest in, then rapidly tapers in outer parts. Genital

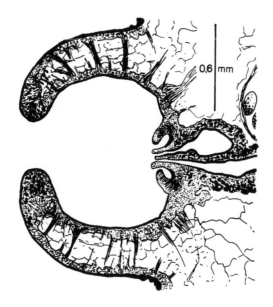

FIGURE 99. Bothriophoron type Näsmark, 1937.

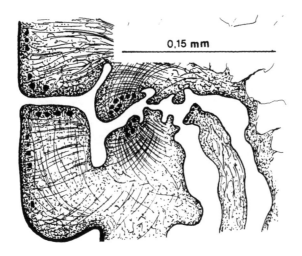

FIGURE 100. Brevisaccus type Eduardo and Javellana, 1987. (Courtesy of S. L. Eduardo and Peralta.)

atrium forming broad, ring-shaped cavity surrounds genital papilla. Well-developed radial muscle fibers of genital atrium present, circular musculature absent.

5. Calicophoron type Näsmark, 1937 (Figure 103)

Genital papilla and genital fold present; they can be exposed on a rectable elevation, called ventral papilla by Näsmark.[647] Genital sphincter moderately developed, sphincter papillae well developed and distinct. Radial muscle fibers well developed. True ventral fold and ventral atrium absent.

6. Clavula type Näsmark, 1937 (Figure 104)

Genital papilla clumsy, usually retracted, with well-defined sphincter papillae. Genital fold thick and broad with strongly developed, compact genital sphincter. Radial fibers well developed. True ventral fold and ventral atrium absent.

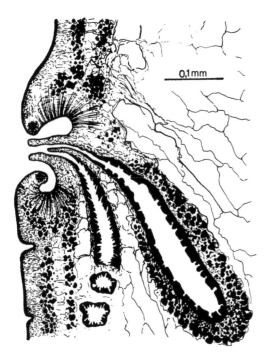

FIGURE 101.   Bubalis type Sey, 1983.

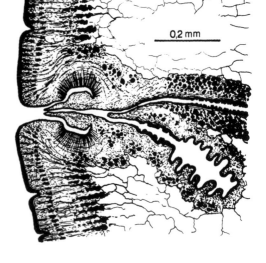

FIGURE 102.   Buxifrons type Näsmark, 1937.

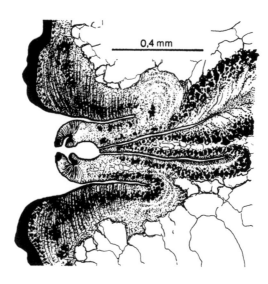

FIGURE 103.   Calicophoron type Näsmark, 1937.

7. Cotylophoron type Näsmark, 1937 (Figure 105)

This type is characterized by presence of genital sucker. Genital papilla long and slender without sphincter papillae. Genital fold thick and broad, without genital sphincter. Radial musculature well developed, longitudinal fibers present.

8. Dawesi type Eduardo, 1982a (Figure 106)

Whole structure characteristically small, genital papilla minute, without sphincter papillae. Genital fold moderately developed with a compact and round genital sphincter. Radial musculature moderately developed.

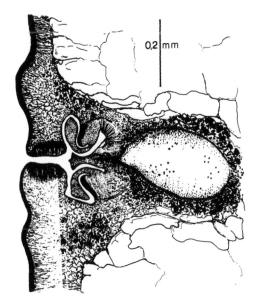

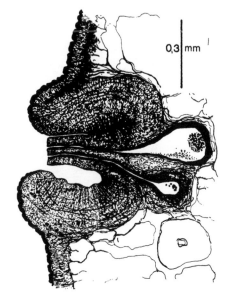

FIGURE 104. Clavula type Näsmark, 1937. FIGURE 105. Cotylophoron type Näsmark, 1937.

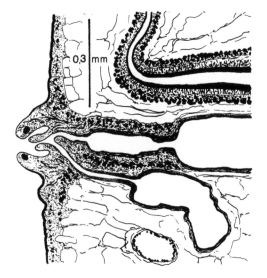

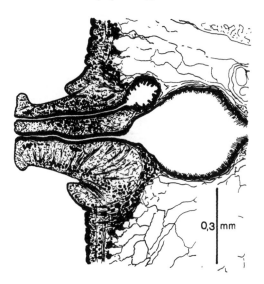

FIGURE 106. Dawesi type Eduardo, 1982. (By permission of Dr. W. Junk Publishers, 1985, Dordrecht.)

FIGURE 107. Deschiensi type Sey, 1984.

9. Deschiensi type Sey, 1984b (Figure 107)

Genital papilla protruding, enormous and thick, with circular muscle units along external margin of genital papilla. Genital fold very small, without genital sphincter. Radial fibers consisting of coarse bundles of fibers.

10. Duplicisphinctris type Sey et Graber, 1980 (Figure 108)

Genital papilla slender, curved without sphincter papillae. Genital fold thick and broad with two sphincters in its inner wall, anterior stronger than posterior one. Radial fibers poorly developed, longitudinal muscle fibers well developed.

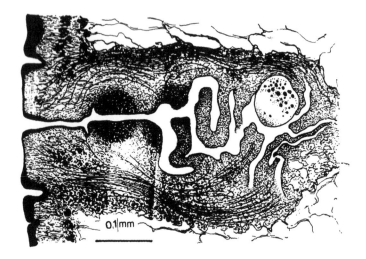

FIGURE 108. Duplicisphinctris type Sey and Graber, 1980. (From Sey and Graber, 1981.)

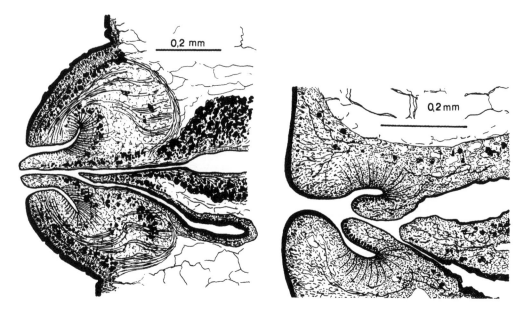

FIGURE 109. Epiclitum type Näsmark, 1937.   FIGURE 110. Explanatum type Näsmark, 1937.

11. Epiclitum type Näsmark, 1937 (Figure 109)

Characterized by cut-shaped outline, with strongly developed radial and longitudinal fibers which give it a suckerlike appearance without forming true sucker. Genital papilla thick, genital fold broad and stout. Genital sphincter and sphincter papillae absent. Ventral fold and ventral atrium absent.

12. Explanatum type Näsmark, 1937 (Figure 110)

Characteristically small in relation to body size, displays one of the smallest terminal genitalia. Genital papilla short with sphincter papillae consisting of loosely packed but distinct muscle units along its outer margin. Genital fold broad and clumsy without genital sphincter. Radial musculature of genital atrium well developed.

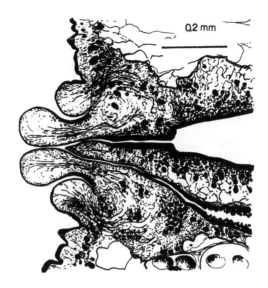

FIGURE 111. Gemellicotyle type Sey, 1987. (By permission of the Editor, Dr. B. S. Chauhana Commemoration Volume, 1975, Bhubaneswar.)

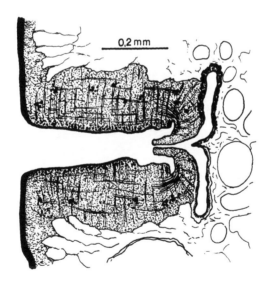

FIGURE 112. Gigantoatrium type Näsmark, 1937. (After Näsmark, 1937.)

13. Gemellicotyle type Sey, 1987 (Figure 111)

Genital papilla voluminous and thick with weakly developed sphincter papillae. Genital fold weakly developed without sphincter; genital atrium straight. Radial musculature poorly developed.

14. Gigantoatrium type Näsmark, 1937 (Figure 112)

Ventral atrium enormously developed, genital papilla inconsiderable, lying at bottom of ventral atrium. Genital papilla provided with slightly developed but distinct sphincter papillae. Ventral and genital sphincters absent, but they possess well-developed radial musculature with some longitudinal fibers. Genital fold absent.

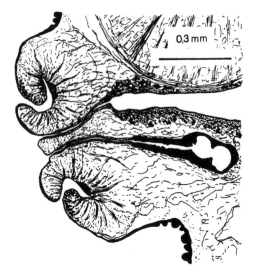

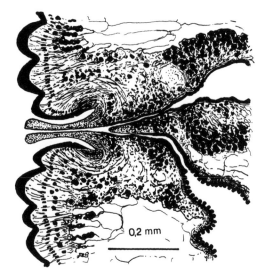

FIGURE 113.   Gigantocotyle type Näsmark, 1937.   FIGURE 114.   Gracile type Näsmark, 1937.

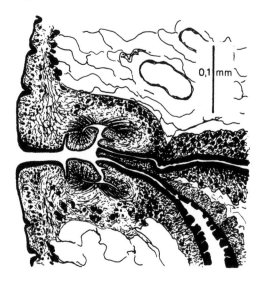

FIGURE 115.   Hippopotami type Näsmark, 1937.

15. Gigantocotyle type Näsmark, 1937 (Figure 113)

Genital papilla voluminous with moderately developed sphincter. Genital fold enormous, with genital sphincter composed of loosely packed muscle units. Radial musculature well developed. Ventral fold and ventral atrium absent.

16. Gracile type Näsmark, 1937 (Figure 114)

Genital papilla shows variable appearance, usually slender, with some poorly developed radial fibers and without sphincter papillae. Genital papilla absent.

17. Hippopotami type Näsmark, 1937 (Figure 115)

Genital papilla small without sphincter papillae. Genital fold well developed, curving inward, without genital sphincter and with strongly developed radial musculature. Genital-atrial-radial layer well developed. True ventral fold and ventral sphincter absent.

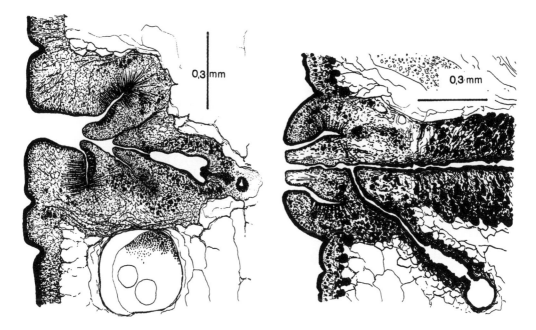

FIGURE 116.   Ichikawai type Näsmark, 1937.   FIGURE 117.   Leydeni type Näsmark, 1937.

18. Ichikawai type Näsmark, 1937 (Figure 116)

Genital papilla thick and clumsy with strongly developed sphincter papilla. Genital fold broad without genital sphincter. Genital atrium straight. Genital-atrial-radial layer well developed. Ventral fold and ventral sphincter absent.

19. Leydeni type Näsmark, 1937 (Figure 117)

Genital papilla thick without sphincter papillae. Genital fold well developed, genital atrium straight. Radial fibers well developed, forming coarse bundles. Ventral fold and ventral atrium absent.

20. Liorchis type Näsmark, 1937 (Figure 118)

Genital papilla slender, protruding, with sphincter papillae consisting of a loosely packed row of muscle units. Genital fold thick and broad without genital sphincter. Genital-atrial-radial layer evenly distributed coarse fibers. Ventral sphincter and ventral fold absent.

21. Macrosphinctris type Sey et Graber, 1979 (= Schistocotyle Sey et Graber, 1979) (Figure 119)

Characterized by presence of genital sucker. Genital papilla slender without sphincter. Anterior and internal part of genital fold thick and broad with strongly developed genital sphincter. Radial fibers strongly developed.

22. Mancupatus type Sey, 1983b (Figure 120)

Genital papilla slender with weakly developed sphincter. Genital fold embedded in tegumental wall and with moderately developed, round genital sphincter. Genital-atrial-radial layer well developed.

23. Microatrium type Näsmark, 1937 (Figure 121)

Characterized by poorly developed structural elements. Genital papilla poorly developed and retracted. In genital atrium, all kinds of musculature absent.

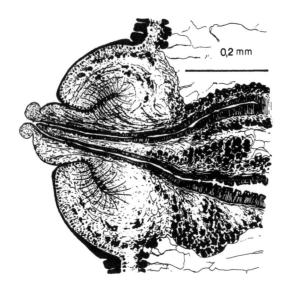

FIGURE 118.   Liorchis type Näsmark, 1937.

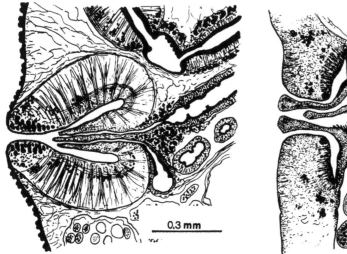

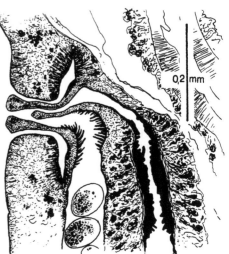

FIGURE 119.   Macrosphinctris type Sey and Graber, 1979.   FIGURE 120.   Mancupatus type Sey, 1983.

24. Microbothrium type Näsmark, 1937 (Figure 122)

Genital papilla short, slender, with well-developed sphincter papillae. Genital fold well defined, with well-developed and compact genital sphincter. Radial musculature well developed. Ventral fold and ventral atrium absent.

25. Minutum type Näsmark, 1937 (Figure 123)

Genital papilla short and thick with sphincter papillae, unevenly developed but distinct. Position of genital papilla may be deeper or shallower. Genital atrium with bundles of strongly developed radial muscle units. Genital fold present, genital sphincter absent.

26. Orthocoelium type Eduardo, 1982a (Figure 124)

Genital papilla stout, with well developed sphincter papillae. Genital fold thick and

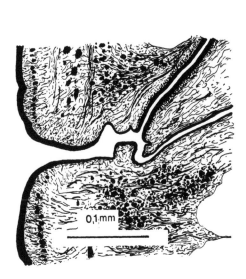

FIGURE 121.   Microatrium type Näsmark, 1937.

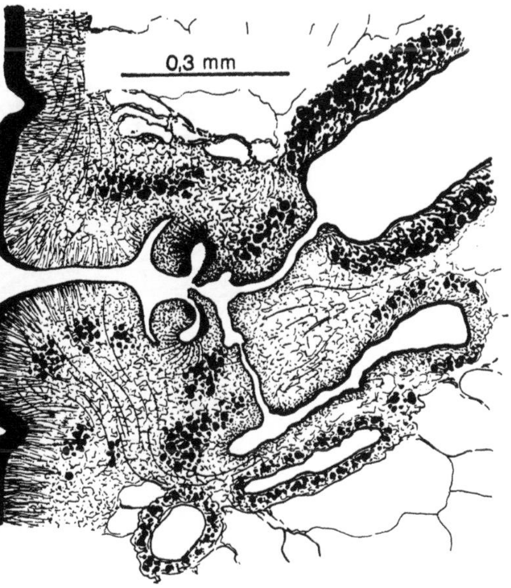

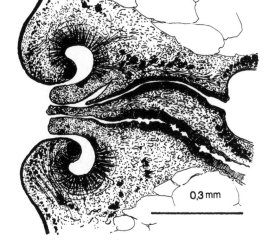

FIGURE 122.   Microbothrium type Näsmark, 1937.

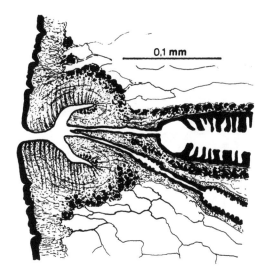

FIGURE 123.   Minutum type Näsmark, 1937.

FIGURE 124.   Orthocoelium type Eduardo, 1982.

broad curved inwards, with strongly developed and compact genital sphincter. Genital-atrial-radial layer strongly developed. Ventral fold and ventral atrium absent.

27. Pertinax type Sey, 1987 (Figure 125)

Genital papilla stout and broad, with poorly developed sphincter papillae. Genital fold and genital sphincter absent. Radial musculature poorly developed.

28. Pisum type Näsmark, 1937 (Figure 126)

Genital papilla short and slender with well developed sphincter papillae. Genital fold usually broad and clumsy, without genital papilla. Genital atrium straight. Genital-atrial-radial layer moderately developed.

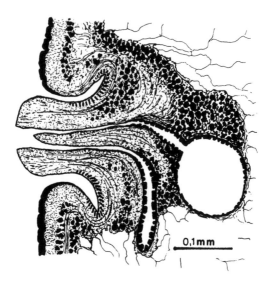

FIGURE 125.   Pertinax type Sey, 1987. (By permission of the *Parasitol. Hung.*, 1987, Budapest.)

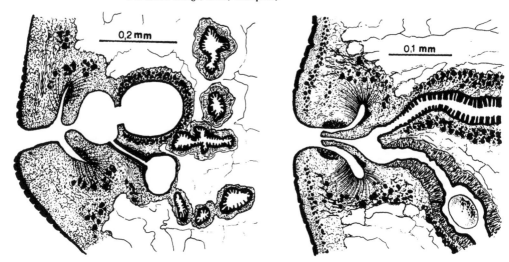

FIGURE 126.   Pisum type Näsmark, 1937.   FIGURE 127.   Scoliocoelium type Näsmark, 1937.

29. Scoliocoelium type Näsmark, 1937 (Figure 127)

Genital papilla slender with moderately developed sphincter. Genital fold well defined, genital sphincter well developed. Radial fibers well developed. True ventral atrium and ventral fold absent.

30. Sellsi type Näsmark, 1937 (Figure 128)

Terminal genitalium composed of genital sucker. Genital papilla thick, broad without sphincter. Genital fold enormous without genital sphincter. Radial musculature well developed. Genital atrium reduced.

31. Streptocoelium type Näsmark, 1937 (Figure 129)

Genital papilla clumsy, well developed with sphincter. Genital fold fairly distinct with strongly developed sphincter. Circular musculature forms a continuous layer throughout. Radial and longitudinal musculature well developed.

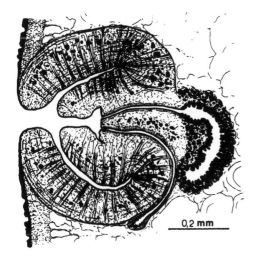

FIGURE 128.   Sellsi type Näsmark, 1937.

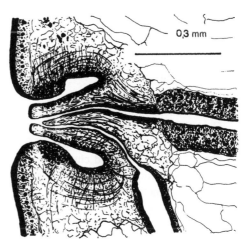

FIGURE 129.   Streptocoelium type Näsmark, 1937.

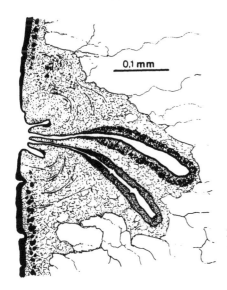

FIGURE 130.   Wagandi type Näsmark, 1937.

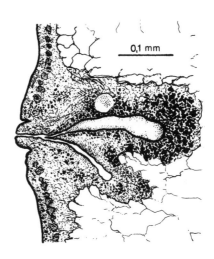

FIGURE 131.   Zygocotyle type Sey, 1987. (By permission of the *Parasitol. Hung.*, 1987, Budapest.)

32. Wagandi type Näsmark, 1937 (Figure 130)

Genital papilla and genital fold present. Genital papilla slender, genital fold clumsy. Sphincter papillae moderately developed. Radial muscle fibers moderately developed; they have some bands of muscle fibers, and they are thinly distributed.

33. Zygocotyle type Sey, 1987 (Figure 131)

Characterized by poorly developed structural elements, both musculature and size. Genital papilla small, with some longitudinal fibers. Circular and radial fibers entirely absent. True genital fold(s) absent, but body wall forms a muscular area without papillalike appearance.

KEY TO TYPES OF TERMINAL GENITALIUM

1. Genital papilla and genital fold present 2
   Genital papilla, genital and ventral folds present .... Bothriophoron
2. Genital sucker present 3
   Genital sucker absent 5
3. Genital sucker without sphincter 4
   Genital sucker with sphincter .... Macrosphinctris
4. Genital papilla long and slender .... Cotylophoron
   Genital papilla thick and clumsy .... Sellsi
5. Sphincter papillae and genital sphincter present 6
   Genital sphincter present 15
   Sphincter papillae present 17
   Circular muscle units absent 24
6. Genital papilla with simple sphincter 7
   Genital papilla with double sphincter .... Duplicisphinctris
7. Sphincter papillae and genital sphincter distinct 8
   Sphincter papillae and genital sphincter continuous .... Streptocoelium
8. Terminal genitalium enormous .... Gigantocotyle
   Terminal genitalium not enormous 9
9. Protrudable ventral papilla present .... Calicophoron
   Protrudable ventral papilla absent 10
10. Genital fold situated freely 11
    Genital fold embedded in parenchyma of ventral pouch .... Mancupatus
11. Genital sphincter enormously developed .... Clavula
    Genital sphincter not enormously developed 12
12. Genital fold in body wall 13
    Genital fold not in body wall 14
13. Genital papilla thick and broad, sphincter papillae relatively well developed ....
    .... Orthocoelium
    Genital papilla long and slender, sphincter papillae slightly developed ....
    .... Scoliocoelium
14. Fibers of sphincter papillae compact .... Microbothrium
    Fibers of sphincter papillae loosely packed .... Brevisaccus
15. Genital fold small, poorly developed radial fibers, genital sphincter small but
    distinct .... Dawesi
16. Genital fold well developed, radial muscle fibers strongly developed, genital sphincter
    insignificant .... Bubalis
17. Terminal genitalium enormously developed .... Gigantoatrium
    Terminal genitalium not enormous 18
18. Sphincter papillae moderately developed 19
    Sphincter papillae well developed .... Ichikawai
    Sphincter papillae poorly developed .... Liorchis
19. Genital fold present 20
    Genital fold absent .... Pertinax
20. Muscle fibers of circular units loosely packed 21
    Muscle fibers of circular units compact .... Wagandi
21. Genital papilla small, sphincter at base or along exterior margin of sphincter
    papillae 22
    Genital papilla large, protrudable, muscle units of sphincter papillae in some rows along
    exterior margin of sphincter papillae .... Deschiensi
22. Genital fold and genital papilla differently developed 23

Radial fibers poorly developed... Explanatum
Radial fibers well developed.. Minutum
23. Genital fold small, genital papilla voluminous........................ Gemellicotyle
    Genital fold thick and broad, genital papilla small............................ Pisum
24. Genital fold present... 25
    Genital fold absent... 26
25. Terminal genitalium cup-shaped in outline showing suckerlike configuration
    ....Epiclitum
    Genital folds small, bend inward with strongly developed radial fibers .. Hippopotami
    Genital folds well developed bend outward with moderately developed radial fibers ..
    .... Leydeni
26. Genital papilla poorly developed.. 27
    Genital papilla well developed... 28
27. Genital papilla at base of canal formed by wall of terminal genitalium ... Microatrium
    Genital papilla and its surrounding area elevated over body surface .... Zygocotyle
28. Radial musculature strongly developed ....Buxifrons
    Radial muscle fibers few... Gracile

1.5.3. ACETABULUM AND KEYS TO ITS TYPES

The ventral sucker or acetabulum is usually a large, stout muscular organ at the posterior end of the body. It exhibits significant variability both in size and appearance, depending on the groups of amphistomes. Of the structural elements, the muscle fibers are the most important. They may be circular, longitudinal, radial, or oblique. The circular muscle units running along the outer and inner margin of acetabulum show a characteristic arrangement on median sagittal section. Diagnosis of structural elements of acetabulum was realized by Näsmark[647] and further details were added by Reinhardt,[772] Eduardo,[271] and Sey.[849,850] The schematic representation of a general type of acetabulum (Figure 132) includes all parts necessary for characterization and differentiation.

KEY TO ABBREVIATIONS

Acetabulum (Figure 132)

bp   — basal part
dec   — dorsal external circular
dec_1 — dorsal external circular outer
dec_2 — dorsal external circular inner
dic   — dorsal internal circular
el   — exterior longitudinal
mp   — marginal part
mec   — middle exterior circular
of   — oblique fibers
rf   — radial fibers
vec   — ventral exterior circular
vec_1 — ventral exterior circular outer
vec_2 — ventral exterior circular inner
v.i.c. — ventral internal circular

Types of acetabulum, based on gross morphological and histomorphological features, were divided by Näsmark[647] into three groups and a possible fourth with uncertain position. A comprehensive survey of a great number of different types of acetabula indicates that the

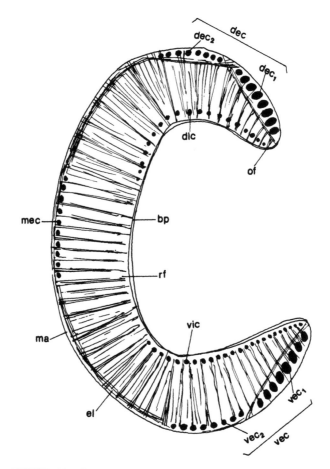

FIGURE 132.   Schematic representation of a general type of acetabulum
(median sagittal section) showing details which are necessary for diagnosis.
(By permission of the *Parasitol. Hung.*, 1987, Budapest.)

fourth group of Näsmark's classification can be placed into group II because of the shape
and arrangement of musculature. Accordingly, three types of groups of acetabulum can be
differentiated.

1.5.3.1. Acetabulum of Group I

Acetabulum, with various forms of appearance, differently developed acetabular cavity
and musculature. Some of them similar to suctorial disk or suckling depression rather than
to sucker. Thickness of acetabulum usually moderate, but sometimes poorly or well developed. Characterized by poorly developed musculature: number of dorsal (d) and ventral (v),
exterior (e) and interior (i) circular (c) muscle units may be equal or dissimilar. In the latter
case, d.e.c. series has a greater number of units than that of v.e.c. series. D.e.c. series does
not divide into d.e.1 and d.e.2 circular. Radial musculature usually weakly developed but
in some types it may be strongly developed along marginal edge. Ventral longitudinal series
is unequally developed: typically more developed ventrally but rarely can be found symmetrically (dorsal and lateral) or dorsally only. Tissue of acetabulum usually vesicular.

1.5.3.1.1. Types of Acetabulum of Group I
1. Asper type Sey, 1987 (Figure 133)*

Size enormous, with papillae in its inner surface. With poorly developed acetabular wall

---

*   Types of acetabulum (median sagittal section) are shown in Figures 133 to 184.

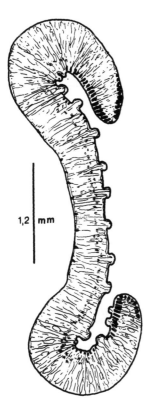

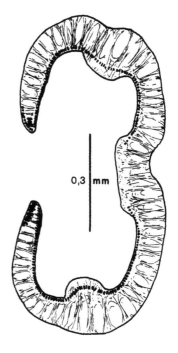

FIGURE 133.   Asper type Sey, 1987. (By permission of the *Parasitol. Hung.*, 1987, Budapest.)

FIGURE 134.   Basidiodiscus type Sey, 1987. (By permission of the *Parasitol. Hung.*, 1987, Budapest.)

and musculature, similar to sucking disk rather than to acetabulum. D.e.c. and v.e.c. with same number and development. D.i.c. and v.i.c. with same number and development of units. Exterior units bigger than those inward. M.e.c. present. Radial fibers poorly developed, exterior longitudinal fibers weakly developed but distinct.

2. Basidiodiscus type Sey, 1987 (Figure 134)

Small in size with some clumsy papillae in its inner surface. Wall of acetabulum narrow. D.e.c. and v.e.c. similar in number but v.e.c. more strongly developed. D.i.c. and v.i.c. series consist of muscle units of similar number and development. M.e.c. units present. Radial fibers poorly developed.

3. Brevicaecum type, Sey, 1987 (Figure 135)

Moderate in size, wall of acetabulum thin. Circular musculature in dorsal and ventral parts of acetabulum show considerable differences. D.e.c. series consists of only a single row of units which are numerous and similar in size. D.i.c. series is well developed with muscle units decreasing inward. Circular musculature in both v.e.c. and v.i.c. series divide clearly into two parts: exterior series with longer units and interior series with much shorter ones. Exterior longitudinal fibers strongly developed along the whole marginal part of acetabulum. Radial muscle fibers moderately developed.

4. Chiostichorchis type Sey, 1987 (Figure 136)

Well developed in size. Wall of acetabulum moderately developed. Characterized by presence of fewer muscle units in dorsal part of acetabulum than that of ventral part. Units

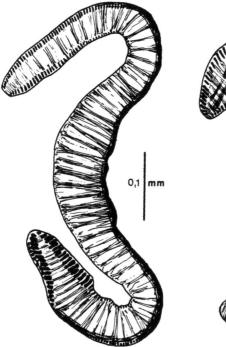

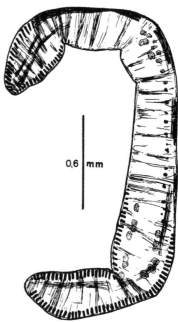

FIGURE 135. Brevicaecum type Sey, 1987. (By permission of the *Parasitol. Hung.*, 1987, Budapest.)

FIGURE 136. Chiostichorchis type Sey, 1987. (By permission of the *Parasitol. Hung.*, 1987, Budapest.)

of d.e.c. series are few, equal in size. V.e.c. series consists of a great number of units with sizes decreasing inward. Number of units in d.e.c. smaller than those in v.i.c. Exterior longitudinal fibers more or less well developed in dorsal and marginal parts of acetabulum. Radial musculature moderately developed, forming some bundles of fibers. M.e.c. present.

5. Choerocotyloides type Sey, 1987 (Figure 137)

Enormous in size, with well developed musculature and moderately developed acetabular wall and acetabular cavity. D.e.c. and v.e.c. series consist of a single row of units, comprising a few, weakly developed units. Ventral part of acetabulum with well developed, oblique fibers. Radial fibers well developed, forming closely packed coarse bundles. D.i.c. and v.i.c. series similar in size and in number. Exterior longitudinal fibers moderately developed. M.e.c. present.

6. Cladorchis type Näsmark, 1937 (Figure 138)

Moderate in size, wall of acetabulum well developed. D.e.c. consists of a longer exterior series, it is followed by an empty space and continued with some more circular units. V.e.c. series forms a compact spot composed of 6 or 7 muscle units. The units of d.i.c. and v.i.c. series are similar in number and size. Ventral exterior longitudinal fibers strongly developed, an isthmus divides them into two parts. M.e.c. series weakly developed, composed of regularly spaced units. Radial fibers form coarse bundles.

7. Cleptodiscus type Sey, 1987 (Figure 139)

Small in size, with symmetrically arranged but poorly developed circular musculature. Wall of acetabulum thin. Number of units of d.e.c. and v.e.c. series more or less similar in number and in size. Number of d.i.c. and v.i.c. series numerous, exterior units bigger,

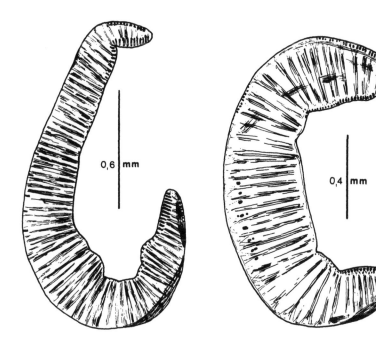

FIGURE 137. Choerocotyloides type Sey,
1987. (By permission of the *Parasitol. Hung.*,
1987, Budapest.)

FIGURE 138. Cladorchis type Näsmark,
1937.

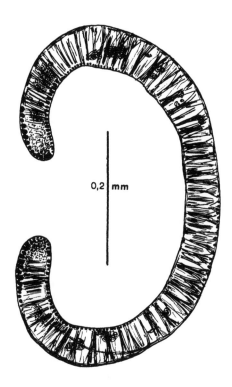

FIGURE 139. Cleptodiscus type Sey, 1987.
(By permission of the *Parasitol. Hung.*, 1987,
Budapest.)

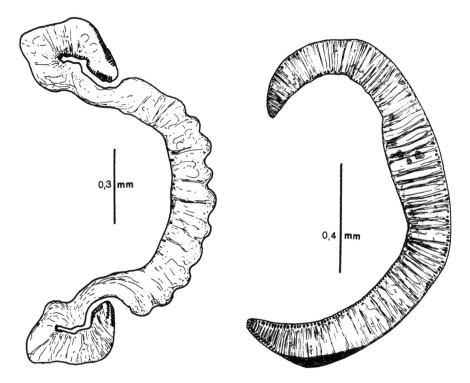

FIGURE 140. Dadayius type Sey, 1987. (By permission of the *Parasitol. Hung.*, 1987, Budapest.)

FIGURE 141. Dermatemytrema type Sey, 1983.

interior decreasing inward. M.e.c. moderately developed. Radial musculature well developed, fibers form smaller bundles.

8. Dadayius type Sey, 1987 (Figure 140)

Moderate in size with weakly developed musculature and thin acetabular wall. Marginal line of acetabulum concave, basal line convex and undulating. This wavy line is the consequence of the lineal protuberances situated on the inner surface of acetabulum, seen on median sagittal section. Units of d.e.c. and v.e.c. as well as d.i.c. and v.i.c. are small in number and inconsiderably developed. Radial musculature weakly developed.

9. Dermatemytrema type Sey, 1983a (Figure 141)

Moderate in size with poorly developed musculature. Units of d.e.c. series are smaller in number than those in d.i.c. series, but they are similar, slightly developed. Units of v.e.c. and v.i.c. are somewhat more developed and are numerous. External longitudinal fibers well developed in ventrolateral region. M.e.c. series consists of numerous muscle units, weakly developed. Radial muscle fibers regularly spaced, slightly developed.

10. Dilymphosa type Sey, 1987 (Figure 142)

Moderate or well developed in size. Acetabular wall moderately developed. D.e.c. series consists of a single row of muscle units with more or less equally developed, numerous muscle units. V.e.c. series short with some muscle units at acetabular extremity. D.i.c. and v.i.c. series are more or less similar in number and size. Exterior longitudinal fibers well developed in ventrolateral region. Radial musculature well developed forming coarse bundles of fibers.

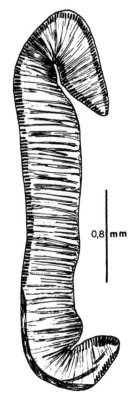 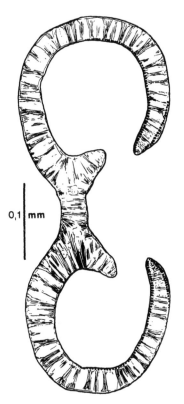

FIGURE 142. Dilymphosa type Sey, 1987. FIGURE 143. Diplodiscus type Näsmark, 1937.

11. Diplodiscus type Näsmark, 1937 (Figure 143)

Small in size with accessory sucker or muscular plug, poorly developed musculature, and thin acetabular wall. Acetabular cavity well developed. Numbers and arrangement of d.e.c., v.e.c. muscle series and d.i.c., v.i.c. muscle series are similar. Number of muscle units of internal series is higher than that of exterior series. M.e.c. present, a few. Exterior longitudinal fibers normally developed. Radial muscle fibers well developed, forming loosely spaced bundles.

12. Gemellicotyle type Sey, 1987 (Figure 144)

Enormous in size with poorly developed musculature and moderately developed acetabular wall. In its center there is a large prominence surrounded by irregularly spaced papillae. D.e.c. and v.e.c. consist of a single row of units. Units of d.e.c. and of v.e.c. are few, small, and equal in size. Number in series d.i.c. and v.i.c. is high; they are small and equal in size. Radial fibers weakly developed. External longitudinal fibers poorly developed.

13. Gigantheus type Sey, 1987 (Figure 145)

Enormous in size with symmetrically arranged musculature. Wall of acetabulum strongly developed. Units of d.e.c. and of v.e.c. and their development are more or less similar. D.i.c. and v.i.c. are similar in number and in size. Exterior longitudinal fibers in dorsal and ventral parts of acetabulum equally developed. M.e.c. moderately developed. Radial fibers moderately developed and distributed evenly.

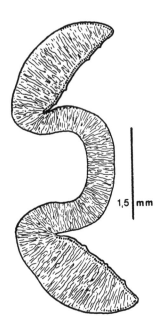

FIGURE 144. Gemellicotyle type Sey, 1987. (By permission of the Editor, Dr. B. S. Chauhana Commemoration Volume, 1975, Bhubaneswar.)

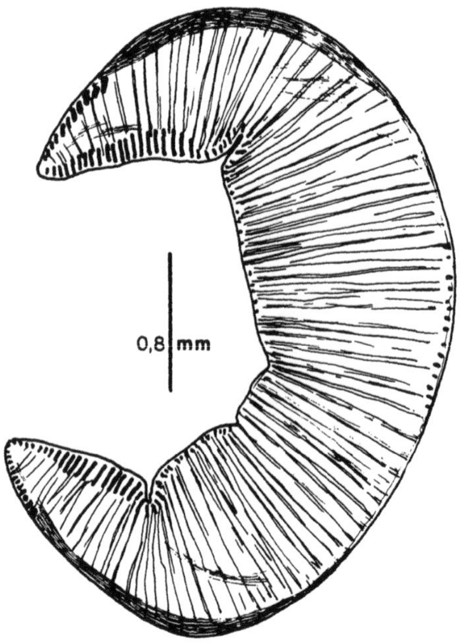

FIGURE 145. Gigantheus type Sey, 1987. (By permission of the *Parasitol. Hung.*, 1987, Budapest.)

14. Megalodiscus type Sey, 1983a (Figure 146)

Well developed in size with muscle plug, poorly developed musculature, and with moderately developed acetabular wall and cavity. Muscle units of d.e.c., v.e.c. and of d.i.c., v.i.c are similar in size, number, and arrangement. M.e.c. absent. Exterior longitudinal fibers normally developed. Radial fibers poorly developed forming loosely packed bundles.

15. Microrchis type Sey, 1978 (Figure 147)

Well developed in size. Wall of acetabulum weakly developed. D.e.c. consists of a single group of units; it is longer than v.e.c. units slightly developed and almost equal in size. The units of d.i.c. and v.i.c. series are similar in number and in size. Ventrolateral, longitudinal muscle fibers form muscular pad. M.e.c. moderately developed. Radial fibers moderately developed, forming small bundles of fibers.

16. Nematophila type Sey, 1973b (Figure 148)

Enormous in size with well-developed musculature, wall and cavity of acetabulum well developed. D.e.c. series composed of numerous, relatively small muscular units. Number of v.e.c. muscle units few. Numbers of units of d.i.c. and v.i.c. series are similar in size and in arrangement. Muscle units of the middle part of the series are the longest, and they gradually decrease both outside and inside. M.e.c. poorly developed. External longitudinal fibers strongly developed along the margin of lower part of the acetabulum. Radial musculature well developed, bundles of fibers are packed.

17. Neocladorchis type Sey, 1986 (Figure 149)

Moderate in size, with moderately developed acetabular wall and acetabular cavity. Units in d.e.c. series are numerous; in v.e.c. series there is a 100-μm-long interruption without muscle units. D.i.c. and v.i.c. series are similar in number and arrangement. M.e.c. series absent. Exterior longitudinal fibers normally developed. Radial fibers poorly developed.

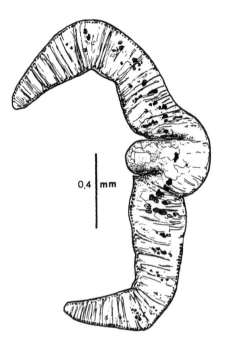

FIGURE 146.   Megalodiscus type Sey, 1983.

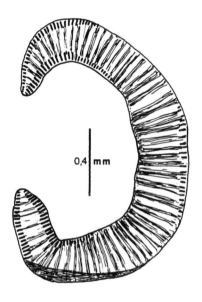

FIGURE 147.   Microrchis type Sey, 1987. (By permission of the *Parasitol. Hung.*, 1987, Budapest.)

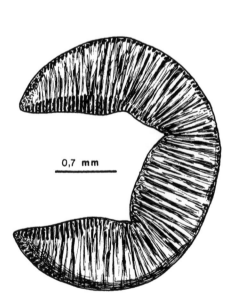

FIGURE 148.   Nematophila type Sey, 1973.

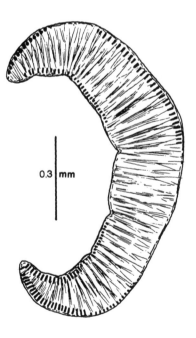

FIGURE 149.   Neocladorchis type Sey, 1986.

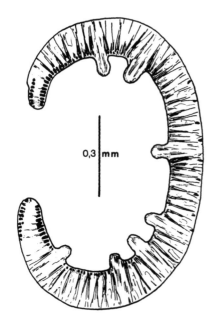 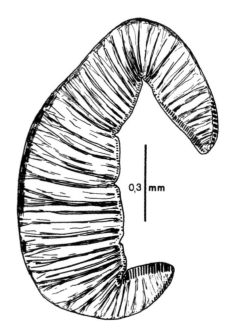

FIGURE 150.   Pfenderius type Näsmark, 1937.   FIGURE 151.   Pseudocladorchis type Näsmark, 1937.

18. Pfenderius type Näsmark, 1937 (Figure 150)

Moderate in size with papillae around its inner surface. Wall of acetabulum narrow. D.e.c. and v.e.c. series weakly developed and their number practically the same. D.i.c. and v.i.c. show similar arrangement and units in number; number of units numerous, exterior units longer, their measurements decrease inward. M.e.c. few. Radial fibers moderately developed, exterior longitudinal fibers poorly developed but distinct.

19. Pseudocladorchis type Näsmark, 1937 (Figure 151)

Moderate in size, with poorly developed circular musculature and with relatively thick wall of acetabulum. Number of units of d.e.c. and v.e.c. series are similar in number and size. Of the ventral circular series, exterior units of v.i.c. are longer than the rest of v.i.c. series. M.e.c. few. Exterior longitudinal fibers weakly developed. Radial muscle fibers well developed, forming loosely spaced bundles.

20. Pyriformis type Sey, 1987 (Figure 152)

Well developed in size, with poorly developed musculature and acetabular cavity. Acetabular wall well developed. Units of d.e.c. and v.e.c. are few, and they are small. Muscle units of d.i.c. and v.i.c. are located symmetrically. External longitudinal fibers normally developed. Radial muscle fibers weakly developed.

21. Sandonia type Sey, 1987 (Figure 153)

Small in size with moderately developed musculature and acetabular wall and with well-developed acetabular cavity. Muscle units of d.e.c. series are few, and they are small. Number of units of v.e.c. moderately developed; sometimes they are situated in double rows. At the tip of the ventral part of acetabulum there is a concentration of muscle units. D.i.c. and v.i.c. series are similar in size and number, and they show similar arrangement. Exterior longitudinal fibers normally developed. Radial muscle fibers moderately developed forming loosely packed bundles.

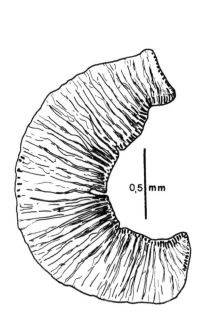

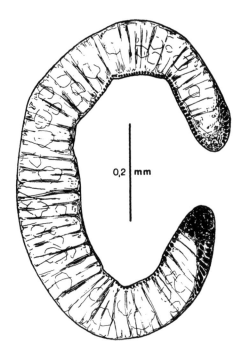

FIGURE 152.   Pyriformis type Sey, 1987. (By permission of the *Parasitol. Hung.*, 1987, Budapest.

FIGURE 153.   Sandonia type Sey, 1987. (By permission of the *Parasitol. Hung.*, 1987, Budapest).

22. Schizamphistomum type Näsmark, 1937 (Figure 154)

Well developed in size, with moderately and well-developed musculature, wall of acetabulum, and acetabular cavity. Number of units of d.e.c. numerous, and they unite with
m.e.c. Number of v.e.c. is small, d.i.c. and v.i.c. series are similar in size and in number,
their units are without joining in middle basal part of acetabulum. Exterior longitudinal
fibers normally developed. Radial fibers well developed forming closely spaced bundles.

23. Solenorchis type Sey, 1980a (Figure 155)

Enormous in size; acetabular wall moderately developed. D.e.c. series consists of only
a single group of units which are moderately developed, and the number of units is not too
high. V.e.c. divided into two series by a well developed muscular pad. Number of units in
exterior series is smaller than that of interior series. Exterior longitudinal fibers well developed in ventrolateral region. Radial musculature well developed, fibers situated close to
one another, forming bundles. M.e.c. moderately developed.

24. Spinolosum type Sey, 1987 (Figure 156)

Well developed in size, with moderately developed musculature, acetabular wall, and
with well developed acetabular cavity. D.e.c. series is a single row, consisting of a great
number of muscle units. D.e.c. series unites with m.e.c. muscle units. Number of v.e.c.
muscle units few. D.i.c. and v.i.c. unit series are similar in size and number forming a
continuous series along the basal line of acetabulum. Exterior longitudinal fibers well developed. Radial fibers well developed and form coarse bundles of fibers, in dorsal and ventral
parts of acetabulum. These strong bundles show somewhat oblique arrangement. In dorsal
and ventral parts of acetabulum.

25. Stichorchis type Näsmark, 1937 (Figure 157)

Enormous in size with moderately developed musculature, wall and cavity of acetabulum
well developed. D.e.c. series consists of a great number of muscle units, including a small,

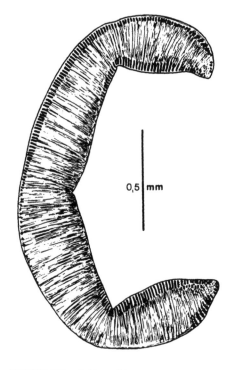

FIGURE 154. Schizamphistomum type Näsmark, 1937.

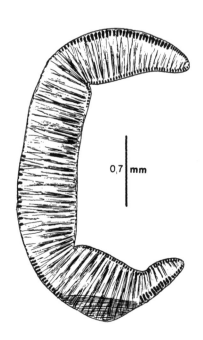

FIGURE 155. Solenorchis type Sey, 1980.

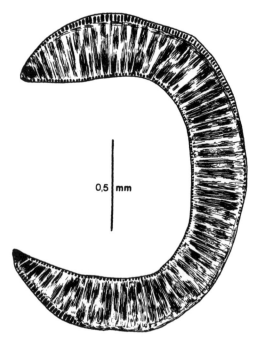

FIGURE 156. Spinolosum type Sey, 1987. (By permission of the *Parasitol. Hung.*, 1987, Budapest.)

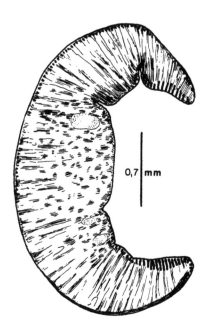

FIGURE 157. Stichorchis type Näsmark, 1937.

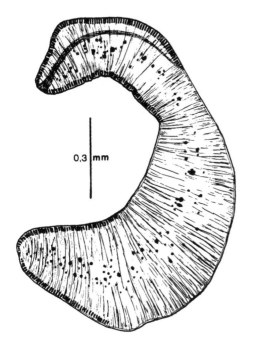

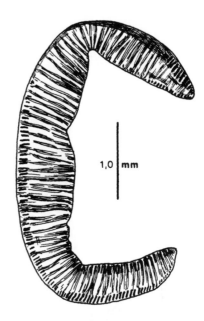

FIGURE 158. Stunkardia type Sey, 1987. (By permission of the *Parasitol. Hung.*, 1987, Budapest.)

FIGURE 159. Taxorchis type Sey, 1987. (By permission of the *Parasitol. Hung.*, 1987, Budapest.)

empty area near the beginning of the series. V.e.c. series is small in number and size. Number, size, and arrangement of d.i.e. and v.i.e. are similar to one another. Exterior longitudinal fibers somewhat more developed in ventrolateral part of acetabulum. Radial musculature moderately developed, obliquely running fibers situated in the middle and basal parts of acetabulum. M.e.c. present.

26. Stunkardia type Sey, 1987 (Figure 158)

Moderate in size, wall of acetabulum strongly developed. Number of units of d.e.c. numerous, evenly developed; number of units of v.e.c. is somewhat smaller but they are similarly developed to d.e.c. units. Number of d.e.c. is somewhat more than that of v.i.c., but units are equal in size. M.e.c. probably absent. Radial musculature poorly developed; fibers are evenly and loosely spaced. In dorsal part of acetabulum, well developed longitudinal muscle fibers present.

27. Taxorchis type Sey, 1987 (Figure 159)

Enormous in size. D.e.c. series is a short row of units with well-developed muscle elements. V.e.c. series consists of a great number of units, more developed in ventrolateral region. Outside of d.i.c. and v.i.c. series, units of the former series are small in size as well as in number compared with that of d.i.c. Exterior longitudinal fibers well developed in dorsolateral region. Radial musculature well developed, fibers coarse and form bundles.

28. Venezuelensis type Sey, 1987 (Figure 160)

Well developed in size, musculature more or less symmetrically developed and arranged. Wall of acetabulum well developed. Number of units of d.e.c. series small in number, with longer units in the middle of the series; v.e.c. are few and universally smaller. Units of d.i.c. and v.i.c. are similar in size and number. M.e.c. series poorly developed with several units. External longitudinal fibers strongly developed both along dorsal and ventral parts of

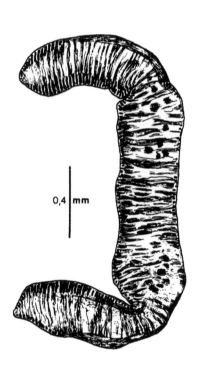

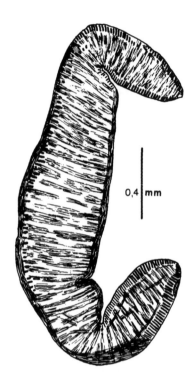

FIGURE 160.   Venezuelensis type Sey, 1987. (By permission of the *Parasitol. Hung.*, 1987, Budapest.)

FIGURE 161.   Wardius type Sey, 1987. (By permission of the *Parasitol. Hung.*, 1987, Budapest.)

acetabulum. Radial musculature strongly developed; muscle fibers form coarse, irregularly spaced bundles.

29. Wardius type Sey, 1987 (Figure 161)

Well developed in size. Acetabular wall moderately developed. The units of d.e.c. series unevenly developed; they are smaller but greater in number than are muscle units of v.e.c. (d.i.c. and v.i.c. series are similar in number and size). External and internal longitudinal fibers strongly developed, and in ventral side longitudinal fibers form a muscular pad. Radial fibers strongly developed and form irregular coarse bundles. M.e.c. very well developed, stretching along the whole marginal side.

30. Zygocotyle type Sey, 1975 (Figure 162)

Moderate in size. Wall of acetabulum moderately developed. D.e.c. series consists of well-developed units. Number of v.e.c. series is smaller, but they are longer than that of d.e.c. series. Number of d.i.c. and v.i.c. is the same, but the units of the latter are somewhat longer, especially of exterior units. External longitudinal fibers are rather developed in ventral part of acetabulum and well developed on lateral side. Radial musculature poorly developed, more or less evenly spaced weak fibers.

KEY TO TYPES OF ACETABULUM OF GROUP I

1. Circular muscle series symmetrical in number and arrangement both in dorsal and ventral halves.. 2

   Number of muscle units of d.e.c. series greater in number than in v.e.c. series..... 9

   Muscle units in v.e.c. greater in number than in d.e.c............................. 22

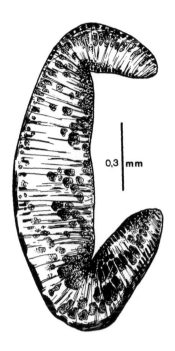

FIGURE 162. Zygocotyle type Sey, 1975.

15. D.e.c. units a few ....Choerocotyloides

    D.e.c. units numerous ....Microrchis

16. Longitudinal muscle fibers moderately developed.........................Dilymphosa

    Both external and internal longitudinal fibers strongly developed............. Wardius

17. D.e.c. continuous.. 18

    D.e.c. not continuous, interrupted by an empty space .... Stichorchis

18. Acetabular tissue slightly vesicular 19

    Acetabular tissue strongly vesicular.. Zygocotyle

19. Number of d.e.c. enormous.. 20

    Number of d.e.c. not enormous.. 21

20. D.i.c. and v.i.c. units big in size ....Schizamphistomum

    D.i.c. and v.i.c. units small in size...Spinolosum

21. V.e.c. and v.i.c. units bigger than that of dorsal circular series .... Brevicaecum

    D.e.c. and v.e.c. units equal in size......................................Stunkardia

22. V.e.c. series continuous 23

    V.e.c. series interrupted by strongly developed oblique fibers.............Solenorchis

    V.e.c. series interrupted by an empty space .... Neocladorchis

23. Dorsal oblique muscle fibers present... 24

    Dorsal oblique fibers absent 25

24. Ventral rim of acetabulum with muscular elements .... Sandonia

    Ventral rim of acetabulum without muscular elements....................Cleptodiscus

25. D.e.c. units large in size...Taxorchis

    D.e.c. units medium in size ....Chiostichorchis

1.5.3.2. Acetabulum of Group II

Acetabulum with typical suckerlike forms, well-developed musculature, well-developed acetabular wall and acetabular cavity. Dorsal exterior circular series often divides into two parts (d.e.c.1 and d.e.c.2), if d.e.c. consists of a single row; in this case it is not so long as in the acetabulum group I. Radial musculature well or strongly developed, exterior longitudinal series normally developed. Tissue of acetabulum is not vesicular.

1.5.3.2.1. Types of Acetabulum of Group II

1. Buxifrons type Näsmark, 1937 (Figure 163)

Small in size, with well-developed musculature, well developed acetabular cavity and wall. D.e.c. and v.e.c. series consist of a single group of units; units of the middle part are the longest. D.i.c. and v.i.c. series are similar in size and arrangement. Radial fibers are well developed, forming regularly spaced bundles. Exterior longitudinal fibers well developed, especially in dorsal and ventral parts of acetabulum. M.e.c. present.

2. Carmyerius type Sey, 1983b (Figure 164)

Well developed in size, with well-developed musculature and acetabular wall and moderately developed acetabular cavity. D.e.c., v.e.c. and d.i.c., v.i.c. series consist of a single group of units. D.e.c. series is similar in size and arrangement. Number of interior series is somewhat greater than that of exterior series. Radial fibers moderately developed, forming loosely packed bundles. Exterior longitudinal fibers well developed. M.e.c. present.

3. Calicophoron type Näsmark, 1937 (Figure 165)

Enormous in size, with well-developed musculature, acetabular wall, and acetabular cavity. D.e.c. and v.e.c. series consist only of a single group of units. D.i.c. and v.i.c. series are similar in size; longest units are in middle part of series. Units in interior series are greater in number than those in exterior series. Sometimes, irregularly spaced units in

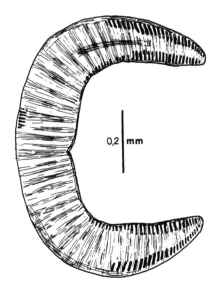

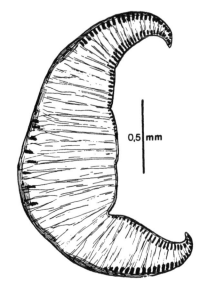

FIGURE 163. Buxifrons type Näsmark, 1937.   FIGURE 164. Carmyerius type Sey, 1983.

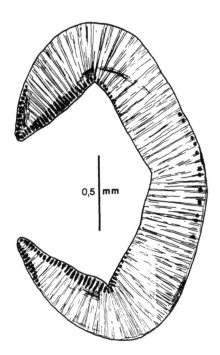

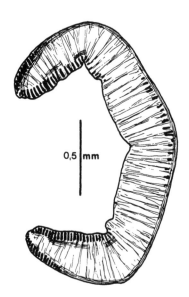

FIGURE 165. Calicophoron type Näsmark, 1937.   FIGURE 166. Cotylophoron type Näs-mark, 1937.

d.e.c.2. region are found, but they cannot be regarded as a true d.e.c.2 series. Radial fibers fairly well developed. M.e.c. units composed of loosely spaced units. Exterior longitudinal fibers are small in number but distinct.

4. Cotylophoron type Näsmark, 1937 (Figure 166)

Well developed in size, with well-developed musculature, acetabular wall, and acetabular

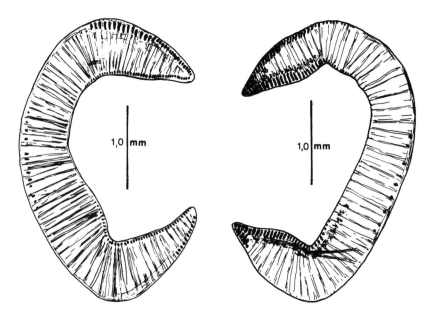

FIGURE 167.   Duplicitestorum type Näsmark, 1937.

FIGURE 168.   Explanatum type Näsmark,
1937.

cavity. D.e.c. and v.e.c. series consist of a single group of units. A well-developed band
of oblique fibers run along the interior border of d.e.c. series. D.i.c. and v.i.c. series are
similar in size, units gradually diminish inward. Radial fibers well developed. M.e.c. consists
of a single group of units. Exterior longitudinal fibers well developed.

5. Duplicitestorum type Näsmark, 1937 (Figure 167)

Enormous in size with moderately developed musculature, acetabular wall and well-
developed acetabular cavity. D.e.c. series consists of two groups of units; d.e.c.2 units are
longer in size and greater in number than d.e.c.1 units. D.i.e. and v.i.e. series composed
of small units. Radial fibers well developed, oblique fibers few and insignificant. Exterior
longitudinal fibers normally developed. M.e.c. units consist of regularly placed units.

6. Explanatum type Näsmark, 1937 (Figure 168)

Size enormous, with well-developed musculature, acetabular wall, and acetabular cavity.
D.e.c. consists of two groups of units, d.e.c.1 units are more strongly developed than d.e.c.2
units. V.e.c., d.i.c., and v.i.c. series consist only of a single group of units, diminishing
in size inward. M.e.c. series consists of irregularly spaced units. Radial fibers well developed,
oblique fibers few, inconsiderable. Exterior longitudinal fibers poorly developed but distinct.

7. Fischoederius type Sey, 1983b (Figure 169)

Moderate in size, with moderately developed musculature and acetabular cavity and
well-developed acetabular wall. D.e.c. series consists of two groups of units, separated by
thin longitudinal fibers. D.e.c. and v.i.c. series are similar in number and size. Radial
musculature well developed. Exterior longitudinal fibers strongly developed. M.e.c. consists
of irregularly spaced units.

8. Gastrothylax type Näsmark, 1937 (Figure 170)

Enormous in size, with well-developed musculature and acetabular wall and moderately
developed acetabular cavity. D.e.c. and v.i.c. series consist only of a single group of units.

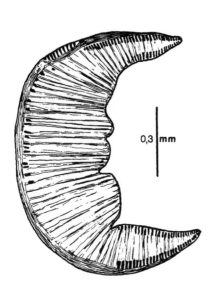

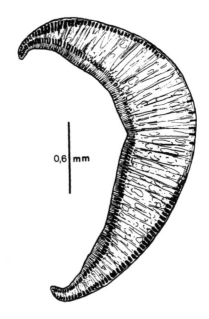

FIGURE 169.   Fischoederius type Sey, 1983.

FIGURE 170.   Gastrothylax type Näsmark, 1937.

Number of units of series is numerous, longest units in middle part of series, diminishing both inward and outward. Radial fibers are well developed, forming closely packed bands. Exterior longitudinal fibers moderately developed. M.e.c. series consists of irregularly spaced units.

9. Gigantocotyle type Eduardo, 1982a (Figure 171)
Enormous in size, with moderately developed musculature, well-developed acetabular wall and acetabular cavity. D.e.c. and v.e.c. divide into two groups of units; d.e.c.1, d.e.c.2 and v.e.c.1, v.e.c.2, respectively. Exterior units of both series are well developed. Number of units in d.e.c.2 is many more than are in d.e.c.1. In v.e.c.2, there are few and small units. D.i.c. and v.i.c. series consist of similar units in size, gradually diminish in size inward. M.e.c. series composed of irregularly packed units. Radial fibers well developed, oblique fibers few and insignificant. Exterior longitudinal fibrs moderately developed but distinct.

10. Nilocotyle type Näsmark, 1937 (Figure 172)
Small in size, with moderately developed musculature, acetabular cavity, and acetabular wall. D.e.c. and v.e.c. consist of a single group of units. Number of units in interior series is greater in number than that of exterior series. Radial fibers well developed, forming loosely packed, weak bundles. Exterior longitudinal fibers well developed. M.e.c. present.

11. Paramphistomum type Nsmark, 1937 (Figure 173)
Well developed in size, with moderately developed acetabular wall and well-developed acetabular cavity. D.e.c. series consists of two groups of muscle units; d.e.c.1 units are well developed; d.e.c.2. units are smaller in size but greater in number than d.e.c.1. units. Exterior units of d.i.c. and v.i.c. series usually longer and gradually decrease inward. M.e.c. series composed of irregularly spaced units. Exterior longitudinal fibers moderately developed. Radial fibers well developed; oblique fibers are few and weakly developed.

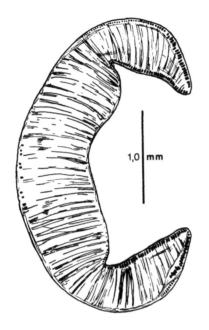

FIGURE 171.   Gigantocotyle type Eduardo, 1982.

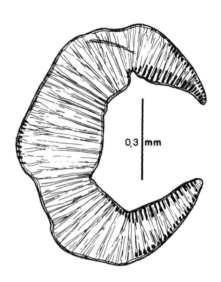

FIGURE 172.   Nilocotyle type Näsmark, 1937.

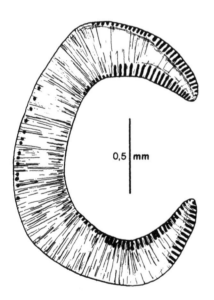

FIGURE 173.   Paramphistomum type Näsmark, 1937.

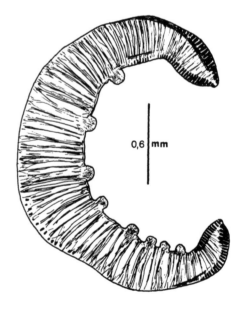

FIGURE 174.   Pertinax type Sey, 1987. (By permission of the *Parasitol. Hung.*, 1987, Budapest.)

12. Pertinax type Sey, 1987 (Figure 174)

Well developed in size, with papillae on the inner surface of acetabulum; musculature moderately developed, acetabular wall and cavity well developed. Circular series consists of a single group of units. D.e.c. series is separated by longitudinal fibers from inside. Units of interior series are similar in size. Radial fibers strongly developed, forming closely spaced coarse bundles. M.e.c. series consists of irregularly spaced units. Exterior longitudinal fibers moderately developed.

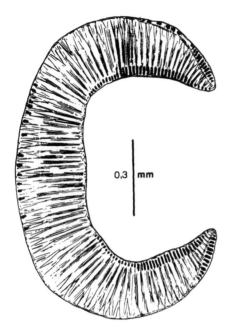
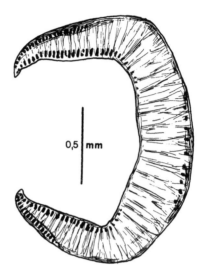

FIGURE 175.   Pisum type Näsmark, 1937.   FIGURE 176.   Stephanopharynx type Näsmark, 1973.

13. Pisum type Näsmark, 1937 (Figure 175)

Moderate in size, with well-developed musculature, strongly developed acetabular wall, and poorly developed acetabular cavity. D.e.c. series consists of two groups of units, d.e.c.1 and d.e.c.2, the latter including some small units. D.e.c.1 and v.e.c. are similar in size. D.i.c. and v.i.c series situated symmetrically; number of units of these series is greater than that of exterior series. Radial musculature very strongly developed, forming closely spaced bundles. M.e.c. series consisting of closely packed units. Exterior longitudinal fibers moderately developed.

14. Stephanopharynx type Näsmark, 1937 (Figure 176)

Well developed in size with moderately developed musculature and acetabular wall and well-developed acetabular cavity. Circular series consists of a single group of units. Units in interior series are greater in number than those of exterior ones. Radial fibers moderately developed forming weak bundles of fibers. Number of oblique fibers: a few. Exterior longitudinal fibers well developed.

15. Streptocoelium type Näsmark, 1937 (Figure 177)

Small in size, with well-developed musculature, moderately developed acetabular wall and cavity. D.e.c. and v.e.c. series consist of a single group of units. They are similar in size and number. D.i.c. and v.i.c. series are also similar in size and arrangement. M.e.c. series usually consists of a few, irregular units. Radial fibers are well developed, forming loosely spaced bundles. Exterior longitudinal fibers composed of a few but well-developed fibers.

16. Symmeri type Näsmark, 1937 (Figure 178)

Enormous in size, with moderately developed musculature and acetabular wall and well-developed acetabular cavity. D.e.c. series consists of two groups of units; d.e.c.1 series is greater in number and bigger in size than d.e.c.2 units are. D.i.c. and v.i.c. series well

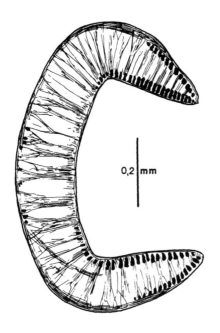 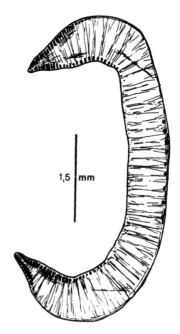

FIGURE 177. Streptocoelium type Näsmark, 1937.

FIGURE 178. Symmeri type Näsmark, 1937.

developed, forming a single group of units. M.e.c. series consists of loosely spaced units. Radial fibers well developed, oblique fibers few and weakly developed. Exterior longitudinal fibers poorly developed but distinct.

KEY TO TYPES OF ACETABULUM OF GROUP II

1. Acetabulum without papillae on inner surface 2
   Acetabulum with papillae on inner surface................................. Pertinax
2. Acetabulum enormous 3
   Acetabulum not enormous 7
3. Ventral exterior circular series divided into two groups................ Gigantocotyle
   Ventral exterior circular series not divided into two groups......................... 4
4. Dorsal exterior circular series divided into two groups............................ 5
   Dorsal exterior circular not divided into two groups 6
5. Dorsal exterior circular 2 units smaller in size but greater in number than dorsal exterior circular 1 .... Explanatum
   Dorsal exterior circular 2 units longer in size and greater in number than dorsal exterior circular 1 .... Duplicitestorum
   Dorsal exterior circular 2 units smaller in size and number than dorsal exterior circular 1 unit .... Symmeri
6. Dorsal exterior circular series small in number .... Calicophoron
   Dorsal exterior circular series consists of a great number of units ....Gastrothylax
7. Dorsal exterior circular series divided into two groups, d.e.c.1 and d.e.c.2 8
   Dorsal exterior circular series not divided into two groups..:...................... 9
8. Dorsal exterior circular 2 units smaller in size but greater in number than d.e.c.1, acetabular wall moderately developed............................Paramphistomum
   Dorsal exterior circular 2 units smaller in size and number than d.e.c.1, *irregularly spaced*, acetabular wall well developed .... Pisum

Dorsal exterior circular 2 units smaller in number than d.e.c.1 and separated from each other by longitudinal fibers, acetabular wall strongly developed ....Fischoederius
9. Acetabulum small in size 10
   Acetabulum well developed in size 11
10. Radial fibers well developed, size of units in each circular series is greater at exterior end and diminish interiorly .... Streptocoelium
    Radial fibers well developed, units of circular series are about the same in size than in interior series ....Buxifrons
    Radial fibers well developed, units of middle part of interior series bigger than those in circular series .... Nilocotyle
11. Acetabular wall moderately developed..12
    Acetabular wall strongly developed.......................................Carmyerius
12. Well-developed band of oblique fibers bordering on the inner margin of d.e.c. series, d.e.c. series consists of units in small number.........................Cotylophoron
    Poorly developed band of oblique fibers bordering on the inner margin of d.e.c. series, d.e.c. numerous, acetabular rim tapers .... Stephanopharynx

1.5.3.3. Acetabulum of Group III

Acetabulum has either typical forms or forms with usually more developed basal than marginal parts. Acetabular musculature and acetabular wall well developed. Dorsal exterior circular series longer or shorter but without division into two parts. Acetabular tissue not vesicular.

1.5.3.3.1. Types of Acetabulum of Group III
1. Brumptia type Näsmark, 1937 (Figure 179)

Enormous in size with well-developed musculature, acetabular wall, and acetabular cavity. Number of units in d.e.c. series greater in number than they are in v.e.c. series. Exterior of d.e.c. series consists of larger units, and they gradually diminish inward. D.i.c. and v.i.c. series are similar in number and size. Radial fibers well developed forming regularly spaced bands. Exterior longitudinal fibers moderately developed but distinct. M.e.c. present.

2. Gastrodiscus type Näsmark, 1937 (Figure 180)

Well developed in size, with well-developed musculature and well-developed acetabular wall and acetabular cavity. Dorsal and ventral circular series consist only of a single group of units. D.e.c. series is greater in number than those in v.e.c. inward. D.i.e. and v.i.e. are similar in size and arrangement; exterior of interior series larger and gradually diminish inward. Radial fibers strongly developed, forming coarse bands. M.e.c. consists of well-developed units. Oblique fibers small in number but distinct. Exterior longitudinal fibers well developed.

3. Hawkesius type Sey, 1985a (Figure 181)

Moderate in size, with well-developed musculature, acetabular wall, and acetabular cavity. D.e.c. and v.e.c. series consist only of a single group of units. Strongest units in middle part of ventral circular series. Radial fibers well developed forming regularly spaced bands. Oblique fibers small in number and poorly developed. Exterior longitudinal fibers moderately developed but distinct. M.e.c. present.

4. Homalogaster type Sey, 1984b (Figure 182)

Well developed in size, with well-developed musculature, acetabular cavity and moderately developed acetabular wall. D.e.c. and v.e.c. series consist of a single group of units,

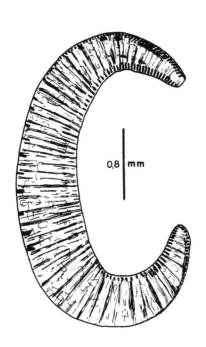

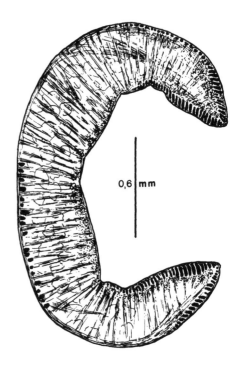

FIGURE 179.   Brumptia type Näsmark, 1937.

FIGURE 180.   Gastrodiscus type Näsmark, 1937.

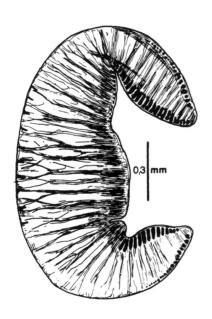

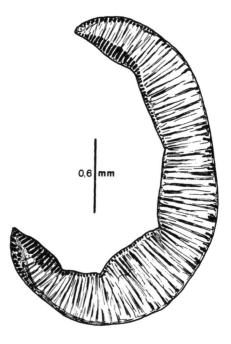

FIGURE 181.   Hawkesius type Sey, 1985.

FIGURE 182.   Homalogaster type Sey, 1984.

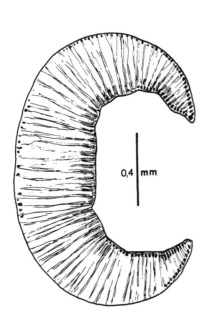

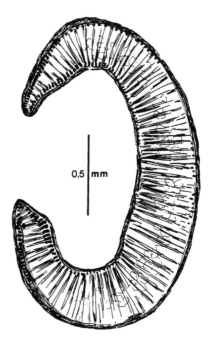

FIGURE 183.   Pseudodiscus type Näsmark,
1937.

FIGURE 184.   Watsonius type Näsmark, 1937.

comprising a small number of units. Units of middle part of d.e.c. and v.e.c. series are longer than exterior and interior ones. Radial fibers well developed, forming irregularly spaced bands. Oblique fibers few and poorly developed. Exterior longitudinal fibers normally developed. M.e.c. present.

5. Pseudodiscus type Näsmark, 1937 (Figure 183)

Well developed in size, with moderately developed circular musculature and with well-developed radial musculature, acetabular wall, and acetabular cavity. D.e.c. and v.e.c. series consist of a single group of units, the former is greater in number than the latter. D.i.c. and v.e.c. series are similar in size and arrangement, but exterior of v.i.c. longer than those in d.i.c. Radial fibers well developed forming regularly spaced bands. Exterior longitudinal fibers well developed. M.e.c. has irregularly spaced fibers.

6. Watsonius type Näsmark, 1937 (Figure 184)

Well developed in size, with well-developed musculature, acetabular wall, and acetabular cavity. External longitudinal fibers strongly developed. D.e.c. and v.e.c. series consist of a single group of units; d.e.c. series greater in number than are v.e.c. series. Dorsal and ventral internal series are similar in size and number. Radial fibers well developed forming unevenly spaced bands.

KEY TO TYPES OF ACETABULUM OF GROUP III

1. Measurement of units of dorsal exterior circular series varies evenly 2
   Exterior of dorsal exterior circular series definitely larger in size than units of
   interior .... Brumptia
2. Units of dorsal exterior circular series greater in number than units in ventral internal
   circular series 3
   Units of middle part of ventral interior circular series strongly developed 5

3. Exterior longitudinal fibers strongly or well developed 4
    Exterior longitudinal fibers poorly developed .... Pseudodiscus
4. Exterior of ventral circular series strongly developed ....Gastrodiscus
    Exterior of ventral circular series moderately developed .... Watsonius
5. Acetabular wall and basal part of acetabular well developed...............Hawkesius
    Acetabular wall and basal part of acetabulum poorly developed .... Homalogaster

2. ECOLOGY OF THE AMPHISTOMES

2.1 LIFE-HISTORY PATTERN

The amphistomes have a dixenous type of developmental cycle, and, accordingly, they develop in a single intermediate host (usually freshwater or occasionally marine mollusks). In general, cercariae encyst on vegetation or other objects in their vicinity. Five larval stages (miracidium, sporocyst, redia, cercaria, and adolescaria) are involved in their ontogenetic cycle, and they will be reviewed below under the following three headings: (1) preparasitic stage, (2) intramolluscan stage, and (3) development in definitive host.

2.1.1. PREPARASITIC STAGE
Egg and Embryonic Development
The eggs of amphistomes belong to the ones which are larger in size. They are oval or spherical in shape and enveloped with transparent thin (developed in freshwater) or thick (developed in marine water) eggshell. Occasionally, it is membranaceous (*Colocladorchis, Dadaytrema*, etc.). The anterior tapering tip with a tiny operculum (eggs without operculum were, however, observed in *Helostomatis bundelkhandensis* and *Schizamphistomoides prescotti* by Agarwal and Agrawal[6] and Agrawal,[8] respectively) and the antiopercular part usually end in a knoblike, polar thickening (Figure 185). The color of the freshly deposited eggs is yellow-green or white. The amphistomes, with regard to the developmental level of the embryos, may be oviparous (most amphistomes of homoiotherm vertebrates), ovoviviparous (amphistomes of lower vertebrates), and viviparous (some amphistomes of fish and reptiles. Embryonic development is affected by various environmental factors, of which temperature seems to be the most important under natural conditions. The life history of amphistomes has been studied in several dozen species (see the biological information under the given species), but the majority of examinations have been carried out at room temperatures which changed with the seasons. Under controlled temperatures, embryonic development was also studied in several species. *Calicophoron microbothrium*,[229,296,536] *C. daubneyi*,[833-835,842] *C. calicophoron*,[258] *C. microbothrioides*,[259,809] *Paramphistomum cervi*,[108,505,848] *P. ichikawai*,[253,488,348] *P. leydeni*,[471] *P. hiberniae*,[468] *Gigantocotyle explanatum*.[883] *Orthocoelium scoliocoelium*,[635] *Gastrothylax crumenifer*,[651] and *Gastrodiscus aegyptiacus*,[582] to mention only a few. These observations showed a more or less similar tempo in embryo formation and similar period in incubation even if they are phylogenetically removed species. The lower temperature at which the cleavage commenced at 4°C. At this temperature, eggs preserved their viability for 3 to 4 months (*C. daubneyi*).[842] With increasing temperatures the incubation period changes proportionally with the degree of change in temperature.

The optimal temperature seems to be about 25 to 27°C at the species studied. The upper limit of the temperature is 37 to 40°C. During incubation at 27°C, no significant change can be observed in the first 5 to 6 d, except for the growth of the embryo and its oval elongation. On the 6th and 7th days, certain structural parts (terebratorium, apical glad, penetration glands) can be detected. On the 8th day the flame cells begin to beat, and the formation of the germinal tissue also takes place during this period. With the growth of the embryo, the vitelline cells gradually decrease in number, and the place of the vitelline cells embracing the embryo is now occupied by two large vacuoles and the so-called mucoid plug[842] (Figure 186).

Hatching begins on the 9th and 10th days and continues for another 5 or 6 d or more. It is rather complicated, a process not yet clear in every detail. On examining the process of the hatching of *C. daubneyi*,[835] it was found that the process consists of two well-visible phases: (1) opening of the operculum (Figure 187) and (2) emergence of the micracidium

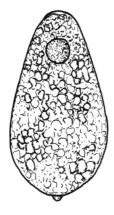

FIGURE 185.   Egg with zygote, vitelline cells, and polar thickening.

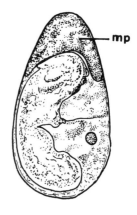

FIGURE 186.   Embryonated egg containing mucoid plug.

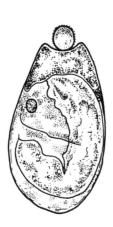

FIGURE 187.   Opening of opercular lid.

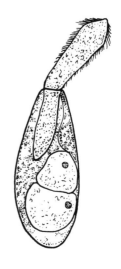

FIGURE 188.   Emergence of miracidium.

(Figure 188). When the egg culture with fully developed micracidia was stimulated with light, a marked change was observed both in the structure of the mucoid plug and in the activity of the miracidia. The mucoid plug enlarged and became opaque, and the interior pressure, which appeared by endosmosis and miracidial activity, pushed off the operculum. The apical part of the miracidium pressed quickly through the opercular opening, and the rest of the body continued to emerge with active body extensions and contractions coupled with vigorous beating of the cilia. In the process of hatching, the role of the enzymatic effect of the unicellular miracidial glands cannot be excluded. Namely, the formation of the mucoid plug is not observable in eggs of several other species of amphistomes. In these cases, the operculum is touched directly with the apical part of the miracidium. The relationship between temperature, development, and hatchability of *C. microbothrium* is demonstrated by Fagbemi[296] (Figure 189).

Miracidium

The emerged and freely swimming miracidia are pyriform or bullet shaped, without eyespots except for *Heronimus*, and they are covered with ciliated epidermal cells. Thapar

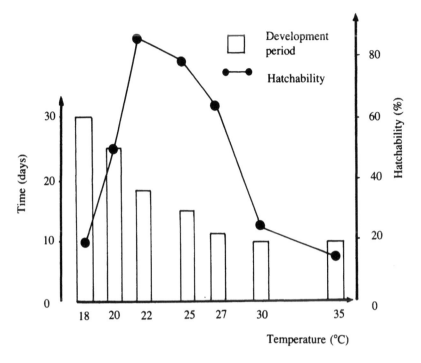

FIGURE 189.   Effect of temperature on the development and hatchability of *Paramphis-tomum* (= *Calicophoron*) *microbothrium*. (Courtesy of B. O. Fagbemi.)

and Singha[968] have reported the presence of the eyespots in miracidia of *Olveria indica*, but this observation later proved to be an artifact.[948] The species studied were 140 to 360 by 40 to 120 μm on average. The freshly hatched miracidia swam rapidly and moved in various directions at the bottom of the dish or at the surface layer of the water. Their life span under normal conditions (natural water bodies and room temperature) varied from 6 to 24 h.

The structure of the miracidia examined shows a close similarity to that of others with a few exceptions. Accordingly, they represent the special amphistome type of miracidium which is characterized by the number and arrangement of the epidermal cells as well as by its internal structure. The epidermal cells are situated in four tiers and arranged by the formula 6:6:4:2 in species of the genus *Diplodiscus*[111,701,818,833] (Figure 190) or 6:8:4:2 in most amphistome species (Figure 191), and a case of intraspecific variability (6:8:4:2, 6:9:4:2) has been recorded in *Heronimus mollis*.[188,567] The anterior end or apical papilla of the miracidium, now called terebratorium[773] is nonciliated and is perforated by the pore of the openings of the glands in the forebody (Figure 192). After silver nitrate treatment, argentophilic structures can be observed on the terebratorium and along the body surface. On the terebratorium examined face on, these formations appear to be arranged in a bilaterally symmetrical fashion. Their orientation can be given with the axes (A_1, A_2, A_3, or A_4) having been drawn across the groups proper. Along the axes, aggregations of these structures can be monitored (Figure 193). The number and arrangement of the argentophilic structures were described in species *Sandonia sudanensis*,[861] *Diplodiscus subclavatus*,[833] *Calicophoron calicophorum*,[937] *C. daubneyi*,[833,810] *C. microbothrium*,[215] *C. microbothrioides*,[813] *C. phillerouxi*,[16,17] *Paramphistomum ichikawai*,[348,808] *P. leydeni*,[812] and *Gastrodiscoides hominis*.[257]

Two main types of the arrangement can be differentiated: (1) the argentophilic structures are situated along three axes (in most of the examined species, e.g. *C. daubneyi*, *Diplodiscus subclavatus*, Figures 193 and 194: A_{1a}:2; A_{1b}:1; A_{1c}:2; A_{2a}:8; A_{2b}:8; A_{3a}:6; A_{3b}:6) or (2) along four axes (*Calicophoron calicophorum*, Figure 195: A_{1a}:2; A_{1b}:1; A_{1c}:2; A_{2a}:7-9; A_{2b}:8-

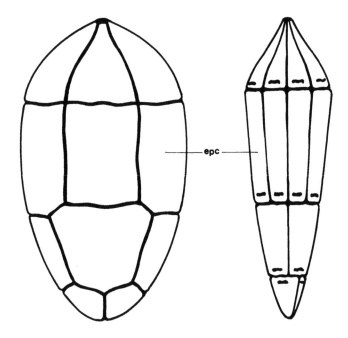

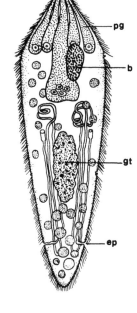

FIGURE 190. Number and arrangement of epithelial cells of miracidium of *Diplodiscus subclavatus*. (By permission of the *Parasitol. Hung.*, 1972, Budapest.)

FIGURE 191. Number and arrangement of epithelial cells, typical for amphistome miracidia. (By permission of the *Parasitol. Hung.*, 1972, Budapest.)

FIGURE 192. Internal structure of miracidium *(Calicophoron daubneyi)*. (By permission of the *Parasitol. Hung.*, 1972, Budapest.)

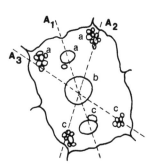

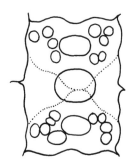

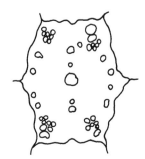

FIGURE 193. Argentophile structure on terebratorium of miracidium *(Calicophoron daubneyi)*. (By permission of the *Parasitol. Hung.*, 1972, Budapest.)

FIGURE 194. Argentophile structures on terebratorium of miracidium *(Diplodiscus subclavatus)*. (By permission of the *Parasitol. Hung.*, 1972, Budapest.)

FIGURE 195. Argentophile structure on terebratorium of miracidium *(Calicophoron calicophorum)*. (After J. P. Swart, 1967.)

9; A_{3a}:4; A_{3b}:4; A_{4a}:2-4; A_{4b}:2-3 and *Gastrodiscoides hominis*). Besides these, anteriorly there are 10 structures between the first and second tiers, two at the junction of the second and third tiers, and a further two between the third and fourth rows (the latter two are openings of the excretory ducts).

Beneath the epidermal cells there is a thin, transparent layer of subepithelium. It is cellular in nature, and its cells can be observed throughout the whole body. The muscle fibers, circular and longitudinal, are found between the epidermal and subepithelial layers which are capable in the miracidia of extensive changes in shape and size.

The internal structures of the miracidia consist of an apical gland, two pairs of penetration glands, a nerve mass, an excretory system, and germinal tissue (Figure 192).

The apical gland, which is often called "primitive gut", is situated in the anterior third of the body. Four nuclei are visible in the posterior part of the gland. The possible function of this gland is that its secretion helps the micracidia in the process of penetration into the snail tissue. There are usually two pairs of penetration glands found in close juxtaposition to the apical gland. The function of these glands might be connected to the hatching of the miracidia (besides the characteristic number of penetration glands, Singh,[883] in *Explanatum explanatum*, and Mukherjee,[633] in *Paramphistomum epiclitum*, reported only one pair of these glands. Dinnik and Dinnik[232] described three glands in the anterior end of the body in *Carmyerius exoporus*).

The nerve mass (or "brain") is situated in the forebody, immediately along the posterior part of the apical gland (Figures 192 and 196). It consists of a mass of fibrillar substance. Two nerves run toward the terebratorium, two other nerves arise laterally and run to the lateral sensory papillae, and a further two nerves are sent off posterolaterally to merge with the body wall near the middle of the animal (Figure 196).

The excretory system shows unified structure in the species studied. It consists of a pair of flame cells situated in the forebody. They open to the excretory ducts which form some loops anteriorly and wind posteriorly and finally empty at the excretory pore (Figures 192 and 196). The latter are situated laterally between the third and fourth tiers of the epidermal cells.

The posterior two thirds of the body is occupied by the germinal activity,[536] which contains a granular mass with embedded germinal cells and embryo balls, and all these together constitute the germinal tissue (Figure 192). It is probable that the "two granulated masses of protoplasm" observed by Dinnik[225] in the miracidia of *Calicophoron phillerouxi* (Figure 197) are a part of the germinal tissue in Lengy's sense. Especially in the posterior part of the germinal tissue, there are several germinal cells or germinal balls. In *Stunkardia dilymphosa*, the germinal cells combine only a single germinal ball, and no other cells or germinal balls were observed.[629] The germinal cells probably originate along the germinal tissue and move centrally and anteriorly where they constitute the embryo ball(s). Deviations from the above-described general constitution can be found in miracidia of *Stichorchis subtriquetrus*,[81,675] *Basidiodiscus ectorchis*, and *Sandonia sudanensis*[861] by the formation of redia in the hindbody instead of germinal tissues (Figure 198).

2.1.2. INTRAMOLLUSCAN LARVAL STAGE

When miracidia reach the vicinity of the susceptible intermediate snail hosts, they become quite agitated and perform definitive attraction to these snails. Moving vigorously they make temporary contacts with various parts of the snails or snail excreta. The best part of the snail body for penetration is the mantle cavity. Here the miracidium applies the terebratorium to the mantle wall and begins a boring movement. Within a short period, the terebratorium commences to embed into the snail's tissue. In general, the complete penetration ceases within 20 to 40 minutes. Observations show[229,536,842] that a major part of the miracidium that has invaded and penetrated the snail fails to develop into sporocysts. After penetration, the miracidium continues its development in the snail tissue and transforms into a sporocyst.

2.1.2.1. Sporocyst

This developmental stage is included in all the examined amphistomes except for *Stichorchis subtriquetrus*,[675] *Basidiodiscus ectorchis*, and *Sandonis sudanensis*.[861] In these

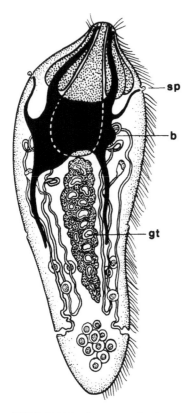
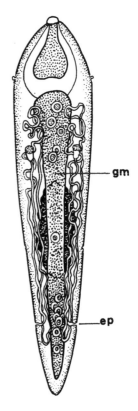
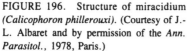

FIGURE 196.   Structure of miracidium (*Calicophoron phillerouxi*). (Courtesy of J.-L. Albaret and by permission of the *Ann. Parasitol.*, 1978, Paris.)

FIGURE 197.   Miracidium of *Calicophoron phillerouxi* (note "two granulated masses of protoplasm"). (Courtesy of N. N. Dinnik and by permission of the *J. Helminthol.*, 1961, London.)

species, miracidia contain fully formed rediae and after penetrations of the miracidia into the snails the rediae are liberated and continue their development. In typical cases, however, the miracidia that have penetrated undergo a series of changes, and the miracidia transform into sporocysts. The changes include: shedding of the epidermal cells, losing some internal structure (terebratorium, apical, and penetration glands), and breaking down the embryo balls into single germinal cells. The shape of sporocysts is usually oval, spherical, saclike, or elongated saccular (Figure 199). Occasionally, it may knoblike (*C. microbothrium*, Figure 200) or like a V (*P. togolense* [= *C. phillerouxi*] Figure 201) or with paired branches along the lateral edges (*Heronimus*, Figure 202). The sporocyst of the latter species, as a unique feature among the amphistomes, retains miracidial eyespots. In the anterior-posterior orientation, the position of the flame cells and the direction of the excretory ducts can be used as they correspond properly with that of the miracidia.

The young sporocysts are similar in structure to the miracidia. They are covered with the miracidial subepithelial layer. Beneath it fine muscle fibers are present. The internal structures include a pair of flame cells and germinal cells which develop into new embryo balls and later into embryonal rediae. In the excretory system of the sporocyst of *Heronimus mollis*, there are several small flame cells around two larger ones. By gradually increasing in size, the sporocysts reach their maturity in 6 to 10 d when liberation of rediae commences. The mature sporocysts contain developing rediae in varying numbers (1 to 80 per sporocyst) and embryo balls. Tandon[949] observed an oblique transverse opening at one end of the sporocyst as a birth pore for emergence of rediae. Several other authors[229,536,848] did not find

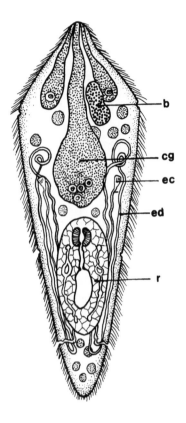

FIGURE 198. Miracidium with a young redia. (By permission of the *Acta Zool. Acad. Sci. Hung.*, 1976, Budapest.)

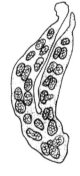

FIGURE 199. Sporocyst of *Calicophoron sukari*. (Courtesy of N. N. Dinnik.)

FIGURE 200. Sporocyst of *Calicophoron microbothrium*. (Courtesy of N. N. Dinnik. and by permission of *J. Parasitology*, 1954, London).

FIGURE 201. Sporocyst of *Calicophoron phillerouxi*. (Courtesy of J.-L. Albaret and by permission of *Ann. Parasitol.*, 1978, Paris.)

FIGURE 202.   Sporocyst
of *Heronimus mollis*. (By
permission of the *J. Parasitol.*, 1960, London.)

any opening of the sporocyst's surface. The liberation of rediae usually takes place through
the rupture made by the escaping rediae. After passing of the rediae this hole is closed and
the damage is not seen later. The shedding of rediae can be continued for several weeks
(e.g., *C. microbothrium*).[229]

2.1.2.2. Redia

This developmental stage has been found in every studied species, except for *Heronimus*
where cercariae develop directly in the sporocyst. The rediae commence their development
from those embryo balls which reorganized anew in the early stage of the sporocysts. They
liberate from sporocysts or rarely from miracidia (e.g., *Sandonia*) in immature forms, and
they continue development in the snail tissue. The rediae usually have elongated bodies and
are sausage shaped (Figure 203) and have a body covered with integument having microvilli,
rarely with locomotor appendages (*Megalodiscus temperatus*, Figure 204), or without them
(in most studied species). The liberated rediae reach their maturity with gradual growing,
and they can be regarded to be in the mature stage when release of cercariae begins. This
period usually occurs by 20 to 25 d after liberation from the sporocyst. At this stage rediae
are found within the body space in the vicinity of the intestine or in the hepatopancreas.
The internal structure of the rediae can be observed in mature rediae, which contain a
digestive system, excretory system, nervous system, and germinal system.

The digestive system consists of a mouth, pharynx, an esophagus, and a rhabdocoele
gut. The mouth is a minute tube leading into the pharynx. The latter is a globular and
strongly muscular organ. The pharynx is surrounded by a fibrous membrane which continues
along the pharynx, too. The esophagus is a short and narrow tube, leading posteriorly to
the gut. The gut is saccular, made up of a single layer of large, rectangular cells. It is longer
in young rediae, in proportion to the body length, than in older ones, but it remains relatively
long (e.g., in *Megalodiscus temperatus*). Along both sides of the pharynx and esophagus
there are drop-shaped cells in various numbers which are termed salivary glands. They are
glandular in nature and contain one[79,949] or two[253,511,830,842] types of glands. The possible
(e.g., histolytic) function of them has not been elucidated. Orlov[675] described a unicellular
cell, one on each side, with unknown function in the rediae of *Stichorchis subtriquetrus*.

The excretory system is composed of three flame cells on each side, situated anteriorly,
midway, and posteriorly. The ducts of the flame cells unite at about the middle part of the
body on each side and proceed through a short common duct into a bladder which opens
on the side of body. Jonathan,[452] in rediae of *Calicophoron ijimai* (= *C. calicophorum*)
and Singh,[884] in rediae of *Srivastavaia indica* (= *Paramphistomum epiclitum*), described
five pairs of flame cells.

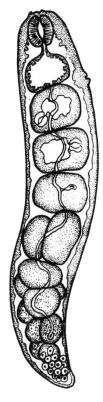

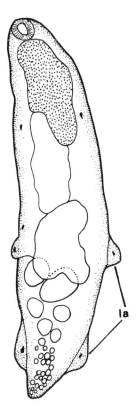

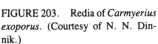

FIGURE 203. Redia of *Carmyerius exoporus*. (Courtesy of N. N. Dinnik.)

FIGURE 204. Redia of *Megalodiscus temperatus* with locomotor appendages. (After W. A. Krull and H. F. Price, 1932.)

The nervous system has a central nerve mass[79] which is located dorsal to the esophagus, similar to that of the miracidium. There are numerous nerve cells in it, but nerves could not be detected.[536]

The germinal system comprises the germinal epithelium, germinal cells, and embryo balls. The germinal epithelium which is located at the posterior end of the germinal cavity produces germinal cells which develop into germinal balls. The germinal balls and cells are located in the posterior half of the body. The newly emerged rediae usually contain 3 to 10 embryo balls. The birth pore can be traced in early development of rediae, but it becomes easily seen and prominent only later, when cercariae have already been situated in the first fifth of its body (Figure 205).

The studies referring to the number of redial generations show that there are species with one redial generation (e.g., *Olveria indica, Gastrothylax crumenifer*) or one daughter redial generation (*Megalodiscus temperatus, Calicophoron daubneyi*) or several redial generations (*Calicophoron microbothrium*). The establishment of the number of the actual redial generations, however, encounters difficulties because sporocysts, rediae both older and younger, might produce rediae in about every 20 d. These redial generations overlap each other, and it is difficult to ascertain the sequence of generations. Moreover, Katkov[471] experimentally demonstrated that the process of change of generations continues until the death of the snails. It was also pointed out that under favorable conditions (23 to 29°C) cercariae develop in rediae at low temperatures (5 to 6°C), newer rediae develop in rediae, and the ratio of rediae to cercariae shifted in favor of rediae. After a certain time, however,

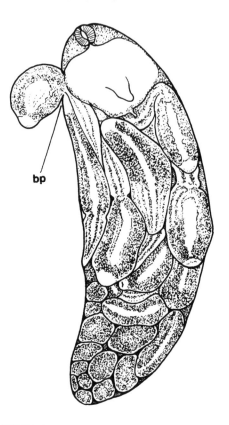

FIGURE 205.   Redia of *Stichorchis subtriquetrus*
with cercaria moves through the birth pore. (After
I. V. Orlov, 1948.)

(about 40 d in the case of *C. microbothrium*)[536] the signs of aging (exhausting of the germinal
epithelium, the presence of only some embryo balls, brownish in color, etc.) can be observed,
and they gradually die off.

2.1.2.3 Cercaria and Adolescaria

These larval stages are always present in the amphistome species investigated (except
for adolescaria of *Heronimus mollis*). The cercariae begin their development from the embryo
balls which are found in the developing rediae. The developing cercariae show a rapid
progress in development, and the first cercariae emerged on an average between 18 (*Carmyerius exoporus*)[232] and 30[229] d after infestation. The liberated cercariae are in an immature
state, and they continue their development in the snail tissue. The developing cercariae
gradually evolve features characteristic of mature cercariae. These traits are so characteristic
of this digenetic group of flukes that we can rightly call them the amphistome type of cercaria
which is easily recognizable and separable from other types of cercariae. The first cercariae
are usually excreted by the infected snail in the 28 to 46 d after exposure of the snail to
miracidia.

The freely swimming cercariae are usually vigorously (*Diplodiscus subclavatus, Calicophoron microbothrium*, etc.) or sluggishly (e.g., *Calicophoron daubneyi*) moving organisms. They have a more or less pear-shaped body and a tail varying in length in relation
to the body length. The body is covered with smooth integument, and it is opaque with
cystogenous rods stretched all over the body within the paranchyma. The body itself may
be heavily pigmented (Cercaria pigmentata Sonsino)[901] or without pigmentation (Cercaria

diplocotylea Pagenstecher)[692] or intermediate in color. Behind the anterior extremity, there is a pair of eyes (except for *Heronimus mollis*) on the dorsal surface just beneath the integument. The acetabulum is usually ventroterminally located in the species investigated. The tail is simple in structure, and occasionally (*Calicophoron microbothrium*[536] and *C. microbothrioides*)[809] it is provided with a fin fold.

The internal structure of the cercariae is composed of the digestive, excretory, nervous, sensory, and reproductive systems.

The digestive system consists of a mouth, pharynx, with or without appendages, and a pair of ceca. The mouth opening is usually terminal, and it is surrounded by a strongly muscular pharynx. In the cercariae of the species having extramural pharyngeal appendages, these diverticules also appear. The pharynx is followed by the esophagus, which is usually a shorter or longer tube with muscular thickening, provided that it is also present in the adult. Ceca are simple tubes along their full length, lateral in position, and terminate at various levels of the posterior body part.

The excretory system consists of a group of flame cells, capillary tubulets, ascending and descending excretory ducts, excretory bladder, and a caudal excretory tube. The flame cells are scattered along the body. Determination of the actual flame cell pattern is difficult due to the opacity and pigmentation. It was revealed completely in *Heronimus mollis*: 2(7 + 7) + (7 + 6);[188] *Diplodiscus subclavatus*: 2 (5 + 5) + (5 + 5) + (5 + 5) + (5 + 5);[74] in other species it was 16 pairs of flame cells,[79] 14 pairs of flame cells (*Calicophoron microbothrium*),[536] 40 pairs of flame cells (*Diplodiscus fischthalicus*);[74] *Cercaria indicae XXIX*: 2(3 + 3 + 3 + 3 + 3);[618] *Cercaria sp.* I, Kerala: 2(3 + 3 + 3); *C. leyteensis no. 32*: 2(3 + 1) + (3 + 2 + 3 + 3),[437] 12 pairs of flame cells (*Calicophoron daubneyi*).[842]

The capillaries of the flame cells are gathered by the ascending excretory ducts originated at about the level of the acetabulum. The ascending excretory ducts in the vicinity of the eyes unite with the branches of the descending excretory ducts which are easily recognizable due to refractory globules. The descending excretory ducts, at about the middle of the body, bifurcate, and the anterior branches unite with each other, while the posterior ones continue posteriorly and terminate at the base of the excretory bladder. The caudal tube is a thin, straight vessel situated in the axis of the tail ending in front of the caudal extremity with or without openings on the surface. The nonepithelial excretory bladder with posterodorsal location is oval or conical in shape and opens on the dorsal surface through the excretory pore.

The nervous system consists of two ganglia connected by a transverse commissure.[79] The ganglia are spaced beneath the eyes, and each of them gives rise to several anterior and posterior nerves. Three pairs of nerves are anterior, two are ventral, one is dorsal, and another dorsal nerve ends at the inner end of the eye. The posterior nerve runs from the ganglion to the cecum at the same side where it disappears. Lengy[536] described sensory bristles along the forebody and the tail of cercariae of *Calicophoron microbothrium*. The greatest number of these bristles was 36 on one individual.

Over the past decade, several papers have reported on the superficial sensory apparatus of various amphistomes detected by the silver technique[74,215,216,809] (Table 1).

The superficial, sensory papillae bearing information may be a useful characteristic in the differentiation of closely related species.

A great number of species of cercariae have been revealed up to now. They have been described either at the larval stages of the developmental cycle of the given species (these cercariae were listed under the amphistome species proper) or independent of snails without the knowledge of the relevant adults (Table 2).

The classification of the amphistome cercariae was done by Cort,[184] and the five species known then were divided into two groups, according to the subfamilies: Paramphistominae (group 1) and Diplodiscinae (group 2). Representatives of group 1 were characterized by

TABLE 1
Superficial Sensory Apparatus of Some Amphistomes

Diplodiscus subclavatus	*Diplodiscus fiscthalicus*	*Calicophoron microbothrium*	*Calicophoron daubneyi*
		Cephalic Papillae	
378—433	269—351	116—152	90—105
		Caudal Papillae	
26—38	32—43	53—65	9—19

TABLE 2
Cercariae and Their Snail Hosts of the Amphistomes Studied

Cercaria	Intermediate host	Locality	Ref.
Cercaria bhaleraoi Mukherjee, 1962	*Indoplanorbis exustus*	India	627,633
C. bareillyi Peter et Srivastava, 1955	*I. exustus*	India	27,717,718
C. chauhani Pandey et Jain, 1971	*I. exustus*	India	700
C. convolutoides Porter, 1938	*Bulinus tropicus*	South Africa	734
C. corti O'Roke, 1917	*Helisoma trivolvis*	U.S.	676
C. diastropha Cort, 1914	*H. trivolvis*	U.S.	184,185,676
C. euphraticus Agrawal, 1971	*Gyraulus convexiusculus*	India	11
C. frondosa Cawston, 1918	*Bulinus schakoi*	South Africa	161,301
C. fraseri Buckley, 1939	*Indoplanorbis exustus*	India	134
C. fursolensis Singh et Malaki, 1963	*Gyraulus convexiusculus*	India	886
C. gyraulusi Peter et Srivastava, 1955	*G. convexiusculus*	India	443,717,718
C. helicorbisi Kumar, Dutt et Jain, 1968	*Helicorbis coenosus*	India	443,514
C. inhabilis Cort, 1914	*Helisoma trivolvis*	U.S.	75,184,185,676
C. indoplanorbisi Peter et Srivastava, 1955	*Indoplanorbis exustus*	India	717,718
C. indicae XXI. Sewell, 1922	*I. exustus*	India	830
C. indicae XXXII. Sewell, 1922	*Bulimus pulchellus*	India	443,446,830
	Amnicola travancorica	Pakistan	442
	Bithynia tentaculata	India	409
C. kareilliensis Mukherjee, 1972	*Bulimus pulchellus*	India	634
C. kylasarmi Rao, 1932	*Indoplanorbis exustus*	India	713,767
C. lewerti Singh, 1957	*I. exustus*	India	882
C. leyteensis no. 32 Ito et Blas, 1977	*Segmentina hemisphaerula, Gyraulus convexiusculus*	Philippines	437
C. leyteensis no. 33 Ito et Blas, 1977	*Segmentina hemisphaerula, Gyraulus convexiusculus*	Philippines	437
C. macroacetabulata Haseeb et Khan, 1982	*G. convexiusculus*	Pakistan	409
C. mathurapurensis Mukherjee, 1962	*Indoplanorbis exustus*	India	627,633
C. mosaica Faust, 1926	*Bulinus fosskalii, B. schakoi*	South Africa	304
C. missouriensis McCoy, 1929	*Helisoma trivolvis*	U.S.	599
C. nervosa Faust, 1922	*Planorbis möllendorfi*	China	303
C. nigrita Fain, 1953	*Ceratophalus mortalensis, Lentorbis junodi, Segmentorbis kinisaensis*	Tanzania	298

TABLE 2 (continued)
Cercariae and Their Snail Hosts of the Amphistomes Studied

Cercaria	Intermediate host	Locality	Ref.
C. obscurior Fain, 1953	Biomphalasia sudanica	Tanzania	298
C. onkari Jain, 1972	Indoplarbis exustus	India	440
C. phulpurensis Tripathi et Srivastava, 1980	I. exustus	India	990
C. sewelli Tripathi et Srivastava, 1980	I. exustus	India	990
C. stelliae Porter, 1938	Bulinus tropicus	South Africa	734
C. sp. I. kerala Mahandas, 1976	Indoplanorbis exustus	India	618
C. udonensis Ito, Papasarathorn et Tongkoon, 1962	I. exustus	Thailand	436
C. umashankari Tripathi et Srivastava, 1980	I. exustus	India	990
C. umhlotia Porter, 1938	Lymnaea natalensis	South Africa	734
C. vaalensis Porter, 1938	Bulinus tropicus	South Africa	734
Cercaria truncatuloides	Lymnae truncatula	Germany	671
Amphistome cercaria	Gyraulus euphraticus	India	11,12
		India	440
	Cleopatra bulimoides	Egypt	297
Paramphistomid cercaria	Bulinus truncatus	Egypt	800
	Planorbis maginentus	Europe	565

the absence of the primary pharyngeal sacs; in species of group 2, the pharyngeal appendages were present. Later, Sewell,[830] emphasized the importance of the pigmentation in the differentiation of the two groups above, and, thus, the cercariae of the first group (Paramphistominae) were named "Pigmentata" and those of the second group (Diplodiscinae) were named "Diplocotylea". After the discovery of newer cercariae, it became evident that neither Cort nor Sewell's classification were adequate because several exceptions have been revealed in both groups.

Jain,[441] while retaining Sewell's nominations, attributed greater importance, (among the character states of the amphistome cercariae) to the structure of the excretory system. Accordingly, the Pigmentata group is characterized by the medially anastomized branches of the ascending main tubes. In the group Diplocotylea, these anastomosed branches do not exist. Besides the main basic structure, another type can be designated among cercariae presently known, based on taking the branches of the ascending main tubes into account.

Hence, amphistome cercariae, described up to now have been assigned to three groups: (1) Cercaria diplocotylea Pagenstecher, 1857 (without pigmentation and mediolateral branches, usually with pharyngeal appendages, main excretory tubes either straight or convoluted [Figures 206 to 208] several cercariae of amphistomes of lower vertebrates); (2) Cercaria intermedia Sey, 1988a (with antero- and posterolateral diverticula of the main excretory tubes, with or without pigmentation, and with pharyngeal appendages; Cercaria kareilliensis, cercaria of Zygocotyle lunata, of Wardius zibethicus and of Pseudodiscus collinsi; [Figures 209 and 210]; (3) Cercaria pigmentata Sonsino, 1892 (pigmented and with anastomosed mediolateral branches and usually without pharyngeal appendages; several cercariae of amphistomes of mammals; Figure 211).

Of the larval stages of flukes, the cercarial stage has been provided with formal taxonomic names in order to facilitate accurate identification. Such dual nomenclature should be dropped when it becomes known to what trematode species a given cercaria belongs. In amphistomes, the adults have usually been described prior to the recovery of their cercariae. In two species (Alassostomoides parvus, Calicophoron microbothrium), however, the probable adequate cercaria Cercaria inhabilis Cort, 1914, and Cercaria pigmentata Sonsino, 1892, respectively,

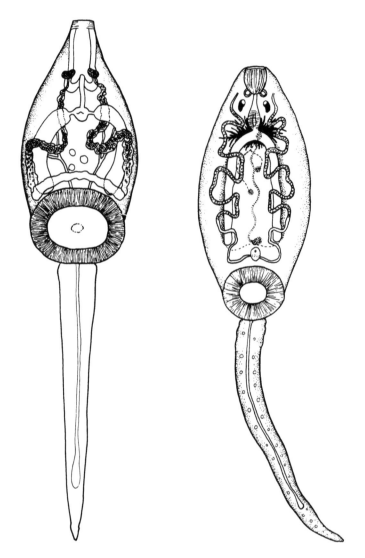

FIGURE 206. Cercaria diplocotylea *(Opisthodiscus diplodiscoides).*
(Courtesy of F. Simon-Vicente et al.
and by permission of the *J. Helminthol.,* 1974, London.)

FIGURE 207. Cercaria diplocotylea
(Cercaria onkari Jain, 1972). (After S.
P. Jain, 1972.)

had been described before recovery of their adults. If these cercariae are really the larval
forms of these adults, the species names should be *Alassostomoides inhabilis* (Stunkard,
1916) Beaver, 1929, and *Calicophoron pigmentatum* (Sonsino, 1892) Odening, Bockhardt
et Gräfner, 1979, respectively.

2.1.2.4. Life Cycle in Definitive Hosts

The amphistomes are parasitic either in the lower part (large intestine, rectum, cecum,
colon) or in the upper one (stomach, rumen, reticulum) of the digestive tube of the definitive
hosts. Species of the latter group are often called ''ruminal amphistomes'' while the former
one ''nonruminal amphistomes''. There is a limited number of experimental observations
on the developmental patterns and pathways of the amphistomes in the definitive hosts. It
is especially true in the case of the amphistomes of the lower vertebrates (Table 3).

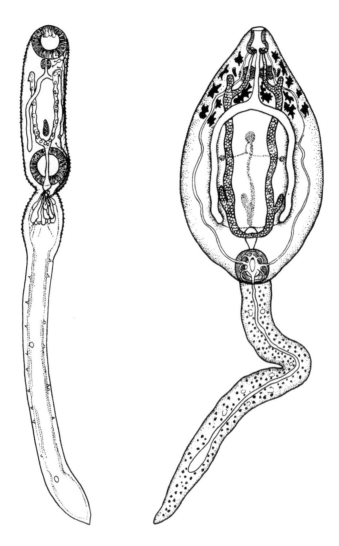

FIGURE 208. Cercaria diplocotylea *(Heronimus mollis).* (By
permission of the *J. Parasitol.,*
1960, Lincoln.)

FIGURE 209. Cercaria intermedia
(Pseudodiscus collinsi). (By permis-
sion of the *J. Helminthol.,* 1960,
London.)

On the basis of relevant information it seems that two basically different life-cycle patterns exist. In the ruminal amphistomes, an intestinal phase of their development is involved. Among these trematodes this is probably a general phenomenon, the conclusion of which can be drawn from experimental data available. In the nonruminal amphistomes, the intestinal phase is missing. They reach their sites of predilection in a short period of time (15 to 24 h); e.g., *Megalodiscus temperatus*[511] *Zygocotyle lunata*[1059] where they usually exist and reach their maturity in various times (Table 3).

The intestinal phase of the rumen flukes might be a varying period of time even in the same species. Dinnik and Dinnik[233] found that *Calicophoron microbothrium* in calves spend 2.5 to 6 weeks in the duodenum. Horak[422] demonstrated that the young specimens of *Calicophoron microbothrium* gradually migrate to the rumen from the 20th day up to the 34th day after infestation.

The longevity of the rumen flukes *Calicophoron microbothrium*[233] and *Paramphistomum cervi*[489] was estimated about 7 and 4 years, respectively. The cecal amphistome, *Zygocotyle lunata*[1059] exceeded about 2 years in experimentally infected ducks.

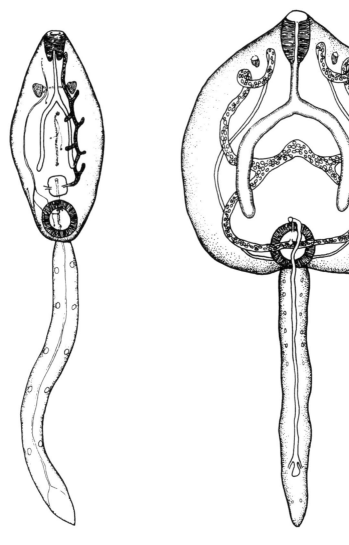

FIGURE 210. Cercaria intermedia *(Wardius zibethicus)*. (Courtesy of K. D. Murrell and by permission of the *J. Parasitol.*, 1965, Lincoln.)

FIGURE 211. Cercaria pigmentata *(Calicophoron daubneyi)*. (By permission of the *Acta Vet. Acad. Sci. Hung.*, 1979, Budapest.)

2.1.3. INTERMEDIATE SNAIL HOSTS AND INTERMEDIATE HOST SPECIFICITY

Of the presently known amphistomes, the life-history pattern has been elaborated completely or partially in 54 species. These examinations show a considerable disproportion as to the species studied and their geographic distribution. There is little information on the intermediate hosts of amphistomes of the lower vertebrates while it is much more abundant in mammalian vertebrates. The most important data on the intermediate snail hosts are summarized in Tables 4 and 5. It seems (Table 4) that intermediate hosts belong to the Mesogastropoda, Basommatophora, and Stylommatophora orders, involving many genera and species mainly of the taxa, Planorbidae, Bulinidae, and Lymnaeidae. Of the planorbid snails used by amphistomes as the most frequent intermediate hosts are the genera: *Armiger*, *Gyraulus*, and *Planorbis*. Bulinid *(Bulinus)* and lymnaeid *(Lymnaea)* snails might often be intermediate hosts of amphistomes. These snails are usually broadly distributed, well adaptive

TABLE 3
Life Cycle in Definitive Hosts

Amphistomes	Definitive host	Prepatent period (d)	Habitat	Ref.
Megalodiscus temperatus	Ranid frogs	60—90	Rectum	511
Diplodiscus subclavatus	Ranid frogs	12	Rectum	355
Opisthodiscus diplodiscoides	Ranid frogs	55	Rectum	879
Progonimodiscus doyeri	Clawed toad	26—30	Rectum	1030
Gastrodiscus secundus	Donkey	98	Large intestine	714
Zygocotyle lunata	Rat	46—61	Cecum	1059
	Duck	41—42	Cecum	1059
Pseudodiscus collinsi	Donkey	90	Large intestine	719
Homalogaster paloniae	Calf	57—94	Cecum	170
Gastrodiscoides hominis	Piglet	74	Colon	257
Fischoederius elongatus	Cow	126	Rumen	630
	Lamb	126	Rumen	12
Gastrothylax crumenifer	Kid	270	Rumen	949
	Buffalo kid	114	Rumen	720
	Calf	90	Rumen	651
Carmyerius mancupatus	Calf	76	Rumen	370
C. exoporus	Calf	100	Rumen	232
Orthocoelium scoliocoelium	Goat	125	Rumen	444
	Goat	173	Rumen	635
O. streptocoelium	Cattle	56	Rumen	253
	Sheep	48	Rumen	253
Explanatum explanatum	Goat	112	Liver	883
Calicophoron calicophorum	Lamb	80—95	Rumen	254
C. daubneyi	Goat	84—91	Rumen	842
C. microbothrium	Calf	100	Rumen	229
	Lamb	89	Rumen	536
	Sheep	71	Rumen	422
	Cattle	56	Rumen	422
	Lamb	69	Rumen	34
C. microbothrioides	Calf	105	Rumen	79
C. phillerouxi	Calf	99—102	Rumen	225
Paramphistomum cervi	Cattle, sheep	103—115	Rumen	507
	Roe deer	87—96	Rumen	507
	Roe deer	85	Rumen	848
P. ichikawai	Lamb	49—51	Rumen	253
	Lamb	42—51	Rumen	488
	Cattle	40—42	Rumen	348
	Sheep	40—41	Rumen	348
P. leydeni	Calf	130	Rumen	654
	Lamb	107—128	Rumen	654
	Cattle	94—96	Rumen	1097
	Calf, lamb	96—97	Rumen	347
	Calf	97—114	Rumen	350

to various environmental conditions both under tropical and continental climates as well as the emergence of the ciliated larvae of the various species of amphistomes is well synchronized with their propagation (mainly in the continental climatic belt). All these and several other factors (historical, biochemical, and behavior of the definitive hosts) can explain their dominance as of intermediate hosts.

Data in Table 5 show that the amphistomes can be arranged into three groups on the basis of the number of snails used. There are 20 species, and they manifest close specificity (they develop in a single snail host). Of these, there is only one species (*Calicophoron daubneyi*) whose strict specificity was also demonstrated experimentally.[811,833] These ex-

TABLE 4

The Higher Molluscan Taxa and Those Genera with the Number of Species which Have Been Recorded as Intermediate Hosts of Amphistomes

	Succinea	Stagnicola	Lymnaea	Physa	Physopsis	Laevipex	Indoplanorbis	Helisoma	Ferrissia	Bulinus	Taphius	Segmentorbis	Segmentina	Planorbis	Menetus	Hippeutis	Helicorbis	Pygmanisus	Gyraulus	Drepanotrema	Choanophalus	Biomphalaria	Bathyomphalus	Armiger	Anisus	Ancylus	Ploiopsis	Cleopatra	Digoniostoma	Physastra
Mesogastropoda																														
Bithyniidae	0	0	0	0	0	0	0	0	0	0	0	0	0	0	0	0	0	0	0	0	0	0	0	0	0	0	0	0	2	0
Paludomidae	0	0	0	0	0	0	0	0	0	0	0	0	0	0	0	0	0	0	0	0	0	0	0	0	0	0	0	2	0	0
Thiaridae	0	0	0	0	0	0	0	0	0	0	0	0	0	0	0	0	0	0	0	0	0	0	0	0	0	0	1	0	0	0
Basommatophora																														
Ancylidae	0	0	0	0	0	0	0	0	0	0	0	0	0	0	0	0	0	0	0	0	0	0	0	0	0	1	0	0	0	0
Planorbidae	0	0	0	0	0	0	0	0	0	0	2	1	4	6	1	1	3	1	14	1	1	2	1	2	11	0	0	0	0	1
Bulinidae	0	0	0	0	2	1	1	3	3	9	0	0	0	0	0	0	0	0	0	0	0	0	0	0	0	0	0	0	0	0
Physidae	0	0	0	3	0	0	0	0	0	0	0	0	0	0	0	0	0	0	0	0	0	0	0	0	0	0	0	0	0	0
Lymnaeidae	0	1	8	0	0	0	0	0	0	0	0	0	0	0	0	0	0	0	0	0	0	0	0	0	0	0	0	0	0	0
Stylommatophora																														
Succineidae	1	0	0	0	0	0	0	0	0	0	0	0	0	0	0	0	0	0	0	0	0	0	0	0	0	0	0	0	0	0

TABLE 5
Intermediate Snail Hosts of Amphistomes

Amphistome species	Intermediate hosts	Locality	Ref.
Heronimus mollis	*Physa gyrina*	U.S.	1005
	P. integra	U.S.	188
	P. sayii	U.S.	1003
Pretestis australianus	*Physastra tetrica*	Australia	32
Amurotrema dombrowskajae	*Anisus acronicus*	U.S.S.R.	403
Megalodiscus temperatus	*Ferrissia fragilis*	U.S.	895
	F. novangliae	U.S.	894
	F. parallela	U.S.	894,895
	Helisoma antrosa, H. campan-ulatum	U.S.	414
	H. trivolvis	U.S.	414, 511, 523, 894, 895, 912
	Laevapex fuscus	U.S.	895
M. microphagus	*Menetus cooperi*	U.S.	283,578
M. ferrissianus	*Ferrissia novangliae*	U.S.	892
Opisthodiscus diplodiscoides	*Ancylus fluviatilis*	Spain	879
Alassostomoides parvus	*Helisoma antrosa*	U.S.	509
	H. trivolvis	U.S.	75,509
Paraibatrema inesperata	*Biomphalaria tenagophila*	Brazil	1001
	B. glabrata	Brazil	
Stichorchis subtiquetrus	*Anisus vortex, Diganiostoma tentaculata, Lymnaea ovata, Succinea putris*	U.S.S.R.	675
	Stagnicola cubensis	U.S.	82
	Lymnaea humilis	U.S.	81
Diplodiscus amphicrus	*Gyraulus heudei*	Vietnam	667
	G. hiemantium	Japan	1073
	G. hiemantium, G. prashadi	Philippines	999
	G. hiemantium	Japan	944
	Segmentina nitidella	Japan	435,944
D. brevicoeca	*Anisus crassilabrum*	Malagasy	778
D. fischthalicus	*Bulinus forskalii*	Togo	74
	Segmentina kinisaensis	Togo	111
D. subclavatus	*Anisus spirorbis, A. vortex, Bathyomphalus contortus, Planorbis marginatus, P. ro-tundatus*	Germany	557
	P. planorbis	Yugoslavia	794
		Hungary	833
		Poland	355
		U.S.S.R.	343
			1007
Catadiscus uruguayensis	*Drepanotrema kermatoides*	Argentina	661
Progonimodiscus doyeri	*Taphius pfeifferi, T. stanleyi*	Zaire	1030
Gastrodiscus aegyptiacus	*Bulinus forskalii*	Sudan	581,582
		Gambia	549
		Loc. not given	549
	Cleopatra bulimoides, C. cy-clostomoides	Egypt	558
G. secundus	*Indoplanorbis exustus*	India	714,715
Zygocotyle lunata	*Helisoma antrosa*	U.S.	1058,1059
	H. campanulata	U.S.	523
	H. trivolvis	U.S.	329,330
Wardius zibethicus	*H. antrosa*	U.S.	637,638
Oliveria indica	*Gyraulus convexiusculus*	India	967
Stephanopharynx compactus	*Bulinus forskalii*	Mauritius	228

TABLE 5 (continued)
Intermediate Snail Hosts of Amphistomes

Amphistome species	Intermediate hosts	Locality	Ref.
	B. cercinus	Mauritius	225
Pseudodiscus collinsi	*Indoplanorbis exustus*	India	716—719
Homalogaster paloniae	*I. exustus*	India	134
	Polypylis hemisphaerula	Japan	169,170
Gastrodiscoides hominis	*Anisus acronicus*	U.S.S.R.	1082,1083
	Helicorbis coenosus	India	256,257
	Lymnaea stagnalis	India	134
Gastrothylax crumenifer	*Armiger crista*	U.S.S.R.	653
	Gyraulus albus	U.S.S.R.	650,653
	G. convexiusculus	India	401,720,949
	G. ehrenbergi	U.S.S.R.	470,870
	Planorbis planorbis	U.S.S.R.	650,653
	P. sieversi	U.S.S.R.	798
Carmyerius exoporus	*Anisus natalensis*	Kenya, Zambia	228
C. gregarius	*Bulinus truncatus, B. forskalii*	Egypt	558
C. mancupatus	*Anisus natalensis*	Zambia	228
	Bulinus liratus, B. mariei	Malagasy	370
C. parvipapillatus	*Physopsis globosus*	Zambia	228
Fischoederius elongatus	*Gyraulus euphraticus, Lymnaea acuminata*	India	830
	L. luteola	India	11, 12, 91, 169, 401, 630, 713, 718, 951
	Gyraulus pulcher	Japan	
Orthocoelium scoliocoelium	*Anisus natalensis*	Kenya	224
	Digoniostoma pulchella	India	439, 444, 626, 627, 635
	Gyraulus convexiusculus	India	631
	Lymnaea luteola, Indoplanorbis exustus	India	401
O. dicranocoelium	*Digoniostoma pulchella*	India	439
O. streptocoelium	*Gyraulus isingi*	Australia	107,253
	G. waterhousei	Australia	107,254
	G. chinensis	Japan	169
Palamphistomum dutti	*G. convexiusculus*	India	992
P. lobatum	*G. convexiusculus*	India	991
Explanatum bathycotyle	*G. convexiusculus*	India	439,629,636
E. explanatum	*G. convexiusculus*	India	401,627,883
Calicophoron calicophorum	*Anisus spirorbis*	U.S.S.R.	870
	Bulinus tropicus	South Africa	376,733,734
	Helicorbis suifunensis	U.S.S.R.	258
	Planorbis kahuika	New Zealand	452
	P. planorbis	U.S.S.R.	478
	Pygmanius pelorius	Australia	254
	Gyraulus scottianus	Australia	784
	G. pulcher	Japan	169
C. daubneyi	*Lymnaea truncatula*	Kenya	226
		Ethiopia	235
		Hungary	833
		Germany	670
		Yugoslavia	543,769
		Bulgaria	340,1015
	L. peregra	Hungary	833
	L. peregra	Germany	670
C. clavula	*Bulinus abyssinicus*	Somalia	897
C. microbothrium	*B. liratus*	Malagasy	750

TABLE 5 (continued)
Intermediate Snail Hosts of Amphistomes

Amphistome species	Intermediate hosts	Locality	Ref.
	B. mariei	Malagasy	368
	B. tropicus	Kenya	229,230,233
		South Africa	938
		Zambia	228
		Ethiopia	364
	B. truncatus, B. forskalii	Egypt	558,562
	B. truncatus	Israel	536
		Iran	23,34
	B. natalensis, B. octoploidus	Ethiopia	364
	Physopsis globosus	Zambia	228
	P. nasutus	Kenya	227,228
C. microbothrioides	Anisus minusculus	U.S.S.R.	259
	Lymnaea cubensis	U.S.	79
	L. humilis	U.S.	79,510,745
	L. truncatula	Bulgaria	809
C. phillerouxi	Bulinus cercinus	Mauritius	225
	B. forskalii	Tanzania	225,235
		Zambia	225
		Togo	16
	B. senegalensis	Gambia	225
C. raja	Physopsis globosus	East Africa	235
C. sukari	Biomphalaria pfeifferi	Kenya	228,229,231,235
		Ethiopia	364
Paramphistomum cervi	Anisus cornetus	U.S.S.R.	465,512
	A. leucostomus	U.S.S.R.	347,489
		Germany	505,506,670
	A. septemgyratus	U.S.S.R.	489
	A. spirorbis	U.S.S.R.	470, 489, 512, 607, 1090
	A. vortex	U.S.S.R.	470,489,512,605
		Germany	100, 505, 506, 507, 669, 822
	Armiger crista	U.S.S.R.	512
		Germany	505,506,669
	A. inermis	U.S.S.R.	512
	Bathyomphalus contortus	U.S.S.R.	470,489,607
		Germany	100, 505, 506, 669, 670
	Choanophalus anophalus	U.S.S.R.	512
	Gyraulus albus	U.S.S.R.	470,512,1090
		Germany	670
	G. ehrenbergi, G. gredleri	U.S.S.R.	512
	Hippeutis complanatus	Germany	505,506
	Planorbis carinatus	U.S.S.R.	1090
	P. planorbis	U.S.S.R.	45, 470, 489, 512, 607, 731, 1090
		Bulgaria	462
		Germany	100, 289, 505, 506, 507, 669, 670, 822, 940
		Turkey	137
		Hungary	848,867
		Italy	36,37
	Segmentina nitida	U.S.S.R.	512,607
		Germany	670
	Indoplanorbis exustus	India	401

TABLE 5 (continued)
Intermediate Snail Hosts of Amphistomes

Amphistome species	Intermediate hosts	Locality	Ref.
P. epiclitum	I. exustus	India	633,884
P. gotoi	Segmentina nitidella	Japan	945
	Gyraulus largielli	Japan	497
P. hiberniae	Anisus leucostomus	England	1067
P. ichikawai	A. centrifugus, A. minusculus	U.S.S.R.	258
	A. spirorbus	Hungary	867
	A. subfilialis	U.S.S.R.	258
	Gyraulus filialis	U.S.S.R.	488
	Helicorbis australiensis	Australia	253
	H. sujfunensis	U.S.S.R.	258,488
	Hippeutis complanatus	U.S.S.R.	470
	Planorbis carinatus	Germany	670
	Segmentina nitida	U.S.S.R.	348
		Germany	670
	S. semiglobosa	U.S.S.R.	258
	S. hemisphaerula	Japan	169
P. leydeni	Anisus leucostoma, A. septem-gyratus	U.S.S.R.	347,654
	A. spirorbis, A. vortex	U.S.S.R.	347,470,654,1098
	Armiper crista	U.S.S.R.	654
	Bath omphalus contortus	U.S.S.R.	344, 347, 470, 654, 1098
	Gyraulus albus	U.S.S.R.	470,654
	G. gredleri	U.S.S.R.	470,1097
	Hippeutis complanatus	U.S.S.R.	654
	Planorbis planorbis	U.S.S.R.	344-347, 470, 471, 517, 654, 1098
		Hungary	858
	Segmentina nitida	U.S.S.R.	344, 347, 654, 1097, 1098
P. liorchis	Helisoma campanulata, H. tri-volvis	Canada	896

amination showed when young snails of *Bulinus truncatus* or *Planorbis planorbis* were exposed to miracidia of this species (which snails might occur together with the usual intermediate host [*Lymnaea truncatula*] even if they picked up the infestations, miracidia were eliminated in the unusual snails, e.g., in *Bulinus truncatus* and *Planorbis planorbis*. The second group of amphistomes which use two intermediate snail hosts comprises 8 species, and the third group of species includes the rest of the amphistomes. Intermediate hosts in the lower case are more than two, and the highest spectra of the intermediate hosts were registered at *Paramphistomum cervi* (18 planorbid snail species). Two species (*Gastrodiscus aegyptiacus, Orthocoelium scoliocoelium*) develop in the phylogenetically most removed hosts (Planorbidae, Paludomidae and Planorbidae, Bithyniidae, respectively).

Comparing the phylogenetic relationships of the intermediate hosts[913] with that of the amphistomes in question, it is difficult to find a definitive congruence between them. Species of the taxon Paramphistomidae, which from an evolutionary point of view are the youngest representations of the amphistomes, can develop in phylogenetically ancient (*Bulinus, Biomphalaria*, etc.) and derived (*Gyraulus, Helisoma*, etc.) species alike. On the other hand, species of the ancient groups of amphistomes (e.g., Diplodiscidae) use snails both phylogenetically ancient (*Taphius, Bulinus*, etc.) and derived (*Gyraulus, Helicorbis*, etc.) snails. In the colonization of the intermediate hosts, seemingly the ecological factors have played a more important role than the phylogenetic closeness of the snails. Specificity, as it is used *in this* sense, that is, as adaptations to a definitive mode of life, means rather an adaptation to a heterogenous but historically united faunistic complex than to species or species groups of snails (e.g., *Stichorchis subtriquetrus*, *Calicophoron microbothrioides*, etc.).

2.2. SCOPE OF DEFINITIVE HOSTS AND HOST SPECIFICITY

The amphistomes are parasitic in altering the number of representatives of each higher grade of vertebrates. The wide range of their distribution and the diversity of the modern forms suggest that the amphistomes could be established in various vertebrates (coaccomodation), and specific parasitic groups of their own evolved, a process which has been accompanied with speciation in different combinations (cospeciation). As it is supposed in Chapter 6, the amphistome-parasitism evolved, similarly to other fluke-host interactions, from ancient rhabdocoele turbellarian stock which in the prevertebrate ages had existed in the form of mollusk-turbellarian relationships. It was the hypothetic Protoamphistoma which was the first amphistomelike group of these flukes representing the developmental line to the modern forms. The Protoamphistome population supposedly had generalized characteristics, a broad distribution and the ability to colonize the adequate groups of the first vertebrates, e.g., fishes. At this time, the life-cycle pattern presently characteristic of amphistomes had evolved and gradually became genetically fixed, which is similar, in principle, in each amphistome studied. Of the fishes and the groups of the subsequently developed vertebrates, there had been potential definitive hosts that actually became those which, by way of the food chain, might have made contact anyway with the adolescariae of the given worm species. By increasing the time of the coaccommodation, characteristic amphistomes of various vertebrate hosts have evolved and become specific to these groups. It should be supposed, however, that besides the specific groups of amphistomes there have been generalized forms with the physiological ability to adapt to groups of vertebrates, which evolved in the time sequence. This is the explanation, on the one hand, of characteristic forms and the presence of groups of amphistomes in the vertebrate grades and of the general distribution of amphistomes in vertebrates as a whole, on the other.

The amphistomes of fishes with a few exceptions are parasitic in species belonging to the major groups (Ostariophysi, Acanthopterygii) of the bony fishes (Teleostei). Fifty-two species have been described from fishes, 20.6% of the total sum of the existing species of the amphistomes. Of the fish-amphistomes, 24 species are parasitic in Cypriniformes, 22 species in Siluriformes, and 15 species in Perciformes (Table 6). Data of Table 6 indicate that amphistomes would rather colonize species of fishes belonging to the first class of freshwater fish,[641] other groups of fishes played an unimportant role as definitive hosts during adaptive radiation of the amphistomes. Some species of amphistomes (*Cleptodiscus, Microrchitrema, Helostomatis*, in part) are also parasitic in marine fishes of the Atlantic and Pacific waters. The great number of species of the Cypriniformes and their feeding behavior (bottom feeding, plant eating, in part) make them preferable definitive hosts. Species of the Cyprinidae and Characidae are the most common hosts of fish amphistomes. Siluriformes is another larger group of species of freshwater fishes. Like the Perciformes, which is also a major host group of the fish amphistomes, they are predatory fish. None of the life-cycle patterns of the parasitic species in Siluriformes and Perciformes is known currently. Supposedly, these fishes pick up infestations by means of adolescariae encysted on the surface of the prey animals. Further fish groups have played insignificant roles as definitive hosts; they are usually harbored with a single amphistome.

The amphistomes of amphibians are parasitic in species of the Anura and Caudata. They are more abundant in species of Ranidae, Bufonidae, and Hylidae. From the amphibians, 36 species of amphistomes have been described to date: 14.1% of the total sum of the existing amphistomes (Table 7). Species of the Ranidae are the most favorable definitive

TABLE 6
The Scope of Definitive Hosts of the Fish-Amphistomes

	Cypriniformes				Siluriformes										Perciformes												Other fish orders						
	Characidae	Citharinidae	Curimatidae	Cyprinidae	Ageneiosidae	Bagridae	Clariidae	Doradidae	Ictaluridae	Loricariidae	Mochochidae	Pangasidae	Pimelodidae	Siluridae	Anabantidae	Centrarchidae	Chaetodontidae	Cichlidae	Kyphosidae	Mastacembelidae	Mugilidae	Percidae	Sciaenidae	Stomateidae	Theraponidae	Sparidae	Atheriniformes	Salmoniformes	Tetraodontiformes	Petromyzoniformes	Ceratodontiformes	Clupeiformes	Mormyriformes
Caballeroinae (5)[a]	0	0	0	5	0	0	0	0	0	0	0	0	0	0	0	0	0	0	0	0	0	0	0	0	0	0	0	0	0	0	1	0	0
Colocladorchiinae (1)	0	0	1	0	0	0	0	0	0	0	0	0	0	0	0	0	0	0	0	0	0	0	0	0	0	0	0	0	0	0	0	0	0
Dadaytrematinae (11)	8	0	0	2	0	0	2	1	0	0	0	0	1	0	0	0	3	1	1	0	0	0	0	0	0	0	0	1	1	1	0	0	0
Diplodiscidae (1)	0	0	0	0	0	0	0	2	0	0	0	0	0	0	1	0	0	0	0	1	0	0	1	1	0	0	0	0	0	0	0	0	0
Helostomatinae (12)	0	0	0	10	1	0	0	0	0	0	0	2	0	0	0	0	0	0	0	0	0	0	0	0	0	0	0	0	0	0	0	0	0
Microrchiinae (2)	2	0	0	0	0	0	0	4	0	0	0	0	0	0	0	0	0	0	0	1	0	0	0	0	0	0	0	0	0	0	0	0	0
Orientodiscinae (5)	0	0	0	3	0	0	1	0	0	0	0	1	0	1	0	0	0	0	0	0	0	0	0	0	0	0	0	0	0	0	0	0	0
Osteochilotrematinae (1)	0	0	0	2	0	0	0	0	0	0	0	0	0	0	0	0	0	0	0	0	0	0	0	0	0	0	0	1	0	0	0	0	0
Prisciamphistominae (2)	0	1	0	0	0	0	0	0	1	0	0	0	0	0	0	6	0	0	0	0	0	0	0	0	0	0	0	0	0	0	0	2	0
Pseudocladorchiinae (5)	2	2	0	0	0	1	0	1	0	1	0	0	1	1	0	0	0	0	0	1	1	1	0	0	1	2	1	0	0	0	0	0	1
Sandoniinae (4)	0	0	0	0	0	0	0	0	0	0	11	0	0	0	0	0	0	0	0	0	0	0	0	0	0	0	0	0	0	0	0	0	0
Travassosininae (1)	4	0	0	0	0	0	0	0	0	0	0	0	0	0	0	0	0	0	0	0	0	0	0	0	0	0	0	0	0	0	0	0	0
Zonocotylidae (2)	0	0	3	0	0	0	0	0	0	0	0	0	0	0	0	0	0	0	0	0	0	0	0	0	0	0	0	0	0	0	0	0	0

[a]   Number of species.

TABLE 7
The Scope of Definitive Hosts of the Amphibian-Amphistomes

	Ambystomidae	Amphiumidae	Plethodontidae	Salamandridae	Sirenidae	Dicamptodontidae	Bufonidae	Discoglossidae	Hylidae	Hyperoliidae	Leptodactylidae	Pelobatidae	Pipidae	Phacophoridae	Pseudidae	Ranidae	Myobotrochidae
Dadaytrematinae (3)[a]	2	0	0	1	0	1	1	0	1	0	0	0	0	0	0	2	0
Diplodiscinae (24)	0	0	0	7	0	0	6	2	3	4	3	1	4	1	3	33	2
Megalodiscinae (6)	3	2	2	2	0	1	3	0	7	0	0	0	0	0	0	19	0
Schizamphistominae (4)	0	0	0	0	1	0	1	0	0	0	0	0	0	0	0	5	0

[a]   Number of species.

TABLE 8
The Scope of Definitive Hosts of the Reptile- and Bird-Amphistomes

	Reptiles											Birds				
	Chelidae	Chelydridae	Chelonidae	Colubridae	Crotalidae	Dermatemydidae	Emydidae	Kinosternidae	Pelomedusidae	Testudinidae	Trionychidae	Anatidae	Phasianidae	Podicipedidae	Recurvirostridae	Scolopacidae
Caballerodiscinae (3)[a]	1	0	0	0	0	1	1	1	0	0	0	0	0	0	0	0
Dadaytrematinae (2)	0	0	0	2	0	0	1	0	0	0	0	0	0	0	0	0
Diplodiscidae (6)	0	0	0	9	1	1	0	0	0	0	0	0	0	0	0	0
Heronimidae (1)	0	2	0	0	0	0	6	7	0	0	0	0	0	0	0	0
Megalodiscinae (1)	0	0	0	1	0	0	0	0	0	0	0	0	0	0	0	0
Nematophilinae (5)	4	0	0	0	0	1	1	1	5	2	0	0	0	0	0	0
Schizamphistominae (16)	1	1	4	0	0	1	15	1	2	0	4	0	0	0	0	0
Zygocotylinae (1)	0	0	0	0	0	0	0	0	0	0	0	17	1	3	2	1

[a]   Number of species.

hosts, due to their diversity, aquatic way of life, and their worldwide distribution. The species of amphibians harboring these amphistomes are mainly predatory animals; hence, the infestation has usually taken place by devouring their own skin which has encysted cercariae or by gulping mud together with adolescariae in the course of hibernation.

In the reptiles (Testudines, Squamata), 34 species of amphistomes have been recovered. They are 13.3% of the total sum of the presently known amphistome species. The majority of the species are parasitic in freshwater turtles, to a lesser extent in marine chelonians and in snakes (Table 8). Of the freshwater turtles, the definitive hosts are mainly those species which are plant eaters or which consume plants occasionally (Emydidae, Pelomedusidae, etc.). Supposedly, infection usually takes place by ingestion of vegetation having encysted cercariae or by means of the prey animals with encysted cercariae on their body surface (certain turtles, snakes).

There is only one species of amphistomes parasitic in birds living in aquatic or in waterside biotopes (Table 8). Cercariae of this species encyst on the aquatic plants, and the infestation is picked up by means of the food materials.

The amphistomes of mammals are parasitic in the taxa of Metatheria and Eutheria and include those groups which might have made contact, by way of their life, with biotopes where the genetically determined life cycle of amphistomes has taken place. Those mammals have become the definitive hosts which were "suitable" to the reproductive strategy of amphistomes. Bearing the life-cycle pattern of these in mind, it becomes obvious that the herbivorous mammals should be the most preferable definitive hosts: Bovidae, Cervidae, etc. (Table 9). Of the species of amphistomes, 133 are parasitic in mammals, 52.3% of the total sum of the hitherto known species. The diversity of the mammal-amphistomes has been affected by two factors: (1) adaptive radiation and speciation of the adequate definitive hosts, mainly herbivorous ones; (2) successful colonization of the anterior part of the digestive tube (forestomach, liver) as an evolutionally new habitat; meanwhile, the retention of the ancient habitat has been preserved by some forms. The adaptive radiation of the amphistomes has been accompanied with intensive speciation, 77.5% of the mammal amphistomes are inhabitants in the stomach of various groups of mammals (Table 9).

Of the data presently available on the host records, it is difficult to draw reliable conclusions as to the definitive host specificity. Our information in this respect is unequal in number and proportion, especially in the case of lower vertebrates. It is difficult to judge whether, e.g., the single host of a given amphistome is an expression of a close specificity or it is the consequence of the sporadic examinations available. With these reservations in mind and accepting the present data on the definitive hosts as the basis of the examination of this topic, it can be said that the final host specificity shows a variable picture. Should we divide the hosts into three categories according to the number of the harboring worms (1, hosts possessing one fluke; 2, hosts possessing two to four flukes; 3, hosts possessing five or more flukes), it can be found that the number of the hosts belonging to the first category (in fishes 74 hosts; in amphibians, 76; in reptiles, 50; and in mammals, 71). Examining the other two categories we find a gradual increasing the frequency of species, especially in the category in which hosts are harbored with more than five fluke species (in fishes there is no host of this category; in amphibians there are 4; in reptiles, 1; and in mammals, 28). The most abundant species of amphistomes are parasitic in cattle and domestic buffalo (66 and 49, respectively). Besides the host-parasite interaction of the hosts belonging to the first category, the most characteristic amphistome fauna is exhibited by the hippopotamus. Of the hitherto known species, 23 amphistomes (8 genera) are specific for this host. Somewhat similar specificity can be observed in amphistomes of the okapi (*Leiperocotyle* spp.).

2.3 EFFECTS OF WORM BURDEN ON DEFINITIVE HOSTS

Adult amphistomes inhabit either the lower part (rectum, cecum) or the anterior portion (rumen, reticulum, liver) of the digestive tube. To attain a simple and easy grouping the species of the former group can be called ruminal amphistomes, the latter nonruminal. There is a basic difference in the life-cycle pattern of the species belonging to one or the other group: viz., the ruminal amphistomes have a longer duodenal stage; in nonruminal species this stage is missing. The pathogenic nature of the amphistomes seems to be associated just with the intestinal phase of these species which, otherwise, are phylogenetically recent and, hence, are regarded as a not yet harmoniously adapted group of amphistomes.

From experimental examinations carried out with the nonruminal amphistomes, e.g., *Zygocotyle lunata*[1059] and *Diplodiscus subclavatus*,[1101] it seems apparent that the worms reach their normal site in 15 to 24 h. These flukes were encountered either in the lumen or on the surface, attached to mucosa, but there were no observations of their cecal embedding. Hence, the reports on pathological infections (*Gastrodiscus aegyptiacus* in horse)[50,114] or amphistomosis (*Homalogaster paloniae* in cattle)[625] of nonruminal amphistomes are not similar either in genesis or in process to that of the clinical picture caused by immature intestinal amphistomes. Fried and Nelson[330] have experimentally demonstrated that the infection of the nonruminal amphistome, *Zygocotyle lunata*, in chicks resulted in a reduction of weight of the cecum and of the cecal content as well as in histopathological changes (disruption of mucosal architecture and loss of villuslike projections).

TABLE 9
The Scope of Definitive Hosts of the Mammal-Amphistomes

	Bovidae	Camelidae	Canidae	Castoridae	Caviidae	Cercopithecidae	Cervidae	Cricetidae	Dasyproctidae	Dugongidae	Elephantidae	Equidae	Giraffidae	Hippopotamidae	Hominidae	Hydrochoeridae	Macropodidae	Muridae	Myocastenidae	Myrmecophagidae	Pongidae	Rhynoceroridae	Suidae	Tapiridae	Tayassuidae	Tragulidae	Trichechidae
Balanorchiidae (1)[a]	1	0	0	0	0	0	2	0	0	0	0	0	0	0	0	0	0	0	0	0	0	0	0	0	0	0	0
Brumptiidae (3)	0	0	0	0	0	0	0	0	0	0	2	0	0	0	0	0	0	0	0	0	0	2	2	0	0	0	0
Chiorchiinae (3)	0	0	0	0	1	0	0	0	0	0	0	0	0	0	0	0	0	2	1	0	0	0	0	0	0	0	3
Cladorchiinae (4)	0	0	0	0	0	0	0	0	1	0	0	0	0	0	0	3	0	0	0	0	0	0	0	1	1	0	0
Gastrodiscidae (2)	1	0	0	0	0	0	0	0	0	0	1	7	0	1	0	0	0	0	0	0	0	1	4	0	0	0	0
Gastrothylacidae (26)	33	0	0	0	0	0	9	0	0	0	0	0	0	0	0	0	0	0	0	0	0	0	0	0	0	0	0
Olveriinae (2)	3	0	0	0	0	0	2	0	0	0	0	0	0	1	0	0	1	0	0	0	0	0	0	0	0	0	0
Orthocoelinae (40)	11	0	1	0	0	0	20	0	0	0	1	0	1	1	0	0	0	0	0	0	0	0	0	0	0	0	0
Paramphistominae (37)	54	1	0	0	0	0	0	0	0	0	1	1	0	0	0	0	0	0	0	0	0	0	0	0	0	0	0
Pfenderiinae (3)	0	0	0	0	0	0	0	0	0	0	1	0	1	0	0	0	1	0	0	0	0	0	0	0	0	0	0
Pseudodiscinae (2)	1	0	0	0	0	0	0	0	0	0	0	2	0	0	0	0	0	0	0	0	0	0	0	0	0	0	0
Solenorchiinae (1)	0	0	0	0	0	0	0	0	0	1	1	0	0	1	0	0	0	0	0	0	0	0	0	0	0	0	0
Stephanopharynginae (1)	12	0	0	0	0	0	0	0	1	0	0	0	0	0	0	0	0	0	0	0	0	0	0	0	0	0	0
Stichorchiinae (2)	1	0	0	2	0	0	3	2	0	0	0	0	0	0	0	0	0	0	0	1	0	0	0	0	0	0	0
Watsoniinae (8)	8	0	0	0	0	7	3	2	0	0	1	0	0	0	1	0	0	4	1	0	2	0	1	0	2	1	0
Zygocotylinae (3)	2	0	0	0	0	0	0	1	0	0	0	0	0	0	0	0	0	2	0	0	0	0	0	0	0	0	0

[a] Number of species.

The amphistomes, inhabitants of forestomach or liver, especially in ruminants (both domesticated and wild) might cause serious disease in massive infestation. The disease caused by amphistomes is usually called paramphistomosis or, more generally, amphistomosis in the relevant literature. It is, however, recommended to nominate it according to the generic name of the species incriminated in the diseases (calicophorosis, paramphistomosis, etc.) due to their higher information content. Various species of amphistomes have been reported as the causative agents of this disease, which is mainly confined to the genera *Calicophoron*, *Cotylophoron*, *Gastrothylax*, *Paramphistomum*, and *Balanorchis*. Along the distributional areas of these species subclinical infestations or acute amphistomosis have often been described.

One of the characteristic features of the life-cycle pattern of the ruminal amphistomes is the involvement of the duodenal stage in their development. The adolescariae encyst in the duodenum, and the juvenile flukes remain there for some weeks. They attach themselves to the mucosa by their ventral sucker, later entering the submucosa. During this time they feed on epithelial cells of the duodenal glands, and the presence of the juvenile forms in the submucosa generates proliferative changes and hypertrophy of the submucosa in several cases. Sharma Deorani and Jain[872] suppose that the involvement of the duodenal stage in the life cycle is a prerequisite to further development, and the still juvenile forms after completing this developmental stage ("limited development") are able to withstand the acid content of the abomasum during their passage to the rumen. The pathological picture, which is the corollary of the settlement of the juvenile forms in the duodenum, is called intestinal amphistomosis and that of the adults in the rumen is ruminal amphistomosis. Before presenting the most important aspects of both of these diseases, some attention should be paid to the amphistomes parasitic in the liver because they are also pathogenetic.

Explanatum explanatum is the only species of the three commonly found in the liver of ruminants. Its life history was studied by Singh,[883] but it is still no fully known as to how these flukes reach their seat of predilection, the bile duct, and whether a longer intestinal phase is involved or not. Kulasiri and Seneviratne[513] and Singh and Kuppuswami[887] suggested a transperitoneal pathway of migration on the basis of gross and histopathological lesions. Arora and Kalra,[35] however, were of the opinion that the route of migration to the liver might possibly be through the diverticulum duodeni. The disease caused by liver amphistomes is called biliary explanatosis. It is characterized by hypertrophy and hyperplasia of biliary epithelium and proliferation of connective tissue in the periportal areas. The bile duct epithelium was sucked in a form of flash-shaped projection. This suggests that the ventral sucker possibly plays a role in causing proliferation and hyperplasia of the epithelium.[35,513,768,887,963,1006]

The most widespread and important disease caused by rumen flukes is intestinal amphistomosis. Since the beginning of the present century, a great number of literary data from various countries of different continents have been accumulated about this disease and about the amphistomes that cause it.

Helminthology

Numerous species of amphistomes have been incriminated as etiologic agents of this disease. Identification of the correct species might be, however, dubious in certain cases, especially in species of *Calicophoron* and *Paramphistomum*, provided it was not based on histological examination of the muscular organs. In this context it should be borne in mind that investigators must face the problem of the identification of the immature flukes which further increases these difficulties. In the last few decades several papers have been published[227,254,836,842,860,863,934] trying to clarify the specific compositions of amphistomes of the endemic areas. These examinations and post-mortem studies on amphistomes derived from the recovered animals show that the most frequent species are the causative agents.

Accordingly, in Europe they are probably *Calicophoron daubneyi, C. microbothrium, Paramphistomum cervi,* and *P. ichikawai*; in Asia, *Paramphistomum epiclitum, P. gracile, Gastrothylax crumenifer, Fischoederius elongatus,* and *F. cobboldi*; in Africa, *Calicophoron microbothrium, C. daubneyi, C. calicophorum*; in Australia, *P. ichikawai, Calicophoron calicophorum*; in North America, *Calicophoron microbothrioides*; in South America, *Balanorchis anastrophus.* These species have the same type of life-history pattern, using various freshwater pulmunate mollusks of the local faunas as intermediate hosts. In heavy infestation they cause acute parasitic gastroenteritis which is accompanied with high morbidity and mortality rates, particularly in young stocks. The different aspects of this disease were studied by Dinnik,[227] Boray,[105,106] Horak,[422] Lengy,[537] and Sey,[857] and an excellent review was given by Horak.[423] The short summarization below is based on these publications.

Clinical Symptoms

The cercariae liberated from the snail usually encyst in a short time, under natural conditions, on vegetation. The adolescariae, picked up by the grazing definitive hosts, encyst in the anterior portion of the intestine, and they can stay there for 2 to 6 weeks. The clinical symptoms manifest 2 to 4 weeks after infection, and they include listlessness and anorexia. Later diarrhea develops when the feces are extremely fluid. The fetid diarrhea can soil the breech and the hind legs. Submandibular edema has been observed and reported several times. Immature flukes are found either on the surface of the mucosa or embedded in the mucosa.

Pathology

Experimental infestations show that about 40,000 juvenile paramphistomes are required to produce acute fatal disease in sheep,[424] and this figure is about 160,000 in cattle.[422] When the gastrointestinal tract is opened after death, immature flukes can be seen attached to the mucosa of the intestine. The latter is hypertrophied and edematous. Numerous erosions caused by the ventral sucker of the flukes are present, and the intestinal mucosa may be hemorrhagic. The histopathological sections show the presence of the young flukes embedded in the mucosa, rarely reaching the muscularis mucosa. Small pieces of mucosa are sucked in filling the major part of the ventral sucker causing necrosis of the cells and resulting in erosions of the villi of the mucosa.

After completing the intestinal phase, the immature flukes move to the rumen, attached to the epithelium and papillae of the ruminal pillar. Adult paramphistomes cause no clinical symptoms. At autopsy the papillae appear anemic, and necrosis can be observed caused by the pressure of the acetabulum of the attached flukes to the epithelium. They reach their maturity in 3 to 3.5 months[229,422] in domestic and 3.5 to 4 months in wild ruminants.[507,867] *Calicophoron microbothrium* grow continuously and reach their maximum size in cattle 5 to 9 months after infection. Their supposed life span is greater than this figure,[233] and from the epizootiological point of view, it plays an important role together with other factors.

Epizootiology

The uptake of a large quantity of adolescariae in a short time period is the prerequisite of manifestation of acute amphistomosis. Ecological conditions and the grazing technique of domestic ruminants might be favored for infection, due to the multiplication of the intermediate hosts and infection of definitive hosts. The peak of the daily egg output, in experimentally infected sheep, was between 12 to 14 hours.[422] In these hours, the livestock generally visit the source of drinking water and thus concentrate the eggs, dropping feces in the waters. Under favorable conditions these water bodies, both under tropical and continental climate, harbor intermediate snail hosts. The high productivity of the snails (e.g., daily output of a single *Planorbis planorbis* was 27.7 eggs;[1101] the same figure in the case of *Bulinus tropicus* was 16.3 eggs.[227] In the course of continuous contamination of the water bodies with eggs by the visiting ruminants and the presence of the susceptible intermediate hosts, there is a possibility that high quantities of adolescariae are yielded. With the beginning of the drier season the natural water bodies decrease, and the previously submerged aquatic plants, harbored with encysted metacercariae, become accessible for grazing ruminants. The outbreaks of acute paramphistomosis are usually confined to the drier season (in Tanganyika;[139] in Hungary;[105] in New Zealand;[1053] in Australia[105]). In South Asian countries outbreaks usually occur from September until January, after the heavy rains, when livestock may graze on pastures which were inundated with water.[24,466,695]

Anthelmintic Treatment and Integrated Control

Although the treatment of the ruminal amphistomosis has no benefit to the affected animals, it may be prophylactic, in effect, to reduce the source of infection for the intermediate hosts. Several already known and recently developed anthelminthics have been used against intestinal and ruminal amphistomosis of domesticated ruminants. Some of them that proved to be the most effective are as follows.

Terenol

It has been tested by several authors against ruminal[175,493,1056] and intestinal[1043] amphistomosis in cattle or both and by Lämmler et al.[524] in goat. Administered at dosages of 65 mg/bw. it was 84.4 to 100% effective against juvenile and adult flukes alike.

Bithionol

Horak,[421] Fedortsenko,[306] and Mereminskiǐ et al.[606] have tried to assess its efficacy against subclinical paramphistomosis in sheep and cattle; and against clinical paramphistomosis in cattle (Ibrović and Levi)[429] and in sheep (Cvetković).[196] In sheep, dose rates of 25 to 100 mg/bw were highly effective against immature and mature amphistomes. Dose rates of 70 mg/bw given alone or combined with 100 mg/bw fenbendazole repeated 48 h later gave more than 90% efficacy against both immature and mature flukes, resulting in a rapid recovery of affected heifers.

Niklosamide

The is one of the anthelminthics which has broad application in field and laboratory usage alike against intestinal amphistomosis. Horak[419,420] found that this drug at the dose rate of 50 mg/bw was highly effective (92.4 to 99.7%) against immature *Calicophoron microbothrium* in artificially infected sheep. Katiyar and Garg[467] made a similar observation using the same drug at the same level of doses in naturally infected sheep. Horak and Clark,[424] Boray,[106] and Rolfe and Boray,[784,785] studying the therapeutic effects of this drug, found remarkable clinical improvement (diarrhea ceased within a few days, appetite returned to normal in 2 or 3 d).

Control measures recommended by various authors[105,106,227,422] against amphistomes aim at the prevention of the losses of the animal production caused by these flukes or to save the animal's life. Amphistomosis has been maintained in a system in which the responsive or infected hosts and susceptible mollusks are present under favorable ecological conditions. It is obvious that a control plan should include integrated measures which prevent the operation of the fluke-snail host-definitive host system. In this context these main regulatory principles have encompassed: (1) elimination of the amphistome infestation by a consistently effective anthelmintic in domestic ruminants and preservation of the pastures from the contamination of the flukes' eggs; (2) decreasing the surface of the natural water bodies along the pasture, prevention of the grazing animals making contact with them, and the recommendation to supply them with clean drinking water; (3) destruction of the foci of the intermediate snail hosts either by drainage or by molluscicides.

In connection with the "economic" importance of the amphistomes, probably their most unique benefit which I ever experienced is the following. Visiting the slaughterhouses in India to collect samples of amphistomes it could often be seen, especially at the Calcutta abattoir, that the bigger specimens were collected by young boys and that chickens were fed on these flukes.

Finally, it is interesting to note that the occurrence of hyperparasitism in amphistomes has been described in two cases. Travassos[958] has found *Atractis trematophila* in ceca of *Nematophila grandis*. Sey and Moravec[868] recovered immature specimens of *Spironoura babei* in cecum of *Amurotrema dombrowskajae*.

3. SYSTEMATICS OF THE AMPHISTOMES

3.1 HISTORY OF THE CLASSIFICATION

The first representative of the amphistomes (*Planaria subclavata*) was described in the 18th Century by Goeze.[351] At the end of the century some 28 species had been recovered and allocated to the genus *Amphistoma* Rudolphi, 1809, *Amphistomum* Nitzsch, 1819, or to other categories. From the beginning of the present century to the 1940s, more than four times as many species were discovered than in the preceding one and a half centuries. The explanation of this can be found not only in the ever-increasing interest to study amphistomes, but also in the notion that amphistomes are not harmless inhabitants of the digestive tube of domestic ruminants. By the 1930s, it was obvious that the rumen flukes have an intestinal phase in their life history which is pathogenetic, causing serious morbidity and mortality. This notion has considerably increased the knowledge of various aspects of these flukes, and newer information has also been contributed to their taxonomy. Nowadays, we have information on the amphistomes of various areas of the globe. The bulk of the literary data has, however, represented a diverse quality of papers which renders them more difficult to evaluate with equal view.

By the turn of the century, it became evident that the species assigned to the genus *Amphistoma* were a heterogeneous assembling of flukes. In his endeavor to establish the homogeneous structure of the taxon, Fischoeder divided the genus *Amphistoma* into various genera (*Paramphistomum, Cladorchis*, etc.),[309-312] and they were transferred to the family Paramphistomidae set up by himself. With the rearrangement of these flukes, Fischoeder established the first system having scientific value of the amphistomes examined. Since that period, several authors have dealt with the classification of amphistomes used up to now. They are Stiles and Goldberger,[915] Stunkard,[921,923] Maplestone,[591] Fuhrmann,[331] Fukui,[334] Travassos,[986] Näsmark,[647] Southwell and Kirshner,[902] Szidat,[941] Skrjabin,[890,891] Baer and Joyeux,[58] and Yamaguti.[1076,1077]

It does not seem to be necessary to analyze separately all these classificational schemes. Instead, the following generalizations can be stated.

1. Systematics of the amphistomes are based on traditional principles and satisfy their purposes. Phylogenetic or biogeographic viewpoints have often been omitted from consideration.
2. Activities of several authors have encompassed the examinations of amphistomes of mammals or, more frequently, of ruminants.[268-278,311,591]
3. The quality of the established systematics was considerably affected by the experiences of the authors with amphistomes and by the available original material for examinations. Fischoeder,[311] Näsmark,[647] and Éduardo,[268] for example, who were thoroughly acquainted with the amphistomes, have stated principles which hold well today.
4. Systematics, based on considerable original material, were established by Fischoeder,[311] Stiles and Goldberger,[915] Maplestone,[591] Fukui,[334] and Näsmark.[647]
5. Systematics, based on literary data, were created by Fuhrmann,[331] Southwell and Kirshner,[902] Szidat,[941] Skrjabin,[890,891] and Baer and Joyeux.[58]
6. Most species were included in the systematics set up by Travassos,[986] Näsmark,[647] Southwell and Kirshner,[902] Fukui,[334] Skrjabin,[890] Baer and Joyeux,[58] and Yamaguti.[1077]

Of the above-mentioned authors, attention should be paid to Fischoeder's,[311] Stunkard's,[923] and Näsmark's[647] activities as their principles applied proved to be correct in the light of later examinations and to Skrjabin's[890] and Yamaguti's[1077,1078] comprehensive works which included the taxonomic and biological information accumulated in the field of amphistomology in past decades, creating the following new systematic schemes of the amphistomes studied.

Fischoeder, 1901—1903
Paramphistomidae
  Paramphistominae
    Paramphistomum, Stephanopharynx, Gastrothylax
  Cladorchinae
    Cladorchis/Taxorchis, Stichorchis, Cladorchis/
    Chiorchis
    Gastrodiscus
    Homalogaster
    ? Balanorchis

Stunkard, 1925
Paramphistomidae
  Paramphistominae
    Paramphistomum, Cotylophoron, Stephanopharynx
  Cladorchinae
    Cladorchis, Taxorchis, Chiorchis, Stichorchis, Pfenderius, Watsonius, Wardius, Pseudocladorchis, Microrchis, Pseudodiscus
  Schizamphistominae
    Schizamphistomum, Alassostoma, Schizamphistomoides, Alassostomoides
  Gastrothylacinae
    Gastrothylax, Carmyerius, Fischoederius
  Zygocotylinae
    Zygocotyle
  Balanorchinae
    Balanorchis
  Diplodiscinae
    Diplodiscus, Catadiscus, Opisthodiscus
  Gastrodiscinae
    Gastrodiscus, Homalogaster, Gastrodiscoides
  Brumptinae
    Brumptia

Näsmark, 1937
Paramphistomidae
  Pseudocladorchinae
    Pseudocladorchis
  Schizamphistominae
    Schizamphistomum, Alassostoma, Schizamphistomoides, Ophioxenus, Alassostomoides, ? Cleptodiscus, ? Dadaytrema
  Stichorchinae
    Stichorchis
  Cladorchinae
    Cladorchis, Taxorchis, Chiorchis, Microrchis
  Pfenderinae
    Pfenderius, Tagumaea
  Diplodiscinae

    Diplodiscus, Opisthodiscus
Zygocotylinae
  Zygocotyle
Balanorchiinae
  Balanorchis
Paramphistominae
  Paramphitomum, Gigantocotyle, Calicophoron, Cotylophoron, Ugandocotyle, Ceylon-ocotyle, Nilocotyle, Buxifrons, Macropharynx
Gastrothylacinae
  Gastrothylax
Brumptiinae
  Brumptia
Watsoniinae
  Watsonius
Gastrodiscinae
  Gastrodiscus, Homalogaster, Gastrodiscoides
Pseudodiscinae
  Pseudodiscus
Stephanopharynginae
  Stephanopharynx

Skrjabin, 1949
Paramphistomatata
  Paramphistomatoidea
    Paramphistomatidae
      Paramphistomum, Cotylophoron, Gigantocotyle, Calicophoron, Ugandocotyle, Cey-lonocotyle, Nilocotyle, Buxifrons, Macropharynx
    Gastrothylacidae
      Gastrothylax, Carmyerius, Fischoederius
  Cladorchoidea
    Cladorchidae
      Cladorchinae
        Cladorchis, Taxorchis, Chiostichorchis
      Pfenderinae
        Pfenderius, Tugumaea
      Balanorchinae
        Balanorchis
      Pseudodiscinae
        Pseudodiscus
      Kalitrematinae
        Kalitrema
      Cleptodiscinae
        Cleptodiscus
      Stichorchinae
        Stichorchis
    Diplodiscidae
      Diplodiscinae
        Diplodiscus, Catadiscus, Megalodiscus, Dermatemytrema
      Opisthodiscinae
        Opisthodiscus
      Schizamphistominae

Schizamphistomum, Chiorchis, Schizamphistomoides, Ophioxenus, Alassosto-
moides, Microrchis, Travassosinia, Dadaytrema, Alassostoma, Neocladorchis
Helostomatinae
Helostomatis, Protocladorchis
Nematophilinae
Nematophila, Orientodiscus
Watsoniinae
Watsonius, Hawkesius
Dadayinae
Dadayius
Zygocotylinae
Zygocotyle, Stunkardia
Nicollodiscinae
Nicollodiscus
Gastrodiscidae
Gastrodiscinae
Gastrodiscus, Homalogaster, Gastrodiscoides
Stephanopharyngidae
Stephanopharynginae
Stephanopharynx
Brumptidae
Brumptinae
Brumptia
Microscaphidioidea
Microscaphidiidae
Microscaphidium, Polyangium, Angiodyctium, Octangioides, Octangium, Deutero-
baris, Hexangitrema, Hexangium, Parabaris
Gyliauchenoidea
Gyliauchenidae
Gyliauchen, Paragyliauchen, Telotrema
Metacetabulidae
Metacetabulum

Yamaguti, 1971
Paramphistomidae
Paramphistominae
Paramphistominae
Paramphistomum
Paramphistomum, Explanatum
Cotylophoron
Calicophoron
Ugandocotyle
Balanorchiinae
Balanorchis
Choerocotyloidinae
Choerocotyloides
Cladorchiinae
Cladorchiini
Cladorchis, Taxorchis, Chiorchis, Pfenderius, Wardius, Chiostichorchis
Olveriini
Olveria

Gastrodiscinae
 Gastrodiscus, Homalogaster, Gastrodiscoides
Orthocoeliinae
 Orthocoelium, Buxifrons, Gigantatrium, Macropharynx, Glyptamphistoma
 Platyamphistoma
 Nilocotyle
  Nilocotyle, Sellsitrema
 Paramphistomoides, Pseudoparamphistoma
Skrjabinocladorchiinae
 Skrjabinocladorchis
Pseudodiscinae
 Pseudodiscus, Choerocotyle, Hawkesius
Solenorchiinae
 Solenorchis, Indosolenorchis
Stephanopharynginae
 Stephanopharynx
Stichorchiinae
 Stichorchis
Watsoniinae
 Watsonius
Zygocotylinae
 Zygocotyle, Stunkardia
Dadaytrematinae
 Dadaytrema, Pseudocleptodiscus, Panamphistomum, Protocladorchis, Basidiodiscus, Amurotrema, Sandonia
Catadiscinae
 Catadiscus
Dermatemytrematinae
 Dermatemytrema
Nematophilinae
 Nematophila
Orientodiscinae
 Orientodiscus, Kachugotrema, Parorientodiscus
Progonimodiscinae
 Progonimodiscus
Diplodiscinae
 Diplodiscus, Pseudodiplodiscus, Megalodiscus, Opisthodiscus, Pseudopisthodiscus
Pseudochiorchiinae
 Pseudochiorchis
Bancroftrematinae
 Bancroftrema
Brevicaecinae
 Brevicaecum
Caballeroiinae
 Caballeroia
Cleptodiscinae
 Cleptodiscus, Neocladorchis
Dadayiinae
 Dadayius
Helostomatinae
 Helostomatis

Schizamphistominae
  Schizamphistomum, Schizamphistomoidea, Alassostoma, Alassostomoides, Halltrema, Ophioxenus, Parachiorchis, Pseudalassostoma, Pseudalassostomoides, Quasichiorchis
Kalitrematinae
  Kalitrema
Macrorchitrematinae
  Macrorchitrema
Microrchiinae
  Microrchis, Travassosinia
Nicollodiscinae
  Nicollodiscus
Pisciamphistomatinae
  Pisciamphistoma
Pseudocladorchiinae
  Pseudocladorchis
Gastrothylacidae
Gastrothylacinae
  Gastrothylax, Carmyerius, Fischoederius
Johnsonitrematinae
  Johnsonitrema
Brumptiidae
  Brumptia
Gyliauchenidae
Gyliaucheninae
  Gyliauchen, Ichthyotrema, Flagellotrema, Leptobulbus, Paragyliauchen
Apharyngogyliaucheninae
  Apharyngogyliauchen

Fischoeder,[311] with the application of the broadest spectra of the comparative morphological characteristics, Stunkard,[923] with the first outline of the rightly interpreted systematic structure, took the direction toward future examinations. Näsmark,[647] similarly in principle, but on quite a different basis, improved the systematics of the amphistomes. It was he who discovered the histomorphological structure of the muscular organs (pharynx, terminal genitalium, acetabulum) seen on median sagittal sections, and he considerably enlarged the scope of our knowledge about the specific characters. These morphological traits laid the foundation not only of new diagnostic principles, but they also created the taxonomic framework in which taxonomic and systematic studies have been done with considerable success.

Näsmark's laborious method has been used since the 1950s. The major part of the great number of papers written since that time have contained histomorphological details. These examinations have successfully contributed to the clarification of the taxonomic states of several doubtful species, and the histomorphological details, combined with other morphological and biological characteristics, are proper tools for analyzing the phylogenetic relationships among amphistomes.

A proposed new systematics, based on the weighted holomorphological traits at hand, is to be given in the following chapter, and its classificational scheme is depicted in the following chart.

Amphistomida/Lühe, 1909/ Odening, 1974
  Heronimata Skrjabin et Schulz, 1932
    Heronimidae Ward, 1917

Heronimus MacCallum, 1902
Zonocotylata Sey, 1988
  Zonocotylidae Yamaguti, 1963
    Zonocotyle Travassos, 1948
Paramphistomata Szidat, 1936
  Cladorchoidea Skrjabin, 1949
    Cladorchiidae Southwell et Kirshner, 1937
      Dadaytrematinae Yamaguti, 1958
        Dadaytrema Travassos, 1931
        Dadayius Fukui, 1929
        Dadaytremoides Thatcher, 1979
        Panamphistomum Manter et Pritchard, 1964
        Neocladorchis Bhalerao, 1937
        Cleptodiscus Linton, 1910
        Macrorchitrema, Vigueras, 1940
        Ophioxenus Sumwall, 1926
    Travassosiniinae Sey, 1987
        Travassosinia Vaz, 1932
    Orientodiscinae Yamaguti, 1971
        Orientodiscus Srivastava, 1938
    Sandoniinae Ukoli, 1972
        Sandonia McClelland, 1957
        Basidiodiscus Fischthal et Kuntz, 1959
        Pretestis Angel et Manter, 1970
        Australotrema Khalil, 1981
    Microrchiinae Yamaguti, 1958
        Microrchis Daday, 1907
        Colocladorchis Thatcher, 1979
        Pseudocladorchis Daday, 1907
        Brevicaecum McClelland, 1957
        Kalitrema Travassos, 1933
        Nicollodiscus Srivastava, 1938
    Osteochilotrematinae Jones et Seng, 1986
        Osteochilotrema Jones et Seng, 1986
    Caballeroiinae Yamaguti, 1971
        Caballeroia Thapar, 1960
        Bancroftrema Angel, 1966
        Platycladorchis Sey, 1985
    Helostomatinae Skrjabin, 1949
        Helostomatis/ Fukui, 1929/ Travassos, 1934
        Protocladorchis Willey, 1935
        Amurotrema Akhmerov, 1959
    Pisciamphistomatinae Yamaguti, 1971
        Pisciamphistoma Yamaguti, 1971
    Megalodiscinae Sey, 1987
        Megalodiscus Chandler, 1923
        Opisthodiscus Cohn, 1904
    Schizamphistominae Looss, 1912
        Schizamphistomum Looss, 1912
        Schizamphistomoides Stunkard, 1925
        Stunkardia Bhalerao, 1931

Alassostoma Stunkard, 1916
Alassostomoides /Stunkard, 1924/ Fuhrmann, 1928
Pseudocleptodiscus Caballero, 1961
Pseudoalassostomoides Yamaguti, 1971
Quasichiorchis Skrjabin, 1949
Lobatodiscus Rhode, 1984
Nemathophilinae Skrjabin, 1949
Nematophila Travassos, 1934
Pseudoalassostoma Yamaguti, 1958
Halltrema Lent et Freitas, 1939
Parachiorchis Caballero, 1943
Caballerodiscinae Sey, 1988
Caballerodiscus Sey, 1988
Elseyatrema Rohde, 1984
Solenorchiinae /Hilmy, 1949/ Yamaguti, 1958
Solenorchis Hilmy, 1949
Chiorchiinae Sey, 1988
Chiorchis Fischoeder, 1901
Chiostichorchis Artigas et Pacheco, 1933
Paraibatrema Ueta, Deberaldini, Cordeiro et Artigas, 1981
Pfenderiinae Fukui, 1929
Pfenderius Stiles et Goldberger, 1910
Stichorchiinae Näsmark, 1937
Stichorchis /Fischoeder, 1901/ Looss, 1902
Cladorchiinae /Fischoeder, 1901/ Lühe, 1909
Cladorchis Fischoeder, 1901
Taxorchis /Fischoeder, 1901/ Stiles et Goldberger, 1910
Diplodiscidae Skrjabin, 1949
Pseudodiplodiscus Szidat, 1939
Catadiscus Cohn, 1904
Dermatemytrema Price, 1937
Diplodiscus Diesing, 1836
Australodiscus Sey, 1983
Progonimodiscus Vercammen-Grandjean, 1960
Gastrodiscidae Stiles et Goldberger, 1910
Gastrodiscus Leukart, 1877
Balanorchiidae Ozaki, 1937
Balanorchis Fischoeder, 1901
Brumptiidae /Skrjabin, 1949/ Yamaguti, 1971
Brumptia Travassos, 1921
Hawkesius Stiles et Goldberger, 1910
Choerocotyle Baer, 1959
Paramphistomoidea Stiles et Goldberger, 1910
Zygocotylidae Sey, 1988
Zygocotylinae Ward, 1917
Zygocotyle Stunkard, 1917
Wardius Barker et East, 1915
Choerocotyloides Prudhoe, Khalil et Yeh, 1964
Oliveriinae Srivastava, Maurya et Prasad, 1980
Oliveria Thapar et Sinha, 1945
Stephanopharynginae Stiles et Goldberger, 1910
Stephanopharynx Stiles et Goldberger, 1910

Pseudodiscinae Näsmark, 1937
  Pseudodiscus Sonsino, 1895
  Macropotrema Blair, Beveridge et Speare, 1979
Watsoniinae Näsmark, 1937
  Watsonius Stiles et Goldberger, 1910
  Homalogaster Poirrier, 1883
  Gastrodiscoides Leiper, 1913
  Skrjabinocladorchis Tchertkova, 1959
Gastrothylacidae Stiles and Goldberger, 1910
  Fischoederius Stiles et Goldberger, 1910
  Velasquezotrema Eduardo et Javellana, 1987
  Carmyerius Stiles et Goldberger, 1910
  Gastrothylax Poirier, 1883
Paramphistomidae Fischoeder, 1901
  Orthocoeliinae Price et McIntosh, 1953
    Orthocoelium /Stiles et Goldberger, 1910/ Price et McIntosh, 1953
    Leiperocotyle Eduardo, 1980
    Paramphistomoides Yamaguti, 1958
    Platyamphistoma Yamaguti, 1958
    Bilatorchis Eduardo, 1980
    Buxifrons Näsmark, 1937
    Gigantoatrium /Yamaguti, 1958/ spelling amended
    Nilocotyle Näsmark, 1937
    Sellsitrema /Yamaguti, 1958/ Eduardo, 1980
    Pseudoparamphistoma Yamaguti, 1958
    Macropharynx Näsmark, 1937
    Palamphistomum Srivastava et Tripathi, 1980
    Gemellicotyle Prudhoe, 1975
    Glyptamphistoma Yamaguti, 1958
  Paramphistominae Fischoeder, 1901
    Ugandocotyle Näsmark, 1937
    Explanatum Fukui, 1922
    Gigantocotyle Näsmark, 1937
    Cotylophoron Stiles et Goldberger, 1910
    Calicophoron Näsmark, 1937
    Paramphistomum Fischoeder, 1901

3.2. ORDER AMPHISTOMIDA (LÜHE, 1909) ODENING 1974

Diagnosis: Digenea. Adults amphistome type or monostome, evolving from amphistome plan; usually conical, strongly muscular in body; hermaphroditic with posteroterminally located, strongly developed acetabulum. Ventral pouch occasionally present. Oral opening is surrounded by well-developed muscular pharynx with or without appendages; esophagus may be muscular (sphincter, muscular thickening, esophageal bulb) or without it; ceca are usually straight, occasionally undulating or sinuous. Testes usually double, rarely single; tandem, diagonal or horizontal; intercecal, occasionally extracecal, round, lobed, or branched in outline. Ovary spherical, posttesticular, intertesticular, rarely pretesticular. Genital pore with or without sucker, terminal genitalium often well developed; cirrus sac and hermaphroditic pouch can be present or absent. Uterine coils mostly intercecal, rarely extracecal. Vitellaria follicular, exceptionally compact mass extending to lateral fields of body, usually well developed; rarely, only a few. Laurer's canal and lymphatic system present. Excretory system stenostome, excretory bladder usually with dorsoventral opening. Parasitic in each group of vertebrates.

Cercariae large, usually encyst freely (adolescariae); amphistome type, with unforked tail. Body either with dense pigmentation (Cercaria pigmentata) or with light pigmentation (Cercaria intermedia) or without pigmentation (Cercaria diplocotylea). A pair of eyespots and several cystogenous cells present (except for Heronimata). Protonephridial system well developed; excretory bladder nonepithelial, newborn cercariae poorly developed. Rediae and sporocyst occasionally absent; rediae without collar. Intermediate hosts usually pulmunate freshwater snails. Miracidia with ciliated epithelial cells in four rows, arranged according to formula 6:8:4:2 or 6:6:4:2 and without pigment spots (except for Heronimata). Eggs large in size, eggshell either thin (freshwater developed) or thick (marine developed), with viviparity, ovoviviparity, or ovoparity. Life-cycle pattern: dixenous type, with one intermediate host.

KEY TO THE SUBORDERS OF AMPHISTOMIDA

1. Adults and cercariae with posteroterminal ventral sucker .... Paramphistomata
   Adults with posteroterminal suckling disk or without ventral sucker 2
2. Adults with suckling disk .... Zonocotylata
   Adults without sucker, cercariae with posteroventral sucker .... Heronimata

3.2.1. HERONIMATA SKRJABIN ET SCHULZ, 1937

Diagnosis: Amphistomida. Body elongated or spatulate. Pharynx small, terminal; esophagus with bulb, ceca straight, extending to posterior extremities. Testes tubular and long on each side between ovarian zones and posterior ends. Cirrus sac spatulate containing a small vesicular seminalis interna; vesicula seminalis externa tubular, winding in median field of anterior extremity. Tubular convoluting pars prostatica surround with prostate cells and a straight ejaculatory duct present. Ovary small, entire in anterior fifth of body. Genital pore situated ventral at level of pharynx; uterine coils form several ascending and descending convolutions before opening into genital pore. Vitellaria tubular, slender, running one on each side of median line along ceca to posterior third of body. Excretory vesicle tubular, long, opening middorsally at level of pharynx.

Cercariae amphistome type with well-developed acetabulum. Eyespot and other pigmentation absent. Excretory vesicle thin-walled, saccate. Encystment and redial stage missing, sporocyst with paired branches, retention of miracidial eyespot. Epithelial cells of miracidia arranged by formula 6:8:4:2 or 6:9:4:2. Miracidia hatch *in utero*, eggs not operculated, shell membranaceous. Parasitic in lung and trachea of freshwater turtles.

Heronimidae Ward, 1917

Diagram: With characters of the suborder
Type genus: *Heronimus* MacCallum, 1902
   Syn. *Aorchis* Barker et Parsons 1914

Type species: *Heronimus mollis* (Leidy, 1856) Stunkard, 1964 (Figure 212)

   Syn. *Heronimus chelydrae* MacCallum, 1902; *Aorchis extensus* Barker et Parsons, 1917;
   Heronimus geomidae MacCallum, 1921; *Heronimus maternum* MacCallum, 1921
      Diagnosis: With characters of the suborder. Body length 12.0—21.0 mm, body width 2.3—4.0 mm. Diameter of pharynx 0.5—0.52 mm, esophagus very short with bulb. Testis tubular and long. Ovary 0.64—0.78 by 0.57—0.66 mm. Ventral sucker absent. Eggs 0.028 by 0.016—0.018 mm with membranaceous shell, containing developing miracidia.
Biology. References: Lynch,[567] Ulmer and Sommer,[1005] Crandall,[188] Ulmer.[1003]
      Preparasitic stage. ① Lynch. Eggs thin shelled, nonoperculated. Size of eggs enlarged

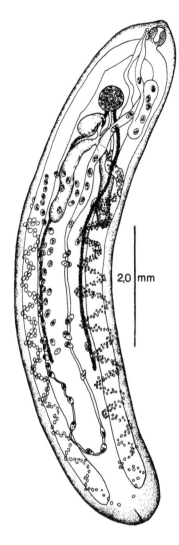

FIGURE 212. *Heronimus mollis* (Leidy, 1856). (After
G. A. MacCallum, 1921.)

as miracidia grow. Living miracidia fusiform with eyespots, measuring 0.30 by 0.10 mm
on average. Body covered with ciliated epithelial cells in four rows, arranged by the
formulae 6:8:4:2 or 6:9:4:2, manifesting intraspecific variation. Inner structure of miracidia
composed of subepithelial and muscular layers, brain mass, apical gland, penetration
glands, sensory papillae, excretory system (with two flame cells and their ducts), and
germinal balls. Life span of miracidia is about $^1/_2$ h in distilled or tap water. ② Crandall.
Fully embryonated eggs, after having been removed from uterus, hatch within 1 to 5 min.
After hatching, miracidia remain active for 4 to 6 h in pond water at room temperature.
Inner structure of miracidia almost completely agreed with that of Lynch's[567] description
as presented above. Penetration into snails usually occurs at tentacle or mantle regions.
Intermediate hosts. ① Crandall: *Helisoma trivolvis; Physa integra*; ② Ulmer and Sommer:
Physa gyrina; ③ Ulmer: *Physa sayii*.

Intramolluscan stage. ① Crandall. Miracidium-sporocyst larval stages have a special
structure without close resemblance to any other trematodes: lateral symmetry, a nearly
constant number of paired branches along the margin of body, retained miracidial eyespots

and special trait of excretory system. Redia stage is absent. Cercariae amphistome type with well-developed acetabulum (Figure 208). Body measuring 0.48—0.50 by 0.16—0.21 mm; tail 0.51—0.60 mm long. Eyespots and other pigmentation absent. Acetabulum 0.90—0.18 mm in diameter, ventroterminal; esophagus deliminating to esophageal bulb, ceca terminating at ventral sucker. Excretory vesicle saccate, between acetabulum and bodytail farrow; excretory vesicle bifurcates anteriorly, just before acetabulum extending to bifurcation. Main excretory tubes without cross-connecting tubes. Cercariae do not emerge from snails and do not encyst; turtles acquire parasites by ingestion of infected snails.

Development in definitive host. ① Crandall. Developing flukes are 1.0 mm long 21 d after infection. Ventral sucker is not recognizable anymore in living specimens. The first mature worms appear in about 90 d postinfection under laboratory conditions. Eggs were not laid and were hatched *in utero* when flukes escaped and reached water.

Habitat: Lung, trachea.

Hosts: *Chelydra serpentina, Claudius angostatus* (Chelydridae); *Chrysemis picta, C. scripta, Emydoidea blandingii, Graptemys pseudogeographica, G. geographica, Pseudemys elegans* (Emydidae); *Kinosternon cruentum, K. panamensis, K. histipes, K. subrubrum, K. leucostomum, Steronotherus odoratus, S. minor* (Kinosternidae).

Distribution: North America — U.S.; Central America — Mexico, Panama, Costa Rica; South America — Brazil.

References: Barker and Parsons,[71] Brenes,[120] Brooks and Mayers,[125] Caballero,[142] Caballero et al.,[151] Crandall,[188] Johnson,[450] Leidy,[529] MacCallum,[570] Stunkard,[922,927] Thatcher,[971] Ward.[1051]

Notes: Data in relevant literature show that MacCallum's species *H. chelydrae* has been indicated as type species. Stunkard[927] pointed out that the above species is specifically identical with *Monostomum molle*: a species described by Leidy.[529] He stated that genus *Heronimus* contains only a single species. Hence, the correct type species and its name is *Heronimus mollis* (Leidy, 1856) Stunkard, 1964. (Figure 212.)

3.2.2. ZONOCOTYLATA SEY, 1987

Diagnosis: Amphistomida. Body elongated, flattened, anterior extremity narrower, posterior wider. Adhesive organ (or rather, suckling disk) flattened, terminal with several transverse ribs. Muscular pharynx terminal; esophagus long, ceca short, extending to ovarian field. Testis single, intercecal, median, entire. Ovary spherical, posttesticular before acetabulum. Genital pore at middle part of esophagus. Uterine coils mainly posterior and dorsal. Vitellaria: two compact masses near inner side of ceca. Eggs with membranaceous shell, viviparity present. Excretory system well developed. Intestinal parasites of freshwater fishes.

Zonocotylidae Yamaguti, 1963

Diagnosis: With characters of the suborder.

Type genus: *Zonocotyle* Travassos, 1949

Syn. *Zonocotyloides* Padilha, 1978

Type species: *Zonocotyle bicaecata* Travassos, 1948 (Figure 213)

Diagnosis: Body length 2.6—5.2 m, body width 0.7—1.3 mm. Adhesive organ 1.29—2.12 by 1.56—2.48 mm; with 11 to 14 transverse ribs. Pharynx 0.19—0.48 mm long; esophagus 0.71—1.56 mm long; ceca short and broad. Testis single, 0.09—0.24 by 0.14—0.22 mm. Ovary: 0.20—0.39 by 0.13—0.37 mm. Vitellaria compact masses, 0.18—0.38 by 0.09—0.13 mm. Eggs 0.070—0.190 by 0.050—0.070 mm, with membranaceous shell.

Biology. Reference: ① Padilha.[691] Miracidia developed in uterus and emerged through genital pore. Newly hatched miracidia pyriform, covered with epithelial cells, situated in four

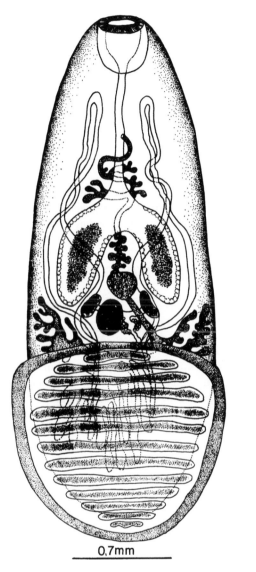

FIGURE 213. *Zonocotyle bicaecata* Travassos, 1948. (After T. Padilha, 1978.)

FIGURE 214. *Zonocotyle haroldtravassosi* (Padilha, 1978). (After T. Padilha, 1978.)

tiers, arranged by formula 6:8:4:2. Size of body: 0.14—0.21 by 0.05—0.12 mm. Inner structure of miracidia composed of two pairs of penetration glands, a single apical gland, a brain mass, a pair of flame cells with their ducts, germinal balls. Life span of miracidia: 8 to 10 h under laboratory conditions.

Habitat: Intestine.

Hosts: *Pseudocurimata gilberti*, *P. elegans*, *P. plumbea* (Curimatidae).

Distribution: South America — Brazil.

References: Kohn et al.,[496] Kohn and Fróes,[495] Padilha,[691] Travassos.[987]

Zonocotyle haroldtravassosi (Padilha; 1978) Kohn, Fernandes, Macedo et Abramson, 1985 (Figure 214)

Syn. *Zonocotyloides haroldtravassosi* Padilha, 1978

Diagnosis: Body length 4.7—7.4 mm, body width 1.6—2.0 mm. Adhesive organ

1.27—1.87 by 1.30—1.56 mm, with 9 or 10 transverse ribs. Pharynx 0.36—0.50 mm long; esophagus 1.43—2.24 mm long; ceca short and broad. Single testis measuring 0.24—0.47 by 0.40—0.70 mm. Ovary irregular in outline 0.29—0.37 by 0.26—0.29 mm. Uterine coils in posterior part of body. Vitellaria: two compact masses, 0.44—0.60 by 0.06—0.08 mm, at level of cecal extremities. Eggs with membranaceous shell, free miracidia in uterine branches.

Biology: Not known yet.

Habitat: Intestine.

Host: *Pseudocurimata gilberti* (Curimatidae).

Distribution: South America — Brazil.

References: Kohn et al.,[496] Padilha.[691]

Notes: Kohn et al.[496] examined specimens of *Z. bicaecata* collected from *Pseudocurimata plumbea* and found that the characters, based on the designation of the genus *Zonocotyloides* by Padilha,[691] proved to be variable in nature. Hence, the latter genus was synonymized with *Zonocotyle* which we can also accept.

KEY TO THE SPECIES OF *ZONOCOTYLE*

1.   Vitellaria compact, united, and V-shaped . . . . *Z. haroldtravassosi*
     Vitellaria compact, separated, elongated . . . . *Z. bicaecata*

3.2.3. PARAMPHISTOMATA SZIDAT, 1936

Diagnosis: Amphistomida. Body usually thick, conical; rarely flattened or divided into two parts; generally with smooth integument; sometimes projections or auricular caudal appendages may be present. Acetabulum large, strongly muscular; terminal, subterminal, or, rarely, ventroterminal. Pharynx terminal, well developed with or without dorsolateral appendages; esophagus of various lengths with or without muscular enlargement; ceca usually straight, with more or less convolutions along their full length. Testes usually double, rarely single; usually intercecal, may be extracecal or cecal, rarely. Male genital duct usually differentiated into tubular seminal vesicle, pars musculosa, pars protatica, and ejaculatory duct; cirrus sac present or absent, hermaphroditic pouch rarely present. Genital pore usually on muscular elevation, in midventral surface, near anterior extremity. Ovary usually posttesticular, seminal receptacle absent, uterine coils mainly intercecal, posttesticular, or dorsal to testes. Vitellaria follicular, exceptionally compact, usually extending in lateral fields of body. Eggs usually numerous, large, thin walled, sometimes with developing miracidium. Laurer's canal present, excretory vesicle usually opening dorsally, subterminally. Lymphatic system developed.

Life-cycle pattern: dixenous type with one intermediate host; cercariae being amphistome type, sporocyst, redial, or adolescarial larval stages rarely absent. Miracidium with 6:8:4:2 or 6:6:8:2 epithelial cell formula. Parasitic in digestive tube of vertebrates.

KEY TO THE SUPERFAMILIES OF PARAMPHISTOMATA

1.   Cirrus sac and pharyngeal appendages present, Laurer's canal and excretory duct without crossing. . . . . Cladorchoidea
     Cirrus sac absent, pharyngeal appendages present or absent; Laurer's canal and excretory duct with or without crossing. . . . . Paramphistomoidea

3.2.3.1. CLADORCHOIDEA SKRJABIN, 1949

Diagnosis: Paramphistomata. Body thick, conical or oval, rarely flattened, and with projecting or auricular appendages posteriorly. Acetabulum terminal or ventroterminal, muscular. Pharynx well developed with dorsolateral appendages, esophagus with various lengths, muscular thickening, sphincter or bulb may be present; ceca usually straight, undulation and branching may occur exceptionally. Testes double, rarely single, intercecal, extracecal, or cecal. Cirrus sac or hermaphroditic pouch present. Ovary usually posttesticular, rarely pretesticular. Genital pore in midventral surface at bifurcation, behind or before it. Uterine coils mainly intercecal. Vitellaria follicular, rarely compact mass, along lateral sides. Eggs usually long, with or without miracidia. Parasitic mainly in fish, amphibians, reptiles; rarely in mammals.

KEY TO THE FAMILIES OF CLADORCHOIDEA

1. Testes double 2
   Testis single .... Diplodiscidae
2. Ovary inter- or posttesticular... 3
   Ovary pretesticular.. Balanochiidae
3. Cirrus sac present... 4
   Hermaphroditic pouch present .... Brumptiidae
4. Primary pharyngeal sacs present .... Cladorchiidae
   Pharyngeal bulb and secondary pharyngeal sacs present .... Gastrodiscidae

3.2.3.1.1. CLADORCHIIDAE Southwell et Kirshner, 1932

Diagnosis: Cladorchoidea. Body elongate oval, pear shaped and uniform, thick, rarely flattened. Projections or auricular appendages posterior. Acetabulum terminal or ventroterminal. Pharynx muscular, with dorsolateral appendages; esophagus with various forms of muscular thickening; ceca straight, undulating or branching in length. Double testes, inter- or extracecal or cecal. Cirrus sac present. Ovary posttesticular. Genital pore at bifurcation, before or behind it. Uterine coils mainly posttesticular. Vitellaria mainly follicular, rarely compact, along lateral sides. Eggs usually large, with or without miracidia. Parasitic in fish, amphibians, reptiles, and, rarely, mammals.

KEY TO THE SUBFAMILIES OF CLADORCHIIDAE

1. One or both testes cecal or extracecal, vitellaria follicular or compact.............. 2
   Testes and uterus intercecal, vitellaria follicular 4
   Testes intercecal, uterus mainly extracecal, vitellaria follicular ...... Travassosiniinae
2. Testes cecal, horizontal 3
   Anterior testis either cecal or extracecal .... Sandoniinae
   Testes extracecal, uterus partly extracecal .... Osteochilotrematinae
3. Ovary intertesticular, vitellaria compact .... Colocladorchiinae
   Ovary posttesticular, testes cecal, vitellaria follicular .... Pseudocladorchiinae
4. Acetabulum simple 5
   Acetabulum with accessory sucker or muscular plug................... Megalodiscine
5. Genital pore bifurcal, slightly prebifurcal or postbifurcal 6
   Genital pore near to pharynx...................................... Helostomatinae
6. Testes preequational 7
   Testes in middle zone... 10
7. Testes horizontal... 8
   Testes tandem... Michrorchiinae
8. Testes round in outline.. 9
   Testes branched.. Cladorchiinae
9. Vitelline follicules posttesticular .... Caballeroiinae
   Vitelline follicules along body sides .... Pisciamphistomatinae

Dadaytrematinae Yamaguti, 1958

Diagnosis: Cladorchiidae. Body elongate oval or conical to lanceolate. Acetabulum ventroterminal. Pharynx terminal, pharyngeal appendages extramural, rarely intramural; esophagus short with bulb; ceca straight extending to acetabulum or before it. Testes intercecal, anteriorly located, tandem or slightly diagonal in position, entire or lobed. Cirrus sac present. Ovary midventral near acetabulum. Genital pore bifurcal, rarely prebifurcal. Uterine coils mainly posttesticular. Vitelline follicules moderately developed, situated in posttesticular part of body. Eggs numerous. Parasites of freshwater and marine fishes.

KEY TO THE GENERA OF DADAYTREMATINAE

1. Testes tandem... 2
   Testes diagonal 5
2. Testes lobed 3
   Testes entire 4
3. Vitelline follicules in posttesticular zone............................ *Neocladorchis*
   Vitelline follicules in ovarian zone.................................... *Dadaytrema*
4. Inner surface of acetabulum simple .... *Cleptodiscus*
   Inner surface of acetabulum with transverse ribs......................... *Dadayius*
5. Ceca end before acetabulum 6
   Ceca end at acetabulum... *Ophioxenus*
6. Vitelline follicules along ceca in ovotesticular zone............................. 7
   Vitellaria reduced, some follicules at cecal ends..................... *Dadaytremoides*
7. Pharyngeal appendages intramural, genital pore bifurcal............. *Macrorchitrema*
   Pharyngeal appendages extramural, genital pore prebifurcal .... *Panamphistomum*

Type genus: *Dadaytrema* Travassos, 1937

Diagnosis: Dadaytrematinae. Body flattened oval. Acetabulum ventroterminal. Pharynx terminal, papillated around mouth opening; appendages extramural; ceca simple, terminating shortly before acetabulum; esophagus short with bulb. Testes lobed, tandem, preequational. Cirrus sac present. Ovary median, before acetabulum. Genital pore postbifurcal. Uterine coils intercecal, posttesticular. Vitelline follicules limited to posterior portion of ceca. Viviparity present. Intestinal parasite of freshwater fishes.

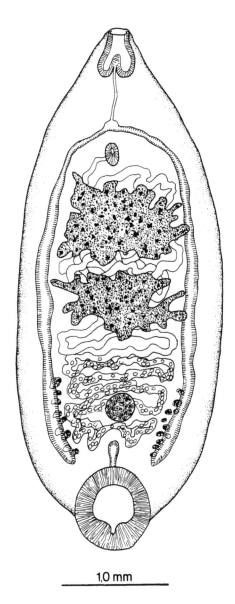

1,0 mm

FIGURE 215. *Dadaytrema oxycephalum* (Diesing, 1836). (After E. Daday, 1907.)

Type species: *Dadaytrema oxycephalum* (Diesing, 1836) Travassos, 1931 (Figure 215)

Syn. *Amphistoma oxycephala* Diesing, 1836; *Chiorchis oxycephalum* (Diesing, 1836) Daday, 1907; *C. papillatus* Daday, 1907; *Dadaytrema elongatum* Vaz, 1932; *D. minima* Vaz, 1932.

Diagnosis. With characters of the genus. Body length 2.2—9.5 mm, body width 0.7—3.5 mm. Acetabulum ventroterminal 0.38—1.76 mm in diameter, Pseudocladorchis type; number of muscle units, d.e.c., 43 − 46; d.i.c., 32 + 20; v.e.c., 36 − 38; v.i.c., 23 + 35 (Figure 151). Pharynx 0.48—1.12 by 0.36—0.47 mm, Subclavatus type (Figure 48); appendages 0.15—0.36 mm long, esophagus 0.83—1.14 mm long with bulb. Testes lobate, tandem, 1.12—1.46 mm in size. Cirrus sac 0.33—0.86 by 0.18—0.52 mm. Ovary 0.26—0.40 mm in diameter. Terminal genitalium Scleroporum type (Figure 76). Eggs 0.070—0.130 by 0.053—0.074 mm; intrauterine miracidia, 0.140—0.190 by 0.085—0.096 mm.

Biology: Not known yet.

Habitat: Intestine.

Hosts: *Prerodora granulosus* (Doradidae); *Pimelodus ornatus* (Pimelodidae); *Colossoma brachypomum, C. bidens, Mylopsus asterias, Mylossoma aureum, Myleus* sp., *Salmo auratus, Salminus maxillosus, Myletes torquatus; Salminus* sp. (Characidae).

Distribution: South America — Brazil, Venezuela, Paraguay.

References: Daday,[197] Diesing,[221] Masi Pallarés et al.,[595] Thatcher,[972] Travassos,[982] Vaz,[1016] Vicente et al.[1033]

Dadayius Fukui, 1929

Diagnosis: Dadaytrematinae. Body elongate conical. Acetabulum large, ventroterminal with transverse wrinkles at its bottom. Pharynx terminal, appendages extramural; esophagus slender, with weakly developed bulb; ceca simple, extending to acetabulum. Testes round, tandem in middle region of body. Ovary median, in front of acetabulum. Genital pore prebifurcal, genital sucker present, cirrus sac not detectable. Uterine coils intercecal. Vitelline follicules lateral, from zone of posterior testis to acetabulum. Intestinal parasite of fishes.

Type species: *Dadayius marenzelleri* (Daday, 1907) Fukui, 1929 (Figure 216)

Syn. *Diplodiscus marenzelleri* Daday, 1907

Diagnosis: With characteristics of the genus. Body length 4.0—6.0 mm, body width 0.7—2.2 mm. Acetabulum 1.20—2.20 mm in diameter, Dadayius type, number of muscle units, d.e.c., 15 or 16; d.i.c., 17—19; v.e.c., 7—9; v.i.c., 35—40 (Figure 140). Pharynx 0.05—0.14 mm in diameter, Megalodiscus type (Figure 42); appendages 0.25—0.30 mm long; esophagus 0.70 mm long, with bulb. Testes 0.66—0.80 mm in diameter, separated by uterine coils. Ovary 0.26—0.30 mm in diameter, terminal genitalium Dadayius type (Figure 73). Eggs 0.140 by 0.060—0.090 mm.

Biology: Not known yet.

Habitat: Intestine.

Host: *Salmo* sp. (Salmonidae).

Distribution: South America — Brazil.

References: Daday,[197] Sey.[856]

Dadaytremoides Thatcher, 1979

Diagnosis: Dadaytrematinae. Body elongated oval. Acetabulum ventroterminal. Pharynx terminal with large appendages situated extramurally; esophagus short, bulb present, ceca slightly undulating, extending to posterior third of body. Testes small, diagonal, irregular in outline; in middle third of body. Cirrus sac present. Ovary mid ventral, at level of cecal extremities. Genital pore postbifurcal. Uterine coils between acetabulum and genital pore. Vitelline follicules small in number, at ovarian level. Viviparity present. Eggs in distal part of uterus contain developing miracidia. Intestinal parasite of freshwater fishes.

Type species: *Dadaytremoides grandistomis* Thatcher, 1979 (Figure 217)

Diagnosis: With characters of the genus. Body length 1.9—4.0 mm, body width 0.7—1.3 mm. Acetabulum 0.35—0.54 mm in diameter. Pharynx 0.22—0.64 mm in diameter, appendages 0.14—0.22 mm in length. Testes 0.19 mm in diameter. Cirrus sac 0.14—0.23 by 0.09—0.16 mm. Ovary 0.10 mm in diameter. Eggs 0.036—0.045 by 0.072—0.090 mm, intrauterine miracidia 0.041—0.062 by 0.095—0.110 mm.

Biology: Not known yet.

Habitat: Intestine.

Hosts: *Chaetostomus leucomeles* (Clariidae), *Astinax fasciatus* (Characidae).

Distribution: South America — Colombia.

Reference: Thatcher.[972]

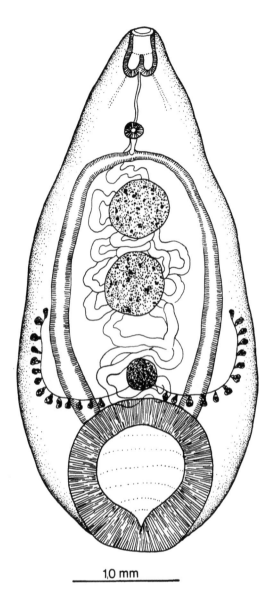

FIGURE 216. *Dadayius marenzelleri* (Daday, 1907). (After
E. Daday, 1907.)

Panamphistomum **Manter et Pritchard, 1964**

Diagnosis: Dadaytrematinae. Body elongate, tapering anteriorly. Acetabulum terminal
with posterior notch. Pharynx well developed, appendages extramural; esophagus short with
strong bulb; ceca extending to ovary. Testes smooth, diagonal near to cecal arch. Cirrus sac
present. Ovary rounded, before acetabulum. Genital pore prebifurcal. Uterine coils mainly
posttesticular. Vitelline follicules in two lateral rows between posterior testis and ovary.
Intestinal parasite of freshwater fishes.

Type species: *Panamphistomum benoiti* **Manter et Pritchard, 1964 (Figure 218)**

Diagnosis. With characters of the genus. Body length 2.4—4.9 mm, body width 0.6—1.4
mm. Acetabulum 0.69 mm in diameter. Pharynx 0.30—0.54 mm, appendages 0.19—0.38
mm long; esophagus 0.9 mm long with bulb. Testes 0.34 mm in diameter. Ovary 0.14 mm
in diameter. Egg 0.147—0.194 by 0.054—0.064 mm.

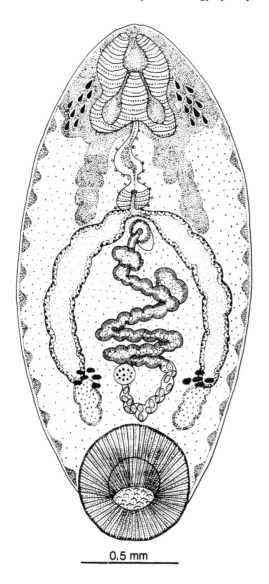

0,5 mm

0,6 mm

FIGURE 217. *Dadaytremoides grandistomis* Thatcher, 1979. (Courtesy of V. E. Thatcher.)

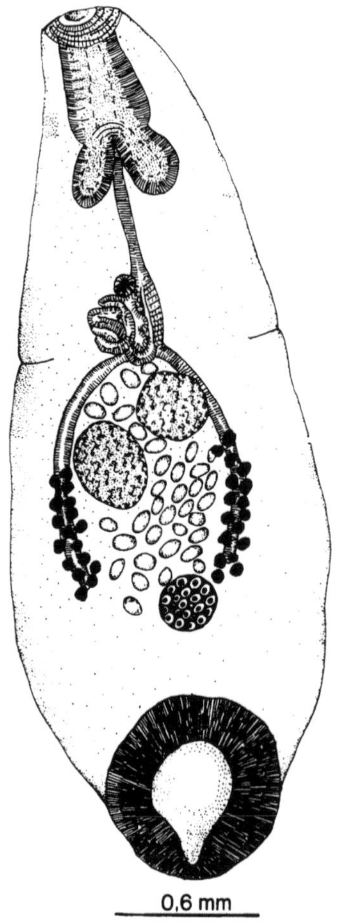

FIGURE 218. *Panamphistomum benoiti* Manter et Pritchard, 1964. (Courtesy of M. H. Pritchard.)

Biology: Not known yet.
Habitat: Intestine.
Hosts: *Haplochromis philander* (Chichlidae); *Clarias lazera* (Clariidae).
Distribution: Africa — Zaire.
Reference: Manter and Pritchard.[589]

Neocladorchis Bhalerao, 1937

Diagnosis: Dadaytrematinae. Body elongated, cylindrical. Acetabulum ventroterminal. Pharynx terminal with well-developed extramural appendages, esophageal bulb present, ceca extending to acetabulum. Testes tandem, lobate, in middle third of body. Cirrus sac present. Ovary submedian, close to acetabulum. Genital pore bifurcal. Uterine coils intercecal between posterior testis and acetabulum. Vitelline follicules along ceca between posterior testis and acetabulum. Intestinal parasites of freshwater fishes.

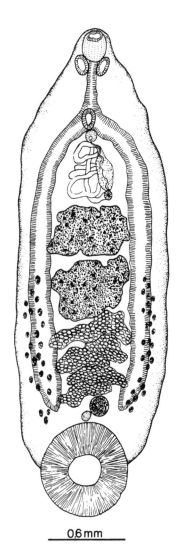

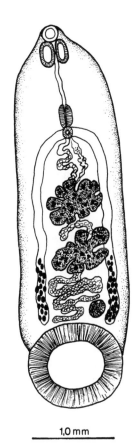

FIGURE 219. *Neocladorchis poonaensis* Bhalerao, 1937. (After G. D. Bhalerao, 1937.)

FIGURE 220. *Neocladorchis multilobularis* Sey, 1986. (By permission of the *Acta Zool.*, 1986, Budapest.)

Type species: *Neocladorchis poonaensis* **Bhalerao, 1937 (Figure 219)**

Diagnosis: Body length 1.8—5.8 mm, body width 0.6—1.4 mm. Acetabulum 0.63—1.07 mm in diameter. Pharynx 0.22—0.33 by 0.24—0.38 mm, appendages 0.22—0.30 by 0.18—0.23 mm; esophagus 0.36—0.68 mm long with bulb. Testes 0.36—0.82 by 0.28—0.67 mm. Ovary 0.13—0.26 mm in diameter. Cirrus sac 0.28—0.33 by 0.17—0.18 mm. Eggs 0.150—0.170 by 0.056—0.060 mm.

Biology: Not yet known.

Habitat: Intestine.

Host: *Barbus dobsoni* (Cyprinidae).

Distribution: Asia — India.

Reference: Bhalerao.[90]

Neocladorchis mutlilobularis **Sey, 1986 (Figure 220)**

Diagnosis. Body length 4.2—5.0 mm, body width 0.1—0.1 mm. Acetabulum 1.12—1.36

mm in diameter, Neocladorchis type, number of muscle units, d.e., 67—69; d.i., 30—32; v.e., 50 + 25—27; v.i., 30—32 (Figure 149). Pharynx 0.20—0.23 mm, Megacotyle type (Figure 41), appendages 0.30—0.32 mm long; esophagus 0.60—0.83 mm long, with bulb; ceca straight, extending to acetabulum. Testes branched, intercecal, tandem; anterior testis 0.41—0.45 by 0.45—0.50 mm, posterior testis 0.35—0.40 by 0.45 mm. Ovary lobate, 0.18—0.20 by 0.15—0.16 mm. Terminal genitalium Scleroporum type (Figure 76). Cirrus sac 0.23 by 0.14 mm. Eggs 0.118—0.136 by 0.053 mm.

Biology: Not yet known.

Habitat: Intestine.

Host: *Spinibarbichthys denticulatus* (Cyprinidae).

Distribution: Asia — Vietnam.

Reference: Sey.[855]

KEY TO THE SPECIES OF *NEOCLADORCHIS*

1. Testes lobed and touch each other in middle line .... *N. poonaensis*
    Testes branched, dendritic, and separated by uterine coils .... *N. multilobularis*

Cleptodiscus Linton, 1910

Diagnosis: Dadaytrematinae. Body elongated oval. Acetabulum ventroterminal. Pharynx with extramural appendages, esophagus short, with strong bulb, ceca slightly sinuous, extending to acetabulum. Testes intercecal, oval, tandem or diagonal, in middle third of body. Cirrus sac present. Ovary submedian, before acetabulum. Genital pore bifurcal. Uterine coils intercecal, mainly posttesticular. Vitelline follicules along posterior portion of intestine. Intestinal parasite of marine fishes.

Type species: *Cleptodiscus reticulatus* Linton, 1910 (Figure 221)

Diagnosis: Body length 1.5—3.7 mm, body width 0.1—0.2 mm, Acetabulum 0.46—0.51 by 0.30—0.65 mm, Cleptodiscus type, number of muscle units, d.e.c., 12—14; d.i.c., 35—36; v.e.c., 12—14; v.i.c., 37—39 (Figure 136).

Biology: Not yet known.

Habitat: Intestine, rectum.

Hosts: *Pomacanthus arquatus, P. aureus (Chaetodontidae).*

Distribution: The Caribbean — Cuba, Puerto Rico, West Indies; North America — U.S.

References: Linton,[553] Overstreet,[684] Vigueras.[1035]

Cleptodiscus bulbosus Hanson, 1955 (Figure 222)

Diagnosis: Body length 3.9 mm, body width 1.2 mm. Acetabulum 0.84 mm in diameter. Pharynx 0.42 m long, with rudimental appendages; esophagus 0.35 mm long. Testes diagonal, right testis 0.21 by 0.27 mm, left testis 0.24 by 0.21 mm. Ovary 0.21 by 0.26 mm. Eggs 0.043—0.061 by 0.031—0.037 mm.

Biology: Not yet known.

Habitat: Intestine.

Host: *Melichthys buniva* (Balistidae).

Distribution: U.S. — Hawaii.

Reference: Hanson.[406]

Cleptodiscus kyphosi Sogandares-Bernal, 1959 (Figure 223)

Diagnosis: Body length 3.2—5.2 mm, body width 0.8—0.9 mm. Acetabulum 0.67 by 0.61 mm, Cleptodiscus type number of muscle units, d.e.c., 11—14; d.i.c., 34—36; v.e.c., 10—14; v.i.c., 36—39 (Figure 139). Pharynx 0.27 mm long, appendages 0.09 mm long;

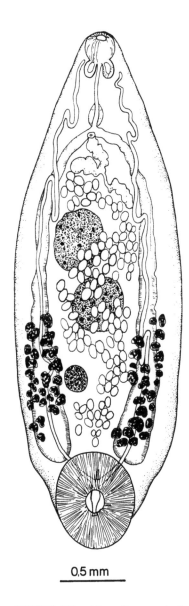

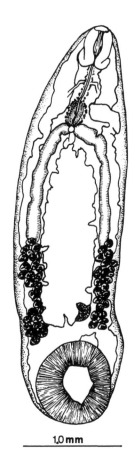

FIGURE 221. *Cleptodiscus reticulatus* Linton, 1910. (After S. Yamaguti, 1971.)

FIGURE 222. *Cleptodiscus bulbosus* Hanson, 1955. (Courtesy of M. L. Hanson.)

esophagus 0.54 mm in length with well developed bulb. Testes tandem, anterior testis 0.360—0.47 by 0.47—0.51 mm, posterior testis 0.38—0.53 by 0.45—0.51 mm. Ovary 0.24 mm in diameter. Eggs 0.056—0.072 by 0.032—0.035 mm.

Biology: Not yet known.

Habitat: Intestine.

Host: *Kyphosus sectatrix* (Kyphosidae).

Distribution: West Indies — Bimini.

References: Sogandares-Bernal,[898] Wotton and Sogandares-Bernal.[1069]

KEY TO SPECIES OF *CLEPTODISCUS*

1. Testes diagonal 2

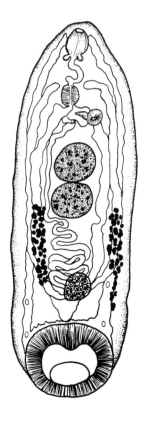

1,0 mm

FIGURE 223. *Cleptodiscus kyphosi* Sogandares-Bernal, 1959. (Courtesy of F. Sogandares-Bernal, and by permission of the *J. Parasitol.*, 1959, Lincoln.)

Testes tandem.. *C. kyphosi*
2. Primary pharyngeal sac rudimental .... *C. bulbosus*
   Primary pharyngeal sacs well developed............................... *C. reticulatus*

Macrorchitrema Vigueras, 1940

Diagnosis: Dadaytrematinae. Body conical. Acetabulum large, ventroterminal. Pharynx terminal with intramural appendages, esophagus sinuous, bulb present, ceca extending to half part of body. Testes tandem, close together, in anterior part of intercecal field. Cirrus sac present. Ovary submedian, preacetabular. Genital pore median, bifurcal. Uterine coils strongly developed, mainly in intestinal field, extending to acetabulum. Vitelline follicules along posterior portion of ceca. Intestinal parasite of marine fishes.

Type species: *Macrorchitrema havanense* Vigueras, 1940 (Figure 224)

Diagnosis: With characters of the genus. Body length 2.0—2.8 mm, body width 0.8—0.9 mm. Acetabulum big, 0.90—1.0 mm in diameter. Pharynx 0.25—0.27 mm long; esophagus 0.35—0.40 mm in length with bulb. Anterior testis 0.56 by 0.29, posterior testis 0.51 by 0.31 mm. Cirrus sac 0.30—0.31 by 0.11—0.12 mm. Ovary 0.14 by 0.17 mm. Eggs 0.060 by 0.038 mm, containing miracidia.

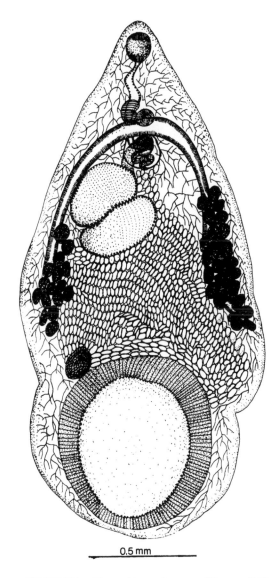

FIGURE 224. *Macrorchitrema havanense* Vigueras, 1940.
(After I. P. Vigueras, 1940.)

Biology: Not yet known.
Habitat: Intestine.
Host: *Holocanthus tricolor* (Chaetodontidae).
Distribution: The Caribbean — Cuba.
Reference: Vigueras.[1034]

Ophioxenus Sumwall, 1926

Diagnosis: Dadaytrematinae. Body elliptical, tapering anteriorly. Acetabulum ventroterminal. Pharynx well developed, appendages extramural, esophagus short, bulb present; ceca extending to acetabulum. Testes entire, diagonal, intercecal. Ovary submedian between posterior testis and acetabulum. Genital pore just bifurcal, cirrus sac weakly developed. Uterine coils winding intercecally among ovary, posterior and anterior testes. Vitelline follicules large between acetabulum and posterior testis, along inner side of ceca. Intestinal parasites of lampreys, frogs, snakes, and turtles.

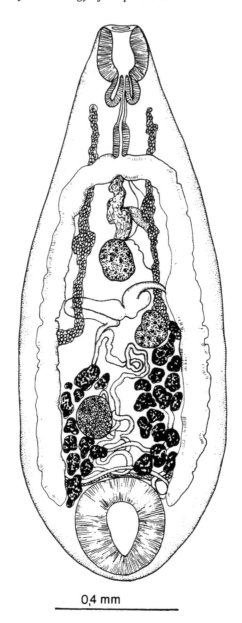

0,4 mm

FIGURE 225.   *Ophioxenus dienteros* Sumwalt, 1926.
(After M. Sumwalt, 1926.)

Type species: *Ophioxenus dienteros* Sumwall, 1926 (Figure 225)

Diagnosis: Body length 1.9 mm, body width 0.6 mm. Acetabulum 0.37 mm in diameter. Pharynx 0.23 by 0.20 mm, appendages 0.09 mm long; esophagus 0.24 mm in length. Testes 0.15 mm in diameter. Ovary 0.15 mm in diameter. Size of eggs is not given.

Biology: Not yet known.

Habitat: Intestine.

Hosts: *Bufo borealis* (Bufonidae); *Thamnophis ordinoides, T. sirtalis* (Colubridae); *Clemmys marmorata* (Emydidae).

Distribution: North America — U.S.

References: Sumwalt,[931] Thatcher.[969]

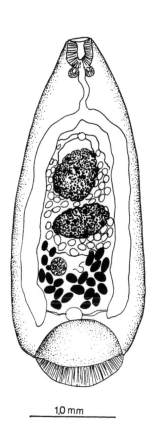

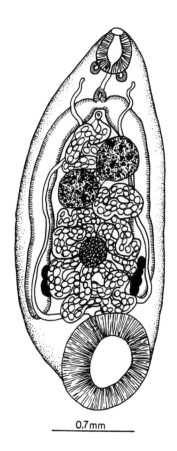

1,0 mm

0,7 mm

FIGURE 226. *Ophioxenus micro-phagus* (Ingles, 1936). (After L. G. Ingles, 1936.)

FIGURE 227. *Ophioxenus micropha-gus* (Ingles, 1936) (Courtesy of M. Bev-erly-Burton.)

Ophioxenos microphagus (Inglis, 1936) Beverly-Burton, 1987 (Figures 226 and 227)

Syn. *Megalodiscus microphagus* Inglis, 1936; *Ophioxenus lampetrae* Beverly-Burton et Morgolis, 1982

Diagnosis: Body length 1.9—5.1 mm, body width 0.8—1.5 mm. Acetabulum 0.59—0.95 by 0.56—1.15 mm. Pharynx 0.25—0.44 by 0.27—0.39 mm, appendages 0.13—0.15 mm long; esophagus 0.26—0.44 mm in length. Anterior testis 0.32—0.60 by 0.19—0.70 mm, posterior testis 0.32—0.60 by 0.14—0.70 mm. Cirrus sac 0.13 by 0.061 mm. Ovary 0.21—0.29 by 0.19—0.32 mm. Eggs 0.110—0.130 by 0.053—0.070 mm.

Biology: Intermediate hosts were found *Menetus cooperi, Gyraulus* sp.[578] and *Menetus cooperi*.[283] Emergence of cercariae from snails was usually sporadic. Daily production was about 20 cercariae. Cercariae encyst on the skin of tadpoles or adult frogs and infestation occurs by ingestion of the sloughed skin or in cases of tadpoles by taking cysts or cercariae directly into the mouth. After infection in spring juveniles they develop rapidly and they retain their reproductive activity during the summer and into the next year.

Habitat: Rectum.

Hosts: Cyclostomata: *Lampetra richardsoni* (Petromyzonidae). Amphibia: *Rana aurora, R. catesbeiana* (Ranidae); *Bufo borealis* (Bufonidae); *Hyla regilla* (Hylidae); *Ambystoma gracile, A. macrodactylum* (Ambystomidae) *Dicamptodon ensatus* (Dicamptodontidae); *Taricha granulosa* (Salamandridae).

Distribution: North America — Canada, U.S.

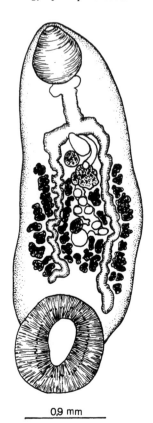

FIGURE 228.   *Ophioxenus singularis* Parker, 1941. (After M. W. Parker, 1941.)

References: Beverly-Burton,[85] Beverly-Burton and Morgolis,[86] Bravo-Hollis,[117] Efford and Tsumura,[283] Ingles,[431] MacCauley,[575] Macy,[578] Moravec.[620]

Ophioxenus singularis Parker, 1941 (Figure 228)

Diagnosis: Body length 3.9 mm, body width 1.3 mm. Acetabulum subterminal, 0.98 by 0.92. Pharynx 0.65—0.53 mm long, appendages small; esophagus 0.52 mm in length. Anterior testis 0.19 by 0.17, posterior testis 0.52 mm in diameter. Ovary 0.18 by 0.15 mm, cirrus sac 0.24 by 0.13 mm. Size of eggs is not given.

Biology: Not yet known.

Habitat: Intestine.

Hosts: *Rana catesbeiana* (Ranidae); *Thammophis sirtalis* (Colubridae).

Distribution: North America — U.S.

Reference: Parker.[702]

KEY TO THE SPECIES OF *OPHIOXENUS*

1. Vitelline follicules intertesticular... 2
    Vitelline follicules along both sides of ceca reaching posterior border of anterior
    testis .... *O. singularis*
2. Vitelline follicules reach anterior border of ovary.................... *O. microphagus*
    Vitelline follicules reach to posterior testis............................. *O. dienteros*

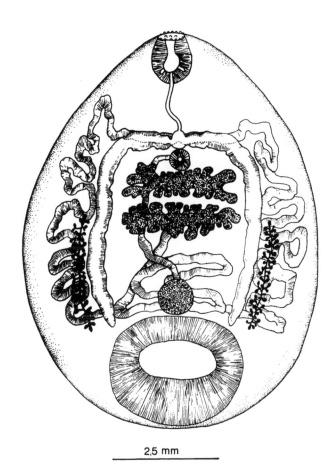

2,5 mm

FIGURE 229. *Travassosinia dilatata* (Daday, 1907). (After E. Daday, 1907.)

Travassosiniinae Sey, 1988

Diagnosis: Cladorchiidae. Body conical or oval with papillae around oral aperture. Acetabulum ventroterminal. Pharynx small, with intramural appendages; esophagus short, with small bulb; ceca simple, terminating preacetabularly. Testes large, tandem, lobate, intracecal and partially cecal. Ovary median, spherical, preacetabular. Genital pore postbifurcal, cirrus sac present. Uterine coils inter- and extracecal. Vitelline follicules extending along posttesticular portion of ceca. Intestinal parasites of freshwater fishes.

Type genus: *Travassosinia* Vaz, 1932

Diagnosis: With characters of the subfamily.

Type species: *Travassosinia dilatata* (Daday, 1907) Vaz, 1932 (Figure 229)

Syn. *Chiorchis dilatatus* Daday, 1907; *C. dilatatus* Viana, 1924

Diagnosis: Body length 9.5—12.0 mm, body width 3.9—4.6 mm. Acetabulum 2.50 mm in diameter. Pharynx 1.12—1.42 by 1.12—1.49 mm, Pseudocladorchis type (Figure 30); appendages 0.22 mm long, esophagus 0.88—1.75 mm long with bulb. Anterior testis 1.35—2.04 by 1.39—2.14 mm, posterior testis 1.39—2.14 by 1.19—2.14 mm. Cirrus sac 0.46 mm in diameter. Ovary 0.56—0.71 mm. Eggs 0.147—0.151 by 0.077—0.084 mm.

Biology: Not yet known.

Habitat: Intestine.

Hosts: *Colossoma brachypomum, Myleus pacu, Colossoma bidens, C. oculus* (Characidae).

Distribution: South America — Brazil, Paraguay.
References: Daday,[197] Vaz,[1016] Vicente et al.[1033]

Orientodiscinae Yamaguti, 1971

Diagnosis: Cladorchiidae. Body elongated fusiform. Acetabulum ventroterminal. Pharynx subterminal, appendages extramural; esophagus short with well developed bulb; ceca sinuous, slightly undulated, terminating shortly before acetabulum. Testes tandem, lobate, in middle part of body. Ovary submedian, near right testis. Genital pore bifurcal or postbifurcal, cirrus sac present. Uterine coils intercecal, transversal convoluted ascending tubes. Vitelline follicules along lateral fields between bifurcation and acetabulum. Intestinal parasites of freshwater fishes.

Type genus: *Orientodiscus* Srivastava, 1938
Diagnosis: With character of the subfamily.

Type species: *Orientodiscus lobatus* Srivastava, 1938 (Figure 230)

Diagnosis. Body length 5.0—7.0, body width 1.1—1.8 mm. Acetabulum 0.68—0.78 mm in diameter. Pharynx 0.22—0.24 by 0.26—0.30 mm, appendages 0.10—0.12 mm in diameter, esophagus 0.44—0.54 mm long. Anterior testis 0.70—1.04 by 1.04—1.14 mm, posterior testis 0.64—1.12 by 1.66 mm. Cirrus sac 0.22—0.30 by 0.12—0.20 mm. Ovary 0.24—0.30 by 0.11—0.12 by 0.06—0.07 mm. Genital pore bifurcal. Eggs 0.110—0.120 by 0.060—0.070 mm.
Biology: Not yet known.
Habitat: Intestine.
Host: *Silundia gangetica* (Siluridae).
Distribution: Asia — India.
Reference: Srivastava.[907]

Orientodiscus jumnai Srivastava, 1938 (Figure 231)

Diagnosis: Body length 5.4 mm, body width 1.2 mm. Acetabulum 0.66 mm in diameter. Pharynx 0.24 by 0.26 mm in size, esophagus 0.32 mm in length. Testes 0.66—0.75 mm in diameter. Ovary 0.24 mm in dimater. Genital pore postbifurcal. Eggs 0.114—0.120 by 0.072—0.076 mm.
Biology: Not yet known.
Habitat: Intestine.
Hosts: *Silundia gangetica* (Siluridae), *Pangasius micronemus* (Pangasidae); *Barbus tambroides, B. daruphani, Cyclocheilichthys apogon* (Cyprinidae).
Distribution: Asia — India, Malaysia.
References: Jones and Seng,[455] Leong et al.,[541] Srivastava.[907]

Orientodiscus fossilis Bilqees, 1972 (Figure 232)

Diagnosis: Body length 2.6—3.2 mm, body width 0.7—1.1 mm. Acetabulum 0.33—0.43 by 0.42—0.46 mm. Pharynx 0.23—0.29 by 0.16—0.33 mm; esophagus 0.06—0.13 mm long. Anterior testis 0.16—0.26 by 0.15—0.24 mm, posterior testis 0.15—0.25 by 0.15—0.24 mm. Cirrus sac 0.18—0.33 by 0.08—0.16 mm. Ovary 0.08—0.12 by 0.13—0.21 mm. Genital pore postbifurcal. Eggs 0.060—0.070 by 0.030—0.035 mm.
Biology: Not yet known.
Habitat: Intestine.
Host: *Heteropneustis fossilis* (Clariidae).
Distribution: Asia — Pakistan.
Reference: Bilqees.[94]

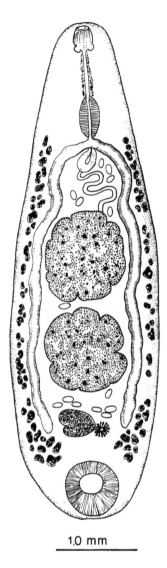

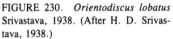

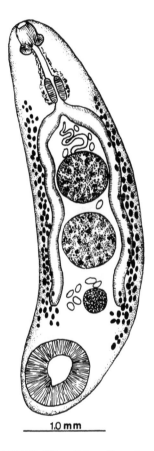

FIGURE 230. *Orientodiscus lobatus* Srivastava, 1938. (After H. D. Srivastava, 1938.)

FIGURE 231. *Orientodiscus jumnai* Srivastava, 1938. (After H. D. Srivastava, 1938.)

***Orientodiscus mastacembeli* Agarwal et Agrawal, 1979 (Figure 233)**

Diagnosis: Body length 2.9—3.0 mm, body width 1.0—1.1 mm. Acetabulum 0.53—0.56 by 0.53—0.57 mm. Pharynx 0.21—0.26 by 0.21—0.25 mm; esophagus 0.16—0.20 by 0.06—0.08 mm long. Right testis 0.23—0.25 by 0.09—0.15 mm; left testis 0.24—0.27 by 0.10—0.15 mm. Cirrus sac 0.16—0.23 by 0.11—0.13 mm. Ovary 0.10—0.17 by 0.12—0.14 mm. Genital pore bifurcal. Eggs 0.120—0.160 by 0.060—0.090 mm.

Biology: Not yet known.

Habitat: Intestine.

Distribution: Asia — India.

Host: *Mastacembelus armatus* (Mastacembelidae).

Reference: Agarwal and Agrawal.[5]

***Orientodiscus orchhaensis* Agarwal et Agrawal, 1980 (Figure 234)**

Diagnosis: Body length 2.0—2.2 mm, body width 0.8—0.9 mm. Acetabulum 0.39—0.45

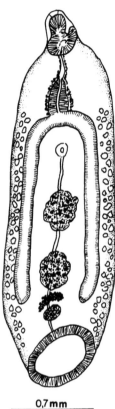

FIGURE 232. *Orientodiscus fossilis* Bilqees, 1972. (Courtesy of F. J. Bilqees.)

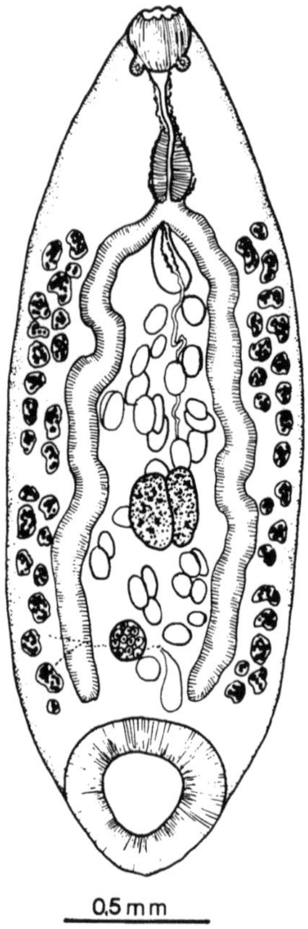

FIGURE 233. *Orientodiscus mastacembeli* Agarwal et Agrawal, 1979. (Courtesy of S. C. Agrawal.)

by 0.54—0.63 mm. Pharynx 0.20—0.24 by 0.20—0.24 mm, esophagus 0.01—0.02 by 0.04—0.06 mm long. Right testis 0.10—0.15 by 0.12—0.16; left testis 0.09—0.12 by 0.13—0.16 mm. Cirrus sac 0.16—0.20 by 0.09—0.17 mm. Ovary 0.09—0.12 by 0.10—0.13. Genital pore bifurcal. Eggs 0.110—0.180 by 0.100—0.120 mm.

Biology: Not yet known.

Habitat: Intestine.

Distribution: Asia — India.

Host: *Mastacembelus armatus* (Mastacembelidae).

Reference: Agarwal and Agrawal.[7]

KEY TO THE SPECIES OF *ORIENTODISCUS*

1. Pharynx with anterior pouch .... *O. fossilis*
    Anterior pouch absent.. 2
2. Vitelline follicules extend from middle of esophageal bulb to anterior margin of acetabulum.. 3
    Vitelline follicules extend from bifurcation to cecal ends 4
3. Testes lobed and tandem.. *O. lobatum*
    Testes entire and tandem.. *O. jumnai*
4. Testes symmetrical and closely attached .... *O. mastacembeli*
    Testes diagonal .... *O. orchhaensis*

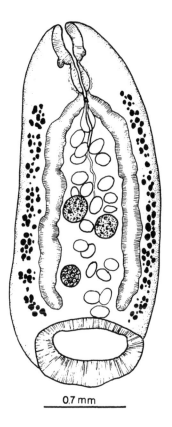

FIGURE 234. *Orientodiscus or-chhaensis* Agarwal et Agrawal, 1980. (Courtesy of S. C. Agrawal.)

Sandoniinae Ukoli, 1972

Diagnosis. Cladorchiidae. Body oval or pear shaped and uniform. Acetabulum terminal or subterminal. Pharynx well developed, primary pharyngeal sacs extramural; esophagus with bulb; ceca simple, extending to acetabular zone or middle part of body. Anterior testis cecal or extracecal, posterior testis intercecal, cirrus sac present. Ovary situated near acetabular zone or before it, genital pore prebifurcal. Uterine coils mainly intercecal and, in part, extracecal. Vitelline follicules posttesticular, rarely confluent posteriorly. Eggs with or without miracidia. Intestinal parasites of freshwater fishes.

KEY TO THE GENERA OF SANDONIINAE

1. Diameter of acetabulum smaller than body width 2
   Diameter of acetabulum longer than body width .... *Basidiodiscus*
2. Testes spherical, diagonal, posterior testis cecal 3
   Testes tandem, lobate, anterior testis cecal .... *Australotrema*
3. Vitelline follicules confluent posteriorly, cecal ends at midzone .... *Pretestis*
   Vitelline follicules not confluent posteriorly, cecal ends at acetabulum ..... *Sandonia*

Type genus *Sandonia* McClelland, 1957

Diagnosis: Sandoniinae. Body oval to elongate. Acetabulum ventroterminal, moderate in

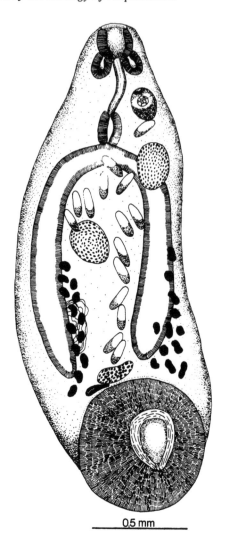

0,5 mm

FIGURE 235.   *Sandonia sudanensis* McClelland,
1957. (Courtesy of F. Moravec.)

size. Pharynx muscular, esophageal bulb well developed; ceca straight, extending to ovarian
zone. Anterior testis cecal, posterior testis cecal or intercecal. Cirrus sac present. Ovary
median before acetabulum. Genital pore at middle part of esophagus. Uterine coils intercecal
and extracecal in part. Vitelline follicules posttesticular. Eggs with miracidia. Intestinal
parasite of freshwater fishes.

Type species: *Sandonia sudanensis* McClelland, 1957 (Figure 235)
   Diagnosis: With characters of the genus. Body length 1.6—3.0 mm, body width 0.7—1.0
mm. Acetabulum 0.54—0.88 by 0.54—0.75 mm, Sandonia type number of muscle units,
d.e.c., 30—32; d.i.c., 66-69; v.e.c., 20—22; v.i.c., 30—32 (Figure 153). Pharynx 0.17—0.29
mm long, Megalocotyle type (Figure 41). Anterior testis 0.26—0.44 by 0.26—0.47 mm,
posterior testis 0.29—0.42 by 0.30—0.44 mm. Cirrus sac 0.22—0.25 by 0.09—0.13 mm.
Ovary 0.16—0.21 by 0.14—0.22 mm. Eggs 0.140—0.160 by 0.086—0.092 mm, contain
developing miracidia.
Biology: Not completely elucidated yet. Sey and Sayed[861] examined embryonic development

of this species collected from fish in the Nile. It was found that the distal part of the uterus contained eggs with fully developed miracidia. After deposition of such eggs, miracidia hatched after a few minutes in water. Miracidia measure 0.24—0.33 by 0.060—0.065 mm. Body covered with ciliated epithelial cells, situated in four rows, arranged along the formula 6:8:4:2. Inner structure composed of two pairs of penetration glands, a single apical gland, a brain mass, a pair of flame cells with their ducts. Posterior part of body occupied by a redia. It is elongated oval, measuring 0.10—0.16 by 0.040—0.070 mm. Oral sucker 0.025 mm in diameter, ceca long. Two pairs of flame cells present. Sporocyst larval stage is lacking.

Habitat: Intestine.

Hosts: *Bagrus docmac* (Bagridae); *Distichodus niloticus, D. rostratus* (Distichodontidae); *Synodontis batensoda, S. budgetti, S. clarias, S. gambiensis, S. membranaceus, S. nigrita, S. ocellifer, S. shall, S. serratus, S. sorex, S. vermiculatus* (Mochocidae).

Distribution: Africa — Egypt, Ghana, Niger, Sudan.

References: Fischthal and Kuntz,[313] Khalil,[481] McClelland,[598] Moravec,[619] Saoud and Wannas,[816] Sey and Sayed,[861] Ukoli.[1002]

Basidiodiscus Fischthal et Kuntz, 1959

Diagnosis: Sandoniinae. Body conical. Acetabulum terminal, large with papilloform projections. Pharynx muscular, esophagus with bulb, ceca simple. Testes slightly oblique, anterior testis extracecal, posterior testis intercecal. Genital pore midventral, near to anterior testis; cirrus sac present. Ovary midventral, posttesticular, postcecal. Uterine coils in area between genital pore and posterior body end. Vitelline follicules lateral, posttesticular. Intestinal parasite of fishes.

Type species: *Basidiodiscus ectorchis* Fischthal et Kuntz, 1959 (Figure 236)

Diagnosis: With characters of the genus. Body length 1.5—2.5 mm, body width 0.4—0.7 mm. Acetabulum 0.97 mm in width, Basidiodiscus type number of muscular units, d.e.c., 16—18; d.i.c., 110—120; v.e.c., 15—17; v.i.c., 113—118 (Figure 134). Pharynx 0.20—0.33 mm long, Megacotyle type (Figure 41); appendages 0.10—0.17 mm, esophagus 0.47 mm long with bulb. Anterior testis extracecal 0.23—0.38 by 0.23—0.35 mm; posterior testis 0.23—0.37 by 0.23—0.41 mm. Cirrus sac 0.11—0.13 by 0.04—0.08 mm. Ovary 0.18 by 0.11 mm. Eggs 0.158—0.178 by 0.093—0.990 mm.

Biology: Not yet completely known. Sey and Sayed[861] found that deposited eggs contained fully developed miracidia. Miracidia contain rediae, thus, sporocyst stage is also lacking in life-cycle pattern. Morphological structure of miracidia of this species was similar to that of miracidia of *Sandonia sudanensis*.

Habitat: Small intestine.

Hosts: *Mormyrus kannume* (Mormyridae); *Synodontis batensoda, S. clarias, S. schall, S. serratus* (Mochocidae).

Distribution: Africa — Egypt, Sudan.

References: Fischthal and Kuntz,[313] Khalil,[481] McClelland,[598] Moravec,[619] Saoud and Wannas,[816] Ukoli.[1002]

Pretestis Angel et Manter, 1970

Diagnosis: Sandoniinae. Body elongated. Acetabulum ventroterminal. Pharynx terminal, with well-developed extramural appendages; esophagus long with bulb; ceca simple and terminate in ovarian field. Testes round, diagonal; anterior testis largely extracecal, posterior testis intercecal. Cirrus sac present. Genital pore pretesticular. Ovary near posterior border of posterior testis. Uterine coils intercecal, mainly posttesticular. Vitelline follicules considerably large, posttesticular. Parasites of freshwater and coastal fishes.

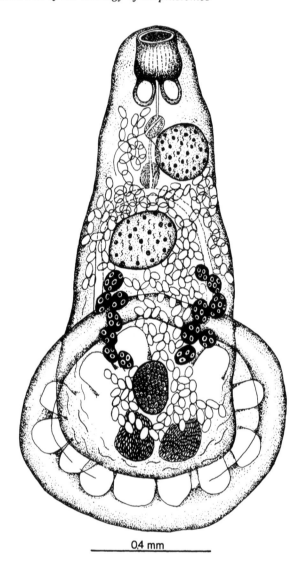

0.4 mm

FIGURE 236.   *Basidiodiscus ectorchis* Fischthal et Kuntz, 1959.
(Courtesy of J. H. Fischthal.)

Type species: *Pretestis australianus* Angel et Manter, 1970 (Figure 237)

Diagnosis: With character of the genus. Body length 0.4—1.1 mm, body width 0.1—0.4 mm. Acetabulum ventroterminal, 0.16—0.25 by 0.19—0.30 mm. Pharynx 0.11 mm long, with extramural appendages; esophagus not visible. Anterior testis 0.06—0.17 by 0.66—0.14 mm, posterior testis 0.06—0.17 by 0.06—0.16 mm. Cirrus sac 0.09 by 0.04 mm. Ovary 0.79—0.84 by 0.53—0.79 mm. Eggs 0.123—0.129 by 0.059—0.065 mm.

Biology: Not completely known. Angel and Manter[32] described the redia and the cercaria (*Cercaria acetabulopapillosa*) from the prosobranch mollusk, *Plotiopsis tetrica* (Thiaridae) which were regarded to be the larval stages of *P. australianus*. These cercariae possess ventroterminal acetabulum, eyespots, and pigmentation, which are characteristic of amphistome type of cercariae. Excretory tube without cross-connecting tubes. A special feature of it is the presence of an acetabular papilla. Cercaria encyst freely, under experimental circumstances on filamentous algae.

Habitat: Digestive tube.

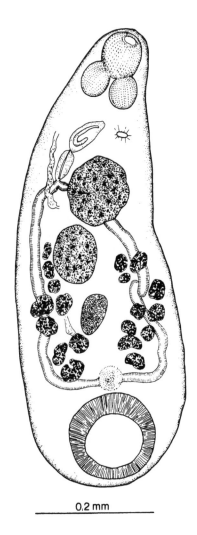

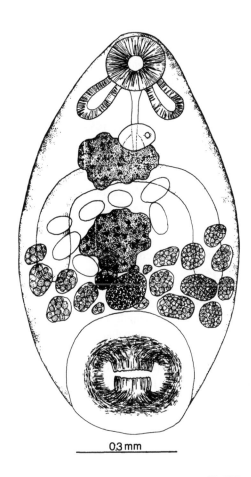

0,2 mm   0,3 mm

FIGURE 237. *Pretestis australianus* Angel et Manter, 1970. (Courtesy of L. M. Angel.)

FIGURE 238. *Australotrema brisbanense* Khalil, 1981. (Courtesy of L. F. Khalil.)

Hosts: *Melanotaenia fluviatilis* (Melanotaeniidae); *Bidyanus bidyanus* (Teraponidae); *Percalatus colonorum, Acanthopagrus butcheri* (Sparidae).

Distribution: Australia.

Reference: Angel and Manter.[32]

Australotrema Khalil, 1981

Diagnosis: Sandoniinae. Body elongate, tapering anteriorly. Acetabulum ventroterminal with a transverse opening and a strong sphincter guarding its opening. Pharynx globular with extramural appendages, esophagus with bulb, ceca straight, terminating at level of ovary. Testes irregularly lobed, tandem or slightly diagonal, anterior testis largely prececal, posterior testis intercecal. Cirrus sac well developed, genital pore prebifurcal. Ovary posttesticular. Uterine coils intercecal. Vitelline follicules fairly large, few in number, confluent posteriorly. Eggs large and few in number. Intestinal parasite of freshwater fishes.

Type species: *Australotrema brisbanense* Khalil, 1981 (Figure 238)

Diagnosis: With characteristics of the genus. Body length 0.8—1.4 mm, body width 0.4—0.9 mm. Acetabulum 0.30—0.50 mm in diameter. Pharynx 0.14—0.21 mm long with bulb, appendages extramural, 0.11—0.16 by 0.06—0.10 mm; esophagus 0.19—0.27 mm long. Anterior testis 0.16—0.23 by 0.15—0.30 mm; posterior testis 0.15—0.29 by 0.13—0.30 mm. Cirrus sac 0.11—0.19 by 0.06—0.09 mm. Ovary 0.05—0.09 by 0.06—0.13 mm. Eggs 0.110—0.150 by 0.080—0.100 mm.

Biology: Not yet known.

Habitat: Intestine.

Host: *Trachystoma petardi* (Mugilidae).

Distribution: Australia.

Reference: Khalil.[482]

Microrchiinae Yamaguti, 1958

Diagnosis: Body elongate. Acetabulum ventroterminal, large. Pharynx terminal, pharyngeal appendages intramural, esophagus short, bulb present. Testes small, diagonal, in middle third of body. Genital pore immediately postbifurcal. Cirrus sac present. Uterine coils intercecal, between bifurcation and cecal ends. Vitelline follicules extending along ceca, in middle third or posterior half of body. Parasites of freshwater fishes.

Type genus: *Microrchis* Daday, 1907

Diagnosis: With characteristics of the subfamily.

Type species: *Microrchis megacotyle* (Diesing, 1836) Daday, 1907 (Figure 239)

Syn. *Amphistoma megacotyle* Diesing, 1836; Daday, 1907; *Chiorchis (Microrchis) megacotyle* Fukui, 1929.

Diagnosis: Body length 5.0—11.0 mm, body width 1.4—4.3 mm. Acetabulum subterminal, 1.12—2.13 mm in diameter, Microrchis type, number of muscle units, d.e.c., 35—37; d.i.c., 57—59; v.e.c., 33—36; v.i.c., 36—38 (Figure 147). Pharynx 0.17—0.26 mm long, Megacotyle type (Figure 41); esophagus 1.37—1.68 mm long with bulb. Anterior testis 0.43—0.55 mm, posterior one 0.32—0.49 mm. Cirrus sac 0.46 by 0.30 mm. Ovary 0.26—0.47 mm in diameter. Terminal genitalium Scleroporum type (Figure 76). Eggs 0.060—0.070 mm in length. Excretory duct and Laurer's canal open jointly at level of ovary.

Biology: Not yet known.

Habitat: Intestine.

Hosts: *Colossoma bidens, Mylosoma aureum* (Characidiae); *Ageneiosus palmitus* (Ageneiosidae).

Distribution: South America — Brazil.

Reference: Daday,[197] Diesing.[221]

Microrchis ferrumequinum (Diesing, 1836) Daday, 1907 (Figure 240)

Syn. *Amphistoma ferrumequinum* Diesing, 1836; *Chiorchis (Microrchis) ferrumequinum*, Fukui, 1929; *Pseudocladorchis ferrumequinum* Vaz, 1932; *P. ferrumequinum* Travassos, 1934; *P. ferrumequinum* Yamaguti, 1958, 1971.

Diagnosis: Body length 5.5—15.0 mm, body width 3.2—5.4 mm. Acetabulum 1.12—2.56 by 1.13—2.18 mm, Microrchis type number of muscle units, d.e.c., 36—38; d.i.c., 55—58; v.e.c., 31—35; v.i.c., 37—39 (Figure 147). Pharynx 1.67—2.34 mm long, Ferrumequinum type (Figure 39); esophagus 2.56—3.22 mm long, with bulb. Testes similar in size, 0.04—0.70 mm in diameter, cirrus sac 0.30—0.33 by 0.15 mm. Ovary 0.57—0.68 by 0.30—0.33 mm. Terminal genitalium Spinolosum type. (Figure 77). Eggs 0.080 by 0.050 mm.

Biology: Not yet known.

Habitat: Intestine.

Hosts: *Platydoras costatus, Pterodoras granulosum, Doras corone, D. dorsalis* (Doradidae).

Distribution: South America — Brazil, Paraguay.

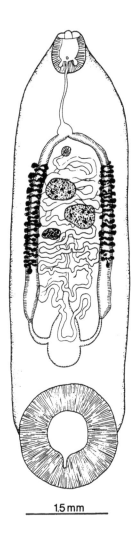

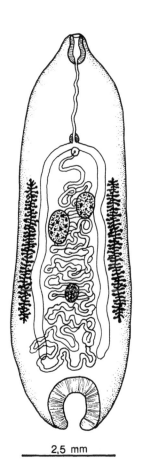

1,5 mm   2,5 mm

FIGURE 239. *Microrchis mega-cotyle* (Diesing, 1836). (After E. Daday, 1907.)

FIGURE 240. *Microrchis ferrumequinum* (Diesing, 1836). (After E. Daday, 1907.)

References: Daday,[197] Diesing,[221] Travassos et al.[989]

Notes: The genus *Microrchis* was designated by Daday[197] for Diesing's[221] species *Microrchis megacotyle* and *M. ferrumequinum*. Skrjabin[890] accepted this taxonomic position for both species, but Yamaguti[1076] transferred the latter species to the genus *Pseudocladorchis*. Both gross morphological and histomorphological features justify retention of *M. ferrumequinum* in this genus.

KEY TO THE SPECIES OF *MICRORCHIS*

1. Ovary near posterior testis; Laurer's canal and excretory duct open jointly ....
.... *M. megacotyle*
Ovary midway between posterior testis and acetabulum; Laurer's canal and excretory duct open separately .... *M. ferrumequinum*

Colocladorchiinae Sey, 1988

Diagnosis: Cladorchiidae. Body flattened, pyriform. Acetabulum large, ventroterminal

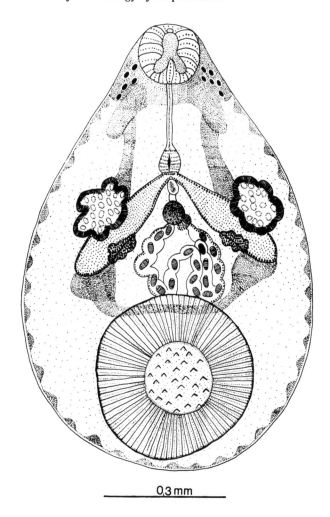

0.3 mm

FIGURE 241.   *Colocladorchis ventrastomis* (Thatcher, 1979). (Courtesy of V. E. Thatcher.)

with minute papillae along inner surface. Pharynx terminal, appendages intramural; esophagus long with bulb; ceca short, extending to midregion of body. Testes horizontal, slightly lobed; cecal and extracecal in part. Ovary intercecal at testicular zone. Genital pore bifurcal, cirrus sac present. Uterine coils between ovary and acetabulum. Vitellaria: two compact masses, at testicular zone, dorsal to ceca. Viviparity present. Intestinal parasite of freshwater fishes.

Type genus: *Colocladorchis* Thatcher, 1979
   Diagnosis: With characters of the subfamily.

Type species: *Colocladorchis ventrastomis* Thatcher, 1979 (Figure 241)
   Diagnosis: Body length 0.7—1.5 mm, body width 0.4—0.8 mm. Acetabulum 0.30—0.58 mm in diameter. Pharynx 0.11—0.19 mm in diameter, esophagus 0.19 mm long with bulb. Testes 0.11—0.22 by 0.094—0.14 mm. Cirrus sac 0.07—0.12 mm long. Ovary 0.50—0.13 mm in diameter. Eggs 0.036—0.049 by 0.060—0.066 mm, intrauterine miracidia 0.053—0.095 by 0.085—0.142 mm.
   Biology: Not yet known.
   Habitat: Intestine.

Host: *Prochilodus reticulatus* (Curimatidae).
Distribution: South America — Colombia.
Reference: Thatcher.[972]

Pseudocladorchiinae (Näsmark, 1937) Yamaguti, 1958

Diagnosis: Cladorchiidae. Body oval, linguiform or conical. Acetabulum terminal or ventroterminal. Pharynx: large or moderate size; appendages extramural, rarely intramural; esophagus short or moderate, bulb mostly present (rarely absent); ceca simple, extending usually to acetabular field. Testes cecal or extracecal horizontal, situated at anterior part of body. Cirrus sac present or absent. Ovary in front of acetabulum, rarely at acetabulum zone. Genital pore bifurcal or prebifurcal. Uterine coils intercecal between ovary and acetabulum. Vitelline follicules along ceca in middle or acetabular zone. Intestinal parasites of freshwater fishes.

KEY TO THE GENERA OF PSEUDOCLADORCHIINAE

1. Acetabulum moderately developed 2
   Acetabulum enormous in size .... *Nicollodiscus*
2. Testes cecal 3
   Testes extracecal, body linguiform with a transverse ridge .... *Kalitrema*
3. Vitellaria posttesticular, confined to ovarian zone .... *Pseudocladorchis*
   Vitelline follicules preovarian .... *Brevicaecum*

Type genus: *Pseudocladorchis* Daday, 1907
  Diagnosis: Pseudocladorchiinae. Body elongate, elliptical. Acetabulum terminal or ventroterminal. Pharynx large, appendages intramural; esophagus long, with muscular thickening; ceca simple, extending to or reaching near acetabulum. Testes cecal, horizontal. Cirrus sac present. Ovary median before acetabulum. Genital pore bifurcal. Uterine coils fill intercecal field. Vitelline follicules along posttesticular portion of ceca. Parasitic in freshwater fishes.

Type species: ***Pseudocladorchis cylindricus* (Diesing, 1836) Daday, 1907 (Figure 242)**
  Syn. *Amphistoma cylindricum* Diesing, 1836; *Pseudocladorchis nephrodorchis* (Daday, 1907) Travassos, Artigas et Pereira, 1928.
  Diagnosis: Body length 4.0—10.0 mm, body width 1.5—3.2 mm. Acetabulum 1.50—2.50 mm in diameter, Pseudocladorchis type, number of muscle units, d.e.c., 43—46; d.i.c., 52—55; v.e.c., 36—38; v.i.c., 56—59 (Figure 151). Pharynx 0.60—1.20 by 0.65—1.30 mm. Pseudocladorchis type (Figure 30), esophagus 0.80—1.70 mm long with muscular thickening. Testes 0.80—1.70 mm in size. Cirrus sac 0.30—0.40 by 0.20—0.30 mm. Eggs 0.100 by 0.060—0.080 mm.
  Biology: Not yet known.
  Habitat: Intestine.
  Hosts: *Mylosoma aureum* (Characidae); *Pimelodus ornatus* (Pimelodidae); *Platydoras costatus* (Doradidae).
  Distribution: South America — Brazil.
  References: Daday,[197] Diesing,[221] Travassos et al.[989]

Pseudocladorchis macrostomus **Daday, 1907 (Figure 243)**
  Syn. *Pseudocladorchis cylindricum* Vaz, 1932
  Diagnosis: Body length 4.0—4.5 mm, body width 0.4—1.5 mm. Acetabulum 1.52—1.68 mm in diameter, Pseudocladorchis type, number of muscle units, d.e.c., 40—45; d.i.c.,

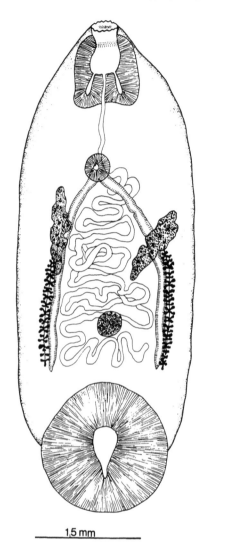

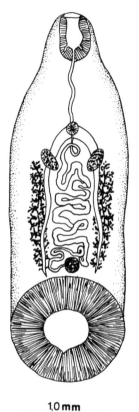

1,5 mm

1,0 mm

FIGURE 242.   *Pseudocladorchis cylindri-cus* (Diesing, 1836). (After E. Daday, 1907.)

FIGURE 243.   *Pseudocladorchis macrostomus* Daday, 1907. (After E. Daday, 1907.)

52—56; v.e.c., 34—37; v.i.c., 54—58 (Figure 151). Pharynx 0.73—0.84 by 0.51 mm, Pseudocladorchis type (Figure 30); esophagus 1.13—1.26 mm long. Testes reniform 0.22 by 0.17 mm. Cirrus sac 0.19 by 0.15 mm. Ovary 0.15—1.18 mm in diameter. Genital pore bifurcal. Size of eggs is not given.

Biology: Not yet known.

Habitat: Intestine.

Host: *Colossoma bidens* (Characidae).

Distribution: South America — Brazil.

Reference: Daday.[197]

KEY TO THE SPECIES OF *PSEUDOCLADORCHIS*

1. Testes branched, ceca terminating in front of acetabulum.............. *P. cylindricus*
    Testes reniform, ceca extending to acetabulum .... *P. macrostomus*

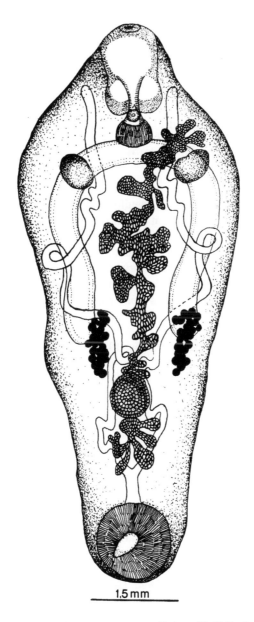

FIGURE 244. *Brevicaecum niloticum* McClelland,
1957. (Courtesy of L. F. Khalil and by permission of
the *J. Helminthol.*, 1963, London.)

Brevicaecum McClelland, 1957

Diagnosis: Pseudocladorchiinae. Body elongate. Acetabulum ventroterminal. Pharynx terminal, appendages extramural; esophagus short with bulb; ceca simple, terminating at middle part of body. Testes horizontal, round or oval, cecal. Cirrus sac seemingly absent. Ovary median, slightly posterior to ceca. Genital pore prebifurcal. Uterine coils in intercecal field, mainly preovarian. Vitelline follicules lateral, at level of cecal ends. Intestinal parasites of freshwater fishes.

Type species: *Brevicaecum niloticum* McClelland, 1957 (Figure 244)

Diagnosis: With characteristics of the genus. Body length 5.7—14.2 mm, body width 1.3—3.8 mm. Acetabulum 1.17—1.96 by 1.27—1.68 mm, Brevicaecum type, number of

muscle units, d.e.c., 46—48; d.i.c., 30—34; v.e.c., 22—24; v.i.c., 20—23 (Figure 132). Pharynx of 0.50—0.81 mm long, Ferrumequinum type (Figure 39); appendages 0.52—0.76 in length; esophagus 0.43—0.79 mm long. Testes 0.23—0.76 by 0.29—0.58 mm. Ovary 0.50—0.80 mm in diameter. Eggs 0.110—0.160 by 0.070—0.090 mm.
Biology: Not yet known.
Habitat: Intestine.
Host: *Citharinus citharinus* (Citharinidae).
Distribution: Africa — Ghana, Nigeria, Sudan.
References: Fischthal and Thomas,[315] Khalil,[480,481] McClelland,[598] Ukoli,[1002]

Kalitrema Travassos, 1933

Diagnosis: Pseudocladorchiidae. Body linguiform, flattened. Transverse ridge present at anterior part of body and a median notch at posterior extremity. Acetabulum ventroterminal, small. Pharynx small, appendages extramural; esophagus short, without bulb; ceca extending to postequatorial zone, far away from acetabulum. Testes irregular in shape, horizontal, extracecal. Cirrus sac seemingly absent. Ovary submedian, at level of cecal end. Genital pore prebifurcal, terminal genitalium suckerlike in appearance. Uterine coils fill intercecal field, postcecal in part. Vitelline follicules along posterior half of ceca. Intestinal parasites of freshwater species.

Type species: *Kalitrema kalitrema* Travassos, 1933 (Figure 245)

Diagnosis: With characteristics of genus. Body length 7.0—8.7 mm, body width 2.6—3.0 mm. Acetabulum 0.76—0.78 mm in diameter. Pharynx 0.36—0.39 mm in diameter, appendages 0.13 mm long: esophagus 0.78—1.32 mm in length. Testes 0.57 by 0.39—0.44 mm in size. Ovary 0.35—0.36 in diameter. Eggs 0.160—0.168 by 0.096—0.110 mm.
Biology: Not yet known.
Habitat: Intestine.
Host: *Plecostomus punctatus* (Loricaridae).
Distribution: South America — Brazil.
Reference: Travassos.[983]

Nicollodiscus Srivastava, 1938

Diagnosis: Pseudocladorchiinae. Body conical. Acetabulum very large, circular, papillose. Pharynx terminal, small; appendages small; appendages extramural; esophagus median in length, with bulb; ceca simple extending to conical end of body. Testes lobed, extracecal, horizontal. Ovary small, near to left cecal end. Genital pore prebifurcal, genital sucker and cirrus sac present. Uterine coils preovarian, intercecal. Vitelline follicules arranged in "U" form, posttesticular and extracecal, confluent between cecal ends. Parasitic in freshwater fishes.

Type species: *Nicollodiscus gangeticus* Srivastava, 1938 (Figure 246)

Diagnosis: With characteristics of the genus. Body length 9.0—13.4 mm, body width 7.0—8.2 mm. Acetabulum 6.24—7.26 mm in diameter. Pharynx 0.76—0.88 by 0.76—0.98 mm; appendages 0.16—0.50 by 0.16—0.24 mm; esophagus 1.32—1.48 mm long. Right testis 1.83—1.97 by 0.88—1.08 mm, left testis 1.54—1.84 by 0.76—1.02 mm. Cirrus sac 0.34—0.46 mm. Ovary 0.46—0.64 by 0.38—0.52 mm. Genital sucker 1.2 mm in diameter. Eggs 0.110—0.153 by 0.052—0.072 mm.
Biology: Not yet known.
Habitat: Intestine.
Hosts: *Silundia gangetica* (Siluridae).
Distribution: Asia — India.
Reference: Srivastava.[907]

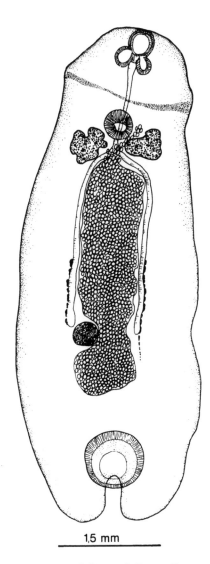

FIGURE 245. *Kalitrema kalitrema* Travassos, 1933. (After L. Travassos, 1933.)

Osteochilotrematinae Jones et Seng, 1986

Diagnosis: Cladorchiidae. Body elongate oval. Acetabulum ventroterminal. Pharynx terminal appendages extramural, esophagus short, with well developed bulb, ceca extending to ovary. Testes extracecal, horizontal, entire or lobate. Cirrus sac present. Ovary median, rounded between cecal end, well apart from acetabulum. Genital pore prebifurcal. Uterine coils radiate from ovary posteriorly and anteriorly, cross ceca laterally. Vitelline follicles along posterior two thirds of ceca. Parasitic in freshwater fishes.

Type genus: *Osteochilotrema* Jones et Seng, 1986

Diagnosis: With characteristics of the subfamily.

Type species: *Osteochilotrema malayae* Jones et Seng, 1986 (Figure 247)

Diagnosis: Body length 4.4—6.8 mm, body width 1.3—2.1 mm. Acetabulum 0.54—0.81 by 0.58—0.83 mm. Pharynx 0.38—0.57 by 0.33—0.58 mm, Ferrumequinum type (Figure 39); appendages 0.17—0.37 by 0.17—0.32 mm; esophagus 1.38 mm long. Testes 0.32—0.39 by 0.33—0.83 mm. Ovary 0.29—0.49 by 0.25—0.45 mm. Eggs 0.071—0.088 by 0.038—0.052 mm.

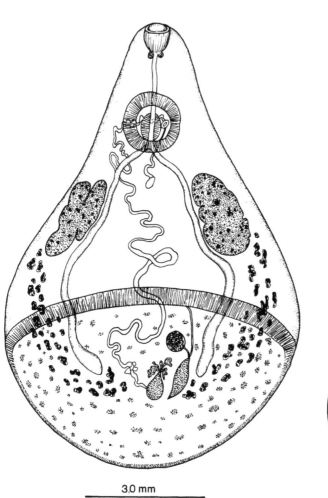

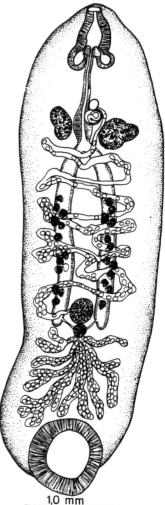

FIGURE 246.   *Nicollodiscus gangeticus* Srivastava, 1938. (After H. D. Srivastava, 1938.)

FIGURE 247.   *Osteochilotrema malayae* Jones et Seng, 1986. (Courtesy of A. Jones and by permission of the *J. Nat. Hist.*, 1986, London.)

Biology: Not yet known.
Habitat: Intestine.
Host: *Osteochilus hasselti, Barbus tramboides* (Cyprinidae).
Distribution: Asia — Malaysia.
Reference: Jones and Seng.[455]

Caballeroiinae Yamaguti, 1971

Syn. Caballeroinae Devaraj, 1972

Diagnosis: Cladorchiidae. Body elongate oval. Acetabulum terminal or ventroterminal. Pharynx small, terminal or subterminal, appendages extramural; esophagus short with or without bulb; ceca simple, somewhat sinuous, extending to acetabulum or before it. Testes horizontal or slightly diagonal, irregular in outline, intercecal in middle third of body. Cirrus sac present. Ovary submedian near to acetabulum or before it. Genital pore bifurcal or postbifurcal. Uterine coils posttesticular, intercecal. Vitelline follicules along posterior third of ceca, sometimes confluent posteriorly. Intestinal parasites of freshwater fishes.

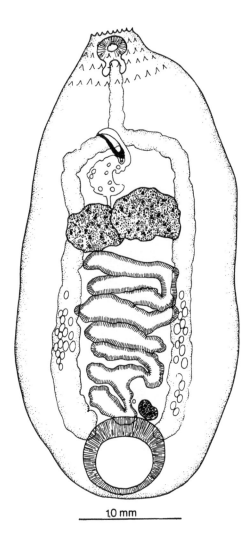

1.0 mm

FIGURE 248. *Caballeroia indica* Thapar, 1960. (After
G. S. Thapar, 1960.)

KEY TO THE GENERA OF CABALLEROIINAE

1. Vitelline follicules cecal, without confluence posteriorly............................ 2
    Vitelline follicules unite posteriorly..................................... *Bancroftrema*
2. Esophagus with bulb.. *Platycladorchis*
    Esophagus without bulb... *Caballeroia*

Type genus: *Caballeroia* Thapar, 1960
  Diagnosis: Caballeroiinae. Body elongate, anterior part with tegumental papillae in rows.
Acetabulum terminal or ventroterminal. Pharynx subterminal, appendages extramural; esophagus short, bulb absent; ceca extending to acetabulum. Testes obliquely diagonal, overlapping
each other medially and slightly overreaching ceca. Ovary median or submedian. Genital
pore bifurcal, cirrus sac present. Uterine coils mainly posttesticular. Vitelline follicules
cecal. Parasitic in freshwater fishes.

Type species: *Caballeroia indica* Thapar, 1960 (Figure 248)
  Diagnosis: Body length 4.6 mm, body width 2.0 mm. Acetabulum 1.23 mm in diameter.

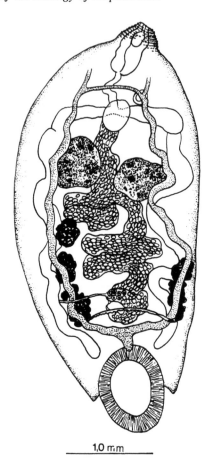

1,0 mm

FIGURE 249. *Caballeroia bhavani* (Achan, 1956). (Courtesy of M. Deveraj.)

Pharynx subterminal, 0.25 mm in diameter, appendages 0.28 mm in length; esophagus 0.82 mm long, without bulb. Right testis 0.70 by 0.45 mm; left testis 0.65 by 0.50 mm. Cirrus sac 0.32 mm long. Ovary 0.25 by 0.22 mm. Eggs 0.080 by 0.050 mm.

Biology: Not yet known.

Habitat: Intestine.

Host: *Cirrhina fulungel* (Cyprinidae).

Distribution: Asia — India.

Reference: Thapar.[966]

Caballeroia bhavani (Achan, 1956) Deveraj, 1972 (Figure 249)

Syn. *Nicollodiscus bhavani* Achan, 1956

Diagnosis: Body length 3.0—7.0 mm, body width 2.2 mm. Acetabulum 1.05 by 0.81 mm. Pharynx 0.17 mm in width, esophagus 0.31 mm long. Right testis 0.54 by 0.48 mm, left testis 0.57 by 0.60 mm. Ovary 0.20 by 0.21 mm. Eggs 0.120—0.194 by 0.072—0.090 mm.

Biology: Not completely elucidated.

Deveraj[211] found that the eggs of *C. bhavani* were operculated with developing miracidia. Miracidia hatched out from the extruded eggs in 2 to 4 h at laboratory temperature. Body length of miracidia 0.144 mm, covered with cilia and move quickly; composed of two pairs of penetration glands, terebratorium and embryos. Miracidia did not show affinity to snails to the genus *Melanodidas* spp.

Habitat: Intestine.
Hosts: *Barbaus hexagonolepis, B. carnaticus* (Cyprinidae).
Distribution: Asia — India.
References: Achan,[2] Devaraj.[211]

KEY TO THE SPECIES OF *CABALLEROIA*

1. Genital pore bifurcal, testes overlapping medially .... *C. indica*
    Genital pore prebifurcal, testes separated by uterine coils................ *C. bhavani*

Bancroftrema Angel, 1966
Diagnosis: Caballeroiinae. Body elongate oval, tapering anteriorly. Acetabulum terminal, large. Pharynx terminal, appendages extramural, esophagus short, slightly sinuous; bulb present; ceca undulating, extending to acetabulum. Testes horizontal, irregular, intercecal, pre-equatorial. Cirrus sac present. Ovary rounded, near acetabulum. Genital pore postbifurcal. Uterine coils posttesticular. Vitelline follicules lateral and median to ceca, confluent anteriorly. Parasitic in lung fishes.

Type species: *Bancroftrema neoceratodi* Angel, 1966 (Figure 250)
Diagnosis: With features characteristic of the genus. Body length 1.7 mm, body width 0.7 mm. Acetabulum 0.39 by 0.52 mm. Pharynx 0.09 mm long, appendages 0.07 mm long, esophagus 0.16 mm in length. Right testis 0.12 by 0.90 mm, left testis 0.16 by 0.19 mm. Cirrus sac 0.46 mm in diameter. Ovary 0.06 by 0.07 mm. Eggs 0.120 by 0.064 mm.
Biology: Not yet known.
Habitat: Intestine.
Host: *Neoceratodus forsteri* (Ceratodidae).
Distribution: Australia.
Reference: Angel.[31]

Platycladorchis Sey, 1986
Diagnosis: Caballeroiinae. Body foliaceous, markedly flattened. Acetabulum terminal or ventroterminal. Pharynx terminal with extramural appendages; esophagus short, with bulb; ceca terminating in front of acetabulum or partially overlapping it. Testes horizontal or slightly diagonal, touching each other. Cirrus sac present, occasionally strongly muscular. Ovary submedian at posterior part of body. Genital pore bifurcal or postbifurcal. Uterine coils preovarian, intercecal. Vitelline follicules postequatorial, lateral, mainly extracecal. Parasitic in freshwater fishes.

Type species: *Platycladorchis microacetabularis* Sey, 1986 (Figure 251)
Diagnosis: Body length 1.8—2.8 mm, body width 0.8—1.1 mm, with rudimental eye-spots. Acetabulum small, 0.20—0.25 mm in diameter. Pharynx 0.15—0.20 mm long; appendages 0.16—0.20 mm in length, esophagus 0.35—0.45 mm long. Right testis 0.20—0.30 by 0.20—0.25 mm, left testis 0.15—0.32 by 0.20—0.25 mm. Cirrus sac 0.32—0.40 by 0.12—0.7 mm. Genital pore postbifurcal. Ovary 0.10—0.12 by 0.05—0.10 mm. Eggs 0.111—0.128 by 0.052—0.063 mm, contain developing miracidia.
Biology: Not yet known.
Habitat: Intestine.
Host: *Spinibarbichthys denticulatus* (Cyprinidae).
Distribution: Asia — Vietnam.
Reference: Sey.[855]

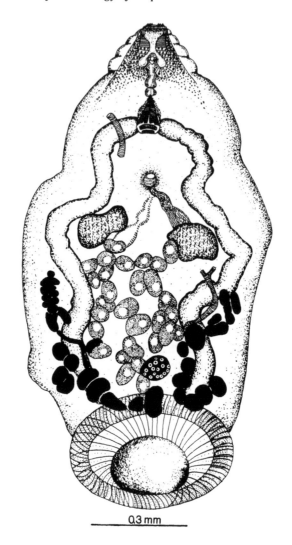

FIGURE 250.   *Bancroftrema neoceratodi* Angel, 1966. (Courtesy of L. M. Angel and by permission of the *J. Parasitol.*, 1966, Lincoln.)

Platycladorchis macroacetabularis Sey, 1986 (Figure 252)

Diagnosis: Immature specimens; body length 2.0—2.2 mm, body width 0.7—1.0 mm. Acetabulum 0.55—0.59 mm in diameter. Pharynx 0.20—0.23 mm long, appendages 0.15—0.17 mm in length; esophagus 0.50—0.54 mm long. Testes lobed, right testis 0.25—0.28 by 0.30—0.32 mm, left testis 0.27—0.29 by 0.25—0.28 mm. Cirrus sac 0.15 by 0.05 mm. Ovary 0.10—0.13 mm in diameter. Genital pore bifurcal.

Biology: Not yet known.

Habitat: Intestine.

Host: *Lissochilus krempfi* (Cyprinidae).

Distribution: Asia — Vietnam.

Reference: Sey.[855]

KEY TO THE SPECIES OF *PLATYCLADORCHIS*

1.   Acetabulum small, cirrus sac large and strongly muscular . . . . *P. microacetbularis*
     Acetabulum large, cirrus sac small and weakly muscular . . . . *P. macroacetabulum*

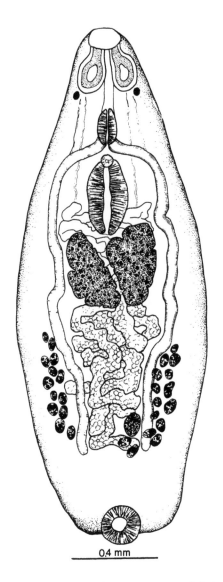

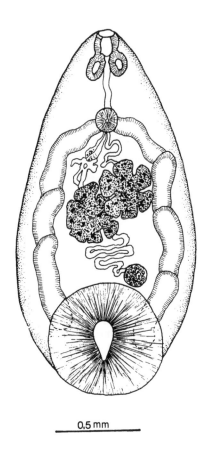

FIGURE 251. *Platycladorchis microacetabularis* Sey, 1986. (By permission of the *Acta Zool.*, 1986, Budapest.)

FIGURE 252. *Platycladorchis macroacetabularis* Sey, 1986. (By permission of the *Acta Zool.*, 1986, Budapest.)

Helostomatinae Skrjabin, 1949

Diagnosis: Cladorchiidae. Body elliptical, elongate, oval. Acetabulum ventroterminal. Pharynx terminal with well-developed extramural appendages; esophagus with bulb; ceca simple, rarely undulating, extending to acetabulum or ending before it. Testes intercecal, horizontal or diagonal. Cirrus sac present. Ovary median or submedian preacetabular. Genital pore prebifurcal, bifurcal or postbifurcal. Uterine coils intercecal, passing between two testes. Vitelline follicules well developed, usually lateral along length of ceca. Parasites of freshwater and marine fishes.

KEY TO THE GENERA OF HELOSTOMATINAE

1. Testes tandem or diagonal 2
   Testes horizontal .... *Helostomatis*

2. Testes lobate . . . . *Protocladorchis*
    Testes round . . . .*Amurotrema*

Type genus: *Helostomatis* (Fukui, 1929) Travassos, 1934b
Diagnosis: Body elliptical, tapering anteriorly. Acetabulum ventroterminal. Pharynx small with well-developed appendages, divided into anterior narrower and posterior expanded parts; esophagus short, with bulb; ceca simple or undulating, extending to acetabulum or ending before it. Testes oval or spherical, entire or lobate, preequatorial. Cirrus sac present. Ovary median or submedian. Genital pore prebifurcal. Uterine coils intercecal passing between two testes. Vitelline follicules lateral, along ceca, variously developed. Parasitic in freshwater and marine fishes.

Type species: *Helostomatis helostomatis* (MacCallum, 1905) Travassos, 1934 (Figure 253)
Syn. *Cladorchis helostomatis* MacCallum, 1905
Diagnosis: Body length 3.0 mm, body width 1.5 mm. Acetabulum 0.39 by 0.24; characterized by its sinuous puckered outline, three spoutlike projections extending backward. Pharynx 0.27 mm long, appendages 0.41 mm long; esophagus 0.51 mm in length, bulb present. Testes 0.39 by 0.24 mm. Ovary 0.13 by 0.06 mm. Eggs 0.140 by 0.060 mm.
Biology: Not yet known.
Habitat: Stomach.
Host: *Helostoma temmincki* (Anabantidae).
Distribution: Asia — Indonesia.
Reference: MacCallum.[571]

Helostomas sakrei Bhalerao, 1937 (Figure 254)
Diagnosis: Body length 1.5 mm, body width 0.7 mm. Acetabulum 0.50 mm in diameter. Pharynx 0.18 by 0.17 mm, appendages 0.21 mm long, esophagus 0.70 mm in length. Testes 0.30—0.32 by 0.13—0.16 mm; cirrus sac 0.09 by 0.05 mm. Ovary 0.13 by 0.08 mm. Eggs 0.080 by 0.042—0.044 mm.
Biology: Not yet known.
Habitat: Intestine.
Host: *Labeo calbasu* (Cyprinidae).
Distribution: Asia — India.
Reference: Bhalerao.[90]

Helostomatis indica Gupta et Verma, 1970 (Figure 255)
Diagnosis: Body length 4.2 mm, body width 1.5 mm. Acetabulum 0.76 by 0.71 mm. Pharynx 0.32 by 0.27 mm, Megalodiscus type, appendages 0.85 by 0.33 and 0.80 by 0.26 mm. Testes 0.44 by 0.35 and 0.38 by 0.33. Cirrus sac 0.16 by 0.10 mm. Eggs 0.120—0.140 by 0.062—0.075 mm.
Biology: Not yet known.
Habitat: Intestine.
Hosts: *Barbus sorana, B. tambroides, Osteochilus hasselti* (Cyprinidae).
Distribution: Asia — India, Malaysia.
References: Gupta and Verma,[397] Jones and Seng,[455] Leong et al.[541]

Helostomatis cirrhini Gupta et Adarsh Kumari, 1970 (Figure 256)
Diagnosis: Body length 3.1 mm, body width 1.1 mm. Acetabulum 0.68 by 0.74 mm. Pharynx 0.28 mm long, esophagus 0.62 mm in length. Right testis 0.19 by 0.14, left testis 0.19 by 0.14 mm. Ovary 0.15 by 0.13 mm. Size of eggs was not given.

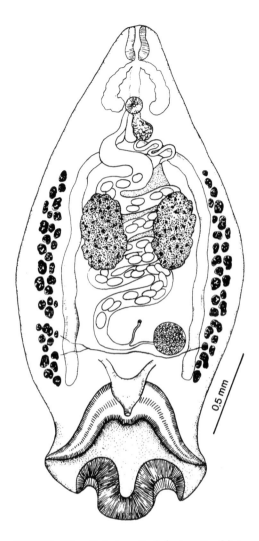

FIGURE 253. *Helostomatis helostomatis* (Mac-Callum, 1905). (After G. A. MacCallum, 1905.)

Biology: Not yet known.
Habitat: Intestine.
Host: *Cirrhina mrigala* (Cyprinidae).
Distribution: Asia — India.
Reference: Gupta and Kumari Adarsh.[393]

Helostomatis muelleri Gupta et Ahmad, 1979 (Figure 257)

Diagnosis: Body length 2.4—3.2 mm, body width 1.1—1.4 mm. Acetabulum 0.59—0.79 by 0.63—0.82 mm. Pharynx 0.27—0.37 mm long, appendages 0.27—0.56 mm long. Right testis 0.13—0.37 by 0.12—0.26 mm, left testis 0.18—0.34 by 0.15—0.36 mm. Cirrus sac 0.14—0.19 mm. Ovary 0.15—0.16 by 0.13—0.21 mm. Eggs 0.110—0.160 by 0.067—0.080 mm.
Biology: Not yet known.
Habitat: Intestine.
Host: *Otolith ruber* (Sciaenidae).
Distribution: Asia — India.
Reference: Gupta and Ahmad.[400]

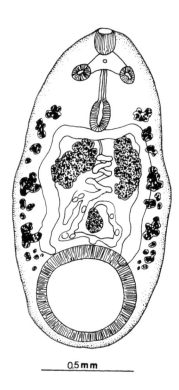
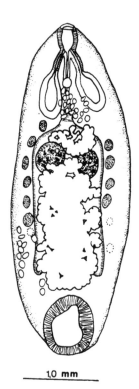

0,5 mm   1,0 mm

FIGURE 254. *Helostomatis sakrei*
Bhalerao, 1937. (After G. D. Bhalerao, 1937.)

FIGURE 255. *Helostomatis indica* Gupta et Verma, 1970.
(Courtesy of A. Jones and by permission of the *J. Nat. Hist.*, 1986,
London.)

Helostomatis bundelkhandensis Agarwal et Agrawal, 1980 (Figure 258)

Diagnosis: Body length 2.0—2.6 mm, body width 0.6—0.9 mm. Acetabulum 0.47—0.67
by 0.46 mm. Pharynx 0.16—0.23 by 0.21—0.23 mm, appendages 0.21—0.24 by 0.02—0.04
mm; esophagus 0.32—0.38 mm long. Right testis 0.19—0.29 by 0.15—0.16 mm, left testis
0.16—0.23 by 0.16—0.17 mm. Cirrus sac 0.30 by 0.16 mm. Ovary 0.08—0.18 by 0.9—0.15
mm. Eggs 0.120 by 0.050 mm, nonoperculated.
Biology: Not yet known.
Habitat: Intestine.
Host: *Mastacembelis armatus* (Mastacembelidae).
Distribution: Asia — India.
Reference: Agarwal and Agrawal.[6]

Helostomatis fotedari Gupta et Tandon, 1983 (Figure 259)

Diagnosis: Body length 3.1 mm, body width 1.3 mm. Acetabulum ventroterminal, 0.70
mm in diameter. Pharynx 0.35 by 0.40 mm, appendages 0.35—0.45 mm long; esophagus
0.47 mm in length, with bulb. Right testis 0.27 by 0.25 mm, left testis 0.35 by 0.08 mm.
Cirrus sac 0.16 by 0.16 mm. Ovary 0.16 by 0.19 mm. Eggs 0.090—0.100 by 0.060—0.080
mm.
Biology: Not yet known.
Habitat: Intestine.
Host: *Stromateus cinereus* (Stromateidae).
Distribution: Asia — India.
Reference: Gupta and Tandon.[398]

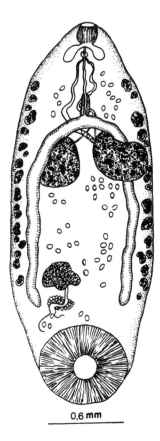

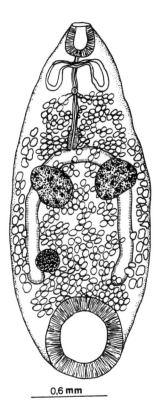

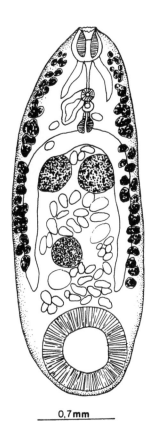

FIGURE 256. *Helostomatis cirrhini* Gupta et Adarsh Kumari, 1970. (After N. K. Gupta and Adarsh Kumari, 1970.)

FIGURE 257. *Helostomatis muelleri* Gupta et Ahmad, 1979. (After N. K. Gupta and Ahmad, 1979.)

FIGURE 258. *Helostomatis bundelkhandensis* Agarwal et Agrawal, 1980. (Courtesy of S. C. Agrawal.)

Helostomatis cyprinorum **Lim et Furtado, 1984. (Figure 260)**

Diagnosis: Body length 1.5—2.0 mm, body width 0.8—1.2 mm. Acetabulum 0.35—0.60 by 0.29—0.64 mm. Pharynx 0.14—0.23 mm long appendages enormously large, 0.50—0.80 mm, esophagus 0.21—0.43 mm in length: ceca with a 3-shaped configuration. Testes 0.85—1.01 mm in diameter. Cirrus sac 0.16—0.19 by 0.04—0.10 mm. Ovary 0.05—0.15 by 0.04—0.10 mm. Eggs 0.110—0.140 by 0.060—0.080 mm.

Biology: Not yet known.

Habitat: Intestine.

Hosts: *Labiobarbus festiva*, *Osteochilus melanopleura* (Cyprinidae).

Distribution: Asia — Malaysia.

Reference: Lim and Furtado.[552]

KEY TO THE SPECIES OF *HELISTOMATIS*

1. Acetabulum puckering in outline.................................... *H. helistomatis*
   Acetabulum not puckering in outline.. 2
2. Primary pharyngeal sacs enormously developed .... *H. cyprinorum*
   Primary pharyngeal sacs not enormously developed 3
3. Vitelline follicules extending to primary pharyngeal sacs or to middle part of esophagus.. 4
   Vitelline follicules extending to middle part of pharynx .... *H. bundelkhandensis*

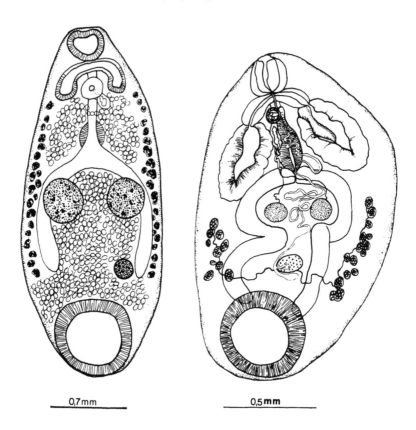

0,7 mm   0,5 mm

FIGURE 259. *Helostomatis fo-*   FIGURE 260. *Helostomatis cyprinorum* Lim
tedari Gupta et Tandon, 1983.   et Furtado, 1984. (Courtesy of S. Lim Lee Hong.)
(After S. P. Gupta and V. L. Tan-
don, 1983.)

Protocladorchis **Willey, 1935**

Syn. *Cladorchis* of MacCallum, 1905; *MacCallumia* Chatterje, 1938

Diagnosis: Helostomatinae. Body oval to elongate, flattened ventrally, convex dorsally. Acetabulum ventroterminal. Pharynx terminal with extramural appendages; esophagus short with bulb; ceca simple, undulating, extending to acetabulum. Testes tandem or diagonal in middle part of body, lobate intercecal. Cirrus sac present, genital pore bifurcal, prebifurcal and postbifurcal. Ovary submedian in posterior half of body. Uterine coils intercecal, post-testicular. Vitelline follicules in lateral fields from testicular or bifurcal zone to acetabulum. Intestinal parasite of freshwater fishes.

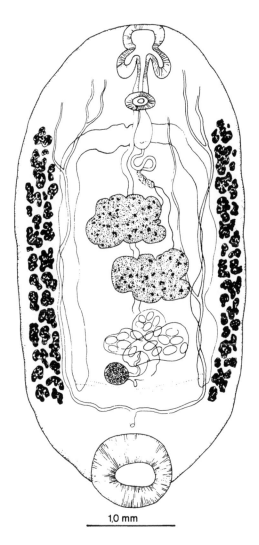

FIGURE 261. *Protocladorchis pangasii* (MacCallum, 1905). (After G. A. MacCallum, 1905.)

Type species: ***Protocladorchis pangasii* (MacCallum, 1905) Willey, 1935 (Figure 261)**

Syn. *Cladorchis pangasii* MacCallum, 1905; *Chiorchis pangasii* (MacCallum, 1905) Fukui, 1929; *Dadaytrema pangasii* (MacCallum 1905) Travassos, 1934

Diagnosis: Body length 2.1—6.0 mm, body width 1.1—3.1 mm. Acetabulum 0.40—1.33 by 0.42—1.43 mm. Pharynx 0.21—0.47 by 0.18—0.47 mm long. Megalodiscus type (Figure 41). Appendages 0.17—0.28 by 0.28 by 0.08—0.19 mm; esophagus 0.09—0.36 mm long. Anterior testis 0.17—0.57 by 0.18—0.53 mm, posterior testis 0.17—0.47 by 0.20—0.53 mm. Ovary 0.11—0.12 mm. Genital pore prebifurcal. Eggs 0.120—0.170 by 0.060—0.090 mm.

Biology: Not yet known.

Habitat: Intestine.

Hosts: *Pangasius nasutus* (Pangasidae); *Barbus daruphani, B. tambroides* (Cyprinidae).

Distribution: Asia — Indonesia, Malaysia.

References: Jones and Seng,[455] Leong et al.,[541] MacCallum,[571] Willey.[1057]

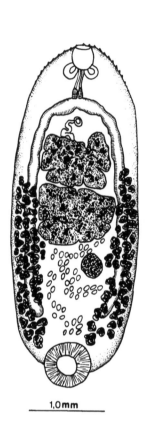 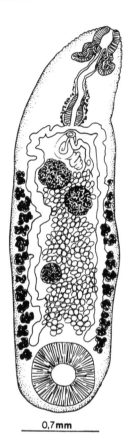

1,0mm   0,7mm

FIGURE 262. *Protocladorchis burmanicus* (Chatterji, 1938). (After R. C. Chatterji, 1938.)

FIGURE 263. *Protocladorchis chinabutae* Jones, 1987. (Courtesy of A. Jones and by permission of the *J. Nat. Hist.*, 1987, London.)

Protocladorchis burmanicus (Chatterji, 1938) Yamaguti, 1954 (Figure 262)

Syn. *Maccallumia burmanica* Chatterji, 1938

Diagnosis: Body length 3.5—7.2 mm, body width 1.4—2.7 mm. Acetabulum 0.40—0.83 by 0.45—0.90 mm. Pharynx 0.21—0.33 by 0.24 mm, appendages 0.15—0.27 mm in diameter; esophagus 0.13—0.62 mm long. Anterior testis 0.76—1.20 by 0.80—1.70 mm, posterior testis 0.80—1.04 by 0.87—1.70 mm. Ovary 0.26—0.46 mm in diameter. Eggs 0.080—0.097 by 0.056—0.060 mm.

Biology: Not yet known.

Habitat: Intestine.

Host: *Pangasius pangasius (Pangasidae)*.

Distribution: Asia — Burma.

Reference: Chatterji.[164]

Protocladorchis chinabutae Jones, 1987 (Figure 263)

Diagnosis: Body length 1.7—4.6 mm, body width 0.6—1.1 mm. Acetabulum 0.28—0.76 by 0.27—0.82 mm. Pharynx 0.15—0.25 by 0.12—0.29 mm; appendages 0.16—0.24 mm long; esophagus 0.30—0.85 mm in length. Testes 0.23—0.40 by 0.23—0.34 mm. Cirrus sac 0.099—0.34 by 0.07—0.19 mm. Ovary 0.09—0.09 by 0.08—0.09 mm. Genital pore bifurcal or immediately postbifurcal. Eggs 0.092—0.100 by 0.052—0.060 mm.

Biology: Not yet known.
Habitat: Intestine.
Host: *Pangasius nasutus* (Pangasidae).
Distribution: Asia — Thailand.
Reference: Jones.[454]

KEY TO THE SPECIES OF *PROTOCLADORCHIS*

1. Testes diagonal 2
   Testes tandem... *P. burmanicus*
2. Vitelline follicules extend from bifurcal zone to cecal ends.............. *P. pangasii*
   Vitelline follicules extend from testicular zone to cecal ends .... *P. chinabutae*

Amurotrema Akhmerov, 1959

Diagnosis: Helostomatinae. Body elongate cylindrical. Acetabulum ventroterminal. Pharynx terminal, with extramural appendages, esophagus with bulb, ceca simple, terminating at considerable distance away from acetabulum. Testes spherical, diagonal, intercecal, in middle part of body. Ovary submedian, round, about half-way between posterior testis and acetabulum. Genital pore near pharyngeal appendages, cirrus sac present. Uterine coils from acetabulum to genital pore in intercecal field. Vitelline follicules lateral, cecal, from zone of posterior testis to ovarian zone. Intestine parasites of freshwater fishes.

Type species: *Amurotrema dombrowskajae* Akhmerov, 1959 (Figure 264)

Diagnosis: With characters of the genus. Body length 8.0 mm body width 0.7—0.2 mm. Acetabulum terminal 0.95 by 0.97 mm. Pharynx 0.63 mm long, Amurotrema type (Figure 36), appendages 0.23 by 0.22 mm, esophagus 1.24 mm in length. Anterior testis 0.67 by 0.75 mm, posterior testis 0.63 by 0.68 mm. Cirrus sac 0.25 by 0.12 mm. Ovary 0.39 by 0.42 mm. Eggs 0.173—0.200 by 0.075—0.094 mm.
Biology: Reference. Gvozdev et al.[403]

Pre-parasitic stage. Eggs laid by adults contained fully developed miracidia. They began to hatch in minutes after falling into water. Miracidia are covered with ciliae and have no eyespot. They measured 0.30 mm in length on an average, the number of epithelial cells is 20, arranged in four rows by the formula 6:8:4:2. The life span of miracidia is usually 20 to 24 h. Intermediate host was *Anisus acronicus* (Planorbidae), percentage infestation was 28.1%, under experimental conditions.

Intramolluscan stage. Sporocyst was not studied. Rediae measuring 1.3 mm in length, containing three well-developed cercariae and several germinal balls. Length of period of intramolluscan stage was about 120 d. Cercariae with large body dimensions (0.70 by 0.33 mm), with eyespots near to anterior extremity. Tail 0.49 mm in length. Cercaria is a Diplocotylea type. After emergence from snails cercariae encysted freely, usually on the surface of submerged aquatic plants. Prepatent period was not given.
Habitat: Intestine.
Hosts: *Ctenopharyngodon idella, Spinbarbichthys denticulatus* (Cyprinidae).
Distribution: Asia — U.S.S.R., Vietnam, China.
References: Akhmerov,[15] Chen et al.,[168] Sey.[854]

Pisciamphistomatinae Yamaguti, 1971

Diagnosis: Cladorchiidae. Body: elongate oval. Acetabulum terminal. Pharynx terminal, appendages intramural, esophagus with bulb, ceca simple, with sinuous walls, terminating away from acetabulum. Testes horizontal or slightly diagonal, pre-equatorial, oval or spherical. Ovary median or submedian in middle part of body. Genital pore bifurcal postbifurcal,

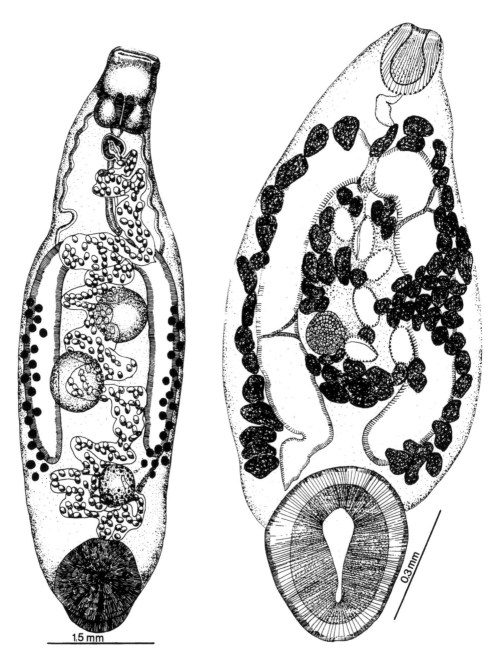

FIGURE 264. *Amurotrema dombrowska-jae* Akhmerov, 1959. (After A. H. Akhmerov, 1959.)

FIGURE 265. *Pisciamphistoma stunkardi* (Holl, 1929). (After S. Yamaguti, 1954.)

cirrus sac present. Uterine coils passing between two testes, with few coils. Vitelline follicules in lateral fields, along ceca or whole length of body. Intestinal parasites of freshwater fishes.

Type genus: *Pisciamphistoma* Yamaguti, 1954

Diagnosis: With characters of the subfamily.

Type species: *Pisciamphistoma stunkardi* (Holl, 1929) Yamaguti, 1954 (Figure 265)

Syn. *Paramphistomum stunkardi* Holl, 1929; *Alassostomoides stunkardi* (Holl, 1929) Skrjabin, 1949

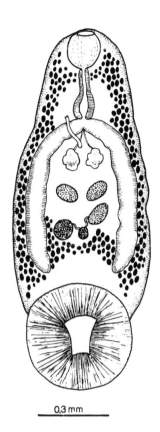

0,3 mm

FIGURE 266. *Pisciamphistoma reynoldsi* Bogitsh et Cheng, 1959. (After B. F. Bogitsh and T. C. Cheng, 1959.)

Diagnosis: Body length 1.4—1.8 mm, body width 0.3—0.7 mm. Acetabulum 0.31—0.54 by 0.27—0.37 mm. Anterior testis 0.10 by 0.05 mm, posterior testis 0.11—0.07 mm. Ovary 0.07 by 0.08 mm. Genital pore postbifurcal. Eggs 0.129 by 0.095 mm, with developing embryo measuring 0.095 by 0.066 mm.

Biology: Not yet known.

Habitat: Intestine.

Hosts: *Chaenobryttus gulosus, Lepomis auritus, L. gibbosus, L. machrochirus, L. megalotis, L. microlophus, Pomoxis nigromaculatus* (Centrarchidae); *Ictalurus natalis* (Ictaluridae); *Etheostoma blennoides* (Percidae), *Esox americanus, E. vermiculatus* (Esocidae).

Distribution: North America — U.S.

References: Aliff et al.,[21] Bangham,[69] Edwards and Nahhas,[282] Haderlie,[404] Holl.[418]

Pisciamphistoma reynoldsi Bogitsh et Cheng, 1959 (Figure 266)

Diagnosis: Body length 1.1—2.2 mm, body width 0.4—0.7 mm. Acetabulum 0.35—0.48 by 0.35—0.46 mm. Pharynx 0.10—0.27 by 0.09—0.27; esophagus 0.21 mm in length. Right testis 0.16—0.27 by 0.06—0.09 mm, left testis 0.06—0.19 by 0.06—0.14 mm. Ovary 0.04—0.11 by 0.05—0.08 mm. Genital pore bifurcal. Eggs 0.090 by 0.040 mm.

Biology: Not yet known.

Habitat: Intestine.

Hosts: *Ictalurus natalis* (Ictaluridae); *Lepomis macrorchis, L. gibbosus* (Centratchidae).

Distribution: North America — U.S.

References: Bogitsh and Cheng,[104] Harmos.[407]

KEY TO THE SPECIES OF *PISCIAMPHISTOMA*

1. Testes horizontal, vitelline follicules confluent anterior and posterior.... *P. reynoldsi*
   Testes tandem, vitelline follicules neither confluent anterior nor posterior ....
   .... *P. stunkardi*

Megalodiscinae Sey, 1987

Diagnosis: Body elongate, conical. Acetabulum terminal with central prominence or accessory sucker. Pharynx terminal, appendages extramural, esophagus short with bulb; ceca simple, terminating at acetabulum or shortly before it. Testes two, tandem, diagonal or horizontal, in middle part of body. Ovary median, spherical. Genital pore bifurcal or post-bifurcal, cirrus sac present. Uterine coils intercecal field, passing between two testes. Vitelline follicules extending along whole or posterior portion of intestine. Parasite of rectum of amphibians, occasionally reptiles.

KEY TO THE GENERA OF MEGALODISCINAE

1. Testes horizontal, vitelline follicules along whole ceca, accessory sucker present .....
   .... *Opisthodiscus*
   Testes obliquely tandem or diagonal, vitellaria posttesticular, central prominence
   present .... *Megalodiscus*

Type genus: *Megalodiscus* Chandler, 1923
   Diagnosis: Body conical. Acetabulum terminal, subterminal, with muscular prominence. Pharynx terminal, appendages extramural, esophagus short with bulb, ceca terminating at or shortly before acetabulum. Testes diagonal or tandem, spherical, intercecal. Ovary median, before acetabulum. Genital pore bifurcal, cirrus sac present. Uterine coils intercecal, passing between two testes. Vitelline follicules posttesticular. Parasites of rectum, occasionally bladder of amphibians and rarely of reptiles.

Type species: *Megalodiscus americanus* Chandler, 1923 (Figures 267 and 268)
   Diagnosis: Body length 3.5—6.0 mm, body width 1.0—2.4 mm. Acetabulum 1.78 mm in diameter, Megalodiscus type number of circular muscle units, d.e.c., 50—52; d.i.c., 90—94; v.e.c., 67—71; v.i.c., 69—86; (Figure 146). Pharynx 0.28—0.45 by 0.3—0.55 mm; Megalodiscus type (Figure 42). Appendages 0.23—0.30 mm long; esophagus 0.41—0.70 mm in length. Anterior testis 0.34—0.65 by 0.32—0.46 mm; posterior testis 0.32—0.75 by 0.38—0.47 mm. Cirrus sac 0.17 by 0.095 mm. Ovary 0.19—0.40 by 0.20—0.35 mm. Terminal genitalium Spinolosum type (Figure 77). Eggs 0.096—0.123 by 0.050—0.058 mm.
Biology: Not yet known.
Habitat: Rectum.
Hosts: *Amphiuma means, A. tridactylum*, (Amphiumidae); *Rana sphenocephala, R. montezumae, R. pipiens* (Ranidae); *Dicamptodon ensatus*, (Dicamptodontidae), *Ambystoma gracile* (Ambystomatidae); *Hyla arborea* (Hylidae); *Taricha granulosa* (Salamandridae).
Distribution: North America — U.S.; Central America — Mexico.
References: Bravo-Hollis,[117] Bravo-Hollis and Caballero Deloya,[119] Chandler,[162] Harwood.[408]

Megalodiscus temperatus (Stafford, 1905) Harwood, 1932 (Figure 269)

Syn. *Megalodiscus ranophilus* Millzner, 1924; *M. temperatus* Harwood, 1932; *M. montezumae* Travassos, 1934; *Diplodiscus temperatus* Stafford, 1905; *D. temperatus* Hunter, 1930; *D. temperatus* Sokoloff and Caballero, 1933.

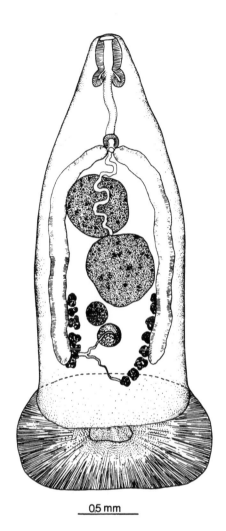

0.5 mm

FIGURE 267. *Megalodiscus americanus* Chandler, 1923. (After A. C. Chandler, 1923.)

FIGURE 268. *Megalodiscus americanus* Chandler, 1923. (median sagittal section).

Diagnosis: Body length 1.2—6.0 mm, body width 1.2—1.4 mm. Acetabulum 1.12—2.15 mm in diameter, Diplodiscus type, number of muscle units, d.e.c., 13—15; d.i.c., 40—44; v.e.c., 12—16; v.i.c., 30—32; m.e.c., 10—12 (Figure 143). Pharynx 0.13—0.50 by 0.18—0.48 mm, Megalodiscus type (Figure 42). Appendages 0.12—0.26 mm in length; esophagus 0.23—0.47 mm long. Anterior testis 0.22—0.52 by 0.17—0.69 mm; posterior testis 0.17—0.53 by 0.19—0.62 mm. Cirrus sac 0.10—0.12 mm in diameter. Ovary 0.13—0.34 by 0.15—0.27 mm. Terminal genitalium Spinolosum type (Figure 77). Eggs 0.083—0.172 by 0.056—0.117 mm, containing developing embryo.

Biology: References. Cary,[159*] Krull and Price,[511] Herber,[413,414] Smith.[894,895]

Preparasitic stage. ① Krull and Price. Eggs were operculated and embryonated when they were deposited. Miracidia pyriform, possessing 20 epithelial cells arranged by the formula 6:8:4:2. A cephalis gland, two pairs of penetration glands, a pair of flame cells

* Cary[159] described observations on the life history of species under the name *Diplodiscus* (= *Megalodiscus*) *temperatus*. Cort, examining Cary's material, found that he had described two different types of cercariae as belonging to this species. Hence Cary's results are not taken into consideration among biological data referring to *Megalodiscus temperatus*.

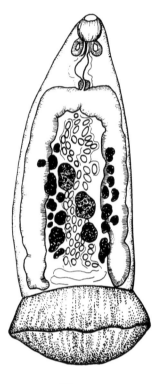

1,9 mm

FIGURE 269. *Megalodiscus temperatus* (Stafford, 1905). (After R. R. Brenes et al. 1959.)

were present. Germ cells formed a large mass which was not attached to the body wall. Intermediate hosts: *Helisoma trivolvis* (Krull and Price, Smith), *Ferrissia fragilis, F. novangliae, F. parallela* (Smith).

Intramolluscan stage. ① Krull and Price. Infestational experiments proved to be negative with the snail *Helisoma trivolvis*, thus, the sporocyst was not examined. Living rediae 0.50 by 0.16 mm in size, posterior end with two pair of appendages (these were not observed by Smith on rediae developed in *F. fragilis*). Pharynx 0.049 by 0.050 mm, gut elongate, sacculate, extending to the anterior pair of appendages. Birth pore situated dorsally about midway between the oral sucker and the anterior appendages. Rediae contain developing cercariae (up to 20). Cercariae amphistomum type belonging to the Diplocotylea group. Body 0.38—0.48 by 0.16—0.22 mm, tails 0.68—0.73 mm in length. Pigmented eyespots dorsal on both sides of esophagus, pigment granules were sometimes massed together. Pharynx together with appendages measuring 0.15—0.14 mm in length; acetabulum 0.11—0.15 mm in diameter. Cystogenous cells containing rodlike structures which were destined to form the cyst. Ceca thick walled, extending near acetabulum. Genital primordia intercecal, in middle part of body. Two main excretory trunks winding anteriorly, without cross-connecting tubes. Tube of tail passes posteriorly and bifurcates near the end of the tails, having no excretory openings. ② Smith. Cercariae were shed from experimentally infected snails (*Helisoma trivolvis, Ferrissia fragilis*) 41 to 45 d post infection with miracidia. Twenty to 30 cercariae emerged daily for a week before the host died.

Development in definitive hosts. ① Krull and Price. Cercariae prefer the pigmented areas of frog and tadpole skin as the site for encystment, and they encyst exceptionally on vegetation or other material. Frogs become infected by devouring their own skin having encysted cercariae. Life span of adolescaria was about 6 or 7 d. Cysts are thin, hyaline, measuring 0.24 by 0.19 mm. Tadpoles usually infected by the adolescariae which were attached loosely to the skin. These adolescariae dropped off and were picked up as particles of food material and swallowed. Prepatent period can occasionally be produced in a month under laboratory conditions, but the usual time is 2 or 3 months. ② Herber. The normal site of these flukes both in young tadpoles and frogs is the rectum, but during metamorphosis of the tadpoles they migrate from rectum to stomach. After metamorphosis, when the young frogs begin to feed on insects and the stomach content becomes acidic, the flukes migrate again to the rectum where pH is 7.1—7.5. Tadpoles can be infected as soon as the mouth becomes open (10-d-old tadpoles). In the heaviest experimental infestation, a tadpole harbored 252 specimens of young worms, all of them in the rectum.

Habitat: Rectum.

Hosts: Amphibia: *Rana aesopus, R. areolata, R. aurora, R. blairi, R. boylii, R. cantabrigensis, R. catesbeiana, R. clamitans, R. grylio, R. montezumae, R. pipiens, R. septentionalis, R. sphenocephala* (Ranidae); *Bufo americanus, B. terrestris, B. woodhousii* (Bufonidae); *Hyla cinerea, H. crucifer, H. chrysoscelis, Pseudacris brimleyi, P. triseriata* (Hylidae); *Ambystoma opacum, A. tigrinum,* (Ambystomidae); *Dicamptodon ensatus* (Dicamptodontidae); *Notophthalmus viridescens* (Salamandridae); *Amphiuma means* (Amphiumidae); *Pseudotriton montanus, Vesmognatus fuscus* (Plethodontidae); Reptilia: *Coluber constrictus* (Colubridae).

Distribution: North America — U.S., Canada; Central America — Mexico, Costa Rica.

References: Bouchard,[109] Bravo-Hollis,[117,118] Brenes,[120] Brenes et al.,[121] Brooks,[123] Catalano and White,[160] Harwood,[408] Ingles,[431] Manter,[586] Parker,[702] Pratt and McCauley,[736] Rankin,[766] Rosen and Manis,[790] Sokoloff and Caballero,[900] Stafford,[912] Ulmer.[1004]

Megalodiscus intermedius (Hunter, 1930) Harwood, 1932 (Figure 270)

Syn. *Diplodiscus intermedius* Hunter, 1930.

Diagnosis: Body length 2.1—3.6 mm, body width 0.4—0.9 mm. Acetabulum 0.70—0.14 mm in diameter. Pharynx 0.22—0.34 mm, appendages 0.17—0.20 mm; esophagus 0.38—0.54 mm long. Anterior testis 0.33—0.40 by 0.21—0.37 mm, posterior testis 0.30—0.35 by 0.19—0.33 mm. Ovary 0.20 mm in diameter. Eggs 0.075—0.100 by 0.030—0.045 mm.

Biology: Not yet known.

Habitat: Rectum.

Host: *Rana catesbeiana* (Ranidae); *Desmagnathus fuscus* (Plethodontidae).

Distribution: North America — U.S.

References: Brandt,[116] Bravo-Hollis,[117] Hunter.[427]

Megalodiscus rankini Bravo-Hollis, 1941 (Figure 271)

Syn. *Opisthorchis americanus* Holl, 1928; *Pseudopisthorchis americanus* (Holl, 1928).

Diagnosis: Body length 1.6—2.0 mm, body width 0.7—1.1 mm. Acetabulum 0.86—1.18 mm in diameter. Pharynx 0.15—0.18 by 0.14—0.18 mm; appendages 0.16—0.18 by 0.11—0.13; esophagus 0.22—0.30 by 0.02—0.04 mm. Anterior testis 0.11—0.14 by 0.11—0.20 mm, posterior testis 0.11—0.15 by 0.12—0.18 mm. Cirrus sac 0.05—0.07 mm in diameter. Ovary 0.13—0.16 by 0.12—0.19 mm. Eggs 0.090—0.110 by 0.046—0.062 mm.

Biology: Not yet known.

Habitat: Rectum.

Hosts: *Acris gryllus* (Hylidae); *Bufo americans* (Bufonidae); *Notophthalmus viridescens* (Salamandridae).

Distribution: North America — U.S.

References: Bravo-Hollis,[117] Brooks and Fusco,[127] Price and Buttner,[746] Rosen and Manis.[790]

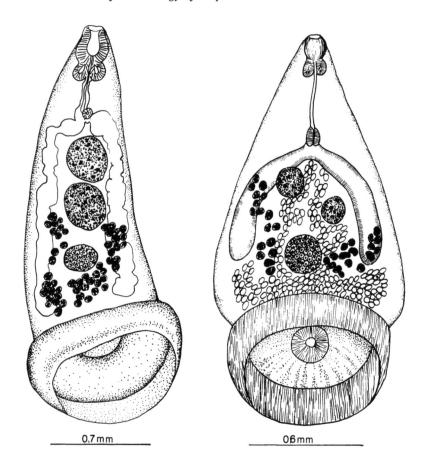

0,7 mm   0,6 mm

FIGURE 270. *Megalodiscus interme-*   FIGURE 271. *Megalodiscus rankini* Bravo-
dius (Hunter, 1930). (Courtesy of M.   Hollis, 1941. (Courtesy of M. Bravo-Hollis.)
Bravo-Hollis.)

Megalodiscus ferrissianus Smith, 1953

Diagnosis: Body length 6.5—7.0, body width 1.5 mm. Anterior and posterior testes similar in size, 0.35—0.38 mm in diameter. Ovary 0.15—0.28 by 0.17—0.38 mm. Genital pore bifurcal. Eggs 0.103—0.119 by 0.058—0.077 mm.

Biology: Reference. Smith.[892]

Preparasitic stage. Eggs contained fully developed miracidia which hatched within 10 to 30 min in water in summertime. Miracidia contained partially developed mother redia. Body of miracidia 0.129 by 0.023 mm when fixed in 10% fromalin. Body covered by 20 epithelial cells, arranged by the formula 6:8:4:2. Of the inner structure of miracidia apical gland, nervous tissue, a pair of flame cells were observed. Posterior two thirds of body filled with a large mass of cells (0.090—0.095 mm) surrounded by a thin membrane. Miracidia penetrated intermediate host through the mantle cavity in a period of 10 h after hatching. Intermediate host: *Ferrissia novangliae*.

Intramolluscan stage: After penetration, miracidia began to enlarge into saclike structure, mother sporocyst. By the third day the sporocyst was 0.150—0.190 mm, and by the fifth day there was a breakdown of all sporocyst structures. Rediae liberated from sporocyst by the end of the 6th or 7th day. First and second generation rediae were similar in shape (sausage- or club-shaped). Mature second-generation rediae were 0.50—1.00 by 0.15—0.20 mm in size with 20 to 40 germ balls and cercarial embryos. Birth pore at level of pharynx, locomotor appendages completely absent. Pharynx 0.040—0.045 mm in diameter, gut short (0.040—0.060 mm long), in excretory system six flame cells present. Cercarial emergence took place around noon and between 7 p.m. and 10 p.m. Body dimension of mature cercariae fixed in formalin: 0.263 by 0.227 mm, tails 0.449 mm in length. Biocellate, body covered with cystogenous cells except in sucker region. Pharyngeal appendages 0.130 mm in length, esophagus bulb present. Developing genital organs are easily seen, testes oblique, 0.043 mm in diameter, ovary 0.034 mm in diameter. Primary collecting vessels of the excretory system with transverse anastomosis; caudal tube with opening. Cercariae encysted on the skin of frogs, adolescariae 0.0250 mm in diameter.

Development in definitive hosts. Cysts were fed to frogs and the first mature worms were recovered 149 days after initial feeding.

Habitat: Rectum.

Hosts: *Rana clamitans, Rana pipiens, Rana castesbeiana* (Ranidae).

Distribution: North America — U.S.

Reference: Smith.[892]

KEY TO THE GENERA OF *MEGALODISCUS*

1. Testes diagonal 2
   Testes tandem.. *M. intermedius*
2. Ceca terminating beyond ovary 3
   Ceca terminating at zone of ovary .... *M. rankini*
3. Vitelline follicules not confluent posteriorly 4
   Vitelline follicules confluent posteriorly .... *M. ferrissianus*
4. Acetabulum Diplodiscus type .... *M. temperatus*
   Acetabulum Megalodiscus type .... *M. americanus*

Opisthodiscus Cohn, 1904

Diagnosis: Megalodiscinae. Body conical or pyriform. Acetabulum large, ventroterminal with additional sucker. Pharynx terminal with extramural appendages, esophagus short with bulb. Ceca simple, extending to acetabulum. Testes two, preovarian, horizontal. Cirrus sac present. Ovary in acetabular or preacetabular zone, median. Genital pore prebifurcal. Uterine coils intercecal, filling area between pharynx and acetabulum. Vitelline follicules lateral, between pharyngeal and acetabular zones, confluent posteriorly. Parasitic in rectum of amphibians.

Type species: *Opisthodiscus diplodiscoides* Cohn, 1904 (Figure 272)

Syn. *Opisthodiscus diplodiscoides nigrivasis* Méhelÿ, 1929.

Diagnosis: With characteristics of the genus. Body length 1.2—4.0 mm, body width 0.7—2.2 mm. Acetabulum 0.6—2.1 mm in diameter, Diplodiscus type, number of muscle units, d.e.c., 21—24; d.i.c., 52—56; v.e.c., 16—18; v.i.c., 53—55; m.e.c., 10—12 (Figure 143). Pharynx 0.38—0.95 by 0.28—0.74 mm, Ferrumequinum type (Figure 39). Appendages 0.32—0.37 mm; esophagus 0.28—0.62 mm long. Right testis 0.11—0.47 by 0.08—0.29 mm, left testis 0.11—0.33 by 0.10—0.47 mm. Cirrus sac 0.19—0.23 by 0.09—0.14 mm. Ovary 0.10—0.29 by 0.09—0.43 mm. Terminal genitalium Spinolosum type (Figure 77). Eggs 0.113—0.132 by 0.056—0.066 mm, containing fully formed miracidia.

Biology: Reference. Simon-Vicente et al.,[879] Martinez-Fernandez et al.[594]

Intramolluscan stage. ① Simon-Vicente et al. Larval stages were observed in artificially infected snail, *Ancylastrum* (= *Ancylus*) *fluviatilis*. Rediae were pyriform, reddish yellow, 0.8 mm in length and 0.18 mm in width. Pharynx large (0.09 mm in diameter), gut short, pear shaped. Birth pore at level of middle cecum. Six to 10 cercariae were present. Cercariae pear shaped, belonging to Diplocotylea group, measuring 0.45 mm in

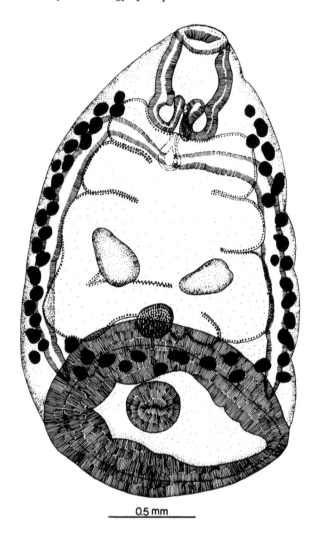

0,5 mm

FIGURE 272. *Opisthodiscus diplodiscoides* Cohn, 1904 (Courtesy of B. Grabda-Kazubska and by permission of the *Acta Parasitol. Pol.*, 1980, Warsaw.)

length and 0.25 mm in width. Tail about 0.65 mm in length. Pharynx with appendages measures 0.26 mm in length. Esophagus 0.07 mm long; cecum extending acetabulum, which is 0.18 mm in diameter. Genital primordia intercecal, postequatorial. Eyespots on the dorsal surface at level of pharyngeal appendages. Cystogenous cells containing bacillary bodies, granules and dark brown pigment. Excretory system consists of two main trunks, without cross-connecting tubes. Tube in tail terminates near the tip, forming elongate swelling, without pore to the surface. Encystment of free-swimming cercariae was mainly induced by the shallowness of the water they were kept in. Cercariae did not show any attraction to frogs found in their vicinity. Cercariae have diurnal periodicity in natural emission. Adolescariae measure 0.35 mm in diameter. Experimental examinations show that infestation presumably happened by ingestion of adolescariae. Prepatent period was about 55 days.

Habitat: Rectum.

Hosts: *Rana dalmatina, R. esculenta, R. iberica, R. perezi, R. ribibunda, R. temporaria* (Ranidae).

Distribution: Europe — Albania, France, Germany, Czechoslovakia, Bulgaria, Spain, Switzerland, Hungary, Romania; Africa — Algeria, Morocco.

References: Benmokhter-Betkonche,[78] Bozhkov,[112] B'tchvarow,[132] Căpuse and Dancău,[158] Cohn,[179] Combes et al.,[181] Dollfus,[240,241] Edelényi,[262] Gassmann,[336] Grabda-Kazubska,[356] Lluch et al.,[554] Méhely,[602] Odening,[663] Prokopič,[751] Sey,[831] Tscherner,[995] Vojkova.[1039]

Schizamphistominae Looss, 1912

Diagnosis: Cladorchiidae. Body conical, elongated oval, fusiform to subcylindrical. Acetabulum ventroterminal with sizes of a wide range. Pharynx terminal with extra- or intramural appendages, esophagus short with bulb, ceca long. Testes tandem or irregular in shape, in anterior or middle third of body. Ovary median, near to acetabulum or before it. Genital pore median, bifurcal or post-bifurcal, cirrus sac, occasionally absent. Uterine coils mainly posttesticular, rarely running between two testes. Vitelline follicules lateral or median to ceca, mainly posttesticular or along whole length of ceca, occasionally confluent posteriorly. Intestinal parasites of reptiles and amphibians.

KEY TO THE GENERA OF SCHIZAMPHISTOMINAE

1. Excretory ducts regular in structure and form some loops 2
   Excretory ducts highly convoluted .... *Schizamphistomum*
2. Testes pre-equatorial, vitellaria various in arrangement 3
   Testes in middle part of body, vitelline follicules extending to posterior testes ....
   .... *Schizamphistomoides*
3. Primary pharyngeal sacs intramural 4
   Primary pharyngeal sacs extramural.. 5
4. Vitelline follicules confluent posteriorly .... *Pseudalassostomoides*
   Vitelline follicules separated posteriorly .... *Lobatodiscus*
5. Ovary posttesticular... 6
   Ovary intertesticular .... *Quasichiorchis*
6. Posterior body end without lateral projection 7
   Body with lateral projection on each side at level of acetabulum .... *Alassostomoides*
7. Vitelline follicules posttesticular 8
   Vitelline follicules along with whole ceca .... *Stunkardia*
8. Testes spherical, obliquely tandem..................................... *Alassostoma*
   Testes lobed, tandem .... *Pseudocleptodiscus*

Type genus: *Schizamphistomum* Looss, 1912

Diagnosis: Schizamphistominae. Body elongated oval or elongate. Acetabulum subterminal, strongly developed with longitudinal aperture. Pharynx subterminal, appendages intramural; esophagus short with bulb; ceca simple, extending near or as far as acetabulum. Testes tandem, entirely irregular in outline. Ovary round, submedian, before acetabulum. Genital pore postbifurcal, cirrus sac present. Uterine coils intercecal between posterior testis and ovary. Vitelline follicules in posttesticular zone, lateral along ceca. Excretory ducts a pair, highly convoluted. Intestinal parasites of chelonians.

Type species: *Schizamphistomum scleroporum* (Creplin, 1844) Looss, 1912 (Figure 273)

Syn. *Amphistoma scleroporum* Creplin, 1844; *Paramphistomum papillosum* MacCallum, 1916; *Schizamphistomoides chelonei* Gupta, 1961; *Schizamphistoma manati* Sokoloff et Caballero, 1932

Diagnosis: Body length 6.7—11.1 mm, body width 2.2—2.8 mm. Acetabulum 1.64—1.88 by 1.31—1.61 mm, Schizamphistomum type number of muscle units, d.e.c., 80—83; d.i.c.,

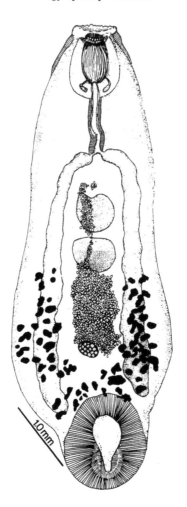

FIGURE 273. *Schizamphistomum scleroporum* (Creplin, 1844). (After F. Groschaft et al. 1977.)

81—84; v.e.c., 20—22; v.i.c., 26—28; m.e.c., 17—19 (Figure 154). Pharynx 0.87—1.22 by 0.89—1.24 mm, Solenorchis type; (Figure 33); esophagus 0.54—1.11 mm long. Anterior testis 0.44—0.83 by 0.7—1.25 mm, posterior testis 0.5—0.71 by 0.75—1.26 mm. Ovary 0.23—0.45 by 0.25—0.44 mm. Cirrus sac 0.66 by 0.40 mm; terminal genitalium Scleroporum type (Figure 76). Eggs 0.090 by 0.060 mm.

Biology: Not yet known.

Habitat: Large intestine.

Hosts: *Chelonia mydas, Eretmochelys imbricata, Caretta caretta* (Cheloniidae).

Distribution: Europe, Australia, North America, Cuba, Trinidad, Puerto Rico.

References: Blair,[97] Groschaft et al.,[377] Looss,[562] MacCallum.[572]

Schizamphistomum taiwanense Fischthal et Kuntz, 1975 (Figure 274)

Diagnosis: Body length 8.6 mm, body width 2.2 mm. Acetabulum 1.70 by 1.40 mm. Pharynx 1.05 by 0.81 mm, esophagus short, 0.61 mm long. Anterior testis 0.73 by 0.74 mm, posterior testis 0.75 by 0.70 mm. Ovary 0.36 mm in diameter. Size of eggs is not given.

Biology: Not yet known.

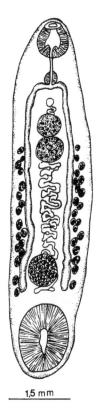

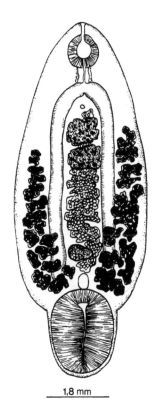

1,5 mm   1,8 mm

FIGURE 274. *Schizamphisto-*   FIGURE 275. *Schizamphisto-*
mum taiwanese Fischthal et Kuntz,   *mum erratum* Blair, 1983. (Cour-
1975. (Courtesy of J. H. Fis-   tesy of D. Blair.)
chthal.)

Habitat: Intestine.
Host: *Chelonia japonica* (Cheloniidae).
Distribution: Aisa — Taiwan.
Reference: Fischthal and Kuntz.[316]

Schizamphistomum erratum Blair, 1983 (Figure 275)

Diagnosis: Body length 6.5—12.9 mm, body width 1.8—3.5 mm. Acetabulum 1.36—2.2
by 1.24—2.02 mm, Dilymphosa type number of muscle units, d.e.c., 142—145; d.i.c.,
50—55; v.e.c., 13—16; v.i.c., 29—32; (Figure 142). Pharynx 0.67—1.30 by 0.67—1.31,
Scleroporum type; (Figure 32) esophagus 0.32—1.02 mm long. Anterior testis 0.30—0.89
by 0.23—1.49 mm. Ovary 0.25—0.59 by 0.20—0.65 mm. Eggs 0.099—0.110 by
0.065—0.110 mm.
Biology: Not yet known.
Habitat: Digestive tract.
Host: *Chelonia mydas* (Cheloniidae).
Distribution: Asia — Sri Lanka, Taiwan, Malaysia; Central America — Puerto Rico.
Reference: Blair.[97]

KEY TO THE SPECIES OF *SCHIZAMPHISTOMUM*

1. Esophagus with strongly developed bulb...2
   Esophagus with weakly developed bulb...................................*S. erratum*

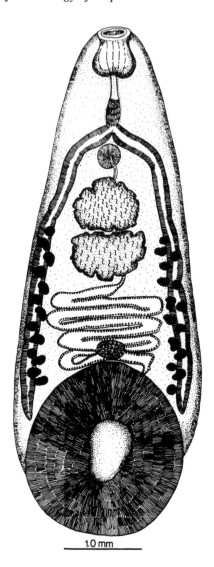

FIGURE 276.   *Schizamphistomoides spinolosus* (Looss, 1901).

2.   Vitelline follicules confluent posteriorly .... *S. scleroporum*
     Vitelline follicules not confluent posteriorly .... *S. taiwanense*

Schizamphistomoides Stunkard, 1925

Diagnosis: Schizamphistominae. Body conical, subcylindrical to fusiform. Acetabulum large, ventroterminal. Pharynx terminal, appendages intramural, or extramural; esophagus short with bulb; ceca simple, slightly winding, terminating at ovarian level. Testes tandem or slightly diagonal, intercecal, lobate. Ovary median, in front of acetabulum. Genital pore bifurcal or postbifurcal, cirrus sac present. Uterine coils between posterior testis and acetabulum. Vitelline follicules extending along outside of ceca, between testes and acetabulum. Intestinal parasites of chelonians.

Type species: *Schizamphistomoides spinolosus* (Looss, 1901) Stunkard, 1925 (Figure 276)

Syn. *Amphistomum spinolosum* Looss, 1901.

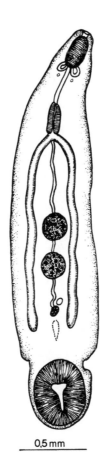

0,5 mm

FIGURE 277. *Schizamphisto-moides constrictus* Price, 1936. (After E. W. Price, 1936.)

Diagnosis: Body length 6.3—7.7 mm, body width 1.7—2.2 mm. Acetabulum 1.74—2.13 by 1.51—1.86 mm, Spinolosum type number of muscle units, d.e.c., 97—108; d.e.c., 60—62; v.e.c., 34—36; v.i.c., 58—61; m.e.c., 16—19 (Figure 156). Pharynx 0.35—0.66 by 0.35—0.59 mm, Spinolosum type (Figure 34); esophagus short 0.69—1.04 mm long. Anterior testis 0.54—0.74 by 0.50—0.89 mm, posteror testis 0.51—0.89 by 0.54—0.96 mm. Ovary 0.25—0.43 by 0.25—0.40 mm. Genital pore postbifurcal, terminal genitalium Spinolosum type (Figure 77). Eggs 0.070—0.075 by 0.038—0.045 mm.
Biology: Not yet known.
Habitat: Intestine.
Host: *Chelonia mydas* (Cheloniidae).
Distribution: Africa — Egypt, Ghana; North America — U.S.; Central America — Panama.
References: Blair,[97] Caballero et al.,[150] Looss,[559] Stunkard.[923]

Schizamphistomoides constrictus Price, 1936 (Figure 277)

Diagnosis: Body length 3.0 mm, body width 0.6 mm. Acetabulum 0.48 by 0.41 mm. Pharynx 0.27 mm long; appendages 0.08 mm long; esophagus 0.17 mm in length. Testes 0.17 mm in diameter. Ovary 0.04 by 0.07 mm. Genital pore bifurcal. Eggs, 0.070—0.075 by 0.038—0.045 mm.
Biology: Not yet known.

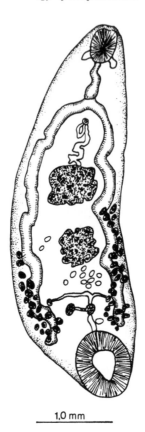

1,0 mm

FIGURE 278. *Schizamphisto-moides prescotti* Agrawal, 1967. (After V. Agrawal, 1967.)

Habitat: Intestine.
Hosts: *Pelomedusa galatea, P. subrufa* (Pelomedusidae).
Distribution: Africa — Tanzania.
Reference: Price.[743]

Schizamphistomoides prescotti Agrawal, 1967 (Figure 278)

Diagnosis: Body length 2.5—7.4 mm, body width 0.5—2.1 mm. Acetabulum 0.46—1.5 by 0.46—1.0 mm. Pharynx 0.60 by 0.50 mm, esophagus 0.16—0.45 mm in length. Anterior testis 0.05—0.50 by 0.06—0.70 mm, posterior testis 0.06—0.48 by 0.07—0.60 mm. Cirrus sac 0.05—0.21 by 0.03—0.08 mm. Ovary 0.05—0.15 by 0.03—0.25 mm. Genital pore just postbifurcal.
Biology: Not yet known.
Habitat: Intestine.
Host: *Hardella thurgi* (Emydidae).
Distribution: Asia — India.
Reference: Agarwal.[8]

KEY TO THE SPECIES OF *SCHIZAMPHISTOMOIDES*

1. Primary pharyngeal sacs extramural.. 2
   Primary pharyngeal sacs intramural .... *S. spinolosus*

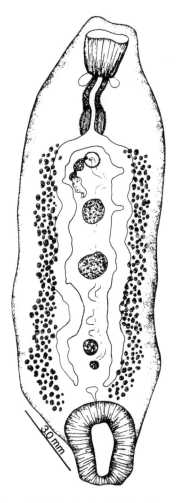

FIGURE 279. *Stunkardia dilymphosa*
Bhalerao, 1931. (After G. Premvati and
M. Agarwal, 1981.)

2. Testes spherical, acetabulum set off from body proper by constriction.. *S. constrictus*
   Testes lobate, acetabulum not set off from body proper by constriction .. *S. prescotti*

Stunkardia Bhalerao, 1931

Syn. *Orientodiscus* Srivastava, 1938 in part, *Schizamphistomoides* Stunkard 1925 in part; *Parorientodiscus* Rohde, 1969; *Kachugotrema* Dwivedi, 1967; *Pseudochiorchis* Yamaguti, 1958; *Neoalassostoma* Farooq, Khanum et Ansar, 1983

Diagnosis: Schizamphistominae. Body elongated. Acetabulum ventroterminal vary in size. Testes tandem, intercecal, middle part of body; spherical, entire or irregular in outline. Ovary median or submedian, before acetabulum. Genital pore bifurcal or postbifurcal, cirrus sac present. Uterine coils occupy space between bifurcation and ovary. Vitelline follicles in cecal or extracecal fields along intestine from bifurcation to acetabulum, occasionally confluent posterior. Intestinal parasites of turtles and amphibians.

Type species: *Stunkardia dilymphosa* Bhalerao, 1931 (Figure 279)

Syn. *Parorientodiscus magnus* Rhode, 1962; *Orientodiscus buckleyi* Siddiqi, 1965; *O. constrictus* Siddiqi, 1965; *O. linguiformis* Siddiqi, 1965; *O. fernandoi* Rohde, 1963; *Stun-*

kardia minuta Palmieri et Sullivan, 1977; *Schizamphistomoides lissemysi* Farooq, 1972; *Neoalassostoma lobata* Farooq, Khanum et Ansar, 1983, *Parorientodiscus lissemysi* Bhutta and Khan, 1975.

Diagnosis: Body length 2.7—20.3 mm, body width 1.5—5.6 mm. Acetabulum 0.52—2.14 by 0.46—3.50 mm, Dilymphosa type, number of muscle units, d.e.c., 126—132; d.i.c., 49—51; v.e.c., 11—13; v.i.c., 29—32; (Figure 142). Pharynx 0.31—2.87 by 0.28—2.1 mm. Stunkardia type (Figure 47); appendages 0.10—0.20 mm long; esophagus 0.22—1.22 mm in length. Anterior testis 0.29—0.36 by 0.21—0.42 mm, posterior testis 0.32—0.37 by 0.29—0.43. Cirrus sac 0.14—0.44 by 0.09—0.61 mm. Ovary 0.10—0.15 by 0.08—0.14 mm. Genital pore just postbifurcal, terminal genitalium Spinolosum type (Figure 77). Eggs 0.093—0.184 by 0.084—0.116 mm.

Biology: Not completely known. Reference. Mukherjee.[629]

Preparasitic stage. Eggs are sticky in nature, colorless, posterior end with asymmetrical thickening, anterior one with operculum. Freshly deposited eggs contain zygote. The major part of embryos attained the maximum size on 5th day of incubation (temperature was not given). Hatching began on the 6th day and the peak of hatching was in the morning hours. Killed miracidia measuring 0.19—0.27 by 0.16—0.21 mm. Body covering with 20 epithelial cells, arranged by the formula 6:8:4:2. Inner structure of miracidia composed of apical gland, two pairs of penetration glands, a total sum of 14 sensory papillae along the body of miracidiaum, excretory system which consists of a pair of flame cells and germinal cells forming a large germinal ball (0.11—0.072 mm on an average).

Habitat: Intestine, rectum.

Hosts: *Lissemys punctata, Irionyx formusus, T. phayrii, T. gangeticus*, (Trionychidae); *Batagur baska, Cyclemis amboinensis, C. dentata, Hardella thurgi, Kachuga tectum, K. sylhetensis, K. kachuga*, Morenia ocellata (Emydidae).

Distribution: Asia — Burma, India, Malaysia, Pakistan.

References: Bhalerao,[87] Chatterji,[163] Farooq,[299] Farooq et al.,[300] Palmieri and Sullivan,[694] Siddiqi.[878]

Notes: *Stunkardia dilymphosa* was described by Bhalerao[87] on the basis of two immature specimens from the freshwater turtle, *Batagur baska*, collected in Burma. Hence, Bhalerao's description might not be complete. Although Chatterji[163] had gone into further detail in the description, his paper has apparently been buried in oblivion. Since that time several related species have been recovered from various turtles of that region (India, Malaysia), and, thus, a group of closely related species has been established. For these species, several newly designated genera were set up.

The gradual increase of the observations referring to the morphology (including histomorphology of the muscular organs) of the species group indicated a close similarity of *S. dilymphosa* with those to be described later on. The genus *Stunkardia* was allocated to the subfamily Zygocotylinae on the basis of the absence of a cirrus sac, the number of lymphatic vessels, and the possession of two lips on each side of the acetabulum.

In the comparative examinations of specimens and the revision of the relevant literary data, Merotra and Gupta[608] and Premvati and Agarwal[737] found that some of the morphological features were erroneously observed, and some of them did not prove to be as stable as believed. Hence, these authors synonymized several species and genera with *S. dilymphosa*. To this list *S. minuta, Schizamphistomoidaes lissemysi, Neoalassostoma lobata*, and *Parorientodiscus lissemysi* were added by the present author. At the same time to the contrary of the authors above, *Kachugotrema* (= *Stunkardia*) *amboinensis* and *Pseudochiorchis* (= *Stunkarida*) *stunkardi* were regarded to be valid on the basis of the differences found in the

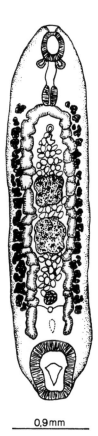

0,9mm

FIGURE 280. *Stunkardia burti*
(Prudhoe, 1944). (Courtesy of S.
Prudhoe and by permission of the
Annu. Mag. Nat. Hist., 1944,
London.)

histomorphological structure of the acetabulum, and they were transferred to this genus.

The validity of the genus *Stunkardia* was questioned by Bhatnagar et al.,[92] and they recommended to supress *Stunkardia* to a synonym of *Zygocotyle*. Their view is unacceptable to the present author due to evidence from histomorphological differences in the species of the genera in question.

Stunkardia burti (Prudhoe, 1944), Sey, 1988 (Figure 280)

Syn. *Chiorchis burti* Prudhoe, 1944

Diagnosis: Body length 3.7—5.1 mm, body width 1.1—1.3 mm. Acetabulum 0.63—0.76 by 0.51—0.64 mm, Cleptodiscus type number of muscle units, d.e.c., 10—12; d.i.c., 32+—36; v.e.c., 11—13; v.i.c., 36—38 (Figure 139). Pharynx 0.42—0.48 by 0.36—0.43 mm, Pseudochiorchis type (Figure 46); appendages 0.13—0.16 mm long; esophagus 0.15—0.21 mm in length. Testes 0.22—0.45 by 0.31—0.67 mm. Ovary 0.12—0.18 mm. Terminal genitalium Spinolosum type (Figure 77). Genital pore just postbifurcal. Eggs 0.150—0.170 by 0.100—0.110 mm.

Biology: Not yet known.

Habitat: Not given.

Distribution: Asia — Sri Lanka.

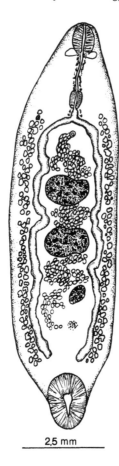 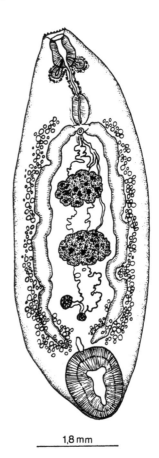

2,5 mm   1,8 mm

FIGURE 281. *Stunkardia amboinensis* (Dwivedi, 1967). (Courtesy of M. P. Dwivedi.)

FIGURE 282. *Stunkardia stunkardi* (Tandon, 1970). (After R. S. Tandon, 1970.)

Host: *Rana hexadactyla* (Ranidae).
Reference: Prudhoe.[754]

Stunkardia amboinensis (Dwivedi, 1967) n. comb. (Figure 281)

Syn. *Kachugotrema amboinensis* Dwivedi, 1967.

Diagnosis: Body length 11.4—12.7 mm, body width 2.5—2.8 mm. Acetabulum 1.09—1.11 mm long, Dilymphosa type, number of muscle units, d.e.c., 60—65; d.i.c., 44-46; v.e.c., 22—23; v.i.c., 44-45; m.e.c., (Figure 142) Pharynx 1.10—1.20 by 0.97—1.00, pharyngeal appendages 0.25—0.26 mm long; esophagus 1.38—1.45 mm long. Anterior testis 1.05—1.20 by 1.77—1.78 mm, posterior testis 0.85—1.10 by 1.80—1.90 mm. Cirrus sac 0.05—0.21 by 0.03—0.25 mm. Ovary 0.30 by 0.60 mm, terminal genitalium Spinolosum type (Figure 77). Eggs 0.160—0.180 by 0.080—0.100 mm.
Biology: Not yet known.
Habitat: Intestine.
Host: *Cyclemis amboinensis* (Emydidae).
Distribution: Asia — India.
Reference: Dwivedi.[260]

Stunkardia stunkardi (Tandon, 1970) n. comb. (Figure 282)

Syn. *Pseudochiorchis stunkardi* Tandon, 1970; *P. lucknowensis* Pandey, 1967 .

Diagnosis: Body length 5.6—11.0 mm, body width 1.8—2.8 mm. Acetabulum 0.83—1.52 by 0.64—1.12 mm, Stunkardia type number of muscle units, d.e.c., 60—65; d.i.c., 44—46; v.e.c., 32—36; v.i.c., 35—40 (Figure 158). Pharynx 0.33—056 by 0.30—0.41 mm, Pseudochiorchis type (Figure 46), appendages 0.18—0.22 mm long; esophagus 0.22—0.51 mm. Anterior testis 0.40—0.48 by 0.48—0.62 mm, posterior testis 0.35—0.50 by 0.48—0.73 mm. Cirrus sac 0.20—0.26 mm long, ovary 0.18—0.22 by 0.15—0.18 mm. Genital pore bifurcal, terminal genitalium Spinolosum type (Figure 77).
Biology: Not yet known.
Habitat: Intestine.
Hosts: *Hardella thurgi, Kachuga kachuga*, (Emydidae); *Lissemys punctata, Trionyx gangeticus* (Trionychidae).
Distribution: Asia — India.
References: Pandey,[698] Tandon.[953]

KEY TO THE SPECIES OF *STUNKARDIA*

1. Pharynx Pseudochiorchis type 2
   Pharynx Stunkardia type .... *S. dilymphosa*
2. Acetabulum Pseudochiorchis type..................................... *S. stunkardi*
   Acetabulum Stunkardia type .... *S. amboinensis*
   Acetabulum Cleptodiscus type .... *S. burti*

Alassostoma Stunkard, 1917

Diagnosis: Schizamphistominae. Body elongated oval. Acetabulum small, terminal. Pharynx with extramural appendages; esophagus short with bulb; ceca with sinuous walls, terminating in front of acetabulum. Testes round, intercecal, pre-equatorial, tandem. Ovary entire oval, little behind posterior testis: genital pore bifurcal, cirrus sac present. Uterine coils preovarian. Vitelline follicules lateral to ceca, posttesticular, penetrating into middle area posteriorly. Intestinal parasites of turtles.

Type species: *Alassostoma magnum* Stunkard, 1917 (Figure 283)

Diagnosis: With characters of the genus. Body length 10.0—12.0 mm, body width 3.0—5.0 mm. Acetabulum 2.00—2.50 by 2.00 mm, Schizamphistomum type number of muscle units, d.e.c., 80—82; d.i.c., 82—84; v.e.c., 21—23; v.i.c., 27—29, m.e.c., 17—19 (Figure 154). Pharynx 0.90—1.35 by 0.60—0.90 mm, Stunkardia type (Figure 47) appendages 0.35—0.60 mm; esophagus 0.60—1.30 mm long. Testes 0.27—0.45 by 0.35—0.9 mm. Ovary 0.27—0.35 by 0.33—0.57 mm. Cirrus sac 0.37 mm long, terminal genitalium Spinolosum type (Figure 77). Eggs 0.100 by 0.130 mm.
Biology: Not yet known.
Habitat: Intestine.
Hosts: *Pseudemys elegans, P. troosti, P. floridana, Chrysemis concirna, C. picta* (Emydidae); *Chelydra serpentina* (Chelydridae).
Distribution: North America — U.S.
Reference: Brooks,[124] Stunkard.[921]

Alassostomoides (Stunkard, 1924) Fukui, 1929

Diagnosis: Schizamphistominae. Body elongate, nearly parallel sides, with small projection at level of middle part of acetabulum. Acetabulum terminal with ventral aperture. Pharynx terminal with extramural or intramural appendages; esophagus short, with bulb; ceca wide, terminating near acetabulum. Testes slightly diagonal or tandem, intracecal near bifurcation, entire or lobed. Ovary median, postequatorial, between acetabulum and posterior

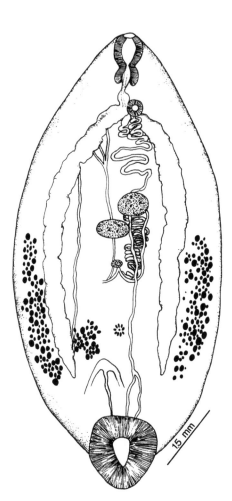

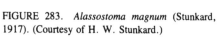

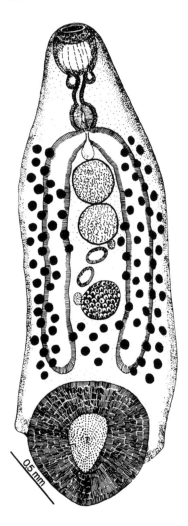

FIGURE 283. *Alassostoma magnum* (Stunkard, 1917). (Courtesy of H. W. Stunkard.)

FIGURE 284. *Alassostomoides parvus* (Stunkard, 1917). (Courtesy of D. R. Brooks and by permission of the *J. Parasitol.*, 1975, Lincoln.)

testis. Genital pore bifurcal, cirrus sac present. Uterine coils between ovary and posterior testis. Vitelline follicules well developed, inter- and extracecal, from bifurcation to acetabulum, confluent posteriorly. Intestinal parasites of amphibians and reptiles.

Type species: *Alassostomoides parvus* **(Stunkard, 1917) Travassos, 1934 (Figure 284)**
Syn. *Alassostoma parvum* Stunkard, 1916, *A.* (*Alassostomoides*) *parvum* Stunkard, 1916.
Diagnosis: Body length 3.0—5.5 mm, body width 0.8—1.6 mm. Acetabulum 0.67—0.20 by 0.64—1.0 mm. Pharynx 0.32—0.64 by 0.27—0.46 mm; appendages 0.06—0.08 mm long; esophagus 0.30—0.60 mm long. Testes 0.28—0.32 by 0.24—0.32 mm. Ovary 0.16—0.20 mm in diameter. Cirrus sac present. Eggs 0.145 by 0.100 mm.
Biology: References. Beaver,[75] Byrd and Reiber.[141]

Intramolluscan stage. ① Beaver. Intermediate host is *Planorbis* (= *Helisoma*) *trivolvis*. Redia 0.97 mm long, very active when obtained from the snail's liver. Two pairs of locomotion appendages present ventrolaterally, the anterior one always bigger than the posterior one. Pharynx 0.06 by 0.04 mm; esophagus short, cecum may extend to anterior locomotion appendage. Three pairs of large flame cells present, excretory pore lateral toward bladder, just behind the posterior locomotion appendage. Birth pore dorsal to anterior part of cecum. Free-swimming cercariae, large, pigmented, with two eyespots lateral to pharyngeal appendages, Diplocotylea type. Body pyriform 0.71 by 0.25 mm in size; tails 1.5 mm long. Pharynx 0.12 mm in diameter, appendages 0.02—0.03 mm long; esophagus short with bulb; ceca terminating near to acetabulum. Acetabulum 0.19 by 0.25 mm. Germinal primordia in postbifurcal median field. Two main excretory trunks form irregular winding tubes, extending to the region of the pharyngeal appendages, without cross-connecting tubes. Caudal tube bifurcate, a short distance from the tail which tips into two bladders with no openings to the outside in mature cercariae; in immature cercariae the tail tube opens to the exterior. Cercariae encyst on the crayfish, *Cambarus propinquus*, and young frogs, *Rana pipiens*. ② Byrd and Reiber. Body of cercariae 1.0 mm in length densely pigmented around the posterior end. Tail is somewhat longer than the body. *Cercaria inhabilis* Cort, 1914 and *Cercaria convoluta* Faust, 1919 are the larval forms of this species. Cysts are found on the outer surface of the intermediate host.

Development in definitive host. ① Beaver. Adolescariae were fed on *Rana castesbeiana* and *Chelydra serpentina* and they developed to adult flukes in both hosts.

Habitat: Rectum.

Hosts: Amphibia: *Rana catesbeiana, R. pipiens* (Ranidae). Reptilia: *Chrysemis picta, C. floridana* (Emydidae); *Sternotherus odoratus* (Kinosternidae); *Chelydra serpentina* (Chelydridae).

Distribution: North America — U.S.

References: Bennett,[80] Brooks,[122] Brooks and Mayers,[125] Guilford,[381] Rausch,[771] Stunkard,[921] Williams.[1062]

Alassostomoides chelydrae (MacCallum, 1918) Yamaguti, 1958 (Figure 285)

Syn. *Paramphistomum cleydrae* MacCallum, 1919

Diagnosis: Body length 2.1—5.5 mm, body width. Acetabulum 0.46—1.20 by 0.42—1.20 mm. Pharynx 0.30—0.80 by 0.22—0.63 mm, appendages small, esophagus 0.63 mm long, with bulb. Anterior testis 0.21—0.38 by 0.12—0.30 mm, posterior testis 0.12—0.40 by 0.12—0.36 mm. Cirrus sac 0.10—0.45 mm long. Ovary 0.06—0.27 mm in diameter. Eggs 0.150—0.180 by 0.100—0.130 mm.

Biology: Not yet known.

Habitat: Rectum.

Hosts: Amphibia: *Rana catesbeiana* (Ranidae); *Bufo americana* (Bufonidae); Reptilia: *Chelydra serpentina* (Chelydridae); *Chrysemys picta, Graptemys pseudogeographica* (Emydidae).

Distribution: North America — U.S.

References: Brooks,[122] Brooks and Mayers,[125] MacCallum.[574]

Notes: Wang[1048] reported on the occurrence of this species in China and recovered them in *Chelonia mydas*. This case report is probably based on misidentification.

Alassostomoides louisianaensis Christian et White, 1973 (Figure 286)

Diagnosis: Body length 3.5—4.1 mm, body width 1.0—1.2 mm. Acetabulum 0.65—0.70 by 0.58—0.62 mm. Pharynx 0.42—0.50 by 0.33—0.36 mm. Anterior testis 0.36—0.42 by 0.36-0.44 mm, posterior testis 0.35—0.40 by 0.35—0.41 mm. Ovary 0.16—0.21 mm in diameter. Genital pore postbifurcal. Eggs 0.110—1.160 by 0.080—0.090 mm.

Biology: Not yet known.

Habitat: Large intestine.

Hosts: *Rana grylio* (Ranidae); *Siren intermedia* (Sirenidae).

Distribution: North America — U.S.

References: Brooks and Buchner,[126] Christian and White.[174]

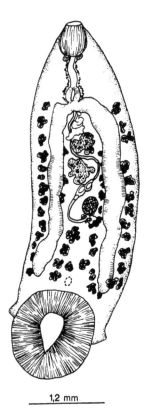

| 1,2 mm | 0,8 mm |

FIGURE 285. *Alassostomoides chelydrae* (MacCallum, 1918). (Courtesy of D. R. Brooks and by permission of the *J. Parasitol.*, 1975, Lincoln.)

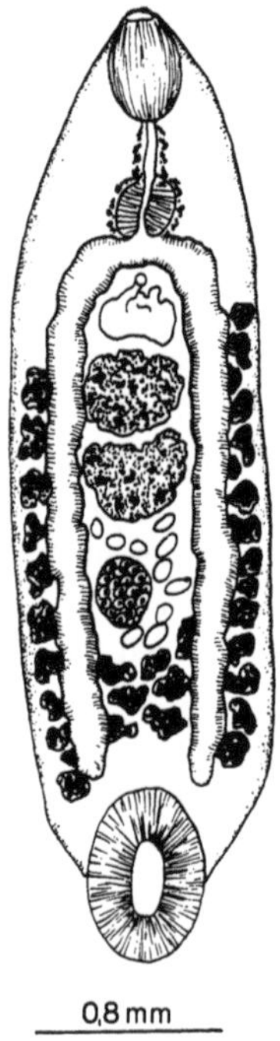

FIGURE 286. *Alassostomoides louisianaensis* Christian et White, 1937. (Courtesy of F. A. Christian and L. L. White.)

KEY TO THE SPECIES OF *ALASSOSTOMOIDES*
(after Christian and White, 1973; modified)

1. Testes diagonal 2
   Testes tandem, irregular in shape .... *A. louisianaensis*
2. Testis round, contiguous, vitelline follicles from posterior testis.......... *A. parvus*
   Testis irregular in shape, not contiguous, vitelline follicles from bifurcation ....
   .... *A. chelydrae*

Pseudocleptodiscus Caballero, 1961
Diagnosis: Schizamphistominae. Body elongate oval. Acetabulum terminal. Pharynx terminal with extramural appendages; esophagus short, ceca undulating and terminate near acetabulum. Testes tandem, irregular in outline, intercecal in middle part of body. Ovary spherical, to right of median line. Genital pore bifurcal, cirrus sac present. Uterine coils mainly posttesticular, passing between two testes. Vitelline follicles extending along ceca from posterior testis to level of acetabulum. Intestinal parasites of turtles.

Type species: *Pseudocleptodiscus margaritae* Caballero, 1961 (Figure 287)
Syn. *Pseudodiplodiscus bravoae* Caballero Rodriguez, 1960; *P. sphaerorchium* (Thatcher, 1963) Yamaguti, 1971; *Dadaytrema sphaerorchium* Thatcher, 1963.
Diagnosis: With characters of the genus. Body length 9.2—10.6 mm, body width 2.50—3.90

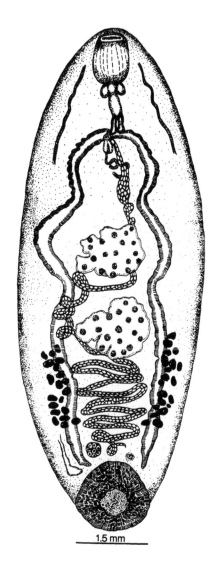

FIGURE 287. *Pseudocleptodiscus margaritae*
Caballero, 1961. (After C. E. Caballero, 1961.)

mm. Acetabulum 0.61—2.20 mm in diameter. Pharynx 0.52—0.70 by 0.24—0.38 mm; appendages 0.24 mm long; esophagus 0.75 mm in length. Anterior testis 0.57—1.14 by 0.70—1.21 mm, posterior testis 0.57—1.05 by 0.70—1.14 mm. Cirrus sac 0.56 by 0.24 mm. Ovary 0.17—0.30 mm in diameter. Eggs 0.070—0.080 by 0.130—0.170 mm, embryonated.

Biology: Not yet known.

Habitat: Intestine.

Host: *Dermatemys mawii* (Dermatemydidae).

Distribution: Central America — Mexico.

References: Caballero,[148] Thatcher.[970]

Notes: Yamaguti[1077] examined the type species of *Pseudocleptodiscus bravoae* (which was described in a thesis by Caballero Rodriquez (1960)) and *Dadaytrema sphaerorchium* and found that they seem to be identical with *Pseudocleptodiscus margaritae*.

Pseudoalassostomoides **Yamaguti, 1971**

Diagnosis: Body elongate, flat. Acetabulum ventroterminal. Pharynx small, terminal with

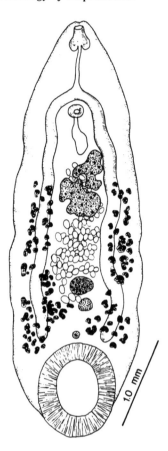

FIGURE 288. *Pseudoalassostomoides rarus* (Rohde, 1963). (Courtesy of K. Rohde.)

extramural appendages; esophagus short with bulb; ceca undulating, terminating near acetabulum. Testes diagonal, deeply indented, pre-equatorial. Ovary submedian, in posterior third of body. Genital pore postbifurcal, cirrus sac present. Uterine coils between ovary and posterior testis. Vitelline follicules posttesticular, along ceca and lateral to them, confluent posteriorly. Intestinal parasites of turtles.

Type species: ***Pseudoalassostomoides rarus*** **(Rohde, 1963) Yamaguti, 1971 (Figure 288)**
Syn. *Chiostichorchis rarus* Rohde, 1963.
Diagnosis: Body length 4.7 mm, body width 1.5 mm. Acetabulum 0.92—0.97 mm in diameter. Pharynx with appendages 0.39 by 0.41 mm, esophagus 0.51 mm long. Anterior testis 0.49 by 0.49 mm, posterior testis 0.39 by 0.49 mm. Ovary 0.16 by 0.25 mm. Eggs 0.130—0.140 by 0.070—0.090 mm.
Biology: Not yet known.
Habitat: Intestine.
Host: *Trionyx* sp. (Trionychidae).
Distribution: Asia — Malaysia.
Reference: Rohde.[782]

Quasichiorchis Skrjabin, 1949

Diagnosis: Schizamphistominae. Body fusiform with rounded body ends. Acetabulum large, ventroterminal. Pharynx small with large extramural appendages; esophagus short,

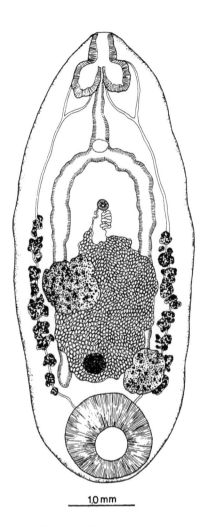

1,0 mm

FIGURE 289. *Quasichiorchis purvisi*
(Southwell et Kirshner, 1937). (After T.
Southwell and A. Kirshner, 1937.)

bulb present. Testes diagonal, postequatorial, posterior testis near acetabulum, irregular in outline. Ovary median, spherical, in front of acetabulum. Genital pore postbifurcal, cirrus sac present. Uterine coils between genital pore and acetabulum. Vitelline follicules extending along ceca, from pretesticular field to cecal ends. Intestinal parasites of turtles.

Type species: ***Quasichiorchis purvisi*** **(Southwell et Kirshner, 1937) Skrjabin, 1949 (Figure 289)**

Syn. *Chiorchis purvisi* Southwell et Kirshner, 1937

Diagnosis: With characters of the genus. Body length 3.0—5.0 mm, body width 0.5—3.0 mm. Acetabulum 1.30 mm in diameter. Pharynx 0.45 mm long; appendages 0.45 mm in length; esophagus 0.80—1.00 mm long. Testes 0.5 mm in diameter. Ovary 0.30 mm in diameter. Eggs 0.100—0.130 by 0.040—0.050 mm.

Biology: Not yet known.
Habitat: Not given.
Host: *Heosemys* (= *Geomyda*) *grandis* (Emydidae).
Distribution: Asia — Malaysia.
References: Balasingam,[62] Southwell and Kirshner.[902]

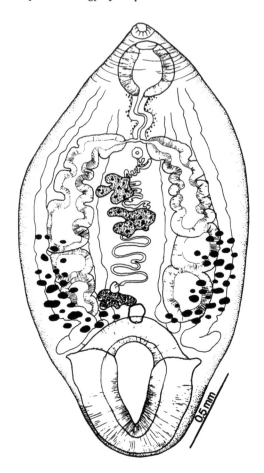

FIGURE 290.   *Lobatodiscus australiensis* Rohde, 1984.
(By permission of Dr. W. Junk Publishers, 1984, Dordrecht.)

Lobatodiscus Rohde, 1984

Diagnosis: Schizamphistominae. Body oval. Acetabulum large, ventroterminal. Pharynx well developed, with intramural appendages; esophagus short, bulb present. Testes slightly diagonal, in middle part of body, lobate. Ovary median, irregular in outline, near acetabulum. Genital pore postbifurcal, no cirrus sac. Uterine coils convoluted, between genital pore and ovary. Vitelline follicules lateral, between posterior testis and anterior margin of acetabulum. Intestinal parasite of turtles.

Type species: *Lobatodiscus australiensis* Rohde, 1984 (Figure 290)
Diagnosis: Schizamphistominae. Body length 2.3—3.0, body width 1.3—1.6 mm. Acetabulum 0.79—0.95 by 0.88—1.0 mm. Pharynx 0.27—0.35 by 0.38—0.42 mm. Anterior testis 0.14—0.19 by 0.24—0.33 mm, posterior testis 0.10—0.16 by 0.20—0.27 mm. Eggs 0.110—0.116 by 0.060—0.066 mm, embryonated.
Biology: Not yet known.
Habitat: Intestine.
Host: *Elseya dentata* (Chelidae).
Distribution: Australia.
Reference: Rohde.[783]
Notes: In the description of the species the presence of an esophageal bulb was not men-

tioned by Rohde.[783] Examining the paratypes it was found that this organ is present in this species.

Nematophilinae Skrjabin, 1949

Diagnosis: Cladorchiidae. Body oval or elongated. Acetabulum terminal or ventroterminal. Pharynx terminal with extra- or intramural appendages; esophagus with elongate bulb or without it; ceca undulating or serpentine, terminating at acetabulum or before it. Testes tandem, lobed or branched, intercecal, in middle or posterior part of body. Ovary median in front of acetabulum. Genital pore bifurcal, cirrus sac present. Uterine coils intercecal, divergently developed. Vitelline follicules small and numerous, extending along ceca from level of esophageal bulb or genital pore to acetabulum. Intestinal parasites of turtles.

KEY TO THE GENERA OF NEMATOPHILINAE

1. Primary pharyngeal sacs extramural.. 2
   Primary pharyngeal sacs intramural .... *Nematophila*
2. Length of pharyngeal appendages shorter than pharynx........................... 3
   Pharyngeal appendages longer than pharynx .... *Pseudalassostoma*
3. Genital pore just behind bifurcation................................. *Parachiorchis*
   Genital pore well behind bifurcation .... *Halltrema*

Type genus: *Nematophila* Travassos, 1934

Diagnosis: Nematophilinae. Body conical, elongated. Acetabulum ventroterminal. Pharynx terminal with intramural appendages; esophagus with elongate bulb; ceca somewhat undulating, terminating in front of acetabulum. Testes tandem, lobed, postequatorial. Cirrus sac present. Ovary submedian, near acetabulum. Genital pore bifurcal. Uterine coils mainly pretesticular, passing between two testes. Vitelline follicules numerous extracecal, stretching from level of esophageal bulb to acetabulum. Intestinal parasites of turtles.

Type species: *Nematophila grandis* (Diesing, 1839) Travassos, 1934 (Figure 291)

Syn. *Amphistoma grande* Diesing, 1839; *Nematophila ovale* Cordero et Vogelsang, 1940; *Paramphistomum argentinum* Cordero et Vogelsang, 1940.

Diagnosis: Body length 16.5—17.8, body width 4.3—5.9 mm. Acetabulum 2.90—3.10 by 2.90—3.30 mm, Nematophila type number of muscle units, d.e.c., 124—128; d.i.c., 23—25; v.e.c., 12—15; v.i.c., 24—26 Figure 148). Pharynx 0.96—1.0 mm long, Nematophila type (Figure 28); esophagus 1.30—1.90 mm in length. Anterior testis 1.38—1.49 by 2.10—2.40 mm, posterior testis 1.10—1.30 by 2.10—2.40 mm. Ovary 1.10—1.30 by 1.40—1.60 mm. Cirrus sac 0.69—0.81 by 0.33—0.35 mm; terminal genitalium Nematophila type (Figure 75). Eggs 0.110—0.120 by 0.059—0.068 mm.

Biology: Not yet known.

Habitat: Intestine.

Hosts: *Geomyda melanosterna* (Emydidae); *Podocnemis expansa, P. unifilis, P. vogli, P. erythrocephala, P. tracaxa* (Pelomedusidae); *Chelus fimbriata, Phrynops geoffroyana, Ph. schöpfii, Ph. hilarii* (Chelidae); *Kinosternon scorpioides* (Kinosternidae); *Kinixys erosa* (Testudinidae).

Distribution: South America — Argentina, Brazil, Bolivia, Paraguay, Venezuela; Central America — Panama.

References: Caballero et al.,[151] Cordero and Vogelsang,[182] Diaz-Ungria,[220] Diesing,[223] Heyneman et al.,[416] Lent and Freitas,[538] Lombardero and Moriena,[556] Masi Pallarés et al.,[595] Sey.[834]

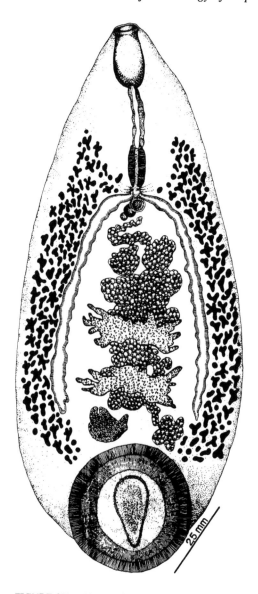

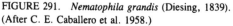

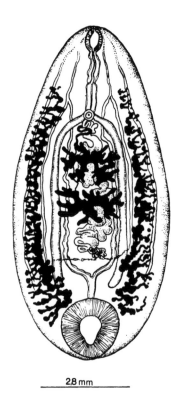

FIGURE 291.   *Nematophila grandis* (Diesing, 1839).
(After C. E. Caballero et al. 1958.)

FIGURE 292.   *Nematophila venezeulensis*
(Cordero et Vogelsang, 1940). (After E. H.
Cordero and E. G. Vogelsang, 1940.)

Nematophila venezuelensis **(Cordero et Vogelsang, 1940) Yamaguti (1958) (Figure 292)**
  Syn. *Alassostoma venezuelensis* Cordero et Vogelsang, 1940.
  Diagnosis: Body length 9.6—10.1 mm, body width 4.3—4.8 mm. Acetabulum 1.50—1.80
by 1.70—1.78 mm, Venezuelensis type number of muscle units, d.e.c., 20—22; d.i.c.,
60—63; v.e.c., 12—14; v.i.c., 62—65; m.e.c., 23—25 (Figure 160). Pharynx 0.66—0.75
by 0.57—0.66 mm, Nematophila type (Figure 28); esophagus 1.8—1.9 mm in length.
Anterior testis 0.86—1.17 by 1.2—1.5 mm, posterior testis 0.90—1.95 by 1.30—1.80 mm.
Cirrus sac 0.60 by 0.11 mm. Ovary 0.45—0.48 by 0.30—0.36 mm. Terminal genitalium
Scleroporum type (Figure 76). Eggs 0.136 by 0.076 mm.
Biology: Not yet known.
Habitat: Intestine.
Host: *Podocnemis* sp. (Pelomedusidae).

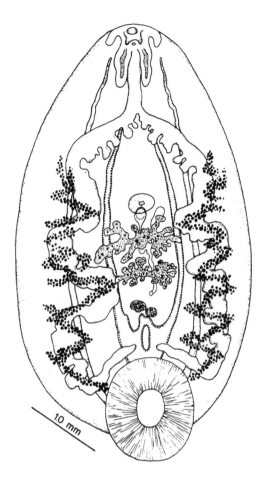

FIGURE 293. *Pseudoalassostoma heteroxenum* (Cordero et Vogelsang, 1940). (After E. H. Cordero and E. G. Vogelsang, 1940.)

Distribution: South America — Venezuela.
Reference: Cordero and Vogelsang.[182]

KEY TO THE SPECIES OF *NEMATOPHILA*

1. Terminal genitalium and acetabulum Nematophila type.................... *N. grandis*
   Terminal genitalium Scleroporum, acetabulum Alassostoma type.... *N. venezuelensis*

Pseudoalassostoma **Yamaguti, 1958**
   Diagnosis: Nematophilinae. Body conical, oval. Acetabulum ventroterminal. Pharynx terminal with extramural appendages; esophagus short, without bulb; ceca serpentine, extending to acetabulum; cecal ends forward medially. Testes tandem, branched, in middle part of body. Cirrus sac present. Ovary median, spherical. Genital pore well behind bifurcation. Uterine coils mainly behind testes. Intestinal parasites of freshwater turtles.

Type species: *Pseudoalassostoma heteroxenum* (Cordero et Vogelsang, 1940) Yamaguti 1958 (Figure 293)
   Syn. *Cladorchis heteroxenus* Cordero et Vogelsang, 1940.
   Diagnosis: With characters of the genus. Body length 5.7 mm, body width 2.9 mm.

Acetabulum 1.3 mm in diameter. Nematophila type number of muscle units, d.e.c., 60—65; d.i.c., 46—48; v.e.c., 38—40; v.i.c., 38—42 (Figure 155). Pharynx 0.30—0.42 mm long, Amurotrema type (Figure 36). Esophagus 0.84 mm in length. Anterior testis 0.54 by 0.78 mm, posterior testis 0.48 by 0.84 mm. Cirrus sac 0.60 by 0.11 mm. Ovary 0.27 by 0.14 mm. Terminal genitalium Scleroporum type (Figure 76). Size of eggs was not given.
Biology: Not yet known.
Habitat: Intestine.
Host: *Podocnemis* sp. (Pelomedusidae).
Distribution: South America — Venezuela.
Reference: Cordero and Vogelsang.[182]

Halltrema Lent et Freitas, 1939

Diagnosis: Nematophilinae. Body, elongated oval. Acetabulum ventroterminal. Pharynx terminal with extramural appendages; esophagus short, without bulb; ceca undulating, terminate near to acetabulum. Testes tandem, deeply lobed, in middle part of body. Ovary median, near posterior testis. Genital pore median, far beyond bifurcation, near anterior testis; cirrus sac present. Uterine coils median in intercecal field. Vitelline follicules numerous, along ceca, beginning before genital pore to cecal ends. Intestinal parasites of freshwater turtles.

Type species: *Halltrema avitellinum* Lent et Freitas, 1939 (Figure 294)

Diagnosis: With characters of the genus. Body length 7.5—11.4 mm, body width 3.5—4.4 mm. Acetabulum 1.31—1.60 mm in diameter. Pharynx 0.63—0.79 mm long, appendages 0.37 mm long; esophagus 1.60—2.10 mm in length. Anterior testis 0.79—1.70 by 1.30—1.80 mm, posterior testis 1.10—1.40 by 1.70—2.20 mm. Cirrus sac 0.26 by 0.30 mm. Ovary 0.20—0.30 mm in diameter. Eggs 0.128—0.136 by 0.064—0.068 mm.
Biology: Not yet known.
Habitat: Stomach and intestine.
Hosts: *Podocnemis expansa* (Pelomedusidae); *Testudo denticulata* (Testudinidae).
Distribution: South America — Brazil, Venezuela.
References: Alho,[20] Caballero et al.,[151] Freitas and Lent,[325] Lent and Freitas.[538]

Parachiorchis Caballero, 1943

Diagnosis: Nematophilinae. Body elongated oval. Acetabulum terminal small. Pharynx terminal with extramural appendages, esophagus short, with bulb; ceca winding, extending to acetabulum. Testes tandem, branched in middle part of body. Ovary median, postequatorial. Genital pore postbifurcal, cirrus sac present. Uterine coils intercecal, dorsal to testes weakly developed. Vitelline follicules along and lateral to ceca from anterior testis to cecal ends, confluent posteriorly. Intestinal parasites of freshwater turtles.

Type species: *Parachiorchis parviacetabularis* Caballero, 1943 (Figure 295)

Diagnosis: With characters of the genus. Body length 2.5—3.9 mm, body width 1.2—1.8 mm. Acetabulum 0.10—0.20 by 0.43—0.56 mm. Pharynx 0.07—0.26 by 0.14—0.22 mm; appendages 0.09—0.10 mm long, esophagus 0.23—0.40 mm in length. Anterior testis 0.16—0.32 by 0.31—0.56 mm, posterior testis 0.18—0.43 by 0.26—0.45 mm. Cirrus sac 0.10—0.26 by 0.08—0.11 mm. Ovary 0.06—0.09 by 0.06—0.10 mm. Eggs 0.088 by 0.060 mm.
Biology: Not yet known.
Habitat: Intestine.
Host: *Dermatemys mawii* (Dermatemydidae).
Distribution: Central America — Mexico.
Reference: Caballero.[145]

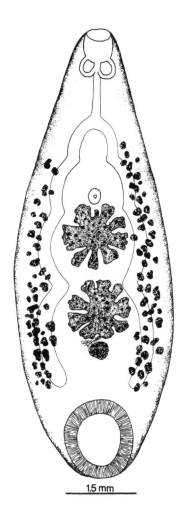

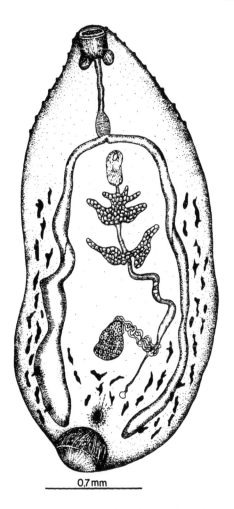

1,5 mm

0,7 mm

FIGURE 294. *Halltrema avitellinum* Lent
et Freitas, 1939. (After J. F. T. Freitas and
H. Lent, 1942.)

FIGURE 295. *Parachiorchis parviacetabu-
laris* Caballero, 1943. (After C. E. Caballero,
1943.)

Caballerodiscinae Sey, 1988

Diagnosis: Cladorchiidae. Body elongated oval. Acetabulum ventroterminal. Pharynx terminal with extramural appendages, esophagus short with well developed bulb; ceca undulating, sinuous, terminating in front of acetabulum. Testes horizontal or slightly diagonal, intercecal, entire or irregular in outline. Ovary submedian, small, situated well apart from acetabulum. Genital pore pre- or postbifurcal, cirrus sac present. Uterine coils mainly posttesticular, passing between two testes. Vitelline follicules extending mostly along posterior portion of ceca. Parasitic in digestive tract of freshwater turtles.

KEY TO THE GENERA OF CABALLERODISCINAE

1. Enormous in size, genital pore prebifurcal .... *Caballerodiscus*
   Size not enormous, genital pore postbifurcal .... *Elseyatrema*

Type genus: *Caballerodiscus* Sey, 1988
Diagnosis: Caballerodiscinae. Body elongated. Acetabulum ventroterminal. Pharynx terminal, with extramural appendages, esophagus short, ceca convoluted, sinuous. Testes

horizontal or slightly diagonal, in middle part of body, lobate or irregular in outline. Cirrus sac present. Ovary submedian, spherical, before acetabulum. Uterine coils posttesticular. Genital pore prebifurcal. Vitelline follicules along posterior part of ceca. Intestinal parasite of freshwater turtles.

Type species: *Caballerodiscus tabascensis* (Caballero et Sokoloff, 1934) Sey, 1988 (Figure 296)

Syn. *Schizamphistomoides tabascensis* Caballero et Sokoloff, 1934.

Diagnosis: Body length 8.0—14.9 mm, body width 2.0—3.7 mm. Acetabulum 1.00—1.50 mm in diameter. Pharynx 0.90—1.10 by 0.40—0.50 mm, appendages 0.35—0.31 by 0.12 mm, esophagus 1.40—1.70 mm long. Right testis 0.24—0.39 by 0.66-0.56, left testis 0.41—0.49 by 0.62—0.66 mm. Ovary 0.20 mm in diameter. Cirrus sac 0.52 by 0.31 mm. Eggs 0.120 by 0.080 mm.

Biology: Not yet known.

Habitat: Large intestine.

Hosts: *Dermatemys mawii* (Dermatemydidae); *Kinosternon leucostomum* (Kinosternidae); *Pseudemys scripta* (Emydidae).

Distribution: Central America — Mexico, Panama.

References: Caballero et al.,[151] Caballero and Sokoloff,[149] Sey,[859] Thatcher.[970]

Caballerodiscus resupinatus (Caballero, 1940) Sey, 1988 (Figure 297)

Syn. *Schizamphistomoides resupinatus* Caballero, 1940.

Diagnosis: Body length 12.6—16.1 mm, body width 4.0—4.4 mm. Acetabulum 1.80—2.10 mm in diameter. Pharynx 0.25—0.48 by 0.48—0.87 mm, appendages 0.87—1.40 mm in length; esophagus 0.78—0.99 by 0.42—0.48 mm. Right testis 0.53—0.85 by 0.62—0.85 mm, left testis 0.62—0.85 by 0.64—0.78 mm. Cirrus sac 0.52 by 0.31 mm. Ovary 0.33 by 0.23—0.72 mm. Eggs 0.140—0.150 by 0.07 mm.

Biology: Not yet known.

Host: *Dermatemys mawii* (Dermatemydidae).

Distribution: Central America — Mexico.

References: Caballero,[143] Caballero et al.,[151] Sey,[859] Thatcher.[970]

KEY TO THE SPECIES OF *CABALLERODISCUS*

1. Testes horizontal, genital pore prebifurcal .... *C. resupinatus*
    Testes slightly diagonal, genital pore at middle part of esophagus ..... *C. tabascensis*

Elseyatrema **Rohde, 1984**

Diagnosis: Caballerodiscinae. Body elongated oval. Acetabulum small, ventroterminal. Pharynx terminal with extramural appendages, esophagus short with bulb; ceca undulating, terminating in front of acetabulum. Testes horizontal, intercecal, preequatorial. Ovary lobate, intercecal. Genital pore well behind bifurcation, cirrus sac present. Uterine coils between testes and ovary, passing between two testes. Vitelline follicles along posterior portion of ceca, terminating at about cecal ends. Intestinal parasites of freshwater turtles.

Types species: *Elseyatrema microacetabulare* **Rohde, 1984 (Figure 298)**

Diagnosis: With characters of the genus. Body length 2.4—3.2 mm, body width 0.8—0.9 mm. Acetabulum 0.33—0.36 by 0.30—0.35 mm. Pharynx 0.24—0.28 by 0.14—0.16 mm; esophagus 0.23—0.36 mm long. Testes 0.24—0.34 by 0.19—0.24 mm. Cirrus sac 0.40 by 0.30 mm. Ovary 0.09—0.13 by 0.08—0.10 mm. Eggs 0.090—0.120 by 0.060—0.080 mm.

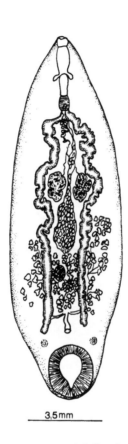

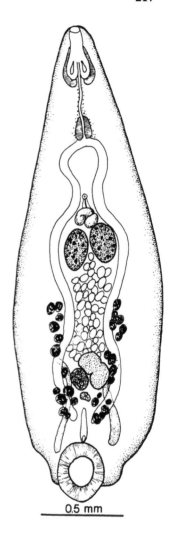

FIGURE 296. *Caballerodiscus tabascensis* (Caballero et Sokoloff, 1934). (After C. E. Caballero and D. Sokoloff, 1934.)

FIGURE 297. *Caballerodiscus resupinatus* (Caballero, 1940). (After C. E. Caballero, 1940.)

FIGURE 298. *Elseyatrema microacetabulare* (Rohde, 1984). (By permission of Dr. W. Junk Publishers, 1984, Dordrecht.)

Biology: Not yet known.
Habitat: Intestine.
Host: *Elseya dentata* (Chelidae).
Distribution: Australia.
Reference: Rohde.[783]

Solenorchiinae (Hilmy, 1949) Yamaguti, 1958

Diagnosis: Cladorchiidae. Body elongate, flattened with transverse terminal lobe. Acetabulum subterminal, with semi lunar muscular cushion anterodorsally. Pharynx subterminal with intramural appendages, esophagus short, bulb weakly developed; ceca sinuous, ribbonlike walls, extending to anterior edge of acetabulum. Testes tandem, transverse oval, in middle third of body, separated by excretory bladder. Ovary posttesticular nearer to posterior testis. Genital pore just behind bifurcation, cirrus sac present. Uterine coils dorsal to testes. Vitelline follicules well developed, extending in lateral fields from pharynx to acetabulum. Parasitic in cecum of dugong.

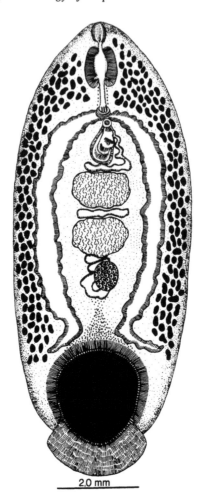

2,0 mm

FIGURE 299. *Solenorchis travassosi*
Hilmy, 1949. (Courtesy of H. Cruz.)

Type genus: *Solenorchis* Hilmy, 1949
   Diagnosis: With characters of the subfamily.

Type species: *Solenorchis travassosi* Hilmy, 1949 (Figure 299)
   Syn. *Solenorchis baeri* Hilmy, 1949; *S. gohari* Hilmy, 1949; *S. naguibmahfousi Hilmy, 1949; Indosolenorchis hirundinaceus* Crusz, 1951; *Zygocotyle* sp. Dollfus, 1958.
   Diagnosis: Body length 8.0—15.7 mm, body width 1.9—4.8 mm. Acetabulum 1.6—3.0 by 1.5—2.9 mm, Solenorchis type, number of muscle units, d.e.c., 66—68; d.i.c., 116—118; v.e.c., 38—42; v.i.c., 92—95; m.e.c., 11-14 (Figure 155). Pharynx 0.79—1.34 mm long; Solenorchis type (Figure 33). Esophagus 0.90—2.12 mm in length. Anterior testis 0.57—2.25 by 0.64—2.2, posterior testis 0.57—2.20 by 0.58—2.34 mm. Ovary 0.30—0.71 by 0.26—0.73 mm. Cirrus sac 0.70—0.75 by 0.51 mm, terminal genitalium Scleroporum type (Figure 76). Eggs 0.12—0.4 by 0.08—0.09 mm.
Biology: Not yet known.
Habitat: Cecum.
Host: *Dugong dugong* (Dugongidae).
Distribution: Africa — Egypt, Israel, Kenya, Djibouti; Asia — Indonesia, Japan, Jemen, Sri-Lanka; Australia, Papua New Guinea.

References: Beck and Forrester,[76] Blair,[96] Crusz,[190] Crusz and Fernand,[193] Dollfus,[239] Hilmy,[417] Sey.[844]

Chiorchiinae, Sey, 1988

Diagnosis: Cladorchiidae. Body elliptical or conical, flattened ventrally and convex dorsally. Acetabulum ventroterminal or ventral. Pharynx terminal or subterminal, appendages extramural or intramural; esophagus short with or without bulb; ceca broad, terminating at acetabulum. Testes tandem, in middle part of body, lobate or entire. Ovary submedian, near acetabulum. Genital pore bifurcal, cirrus sac present. Uterine coils between bifurcation and ovary. Vitelline follicules well developed extending along whole ceca or posterior portion of ceca to acetabulum. Intestinal parasites of mammals.

KEY TO THE GENERA OF CHIORCHIINAE

1. Testes entire in outline... 2
   Testes irregular in outline... *Chiorchis*
2. Vitelline follicules along whole body length........................... *Parabatrema*
   Vitelline follicules in posterior half of body .... *Chiostichorchis*

Type genus: *Chiorchis* Fischoeder, 1901
   Diagnosis: Chiorchiinae. Body elliptical, flattened ventrally and convex dorsally. Acetabulum ventral. Pharynx subterminal, appendages intramural, esophagus long with bulb; ceca extending to acetabulum. Testes tandem, deeply lobed, in middle third of body. Ovary median. Genital pore bifurcal, cirrus sac present. Uterine coils dorsal to testes. Vitelline follicules situated between esophageal bulb and acetabulum. Intestinal parasite of manatees.

Type species: *Chiorchis fabaceus* (Diesing, 1838) Fischoeder, 1901 (Figure 300)

Syn. *Amphistoma fabaceus* Diesing, 1838; *Schizamphistoma manati* Sokoloff et Caballero, 1932.
   Diagnosis: Body length 4.0—14.0 mm, body width 2.5—3.7 mm. Acetabulum 1.25—1.65 mm in diameter. Cladorchis type, number of muscle units, d.e.c., 45—48; d.i.c., 24—26; v.e.c., 5—8; v.i.c., 26—29; m.e.c., 8—10 (Figure 138). Pharynx 0.50—0.65 mm in diameter, Microrchis type (Figure 43), esophagus 2.18—2.48 mm long. Testes 0.40—0.50 by 0.60—0.80 mm. Ovary 0.23—0.32 mm in diameter. Cirrus sac 0.29—0.38 by 0.19—0.22 mm, terminal genitalium Scleroporum type (Figure 76). Eggs 0.140—0.150 by 0.082—0.086 mm.
Biology: Not yet known.
Habitat: Large intestine.
Hosts: *Trichechus inunguis, T. manatus, T. senegalensis* (Trichechidae).
Distribution: Africa — Zaire; South America — Brazil; North America — U.S.
References: Beck and Forrester,[76] Bravo-Hollis and Caballero Deloya,[119] Sokoloff and Caballero,[899] Stunkard.[925]

Chiostichorchis Artigas et Pacheco, 1933

Diagnosis: Chiorchiinae. Body conical. Acetabulum ventroterminal with longitudinal aperture. Pharynx terminal with extramural appendages; esophagus short, without bulb; ceca somewhat sinuous, extending to acetabular zone. Testes tandem, spherical, in middle third of body. Ovary round, at acetabulum. Genital pore bifurcal, genital sucker and cirrus sac present. Uterine coils convoluted, dorsal to ovary. Vitelline follicules lateral from testicular zone to acetabular zone. Parasitic in cecum of rodents.

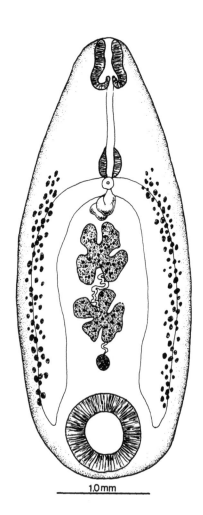 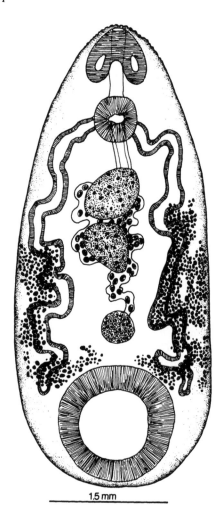

FIGURE 300. *Chiorchis fabaceus* (Diesing, 1838).

FIGURE 301. *Chiostichorchis waltheri* Sprehn,
1932. (After B. B. Babero, 1953.)

Type species: ***Chiostichorchis waltheri* (Sprehn, 1932) Travassos, 1934 (Figures 301
and 302)**

Syn. *Stichorchis waltheri* (Sprehn, 1932); *S. myopotami* Artigas et Pachecho, 1933.

Diagnosis: With characters of the genus. Body length 7.0—9.5 mm, body width 4.5—8.5
mm. Acetabulum 1.81—2.30 mm in diameter, Chiostichorchis type, number of muscle
units, d.e.c., 46—48; d.i.c., 43—45; v.e.c., 42—44; v.i.c., 81—83; m.e.c., 12—14 (Figure
136). Pharynx 0.52—0.92 mm long, Megalocotyle type (Figure 41), appendages 1.02—1.08
mm in length; esophagus 0.29—0.32 mm long. Anterior testis 0.20—0.80 by 1.08—1.24
mm; posterior testis 1.08—1.35 by 1.02—1.62 mm. Ovary 0.25—0.40 by 0.50—0.63 mm.
Terminal genitalium Waltheri type (Figure 79). Eggs 0.150—0.160 by 0.090—0.140 mm.
Biology: Not yet known.

Host: *Myocaster coypus* (Myocasteridae).

Distribution: South America — Brazil, Argentina.

References: Artigas and Pacheco,[38,39] Boero and Boehringer,[102] Grieder,[375] Martinez,[593]
Pereira,[711] Sprehn.[905]

Paraibatrema **Ueta, Deberaldini, Cordeiro et Artigas, 1981**

Diagnosis: Chiorchiinae. Body elongate, flattened, ventral side concave, dorsal convex.

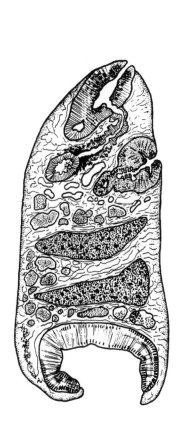

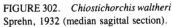

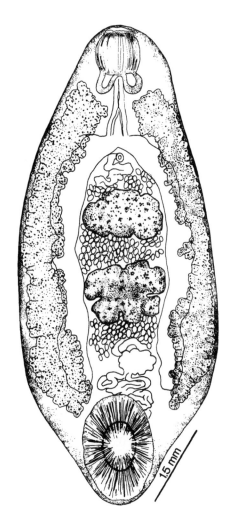

FIGURE 302. *Chiostichorchis waltheri*
Sprehn, 1932 (median sagittal section).

FIGURE 303. *Paraibatrema inesperatum* (Ueta,
Deberaldini, Cordero et Artigas, 1981). (Courtesy
M. T. Ueta et al.)

Acetabulum ventroterminal. Pharynx terminal with extramural appendages, esophagus short
with bulb, ceca long, broad, somewhat sinuous, extending to acetabulum. Testes tandem,
lobate, in middle third of body. Ovary submedian, irregular in outline, at halfway between
posterior testis and acetabulum. Genital pore bifurcal, cirrus sac not observed. Uterine coils
dorsal to testes, between ovary and cecal arch. Vitelline follicules strongly developed, along
lateral side from pharynx to acetabulum. Intestinal parasites of rodents.

**Type species: *Paraibatrema inesperatum* Ueta, Deberaldini, Cordeiro et Artigas, 1981
(Figure 303)**

Diagnosis: With characters of the genus. Body length 7.1—11.4 mm, body width 2.5—4.6
mm. Acetabulum 1.10—1.50 by 1.30—1.80 mm. Pharynx 0.50—0.80 mm in diameter,
appendages 0.30—0.40 mm long; esophagus 0.80—1.40 mm in length. Eggs 0.143—0.158
by 0.095—0.113 mm.

Biology: Not yet known. This species was described based on the material which originated
from artificially infected mice and rats. Cercariae were collected from planorbid snails,
Biomphalaria tenagophila. Prepatent period was 20 d. *Biomphalaria glabrata* was apparently susceptive to this fluke. Larval stages were not studied in details.

Habitat: Large intestine.
Hosts: *Mus musculus* f. domestica (exp.), *Rattus norvegicus* f. domestica (exp.) (Muridae).
Distribution: South America — Brazil.
Reference: Ueta et al.[1001]

Pfenderiinae Fukui, 1929

Diagnosis: Cladorchiidae. Body conical to elongate. Acetabulum ventroterminal, relatively large, sometimes covered with papillae on inner surface. Pharynx terminal with extramural appendages, esophagus short, bulb present, rarely absent, ceca long, undulating, extending to acetabular zone; diverticules may be found along inner line of ceca. Testes horizontal, intercecal or partly cecal, irregular in outline, in middle third of body. Ovary median at acetabulum. Genital pore bifurcal or prebifurcal, cirrus sac strongly developed. Uterine coils posttesticular, passing between two testes. Vitelline follicules extending along cecal from bifurcation to cecal ends. Intestinal parasites in elephants.

Type genus: *Pfenderius* Stiles et Goldberger, 1910
   Syn. *Tagumaea* Fukui, 1926
Diagnosis: With characters of the subfamily.

Type species: *Pfenderius papillatus* (Cobbold, 1882) Stiles et Goldberger, 1910 (Figure 304)

Syn. *Amphistoma papillatus* Cobbold, 1882; *A. papillatus* Fischoeder, 1903, *A. papillatus* Maplestone, 1923
Diagnosis: Body length 4.5—5.6 mm, body width 2.5—2.7 mm. Acetabulum 1.70 by 1.40 mm, Pfenderius type, number of muscle units, d.e.c., 47—49; d.i.c., 24—27; v.e.c., 5—8; v.i.c., 27—29; m.e.c., 8—11 (Figure 150). Pharynx 0.80—1.00 mm long, Megacotyle type (Figure 41), appendages 0.17—0.21 mm long; esophagus 1.00 mm long. Right testis 0.32—0.35 mm long and 0.57—0.63 mm in dorsoventral direction. Cirrus sac 0.48 mm in length, ovary 0.19—0.22 by 0.13—0.16 mm. Terminal genitalium Scleroporum type (Figure 76). Genital pore bifurcal. Eggs 0.120—0.130 by 0.600—0.700 mm.
Biology: Not yet known.
Habitat: Colon.
Host: *Elephas maximum* (Elephantidae).
Distribution: Asia — India, Burma, Malaysia.
References: Fernando,[308] Fukui,[333] Malik et al.,[583] Westhuysen.[1052]

Pfenderius heterocaeca (Fukui, 1926) Yamaguti, 1958 (Figure 305)

Syn. *Tagumaea heterocaeca* Fukui, 1926.
Diagnosis: Body length 3.1—4.8 mm, body width 2.5—3.0 mm. Acetabulum 1.20—1.40 by 1.00 mm in size, Pyriformis type, number of muscle units, d.e.c., 28—35; d.i.c., 29—34; v.e.c., 19—31; v.i.c., 33—48; m.e.c., 6—8 (Figure 152). Pharynx 0.61—0.78 by 0.48—0.50 mm; Pfenderius type (Figure 29); esophagus 1.50 mm in length with bulb. Testes 0.40—0.51 by 0.33—0.42 mm. Cirrus sac 0.62—0.82 by 0.33—0.70 mm. Ovary 0.23—0.33 by 0.20—0.30 mm. Terminal genitalium Scleroporum type (Figure 76). Genital pore prebifurcal. Eggs 0.130—0.140 by 0.070—0.080 mm.
Biology: Not yet known.
Habitat: Intestine.
Host: *Elephas maximus* (Elephantidae).
Distribution: Asia - Burma, India.
References: Bhalerao,[89] Fukui,[333,334] Malik et al.,[583] Westhuysen.[1052]

Pfenderius birmanicus Bhalerao, 1935 (Figure 306)

Diagnosis: Body length 2.1—2.3 mm, body width 2.1 mm. Acetabulum 0.92—1.02 mm

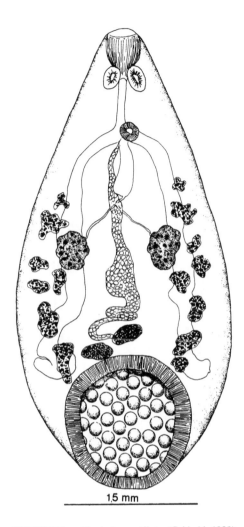

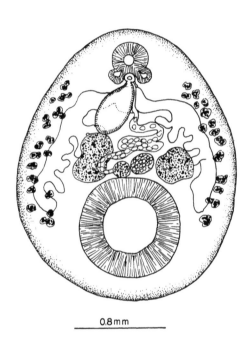

FIGURE 304. *Pfenderius papillatus* (Cobbold, 1882). (After R. Vercruysee, 1950.)

FIGURE 305. *Pfenderius heterocaeca* (Fukui, 1926). (After T. Fukui, 1926.)

in diameter. Pharynx 0.93 mm long, appendages 0.2 mm in length; esophagus 0.52 mm in length with bulb. Testes 0.28—0.33 by 0.31—0.46 mm. Cirrus sac 0.50 by 0.30 mm in size. Ovary 0.25 by 0.20 mm. Genital pore prebifurcal. Eggs 0.160 by 0.090 mm.
Biology: Not yet known.
Habitat: Intestine.
Host: *Elephas maximux* (Elephantidae).
Distribution: Asia — Burma, India.
References: Bhalerao,[89] Westhuysen.[1052]

KEY TO THE SPECIES OF *PFENDERIUS*

1. Acetabulum with papillae on inner surface............................ *P. papillatus*
   Acetabulum without papillae.. 2
2. Ceca divided into anterior broad and posterior short portions......... *P. heterocaeca*
   Ceca diverticulated intervally .... *P. birmanicus*

Stichorchiinae Näsmark, 1937
   Diagnosis: Cladorchiidae. Body elongated, somewhat curved ventrally. Acetabulum ven-

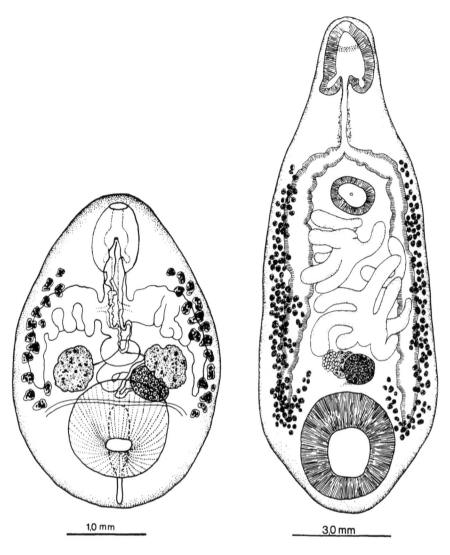

FIGURE 306.   *Pfenderius birmanicus* Bhalerao, 1935. (After G. S. Bhalerao, 1935.)

FIGURE 307.   *Stichorchis giganteus* (Diesing, 1836). (After F. Fischoeder, 1901.)

tral. Pharynx terminal with extramural or intramural appendages, esophagus short, without bulb; ceca voluminous, occasionally sinuous, reaching as far as acetabulum or beyond it. Testes tandem, branched, in middle third of body. Ovary median, before acetabulum. Genital pore pre- or postbifurcal, cirrus sac present. Uterine coils winding, dorsal to testes. Vitelline follicules extending along ceca between bifurcal-testiscular zone to acetabular zone and median between posterior testis and cecal ends. Intestinal parasites of rodents and pigs. Type genus: *Stichorchis* (Fischoeder, 1901) Looss, 1902

Diagnosis: With characters of the subfamily.

Type species: *Stichorchis giganteus* (Diesing, 1836) Travassos, 1922 (Figure 307)

Syn. *Amphistoma giganteus* Diesing, 1836; *Cladorchis (Stichorchis) gigantus* Fischoeder, 1901; *C. giganteus* Maplestone, 1923.

Diagnosis: Body length 16.0—17.0 mm, body width 4.8—6.4 mm. Acetabulum 2.82—3.16 mm in diameter, Giganteus type, number of muscle units, d.e.c., 18—20; d.i.c., 45—47; v.e.c., 14—16; v.i.c., 46—49; m.e.c., 14—16 (Figure 145). Pharynx 1.8—2.0 by 1.5—1.7

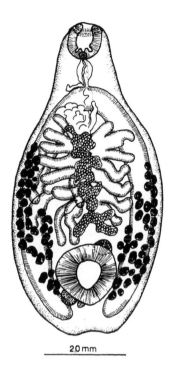

2.0 mm

FIGURE 308. *Stichorchis subtri-quetrus* (Rudolphi, 1814). (After F. Fischoeder, 1901.)

mm, Stichorchis type (Figure 35); appendages 0.84—1.13 mm long; esophagus 2.11—2.56 mm long. Testes 3.54 by 3.23 mm. Cirrus sac 1.12—1.67 mm in dorsoventral direction. Ovary 0.78 mm in diameter. Genital sucker present. Terminal genitalium Cladorchis type (Figure 72). Genital pore postbifurcal. Size of eggs 0.143—0.152 by 0.076—0.084 mm.

Biology: Not yet known.

Habitat: Large intestine.

Hosts: *Sus scrofa* f. domestica (Suidae); *Tayassus tajacu T. albirostris* (Tayassuidae); *Dasyprocta agouti* (Dasyproctidae); *Tamandua longicaudata* (Myrmecophagidae).

Distribution: South America — Brazil, Venezuela, and Trinidad.

References: Diaz-Ungria,[218] Diesing,[221] Fernandez and Travassos,[307] Fischoeder,[311] Freitas and Costa,[327,328] Oliveira Sobrinko,[673] Pinto and Almeida,[728] Stiles and Goldberger,[915] Travassos,[980] Vaz,[1017] Viana,[1032] Vogelsang,[1036] Vogelsang and Rodriguez.[1038]

Stichorchis subtriquetrus (Rudolphi, 1814) Looss, 1902 (Figure 308)

Syn. *Amphistoma subtriquetrus* Rudolphi, 1814; *A. subtriquetrus* Wertsumb, 1823; *Distoma amphistomoides* Bojanus, 1817; *Cladorchis (Stichorchis) subtriquetrus* Fischoeder, 1901; *Paramphistomum castori* Kofoid et Park, 1937.

Diagnosis: Body length 4.0—12.0 mm, body width 2.0—5.0 mm. Acetabulum 2.24—2.67 mm in diameter, Stichorchis type, number of muscle units, d.e.c., 12—40; d.i.c., 20—22; v.e.c., 6—8; v.i.c., 22—25; m.e.c., 20—22 (Figure 157). Pharynx 1.52—1.64 by 0.98—1.14 mm, Stichorchis type (Figure 35). Esophagus 0.51—0.59 mm long. Testes 1.98—2.24 by 1.53—1.76 mm in outline. Ovary 0.63—0.84 mm in diameter. Cirrus sac 0.51—0.56 by 0.21—0.28 mm, terminal genitalium Stichorchis type (Figure 78). Genital pore prebifurcal. Eggs 0.164—0.186 by 0.093—0.118 mm.

Biology: References. Bennett and Humes,[81] Bennett and Allison,[82] Orlov.[675]

Preparasitic stage: ① Bennett and Humes. Deposited eggs contained embryo surrounded by vitelline cells. At the end of the first week-long incubation (in tap water), the embryo was about 0.04 mm in diameter. At about the end of the third week, miracidia were physiologically ready to hatch. Miracidia 0.18 by 0.06 mm. Epithelial cells situated in four rows arranged by the formula 6:8:4:2. Miracidia composed of "primitive gut" (cephalic gland), two large flame cells. Penetration glands were not observed. The posterior half of body was occupied by a single redia. Redia measured 0.11 by 0.06 mm. Intermediate host *Fossaria* (= *Lymnaea*) *parva*. ② Orlov. In eggs, cultivated in river water at 20 to 24°C, miracidia became mature in 3 or 4 weeks. Life span of miracidia was about 14 h measuring 0.17 by 0.06 mm. Hatched miracidia contained rediae, 0.53 by 0.04 mm in size. Intermediate host: *Planorbis vortex, Lymnaea ovata*, rarely *Bithynia tentaculata* and *Succinea putris*. ③ Bennett and Allison. Eggs in lake water, hatched in 20 to 27 d at room temperature.

Intramolluscan stage. ① Bennett and Humes. Mature rediae appeared on the 21st day after penetration of miracidia into snails. Mature rediae measuring 0.6 by 0.3 mm in size; birth pore situated ventrally at level of esophagus, three pairs of flame cells present, containing 15 to 17 daughter rediae in various stages of development. ② Orlov. Rediae, after penetration of the miracidia, reached the liver by the end of the 24th hour. Rediae matured in 12 to 21 d post-infection, and during that period they developed two penetration glands (in contrast with any other one of the known amphistome rediae). Cercariae reached their maturity in 35 to 45 d after infestation, and they emerged during this period. Body of cercariae measuring 28 by 0.22 mm, tail 0.80 by 0.08 mm. Pigmented eyes prominent, pharynx 0.05 mm in diameter, acetabulum 0.08 mm in diameter. Ceca extending to acetabulum. Caudal tube opens laterally near the extremity. Descending trunks undulating without cross-connecting tubes. Life span of cercariae in water up to 18 h. Encystment usually took place within 24 h after emergence. Cyst: 0.12 mm in diameter. Life span of adolescariae was about 3 months. ② Bennett and Allison. Emergence of cercariae took place 50 d post-infection.

Habitat: Intestine, cecum, colon.

Hosts: *Castor fiber, C. canadensis* (Castoridae); *Bos primigenius* f. taurus (Bovidae); *Ondatra zibethica, Microtus arvalis* (Cricetidae).

Distribution: North America — Canada, U.S.; Central America — Mexico; Europe — U.S.S.R., Germany, Poland, Norway, France.

References: Babero,[51] Bakke,[61] Bush and Samuel,[138] Caballero,[146] Erickson,[290] Fischoeder,[311] Joszt,[456] Kiselene and Mitskus,[487] Kofoid and Park,[494] Romashov,[789] Rudolphi,[796] Sey,[837] Wiedmann.[1054]

Cladorchiinae (Fischoeder, 1901) Lühe, 1909

Diagnosis: Cladorchiidae. Body pyriform or elliptical. Acetabulum ventroterminal, ventral, or terminal. Pharynx terminal with extramural appendages; esophagus short without bulb; ceca straight or strongly convoluted. Testes tandem or horizontal, deeply lobate or branched, pre-equatorial or in middle part of body, extends to acetabular zone or terminates before it. Ovary median or submedian. Genital pore bifurcal or postbifurcal, genital sucker and cirrus sac present. Uterine coils intercecal, dorsal to testes. Vitelline follicules lateral from genital pore to acetabulum or posttesticular. Intestinal parasites of mammals.

KEY TO THE GENERA OF CLADORCHIINAE

1. Testes in equatorial zone, uterine coils dorsal to testes, acetabulum ventral ....
   .... *Cladorchis*
   Testes in pre-equatorial zone, uterine coils mainly posterior to testes, acetabulum terminal .... *Taxorchis*

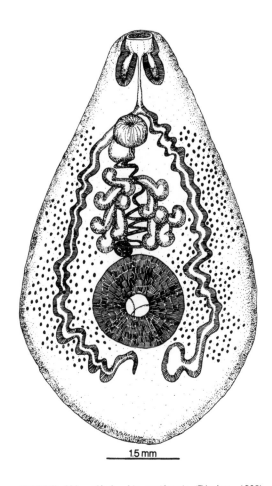

1,5 mm

FIGURE 309. *Cladorchis pyriformis* (Diesing, 1838).
(After F. Fischoeder, 1901.)

Type genus: *Cladorchis* Fischoeder, 1901

Diagnosis: Cladorchiinae. Body conical or pyriform. Acetabulum ventral or ventroterminal; esophagus short, without bulb; ceca long, winding. Testes tandem or horizontal, branched, in middle zone of body. Ovary median at acetabulum. Genital sucker and cirrus sac present. Uterine coils dorsal to testes. Genital pore bifurcal. Vitelline follicules lateral, along ceca. Intestinal parasites of tapirs.

Type species. *Cladorchis pyriformis* **(Diesing, 1838) Fischoeder, 1901 (Figure 309)**

Syn. *Amphistoma pyriforme* Diesing, 1838

Diagnosis: Body length 5.2—12.3 mm, body width 0.9—1.1 mm. Acetabulum 1.53—2.14 mm in diameter, Pyriformis type, number of muscle units, d.e.c., 32—27; d.i.c., 26—32; v.e.c., 17—31; v.i.c., 32—34; m.e.c., 6—8 (Figure 152). Pharynx 0.42—0.53 mm long, Cladorchis type (Figure 38); appendages 1.12—1.26 mm long, esophagus 1.11—1.23 mm in length. Testes 1.42—1.76 by 1.73—1.94 mm in size. Cirrus sac 1.78—1.89 by 2.84—3.19 mm. Ovary 0.48—0.54 mm in diameter. Terminal genitalium Cladorchis type (Figure 72). Eggs 0.14—0.153 by 0.076—0.087 mm.

Biology: Not yet known.

Habitat: Cecum.

Hosts: *Tapirus bairdi* (Tapiridae); *Dasyprocta agouti* (Dasyproctidae)

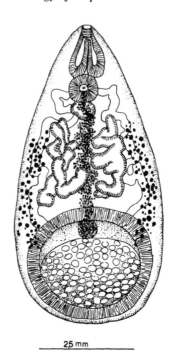

FIGURE 310. *Cladorchis asper*
(Diesing, 1838). (After F. Fischoeder, 1903.)

Distribution: South America — Brazil, Surinam.
References: Diesing,[223] Fischoeder,[311] Näsmark,[647] Prudhoe.[755]

Cladorchis asper (Diesing, 1838) Stiles et Goldberger, 1910 (Figure 310)

Syn. *Amphistoma asperum* Diesing, 1938.

Diagnosis: Body length 4.11—11.6, body width 3.1—3.8 mm. Acetabulum 3.26—4.58 mm in diameter, Asper type, number of muscle units, d.e.c., 15—20; d.i.c., 38—60; v.e.c., 12—18; v.i.c., 42—45; m.e.c., 6—8 (Figure 133), inner surface covered with papillae. Pharynx 0.50—0.80 mm in length, Cladorchis type (Figure 38), appendages 0.80 mm long; esophagus 1.50—1.80 long. Testes 2.00—2.20 by 0.90—1.10 mm in size. Cirrus sac 1.70—1.90 by 1.90—2.2 mm. Ovary 0.40—0.60 mm in diameter. Terminal genitalium Cladorchis type (Figure 72). Eggs 0.140—0.150 by 0.070—0.080 mm.

Biology. Not yet known.

Habitat: Cecum.

Host: *Tapirus bairdi* (Tapiridae).

Distribution: South America — Brazil.

References: Diesing,[222] Fischoeder,[311] Näsmark.[647]

KEY TO THE SPECIES OF *CLADORCHIS*

1. Acetabulum subterminal, inner surface with papillae .... *C. asper*
   Acetabulum ventral, inner surface without papillae .... *C. pyriformis*

Taxorchis (Fischoeder, 1901) Stiles et Goldberger, 1910

Diagnosis: Body elongate to pyriform. Acetabulum terminal. Pharynx terminal with extramural appendages, esophagus short, without bulb; ceca straight or slightly undulating,

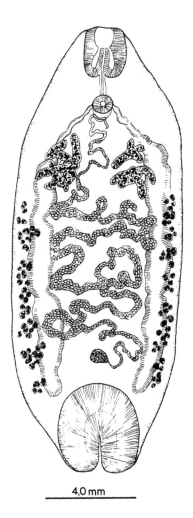

4,0 mm

FIGURE 311. *Taxorchis schistocotyle* (Fischoeder, 1901). (After F. Fischoeder, 1901.)

terminate at acetabulum or before it. Testes horizontal or diagonal, deeply lobed or branched. Ovary median near acetabulum or before it. Genital pore bifurcal or postbifurcal, genital sucker and cirrus sac present. Uterine coils occupying whole intercecal field. Vitelline follicules lateral, posttesticular. Parasitic in cecum of mammals.

Type species: *Taxorchis schistocotyle* (Fischoeder, 1901) Stiles et Goldberger, 1910 (Figure 311)

Syn. *Cladorchis (Taxorchis) schistocotyle* Fischoeder, 1901.

Diagnosis: Body length 19.5—20.5 mm, body width 7.2—8.5 mm. Acetabulum 3.84—4.23 by 2.97—3.38 mm, Taxorchis type, number of muscle units, d.e.c., 40—43; d.i.c., 28—31; v.e.c., 140—150; v.i.c., 31—33 (Figure 159). Pharynx 2.56—2.98 mm long, Taxorchis type (Figure 49), appendages 1.34—1.57 mm long, esophagus 2.96—3.58 mm in diameter. Testes 3.24—3.83 mm in diameter. Cirrus sac 1.46—1.67 mm long. Ovary 0.87—0.96 mm in length. Terminal genitalium Cladorchis type (Figure 72). Genital pore bifurcal. Eggs 0.143—0.157 by 0.078—0.086 mm.

Biology: Not yet known.

Habitat: Ceca.

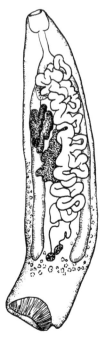

3.4mm

FIGURE 312. *Taxorchis ringueleti* Sutton, 1975. (Courtesy
of C. A. Sutton.)

Hosts: *Tayassus tajacu* (Tayassidae); *Hydrochoerus isthimus, H. hydrochoerus, H. capibara*
(Hydrochoeridae).
Distribution: South America — Brazil, Argentina; Central America — Panama.
References: Faust,[305] Fischoeder,[311] Lombardero and Moriena,[555] Travassos.[985]

Taxorchis ringueleti Sutton, 1975 (Figure 312)
Syn. *Taxorchis caviae* Kawazoe, Cordeiro et Artigas, 1981.
Diagnosis: Body length 17.0—22.0 mm, body width 3.5—4.0 mm. Acetabulum 2.10—2.60
by 1.60—1.90 mm. Pharynx 0.19—1.20 by 1.70—1.60 mm, appendages 1.20 mm long,
esophagus 1.30 mm in length. Testes 3.60 by 1.60 mm in size. Cirrus sac 0.72—1.08 mm
in diameter. Ovary 0.30—0.50 mm in size. Genital pore postbifurcal. Eggs 0.110—0.180
by 0.070 mm.
Biology: Not yet known.
Habitat: Large intestine.
Host: *Cavia aperea* (Caviidae).
Distribution: South America — Argentina, Brazil.
References: Kawazoe et al.,[474] Sutton.[932]

KEY TO THE SPECIES OF *TAXORCHIS*

1. Testes horizontal in pre-equatorial zone..............................*T. schistocotyle*
   Testes diagonal, in middle portion of body ....*T. ringueleti*

Notes: Kawazoe et al.[474] described the species *Taxorchis caviae* from *Cavia aperea* found in Brazil. Apparently, they were unaware of Sutton's[932] paper in which *T. ringueleti* was described from the same host in Argentina. Comparing the form, size, and the topography of the organs of the two species in question and their definitive hosts, we can state that they are closely related and the differences, if any, are within the scope of the individual variations. Hence, *T. caviae* is considered to be a junior synonym of *T. ringueleti*.

3.2.3.1.2. DIPLODISCIDAE Skrjabin, 1949

Diagnosis: *Cladorchidae*. Body conical or pyriform. Acetabulum terminal or ventroterminal with different muscular structures. Pharynx terminal with extra- or intramural appendages; esophagus short or long, with bulb; ceca long or short. Single testis, usually in middle third of body. Ovary posttesticular, usually median. Genital pore bifurcal or prebifurcal, cirrus sac present. Uterine coils usually intercecal, more or less developed. Eggs occasionally embryonated. Intestinal parasites of amphibians, reptiles, rarely fishes.

KEY TO THE GENERA OF DIPLODISCIDAE

1. Acetabulum with accessory sucker 2
   Acetabulum without accessory sucker 3
   Acetabulum with ruffled peduncle . . . . *Progonimodiscus*
2. Ceca terminating at anterior margin of acetabulum . . . . *Diplodiscus*
   Ceca terminating at middle part of body . . . . *Australodiscus*
3. Vitelline follicules in middle zone of body 4
   Vitelline follicules transversal at anterior margin of acetabulum . . . . *Dermatemytrema*
4. Acetabulum constricted into two positions . . . . *Catadiscus*
   Acetabulum without acetabular constriction . . . . *Pseudodiplodiscus*

Type genus: *Diplodiscus* Diesing, 1836
   Diagnosis: Body conical in shape. Acetabulum terminal with or without accessory sucker. Pharynx terminal with extra- or intramural appendages; esophagus short with bulb; ceca straight and wide; extending to acetabular zone. Testis spherical, posttesticular, near acetabulum or before it. Genital pore bifurcal, pre- or postbifurcal, cirrus sac present. Uterine coils intercaecal between ovary and genital pore. Vitelline follicules extending along ceca and sometimes confluent posteriorly. Parasitic in rectum of amphibians.

Type species: *Diplodiscus subclavatus* (Goeze, 1882) Diesing, 1836 (Figure 313)
   Syn. *Planaria subclavata* Goeze, 1982; *Amphistomum subclavatum* Looss, 1892; *Diplodiscus unguiculatus* (Rudolphi, 1819) Diesing, 1836; *Diplodiscus* sp. Honer, 1960; *Diplodiscus subclavatus paludinae* Honer, 1961.
   Diagnosis: Body length 2.1—3.5 mm, body width 1.1—1.7 mm. Acetabulum 0.81—1.36 mm in diameter, Diplodiscus type, number of muscle units, d.e.c., 23—25; d.i.c., 52—55; v.e.c., 17—19; v.i.c., 54—57; m.e.c., 8—10 (Figure 143). Pharynx 0.34—0.43 mm long, Subclavatus type (Figure 48); esophagus 0.14—0.25 mm in length. Testis 0.37—0.39 by 0.28—0.42 mm. Cirrus sac 0.18—0.20 by 0.01—0.01 mm. Ovary 0.18—0.21 by 0.13—0.15 mm. Terminal genitalium Scloeroporum type (Figure 76). Genital pore bifurcal. Eggs 0.123—0.138 by 0.082—0.097 mm.
   Biology: References. Lang,[521] Looss,[557] Rozman,[794] Sey,[833] Salami-Cadoux and Grégorio,[806] Bourgat and Kulo,[111] Bayssade-Dufour et al.,[74] Grabda-Kazubska.[355]
   Preparasitic stage. ① Looss. Eggs were fully embryonated when deposited. Eggs contain 15—20 yolk cells which liquified during the process of cleavage, and vacuoles appear and form an elongate space. Beneath the embryonic epidermal layer longitudinal

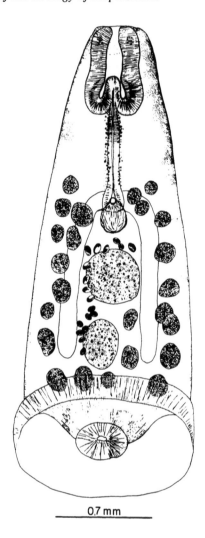

0,7 mm

FIGURE 313.   *Diplodiscus subclavatus*
(Goeze, 1882). (By permission of the *Parasitol. Hung.*, 1983, Budapest.)

and circular muscle elements developed. At this time the flame cells begin to operate,
the nervous system and the central body cavity can be observed. Germ cells and the
formed germinal balls are connected with the body wall. Intermediate hosts: *Planorbis
nitidus, P. vortex, P. marginatus, P. rotondatus, P. spirorbis, P. contortus*; ② Bourgat
and Kulo. *Bulinus forskalii, Segmentorbis kinisaensis.*

   Intramolluscan stage. ① Looss. After penetration of miracidia into snails, they begin
to develop into sporocyst in the liver of the snail. Proliferation of sporocyst by fusion,
as it was observed for other trematodes, did not occur in this species. Rediae escape from
the sporocyst in an early developmental stage (0.22 mm long). The number of rediae
produced by a sporocyst may be up to 20, the propagative period lasting a few months.
Rediae have three pairs of flame cells, excretory pore near posterior extremity; moderately
large germ balls fill up the body cavity. First cercariae appeared 8 weeks after infestation.
Eyespot was situated in each lateral side at level of esophagus. Emergence of cercariae
mostly in the morning hours with a life span of about 28 h in water. They often encyst
on the shell of snails; the cysts are deposited in the bottom zone, and the definitive hosts

become infected by ingestion of the latter. ② Lang. Cercariae usually encyst on the skin of frogs; dark spots are preferable for encystation. Infestation picked up by devouring the cast off skin. ③ Rozman. Body measurement of cercaria 0.8 by 0.39 mm. ④ Bourgat and Kulo. Body of cercaria measuring 0.33 mm long, tail 0.95 mm in length. Acetabulum 0.16 by 0.11 mm, adolescarial cyst 0.3 mm in diameter. ⑤ Bayssade-Dufour et al. The number of superficial sensory apparatus was 378—433 around the cephalic region, 74 along pharynx and appendages, 130 on the body, 19—24 around the acetabulum, and 52—76 along the tail (Table 1). On the basis of the superficial sensory apparatus, Bayssade-Dufour et al. regarded the African "*D. subclavatus*" to be *D. fischthalicus*. ⑥ Grabda-Kazubska. Rediae are small, sausagelike without collar or locomotor projections, measuring 1.0—1.3 by 0.20—0.38 mm in size. Well-developed pharynx leads into saclike intestine. The body cavity is filled with germinal balls of various size. The excretory system consists of three flame cells at each side of the body. Body of fully developed cercariae measures 0.44—0.59 by 0.20—0.28 mm. Pharynx 0.12—0.13 by 0.08—0.09 mm in size; esophagus with well developed bulb, intestinal ceca extending to anterior part of acetabulum. Tail is much longer than that of the body, measuring 0.97—1.1 mm in length at its base provided with 4 or 5 transparent cells. In anterior part of body two well developed eyespots are present. The integument is opaque due to numerous cytogenous cells. Reproductive system composed of primordia of single testis and ovary in posterior half of body. The excretory system consists of a great number of flame cells and lacunar system. Capillaries run to collecting ducts which at level of bifurcation join the main lateral ducts. These ducts at the level of eyespots turn abruptly posteriorly and open into the vesicle. It is saclike and opens outside by a minute duct. Cross-connecting tubes between the main tubes absent. The main excretory tubes contain big excretory granules. The single dilated caudal tube extending near the apex of the tail and the excretory pores are situated here. Adolescariae are flattened, oval in outline, covered with thin cyst wall. Cyst dimensions: 0.31—0.37 by 0.28—0.32 mm. They encyst on frog's skin or directly in water.

Development in definitive hosts. As to the mode of infestation of the final hosts Looss[557] was of the opinion that the cercariae encyst freely in water, accumulate on the bottom, and are swallowed by frogs with mud during hibernation. Lang[521] came to the conclusion that infestation is picked up by ingestion of the sloughed skin with encysted adolescariae. These questions were studied by Grabda-Kazubska[355] in full, and her observations coincided with correctness of Lang's statement and to a lesser extent with Looss's opinion. Although cercariae may encyst freely in water, their longevity is only a few days; hence, they might not be accumulated in great numbers. Cercariae have never shown affinity to the tadpole skin; they are usually infected by taking up and ingesting adolescariae. Adolescariae excyst in the intestine and reach their site of predilection in 48 h. In the rectum they reach their maturity in 18—48 h in juvenile frogs, and egg production was observed in infected tadpoles as early as 12 d after infestation.

Habitat: Rectum.

Host: *Triturus alpestris, T. helveticus, T. cristatus, T. vulgaris,* (Salamandridae); *Bombina bombina, B. variegata* (Discoglossidae); *Pelobates fuscus* (Pelobatidae); *Rana esculenta, R. tigrina, R. macronensis, R. perezi, R. chensinensis, R. galomensis, R. lessonae, R. ridibunda, R. dalmatina, R. temporaria* (Ranidae); *Bufo bufo, B. regularis, B. viridis* (Bufonidae); *Hyla arborea* (Hylidae) and probably, as a result of secondary transmission: *Esox lucius* (Esocidae); *Natrix natrix, N. chrysorga* (Colubridae); *Emys orbicularis* (Emydidae); *Vipera berus* (Viperidae); *Alopex lagopus* (Canidae).

Distribution: Eurasia — Albania, Czechoslovakia, Bulgaria, France, Germany, Hungary, Holland, Italy, Poland, Rumania, U.S.S.R., China, Switzerland, Spain, Yugoslavia; Africa — Tunisia.

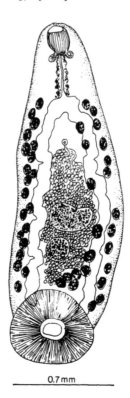

FIGURE 314. *Diplodiscus am-phichrus* Tubangui, 1933.

References: André,[30] Bailenger and Chanseau,[59] Baruš et al.,[72] Bourgat and Kulo,[111] B'tchvarov,[131—133] Căpuşe and Dancău,[158] Chen et al.,[168] Chiriac and Malcoci,[172] Chiriac and Udrescu,[173] Diesing,[221] Dollfus,[241] Dubinina,[248,249] Edelényi,[262] Fukui,[334] Kalabe-kov,[461] Kozák,[504] Lluch et al.,[554] López-Neyra,[563] Maeder et al.,[579] Markusheva and Sagalovich,[592] Milogradova and Spasskiĭ,[614] Näsmark,[647] Odening,[662] Petriashvili,[721] Pike,[726] Prokopič,[751] Prokopič and Křivanec,[753] Ryzhikov et al.,[801] Salami-Cadoux and Grégo-rio,[806] Sey,[831,849] Stoĭkova,[971] Szulc,[943] Taranko-Tulecka,[960] Vojkova,[1039,1040] Volna-Ná-bělkova.[1041]

Notes: In the literature of the authorship of *D. subclavatus* has often been attributed to Pallas.[693] Looss[557] convincingly demonstrated that the species described by Goeze,[351] and not the one mentioned by Pallas,[693] is equal to *D. subclavatus*, as is known at present. There are several reports on the occurrence of *D. subclavatus* in African countries south of the Sahara. Bayssade-Dufour et al.,[74] comparing the excretory system, the argentophilic structure of the cercarial tail, and the body surface and biological properties of the European and African *D. subclavatus*, found that there are important differences in number and arrangement of these structures, especially in those situated along the tail, their reproductive biology, development of embryo and intermediate hosts. They concluded that the species described under the name *D. subclavatus*, south of the Sahara, is identical with *D. fischthalicus*.[850] Hence, the host and distributional data referring to *D. subclavatus* in this area are listed under the name *D. fischthalicus* Meskal, 1970.

Diplodiscus amphichrus Tubangui, 1933 (Figure 314)

Syn. *Diplodiscus sinicus* Li, 1937; *D. amphichrus var. japonicus* Yamaguti, 1936; *D. japonicus* Yamaguti, 1936.

Diagnosis: Body length 1.6—3.4 mm, body width 0.2—0.8 mm. Acetabulum 0.54—1.16 mm in diameter, Diplodiscus type, number of muscle units, d.e.c., 24—26; d.i.c., 49—51; v.e.c., 16—19; v.i.c., 51—57; m.e.c., 8—10 (Figure 140). Pharynx 0.29—0.32 mm long, Subclavatus type (Figure 48), appendages extramural, 0.13 mm in length; esophagus 0.35 mm in length. Testis 0.14—0.14 mm in diameter. Genital pore postbifurcal. Cirrus sac 0.13—0.15 by 0.09—0.12 mm. Terminal genitalium postbifurcal, Sphinolosum type (Figure 77). Eggs 0.113—0.119 by 0.068—0.079 mm.

Biology: Not completely known. References. Tubangui and Masilungan,[993] Yamaguti,[1073] Odening.[665,667]

Preparasitic stage. ① Yamaguti, as *Diplodiscus amphichrus var. japonicus*. Eggs laid not embryonated, mature miracidia develop within 9 to 13 d at 32°C. Miracidia elliptical, 0.11 by 0.05 mm in size, covered with epithelial cells arranged in four rows. Apical gland, a pair of flame cells, and germinal balls are present. Intermediate hosts: *Anisus (Gyraulus) hiemantium; Gyraulus prashadi* (Tubangui and Masilungan); *Gyraulus heudei* (Odening).

Intramolluscan stage. ① Yamaguti. Body of redia 0.26—0.60 by 0.07—0.18 mm when fixed, tapering posteriorly. Pharynx muscular 0.07—0.09 mm in diameter, surrounded posteriorly by salivary glands; cecum saccular. Numerous germ balls found in various stages of development. Three pairs of flame cells present, excretory pore near middle pair of the flame cells. Body of cercariae conical in form, measuring 0.36—0.75 by 0.16—0.35 mm. Around them and the anterior body end there were several, alternating rows of hairs. Two large eyespots situated dorsally at anterior third of body. Parenchymal cells contain fine, brown pigment granules. Cystogenous cells fill with rhabdites. Pharynx 0.54—0.84 mm in diameter, with a pair of appendages; esophagus sometimes long; ceca extending to about posterior third of body. Acetabulum terminal, measuring 0.09—0.16 by 0.13—0.20 mm. Recurrent main excretory vessel enters the acetabulum and divides into six, blind tubules. Tail dorsoterminal, 0.6—0.9 mm long, with a row of large vesicular cells around the caudal tube. Caudal tube, near to the tip of the tail, divides into two branches, opening on the lateral margin of the tail. Secreted cercariae ready to encyst on the objects (green algae, wall of the container, etc.); cysts measuring about 0.27 by 0.25 mm in size. Takahashi's[944] species "*Diplodiscus subclavatus*" developing in *Segmentina mica* and *Planorbis compressus* is equal with this species. ② Tubangui and Masilungan. Cercaria Dilpocotylea type, characterized by the presence of 2 rows of hairlike formation near oral aperture. Body pyriform 0.21—0.24 by 0.2—0.22 mm, tails 0.76—1.00 mm long. Cercariae encyst on solid objects, cysts measuring 0.24—0.25 by 0.2—0.23 mm. ③ Odening. Body measuring 0.41—0.59 by 0.27—0.29 mm; tail 0.56—0.69 mm long. Pharynx 0.12 by 0.09—0.15 mm, acetabulum 0.15—0.18 mm in size.

Habitat: Rectum.

Hosts: *Megalobatrachus japonicus, Triturus pyrrhogaster*, (Salamandridae); *Rana cancrivora, R. cyanophlyctis, R. taipehensis, R. nigromaculata, R. limnocharis, R. tigrina, R. rugosa, Occidozyga lima*, (Ranidae); *Natrix tigrina* (Colubridae).

Distribution: Asia — Philippines, Vietnam, India, Korea, Japan, China.

References: Bhutta and Khan,[93] Chen et al.,[168] Ito,[435] Mukherjee,[631,634] Odening,[665,666] Sey,[849,854] Singh,[881] Tubangui.[997]

Diplodiscus magnus (Srivastava, 1934) Sey, 1988 (Figure 315)

Syn. *Diplodiscus amphichrus var. magnus* Srivastava, 1934; *D. sinicus* Li, 1937; *D. melanosticti*, Yamaguti et Mitunaga, 1943; *D. mehrai* of Pandey et Jain, 1974; *D. mehrai* of Pandey, 1973.

Diagnosis: Body length 2.6—6.7 mm, body width 1.2—2.1 mm. Acetabulum 0.87—1.53 mm in diameter. Pharynx 0.35—0.56 mm long; appendages 0.14—0.28 mm in size. Esoph-

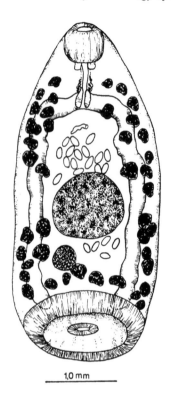

FIGURE 315. *Diplodiscus magnus* (Srivastava, 1934). (After H. D. Srivastava, 1934.)

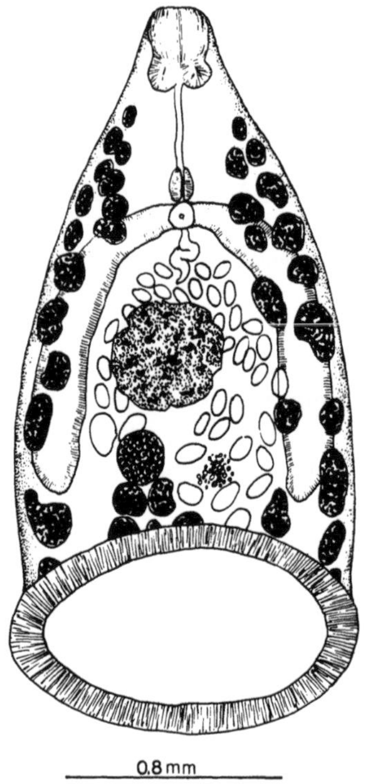

FIGURE 316. *Diplodiscus mehrai* Pande, 1937. (After B. P. Pande, 1937.)

agus 0.13—0.26 mm in length, bulb present. Testis 0.37—0.70 mm in size. Ovary 0.17—0.40 mm in size. Genital pore postbifurcal. Eggs 0.124—0.146 by 0.053—0.712 mm.
Biology: Not yet known.
Habitat: Rectum.
Hosts: *Rana rugulosa, R. tigrina, R. limnocharis, R. cyanophlyctis* (Ranidae); *Bufo melanostrictus* (Bufonidae).
Distribution: Asia — Japan, Taiwan, China, India.
References: Fischthal and Thomas,[314] Fischthal and Kuntz,[317] Li,[551] Srivastava,[906] Zhu et al.[1099]

Diplodiscus mehrai **Pande, 1937 (Figure 316)**

Syn. *Diplodiscus japonicus* (Yamaguti, 1936); *D. amphichrus* of Singh, 1954; of Agarwal, 1966; of Mukherjee, 1966; *D. amphichrus brevis* Nama et Kichi, 1973; *D. amphichrus* of Mukherjee and Ghosh, 1972; *D. minutus* Li et Gu, 1978.

Diagnosis: Body length 2.1—3.2 mm, body width 1.0—1.3 mm. Acetabulum 0.56—0.74 by 0.90—1.14 mm, Diplodiscus type, number of muscle units, d.e.c., 36—38; d.i.c., 42—45; v.e.c., 46—48; v.i.c., 35—37; m.e.c., 8—10 (Figure 143). Pharynx 0.36 mm in length, Megalodiscus type, (Figure 42), appendages 0.28 mm long; esophagus 0.38—0.50 mm long, bulb present. Testes 0.23—0.40 by 0.34—0.45 mm. Cirrus sac 0.18 by 0.16 in size. Ovary 0.16—0.18 mm in diameter. Terminal genitalium Spinolosum type (Figure 77). Genital pore bifurcal. Eggs 0.146 by 0.093 mm in size.
Biology: Not yet known.
Habitat: Rectum.

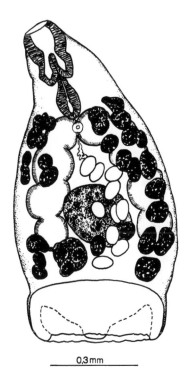 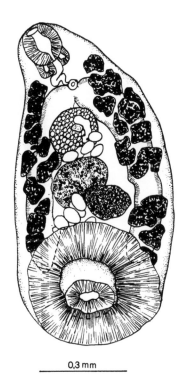

0,3 mm   0,3 mm

FIGURE 317. *Diplodiscus sacculosus* Yuen, 1962. (By permission of the *J. Parasitol.*, 1962, Lincoln.)

FIGURE 318. *Diplodiscus pallascatus* Manter et Pritchard, 1964. (Courtesy of M. Pritchard.)

Hosts: *Rana cyanophlyctis, R. longicrus, R. nigromaculata, R. limnocharis, R. tigrina, R. lima, R. esculanta,* (Ranidae); *Bufo bufo, B. melanostrictus, B. viridis* (Bufonidae); *Paramesotriton deloustali* (Salamandridae); *Psammodynaster pulverulentus* (Colubridae); *Pila globosa* (Pilidae).

Distribution: Asia — China, Japan, Vietnam, Taiwan, India, U.S.S.R.

References: Chen et al.,[168] Fischthal and Kuntz,[317] Kaw,[473] Li,[551] Pande,[696] Petriashvili,[721] Sey.[849,855]

Diplodiscus sacculosus Yuen, 1962 (Figure 317)

Diagnosis: Body length 1.2—1.3 mm, body width 0.71—0.78 mm. Acetabulum 0.56—0.64 by 0.17—0.20 mm. Pharynx 0.21—0.22 by 0.15—0.18 mm; appendages 0.11—0.13 mm long; esophagus 0.24—0.28 mm in length, with bulb. Testis 0.19—0.28 by 0.21—0.34 mm, ovary 0.11—0.14 by 0.14—0.50 mm. Genital pore bifurcal. Eggs 0.100—0.110 by 0.060—0.070 mm.

Biology: Not yet known.

Habitat: Rectum.

Host: *Rana erythraea* (Ranidae).

Distribution: Asia — Malaysia.

Reference: Yuen.[1080]

Diplodiscus pallascatus Manter et Pritchard, 1964 (Figure 318)

Diagnosis: Body length 0.4—1.5 mm, body width 0.2—0.7 mm. Acetabulum 0.27—0.72 by 0.23—0.57 mm. Pharynx 0.11-0.23 by, appendages extramural, 0.28 mm in length, esophagus 0.17—0.28 mm, with bulb. Testis 0.06—0.11 by 0.11—0.33 mm. Cirrus sac

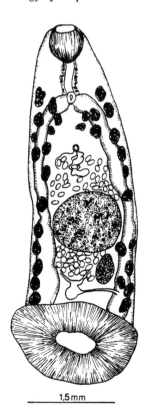

1,5mm

FIGURE 319. *Diplodiscus lali*
K. C. Pandey et K. K. Chakrabarti, 1968.

0.14—0.21 by 0.13—0.19 mm in size. Ovary 0.04—0.16 by 0.06—0.20 mm. Genital pore
postbifurcal. Eggs 0.090—0.130 by 0.064—0.086 mm.
Biology: Not yet known.
Habitat: Intestine.
Host: *Bufo regularis* (Bufonidae).
Distribution: Africa — Zaire.
Reference: Manter and Pritchard.[589]

Dilpodiscus lali Pandey et Chakrabarti, 1968 (Figure 319)

Syn. *Pseudodiplodiscoides pilai* Murty, 1970.
Diagnosis: Body length 3.7—5.8 mm, body width 1.2—1.9 mm. Acetabulum 0.82—1.12
by 1.51—1.83 mm. Pharynx 0.34—0.58 mm long, esophagus 0.53 mm in length, with
bulb. Testis 0.54—1.06 by 0.58—0.85 mm. Ovary 0.25—0.34 by 0.24—0.48 mm. Cirrus
sac 0.18—0.21 by 0.06—0.07 mm. Genital pore postbifurcal. Eggs 0.110—0.130 by
0.060—0.080 mm.
Biology: Not completely elucidated. Reference. Pandey and Agrawal.[701] Eggs laid hatched
within 24 h at room temperature. Miracidia measuring 0.012—0.013 by 0.03—0.05 mm.
Body covered with epithelial cells, arranged in four rows by the formula 6:6:4:2. Miracidia
composed of an apical papilla, two pairs of penetration glands, one pair of flame cells
with openings outside between the third and fourth tiers of epithelial plates. Major part
of the germinal cavity occupied by germinal balls. Miracidia survive 2 h in tap water.
Habitat: Intestine and rectum.

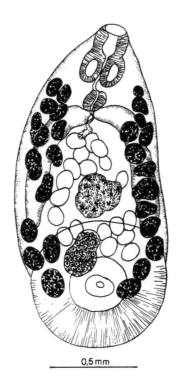

0,5 mm

FIGURE 320. *Diplodiscus brevicoeca*
Richard, Chabaud et Brygoo, 1968. (After
J. Richard et al. 1968.)

Host: *Rana tigrina* (Ranida).
Distribution: Asia — India.
References: Agrawal,[9] Pandey and Chakrabarti,[699] Pandey and Agrawal.[701]

Diplodiscus brevicoeca **Richard, Chabaud et Brygoo, 1968 (Figure 320)**

Diagnosis: Body length 1.4—1.7 mm, body width 0.6—0.9 mm. Acetabulum 0.53 by
0.62—0.80 mm. Pharynx 0.17—0.24 by 0.18—0.25 mm; appendages 0.10—0.15 mm long;
esophagus 0.30—0.40 mm in length with bulb. Testis 0.16—0.33 by 0.15—0.42 mm.
Ovary 0.12—0.18 mm in diameter. Genital pore prebifurcal. Eggs 0.130 by 0.085 mm.
Biology: Not yet completely known. Reference. Richard et al.[778]

Preparasitic stage. Miracidia similar to amphistome type of miracidia. Intermediate
host *Anisus crassilabrum*.

Intramolluscan stage. Body of rediae measuring 0.5—0.8 by 0.16—0.26 mm. Birth
pore situated 0.20 mm from the anterior extremity. Pharynx 0.075—0.090 by 0.085—0.10
mm; cecum 0.16 mm in length. Body cavity contains cercariae in various stages of
development. No locomotion appendages. Body of cercariae 0.32 by 0.31 mm, tails 0.90
mm in length. Acetabulum 0.15 by 0.17 mm, with an accessory sucker, measuring 0.065
mm in diameter. Pharynx with appendages of 0.115 mm length. Eyespots lateral to
pharynx, body strongly pigmented. Cercariae encyst rapidly on the subjects of the environment. Cyst opaque and inner organ not easily seen.
Habitat: Intestine.
Host: *Ptychadena mascareniensis* (Ranidae).
Distribution: Africa — Malagasy.
Reference: Richard et al.[778]

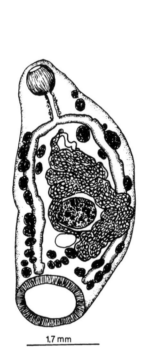

1,7 mm

0,7 mm

FIGURE 321. *Diplodiscus chauhani* Pandey, 1969. (After K. C. Pandey, 1969.)

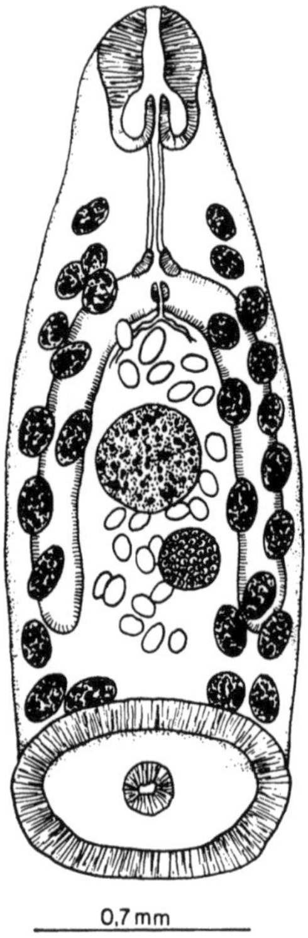

FIGURE 322. *Diplodiscus fischthalicus* Meskal, 1970. (Courtesy of R. Bourgat.)

Diplodiscus chauhani Pandey, 1969 (Figure 321)

Syn. *Diplodiscus* sp. Anjaneyulu, 1967; *D. anjaneyului* Pandey, 1973; *D. anjaneyului* Pandey et Jain, 1974.

Diagnosis: Body length 4.3—6.6 mm, body width 1.5—2.0 mm. Acetabulum 1.35—1.72 by 0.67—1.12 mm. Pharynx 0.45—0.67 mm long, esophagus 0.49—0.78 mm in length, without bulb. Testis 0.62—0.73 by 0.97—1.20 mm. Cirrus sac 0.21—0.25 by 0.25—0.27 mm. Ovary 0.24—0.31 by 0.36—0.45 mm. Genital pore postbifurcal. Eggs 0.100—0.120 by 0.080—0.090 mm.

Biology: Not yet known.

Habitat: Rectum.

Host: *Rana cyanophlictis* (Ranidae).

Distribution: Asia — India.

Reference: Pandey.[697]

Diplodiscus fischthalicus Meskal, 1970 (Figure 322)

Syn. *Diplodiscus magnus* of Fischthal et Thomas, 1968; *D. subclavatus* of Skrjabin, 1916; of Pike, 1979; *D. subclavatus* of Bourgat and Kulo, 1977.

Diagnosis: Body length 1.8—2.7 mm, body width 0.8—0.9 mm. Acetabulum 0.48—0.99 in diameter. Pharynx 0.14—0.23 by 0.16—0.25 mm, Megalodiscus type (Figure 42), appendages 0.08—0.17 mm long, esophagus 0.18—0.29 mm in length with bulb. Testis 0.29—0.52 by 0.29—0.40 mm. Cirrus sac 0.10 mm in diameter. Ovary 0.12—0.18 mm in diameter. Eggs 0.100—0.120 by 0.050—0.080 mm.

Biology: Not yet known.

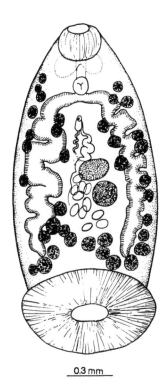

0.3 mm

FIGURE 323. *Diplodiscus nigro-maculati* Wang, 1977. (After P. Q. Wang, 1977.)

Habitat: Rectum.

Hosts: *Rana angolensis, R. cyanophlyctis, R. galamensis, R. occipitalis, R. tigrina, Ptychadema mascareniensis, P. maccarthyensis, Phrynobatrachus accraensis, Ph. pumilio, Aubria subsigillata* (Ranidae); *Hyperolius fusciventris, Phlyctimantis leonardi, Kassina senegalensis, Leptopelis aubryi* (Hyperoliidae); *Bufo regularis* (Bufonidae); *Xenopus muelleri, X. tropicalis* (Pipidae); *Boophis albilaris* (Rhacophoridae).

Distribution: Africa — Ethiopia, Sudan, Zaire, Togo, Cameroon.

References: Bourgat,[110] Bourgat and Kulo,[111] Fischthal and Thomas,[314] Gassmann,[337] Meskal,[609] Pike,[726] Sey,[849] Skrjabin.[889]

Diplodiscus nigromaculati Wang, 1977 (Figure 323)

Diagnosis: Body length 1.6—2.0 mm, body width 0.6—0.9 mm. Acetabulum 0.48—0.56 by 0.88 mm. Pharynx 0.19—0.24 by 0.24—0.27 mm, appendages 0.12—0.13 by 0.11—0.12 mm. Esophagus 0.24—0.26 mm long with bulb. Testis 0.11—0.16 by 0.17—0.25 mm, cirrus sac 0.12—0.17 by 0.06—0.11 mm. Ovarium 0.12—0.17 by 0.12—0.16 mm. Genital pore bifurcal. Eggs 0.118—0.126 by 0.040—0.042 mm.

Biology: Not yet known.

Habitat: Intestine.

Host: *Rana nigromaculata* (Ranidae).

Distribution: Asia — China.

Reference: Wang.[1049]

KEY TO THE SPECIES OF *DIPLODISCUS*

1. Primary pharyngeal sacs extramural.. 2

Pseudodiplodiscus Szidat, 1939

Diagnosis: Diplodiscidae. Body conical. Acetabulum large, subterminal. Pharynx terminal with intramural appendages, esophagus long with bulb; ceca relatively short, terminating at level of ovary. Testis single, middle part of body. Cirrus sac present. Ovary subterminal halfway between testis and acetabulum. Genital pore bifurcal. Uterine coils winding between acetabulum and genital pore, dorsal to testis. Vitelline follicules extending from testicular zone to ovary. Intestinal parasites of fish.

Type species: *Pseudodiplodiscus cornu* (Diesing, 1839) Szidat, 1959 (Figure 324)

Syn. *Amphistoma cornu* Diesing, 1839; *Diplodiscus cornu* (Diesing, 1839) Daday, 1907.

Diagnosis: With characters of the genus. Body length 3.3—4.0 mm, body width 2.0—2.2 mm. Acetabulum 2.00—2.20 mm. Pharynx 0.25—0.42 mm, Pseudodiplodiscus type (Figure 31), esophagus 0.70—0.80 mm long. Testes 0.61—1.00 mm in diameter. Ovary 0.20 mm in diameter. Eggs 0.130—0.140 by 0.060—0.070 mm.

Biology: Not yet known.

Habitat: Intestine.

Hosts: *Doras dorsalis, D. vaca* (Doradidae).

Distribution: South America — Brazil.

References: Daday,[197] Diesing.[223]

Notes: Issa and Ebaid[434] described this species from the Nile fish, *Synodontis membranaceus*. It is obvious that this species was mistaken for the common amphistome of this fish *Sandonia sudanensis*.

Catadiscus Cohn, 1904

Diagnosis: Diplodiscidae. Body conical or pyriform. Acetabulum ventroterminal, constricted into two portions. Pharynx terminal, with extramural appendages, esophagus with bulb: ceca short. Testis simple, intercecal, posttesticular, median or submedian. Ovary oval or spherical, submedian, posttesticular. Genital pore bifurcal, prebifurcal, or postbifurcal, cirrus sac present. Uterine coils between bifurcation and acetabulum, sometimes reaching lateral zone of body. Vitelline follicules large, sometimes appearing in compact shape,

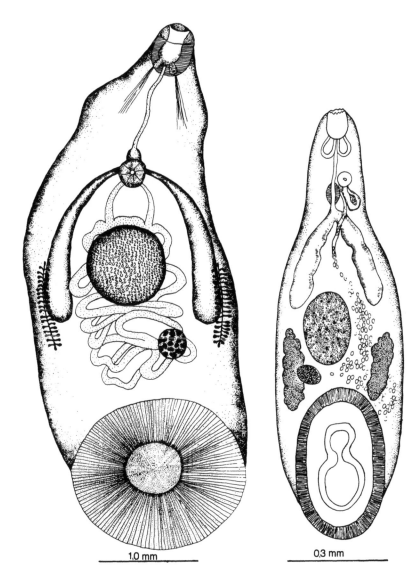

FIGURE 324. *Pseudodiplodiscus cornu* (Diesing, 1839). (After E. Daday, 1907.)

FIGURE 325. *Catadiscus dolichocotyle* (Cohn, 1903). (After F. Mañé-Garsón and A. M. Gortari, 1965.)

lateral, posttesticular, occasionally confluent enteriorly. Intestinal parasites of frogs and snakes.

Type species: *Catadiscus dolichocotyle* **(Cohn, 1903) Cohn, 1904 (Figure 325)**
Syn. *Amphistoma dolichocotyle* Cohn, 1903
Diagnosis: Body length 1.4—15 mm, body width 0.4—0.5 mm. Acetabulum 0.52—0.56 by 0.45—0.48 mm. Pharynx 0.13—0.15 by 0.09 mm; appendages 0.13 mm long; esophagus 0.10—0.13 mm in length with bulb. Testis 0.24—0.28 by 0.20—0.28 mm. Cirrus sac 0.09—0.12 mm. Ovary 0.03 mm in diameter. Genital pore prebifurcal. Eggs 0.070—0.080 by 0.020—0.040 mm.
Biology: Not yet known.
Habitat: Intestine.

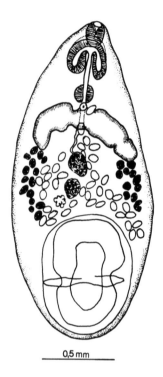

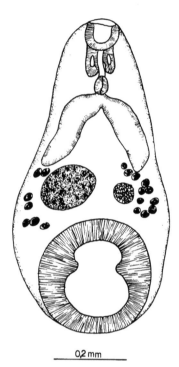

0,5 mm

0,2 mm

FIGURE 326. *Catadiscus cohni* Travassos, 1926. (After L. Travassos, 1926.)

FIGURE 327. *Catadiscus pygmaeus* (Lutz, 1928). (After L. Travassos, 1926.)

Hosts: *Liophis miliaris, Chironius fuscus* (Colubridae).
Distribution: South America — Uruguay.
References: Cohn,[178] Freitas and Lent,[324] Mañé-Garzón and Gortari.[585]

Catadiscus cohni Travassos, 1926 (Figure 326)

Diagnosis: Body length 1.7—2.1 mm, body width 0.8—1.0 mm. Acetabulum 0.06—0.70 by 0.60—0.80 mm. Pharynx 0.12—0.15 mm in diameter, appendages 0.30—0.40 mm in length, esophagus 0.30 by 0.60 mm with bulb. Testis 0.80 by 0.11—0.20 by 0.10 mm. Cirrus sac 0.20 mm long. Ovary 0.07 by 0.01—0.17 mm. Genital pore bifurcal. Eggs 0.080 by 0.040—0.090 by 0.040 mm.
Biology: Not yet known.
Habitat: Large intestine.
Host: *Bufo marinus* (Bufonidae).
Distribution: South America — Brazil.
References: Freitas and Lent,[324] Travassos.[981]

Catadiscus pygmaeus (Lutz, 1928) Freitas et Lent, 1939 (Figure 327)

Syn. *Diplodiscus pygmaeus* Lutz, 1928.
Diagnosis: Body length 1.0 mm, body width 0.6 mm. Acetabulum 0.35 by 0.32 mm. Pharynx 0.12 by 0.17 mm; appendages 0.09 mm long; esophagus 0.15 mm in length with bulb. Testis 0.16 by 0.21 mm. Ovary 0.05 by 0.08 mm. Genital pore bifurcal. Eggs 0.080—0.088 by 0.056 mm.
Biology: Not yet known.
Habitat: Rectum.
Host: *Pseudis paradoxa* (Pseudidae).

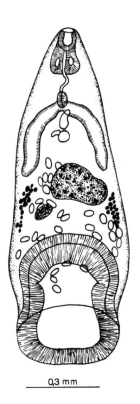

0,3 mm

FIGURE 328. *Catadiscus uru-*
guayensis Freitas et Lent, 1939.
(After J. F. T. Freitas and H. Lent,
1939.)

Distribution: South America — Venezuela.
References: Caballero and Diaz-Ungria,[152] Freitas and Lent,[324] Lutz.[566]

Catadiscus uruguayensis **Freitas et Lent, 1939 (Figure 328)**

Diagnosis: Body length 0.9—2.1 mm, body width 0.4—0.7 mm. Acetabulum 0.40—0.60 by 0.30—0.50 mm. Pharynx 0.10—0.13 by 0.18—0.23 mm, appendages 0.13—0.22 mm long; esophagus 0.22—0.23 mm long with bulb. Testis 0.10—0.25 by 0.15—0.26 mm. Cirrus sac 0.10—0.15 by 0.03—0.07 mm. Ovary 0.08—0.17 by 0.07—0.08 mm. Genital pore bifurcal. Eggs 0.100—0.110 by 0.053—0.055 mm.

Biology: Not yet known. Reference. Núñez[661] collected amphistome cercariae excreted by snail, *Drepanotrema kermatoides* (Planorbidae) and adult flukes from frogs which lived in the same biotope. It was believed that these cercariae belonged to the species, *C. uruguayensis*.

Preparasitic stage. Eggs are ovoid, operculated, and when laid, they contained developed miracidia. Ciliae are found on the anterior and posterior extremities with an empty zone along middle part of miracidia. A large apical gland, four penetration glands, a pair of flame cells, and germinal balls are found within the miracidia. Epithelial cells were not observed.

Intramolluscan stage. Rediae fusiform, without locomotion appendages. Sensory bristles present around mouth. Ceca short, dilated. Three pair of flame cells are present. Body fills up with developing germ cells and cercariae. Body of cercariae measuring 0.21—0.29 by 0.23—0.26; tails 0.55—0.71 mm long. Esophagus long, ceca extending to acetabulum.

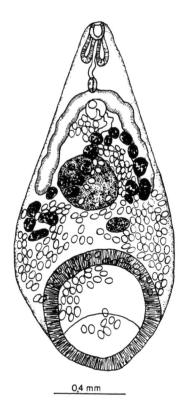

0,4 mm   0,7 mm

FIGURE 329.   *Catadiscus marinholutzi*   FIGURE 330.   *Catadiscus inopina-*
Freitas et Lent, 1939. (After J. F. T. Fre-   *tus* Freitas, 1941. (After J. F. T. Frei-
tias and H. Lent, 1939.)   tas, 1941.)

Two eyespots are present just behind pharynx. Main collecting tubes undulating without cross-connecting tubes. Emerged cercariae encyst on vegetation, adolescariae are about 0.20 mm in diameter.

Habitat: Large intestine.

Hosts: *Leptodactylus ocellatus* (Leptodactylidae); *Lysapsus limellus* (Pseudidae).

Distribution: South America — Brazil, Uruguay.

References: Freitas,[323] Freitas and Lent.[325]

Catadiscus marinholutzi Freitas and Lent, 1939 (Figure 329)

Diagnosis: Body length 1.6—2.4, body width 1.0—1.2 mm. Acetabulum 0.70—0.90 by 0.53—0.98 mm. Pharynx 0.08—0.10 mm long; appendages 0.20—0.25 mm in length; esophagus 0.28—0.33 mm long with bulb. Testis 0.28—0.50 by 0.33—0.55 mm. Cirrus sac 0.15—0.18 by 0.17—0.25 mm. Ovary 0.10—0.17 by 0.12—0.15 mm. Genital pore bifurcal. Eggs 0.110—0.120 by 0.050—0.070 mm.

Biology: Not yet known.

Habitat: Large intestine.

Hosts: *Leptodactylus ocellatus, L. caliginosus* (Leptodactylidae); *Leptodeira septentrionalis, Thamnophis sauritus, Coniophanes quinqveriktatus* (Colubridae).

References: Freitas and Lent,[324] Thatcher.[971]

Catadiscus inopinatus Freitas, 1941 (Figure 330)

Diagnosis: Body length 2.5—3.9 mm, body width 1.2—1.7 mm. Acetabulum 0.65—0.88 by 0.53—0.78 mm. Pharynx 0.09—0.11 mm long, appendges 0.26—0.27 mm long; esoph-

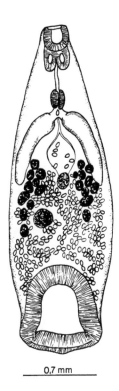

FIGURE 331. *Catadiscus frei-taslenti* Ruiz, 1943. (After J. Ruiz, 1943.)

agus 0.39—0.43 mm long with bulb. Testis 0.47 by 0.40 mm in size. Cirrus sac 0.14—0.19 by 0.14—0.21 mm. Ovary 0.17—0.23 by 0.16—0.17 mm. Genital pore bifurcal, eggs 0.084—0.100 by 0.050—0.055 mm.

Biology: Not yet known.

Habitat: Large intestine.

Host: *Leptodactylus ocellatus* (Leptodactylidae).

Distribution: South America — Brazil, Paraguay.

References: Freitas,[321] Lent et al.,[539] Masi Pallarés et al.[595,596]

Catadiscus freitaslenti Ruiz, 1943 (Figure 331)

Diagnosis: Body length 3.1—3.1 mm, body width 0.9—1.0 mm. Acetabulum 0.79—0.87 by 0.77—0.79 mm. Pharynx 0.20—0.21 by 0.28—0.29 mm, appendages 0.30 mm in length; esophagus 0.36—0.41 mm long with bulb. Testis 0.11 mm in diameter, two vasa efferentia present. Cirrus sac 0.15—0.18 by 0.13—0.17 mm. Ovary 0.16—0.19 mm in size. Genital pore bifurcal. Eggs 0.070—0.080 by 0.033—0.050 mm.

Biology: Not yet known.

Habitat: Large intestine.

Hosts: *Liophis miliaris* (Colubridae); *Bufo paracnemis* (Bufonidae); *Leptodactylus ocellatus*, *L. tiphonius* (Leptodactylidae).

Distribution: South America — Brazil, Paraguay.

References: Lent et al.,[539] Masi Pallarés et al.,[595,596] Ruiz.[797]

Catadiscus mirandai Freitas, 1943 (Figure 332)

Diagnosis: Body length 2.9 mm, body width 0.8 mm. Acetabulum 0.55 by 0.46 mm. Pharynx 0.11 mm long, appendages 0.24 mm long; esophagus 0.33 mm long with bulb.

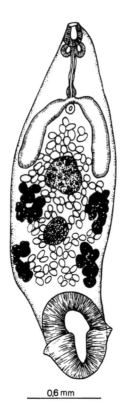 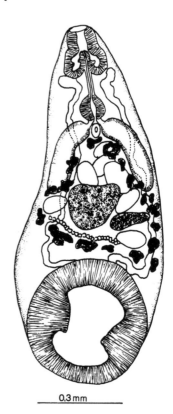

0,6 mm   0,3 mm

FIGURE 332. *Catadiscus mir-*   FIGURE 333. *Catadiscus rodri-*
andai Freitas, 1943. (After J. F.   *guezi* Caballero, 1955. (After C. E.
T. Freitas, 1943.)   Caballero, 1955.)

Testis 0.25 mm in diameter. Cirrus sac 0.10 mm in diameter. Ovary 0.20 by 0.18 mm.
Genital pore bifurcal. Eggs 0.110—0.120 by 0.061—0.078 mm.
Biology: Not yet known.
Habitat: Large intestine.
Host: *Pipa carvalhoi* (Pipidae).
Distribution: South America — Brazil.
Reference: Freitas.[322]

Catadiscus rodriguezi Caballero, 1955 (Figure 333)

Diagnosis: Body length 1.2—1.3 mm, body width 0.3—0.5 mm. Acetabulum 0.36—0.46
by 0.38—0.41 mm. Pharynx 0.07—0.07 by 0.03—0.05 mm; appendages 0.18—0.19 mm
in length; esophagus 0.15—0.17 mm long with bulb. Testis 0.13—0.16 by 0.16—0.19 mm,
cirrus sac 0.09—0.10 by 0.04—0.06 mm. Ovary 0.03—0.05 by 0.11—0.12 mm. Genital
pore bifurcal, cirrus sac 0.09—0.10 by 0.04—0.06 mm. Eggs 0.106—0.118 by 0.068—0.072
mm.
Biology: Not yet known.
Habitat: Rectum.
Host: *Leptodactylus pentadactylus* (Leptodactylidae).
Distribution: Central America — Panama.
Reference: Caballero.[147]

Catadiscus propinquus Freitas et Dobbin, 1956 (Figure 334)

Diagnosis: Body length 1.1—1.6 mm, body width 0.6—1.1 mm. Acetabulum 0.43—0.58

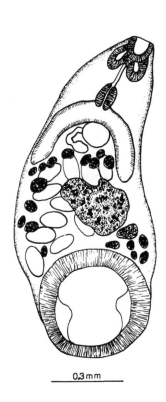 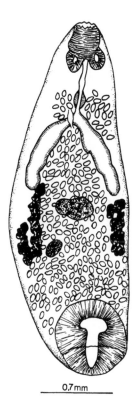

FIGURE 334. *Catadiscus propinguus* Freitas et Dobbin, 1956. (After J. F. T. Freitas and J. E. Dobbin, 1956.)

FIGURE 335. *Catadiscus corderoi* Moné-Garsón, 1956. (After F. Mañé-Garsón, 1958.)

by 0.43—0.53 mm. Pharynx 0.09—0.16 mm long, appendages 0.15—0.22 mm in length, esophagus 0.15—0.25 mm long, with bulb. Testis 0.13—0.79 by 0.17—0.36 mm. Cirrus sac 0.09—0.14 by 0.06—0.13 mm. Ovary 0.04—0.08 by 0.06—0.13 mm. Genital pore bifurcal. Eggs 0.118—0.155 by 0.059—0.088.
Biology: Not yet known.
Habitat: Large intestine.
Host: *Rana palmipes* (Ranidae).
Distribution: South America — Brazil.
References: Dobbin,[236] Freitas and Dobbin.[326]

Catadiscus corderoi Mañé-Garzón, 1956 (Figure 335)

Diagnosis: Body length 3.1 mm, body width 0.9 mm. Acetabulum 0.70 by 0.39 mm. Pharynx 0.30 mm in diameter; appendages 0.14 mm in length; esophagus 0.22 mm long, with bulb. Testis 0.12 by 0.18 mm. Cirrus sac 0.07 by 0.09 mm. Ovary 0.12 by 0.06 mm. Genital pore bifurcal. Eggs 0.057—0.060 by 0.02—0.03 mm.
Biology: Not yet known.
Habitat: Intestine.
Host: *Pseudis meridionalis* (Pseudidae).
Distribution: South America — Uruguay.
Reference: Mañé-Garzón.[584]

Catadiscus eldoradiensis Artigas et Perez, 1964 (Figure 336)

Diagnosis: Body length 2.1—3.0 mm, body width 0.8—1.0 mm. Acetabulum 0.60—0.75

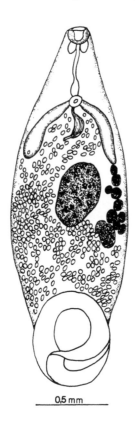

0,5 mm

0,7 mm

FIGURE 336. *Catadiscus el-doradiensis* Artigas et Perez, 1964. (After P. T. Artigas and M. D. Perez, 1964.)

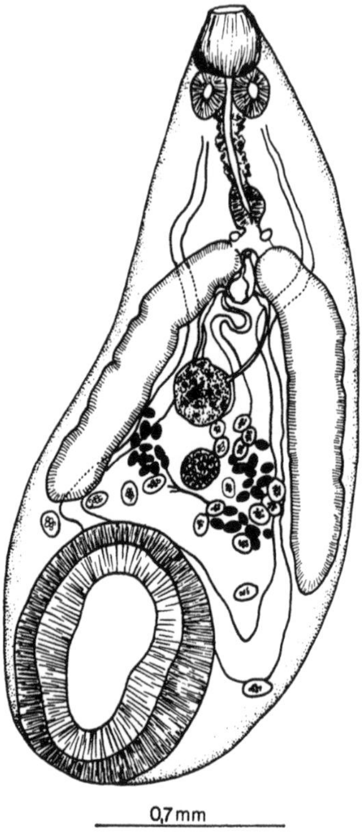

FIGURE 337. *Catadiscus longicae-calis* Poumarau, 1965. (By permission of the Physis, 1965, Buenos Aires.)

by 0.56—0.87 mm; pharynx 0.10—0.18 by 0.16—0.25 mm, appendages 0.10—0.13 by 0.10—0.12 mm; esophagus 0.25—0.34 mm long, with bulb. Testis 0.35—0.60 by 0.22—0.41 mm. Cirrus sac 0.16—0.42 by 0.11—0.12 mm. Ovary 0.10—0.20 by 0.08—0.14 mm. Genital pore bifurcal. Eggs 0.620—0.820 by 0.300—0.320 mm.
Biology: Not yet known.
Habitat: Large intestine.
Host: *Leptodactylus ocellatus* (Leptodactylidae).
Distribution: South America — Brazil.
Reference: Artigas and Pérez.[40]

Catadiscus longicaecalis Poumarau, 1965 (Figure 337)
Diagnosis: Body length 2.8 mm, body width 1.1 mm. Acetabulum 0.90 by 0.76 mm; pharynx 0.13—0.19 by 0.12—0.20 mm, appendages 0.22 mm long, esophagus 0.34 mm in length, with bulb. Testis 0.23 by 0.16 mm. Cirrus sac 0.14 by 0.05 mm. Ovary 0.13 by 0.11 mm. Genital pore bifurcal. Eggs 0.090—0.100 by 0.052—0.065 mm.
Biology: Not yet known.
Habitat: Large intestine.
Hosts: *Bothrops neuwiedi* (Crotalidae); *Xenodon merremii, Lystrophis d'orbignyi* (Colubridae).
Distribution: South America — Argentina.
Reference: Poumarau.[735]

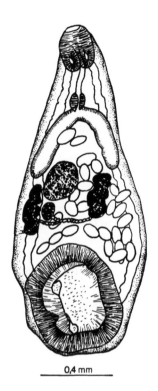

FIGURE 338. *Catadiscus rochai*
Correa et Artigas, 1978. (After
A. A. S. Correa and P. T. Artigas,
1978.)

Catadiscus rochai Correa et Artigas, 1978 (Figure 338)

Diagnosis: Body length 1.8 mm, body width 0.7 mm. Acetabulum 0.49 by 0.60 mm. Pharynx 0.14 mm long, appendages 0.18 mm in length, esophagus 0.38 mm long, with bulb. Testis 0.17 by 0.22 mm. Ovary 0.10 by 0.07 mm. Eggs 0.110 by 0.050 mm.

Biology: Not yet known.

Habitat: Intestine.

Host: *Dromicus typhlus* (Colubridae).

Distribution: South America — Brazil.

Reference: Correa and Artigas.[183]

KEY TO THE SPECIES OF *CATADISCUS**

1. Ceca short 2
   Ceca long .... *C. longicoecalis*
2. Vitelline follicules symmetrical 3
   Vitelline follicules assymetrical .... *C. eldoradiensis*
3. Vitelline follicules confluent anteriorly 4
   Vitelline follicules not confluent anteriorly 5
4. Pharynx/acetabulum ratio bigger than 1:3 .... *C. marinholutzi*
   Pharynx/acetabulum ratio smaller than 1:3 .... *C. propinquus*
5. Acetabulum/body length ratio bigger than 1:4 6
   Acetabulum/body length ratio smaller than 1:4 7

* *Catadiscus bathracorum* Cordero, 1926 was designated only without description.

6. Vitelline follicules extending to cecal zone and partly pretesticular, lateral...........
   ....*C. inopinatus*
   Vitelline follicules extending to cecal end, reaching anterior margin of testes,
   lateral .... *C. corderoi*
   Vitelline follicules extending to cecal end, reaching anterior end of testis, confluent
   posteriorly.. *C. rodriguezi*
   Vitelline follicules postcecal, posttesticular, lateral.................... *C. mirandai*
7. Eggs bigger than 0.1 mm 8
   Eggs smaller than 0.1 mm 9
8. Testis and ovary separated by uterine coils....................... *C. uruguayensis*
   Testis and ovary continuous.. *C. rochai*
9. Diameter of acetabulum smaller than 0.5 mm................................... 10
   Diameter of acetabulum bigger than 0.5 mm................................... 11
10. Esophagus shorter than pharynx, vitelline follicules extending to cecal end, pharynx
    body length ratio 1:3.8 .... *C. pygmaeus*
    Esophagus longer than pharynx, vitelline follicules postcecal pharynx/body length ratio
    1:5.1 .... *C. dolichocotyle*
11. Vas deferens single, vitellaria postcecal.................................. *C. cohni*
    Vas deferens double, vitelline follicules extending to about middle part of body.....
    .... *C. freitaslenti*

Dermatemytrema Price, 1937

Diagnosis: Diplodiscidae. Body pyriform, with a ridgelike collar near anterior extremity. Acetabulum terminal, somewhat trifoliate. Pharynx terminal, with extramural appendages, esophagus long with bulb; ceca short extending to vitelline follicules. Testis single, submedian, cecal. Cirrus sac present. Ovary median, intercecal, near acetabulum. Genital pore prebifurcal. Uterine coils wide, intercecal, in middle field. Intestinal parasites of freshwater turtles.

Type species. *Dermatemytrema trifoliata* Price, 1937 (Figure 339)

Diagnosis: With characters of the genus. Body length 1.9—2.0 mm, body width 0.2 mm. Acetabulum 0.66—0.72 by 0.73—0.80 mm, Dermatemytrema type, number of muscle units, d.e.c., 14—16; d.i.c., 42—45; v.e.c., 13—16; v.i.c., 42—46; m.e.c., 30—35 (Figure 141). Pharynx 0.67—0.7 mm long, Subclavatus type (Figure 48); appendages 0.09—0.10 mm long; esophagus 0.50—0.60 mm in length with bulb. Testis 0.08—0.08 by 0.05—0.08 mm. Cirrus sac 0.13—0.17 by 0.08—0.12 mm. Ovary 0.08—0.09 by 0.07—0.07 mm. Terminal genitalium Spinolosum type (Figure 77). Eggs 0.140 by 0.072—0.080 mm.
Biology: Not yet known.
Habitat: Stomach.
Host: *Dermatemys mawii* (Dmeratemydidae).
Distribution: Central America — Mexico.
References: Caballero,[145] Price and Buttner,[746] Sey,[849] Thatcher.[970]

Australodiscus Sey, 1983

Diagnosis: Diplodiscidae. Body pyriform in shape. Acetabulum terminal, appendages extramural; esophagus long, with well-developed bulb; ceca extending to ovarian zone. Testis intercecal, in middle third of body, irregular in outline. Ovary small, posttesticular, at level of cecal extremities. Genital pore just behind bifurcation, cirrus sac present. Uterine coils intercecal and postcecal. Vitelline follicules lateral, along ceca, confluent posteriorly. Intestinal parasites of frogs.

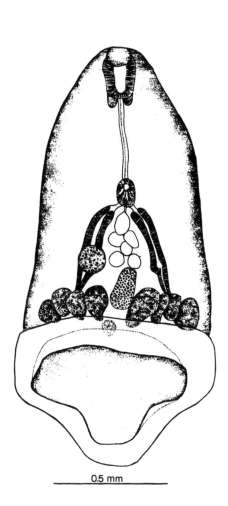

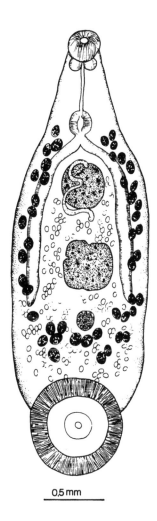

FIGURE 339. *Dermatemytrema trifoliata* Price, 1937. (After E. W. Price, 1953.)

FIGURE 340. *Australodiscus megalorchus* (Johnston, 1912). (After S. J. Johnston, 1912.)

Type species: *Australodiscus megalorchus* **(Johnson, 1912) Sey, 1983 (Figure 340)**

Syn. *Diplodiscus megalorchus* Johnston, 1912; *D. microrchus* Johnston, 1912.

Diagnosis: With characters of the genus. Body length 3.0—4.0 mm, body width 1.25—1.50 mm. Acetabulum 1.08 mm in diameter. Pharynx 0.31 mm in diameter, appendages 0.33 mm long; esophagus 0.52 mm in length. Eggs 0.130 by 0.060 mm.

Biology: Not yet known.

Habitat: Rectum.

Hosts: *Hyla aurea, H. ewingii* (Hylidae); *Limnodynaster peronii, L. tasmaniensis* (Myobatrachidae).

Distribution: Australia.

References: Brace et al.,[113] Johnston,[451] Sey.[849]

Progonimodiscus **Vercammen-Grandjean, 1960**

Diagnosis: Diplodiscidae. Body conical, tapering anteriorly. Acetabulum ventroterminal, large with riffled peduncle. Pharynx terminal with extramural appendages, esophagus short with bulb; ceca short. Testis single, transverse elongate median almost equatorial, two vasa efferentia present. Cirrus sac present. Ovary ovoid near to acetabulum. Genital pore at

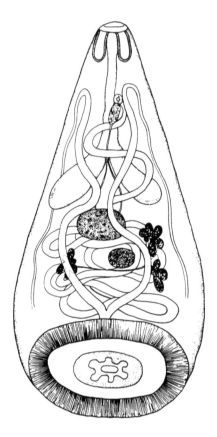

FIGURE 341.   *Progonimodiscus doyeri* (Ortlepp, 1926). (By permission of the *J. Helminthol.*, 1973, London.)

esophagus. Uterine coils transverse in position, reaching lateral margin of body, between
acetabulum and genital pore. Vitelline follicules form lateral clusters between ceca and
acetabulum. Intestinal parasites of frogs.

Type species: ***Progonimodiscus doyeri* (Ortlepp, 1926) Vercammen-Grandjean, 1960**
(Figure 341)

Syn. *Diplodiscus subclavatus* Grobbelaar, 1922; *D. doyeri* Ortlepp, 1926; *O. doyeri*
Ortlepp, 1926; *O. doyeri victoriani* Vercammen-Grandjean, 1960.

Diagnosis: With characters of the genus. Body length 1.9—3.7 mm, body width 1.0—1.5
mm. Acetabulum 0.64—1.03 by 0.93—1.37 mm. Pharynx 0.23—0.38 mm long, Megacotyle type, (Figure 41), appendages 0.15—0.21 by 0.13—0.18 mm; esophagus 0.38—0.42
mm long with bulb. Testis 0.25—0.30 by 0.30—0.41 mm. Ovary 0.16 by 0.20 mm. Eggs
0.112—0.136 by 0.072—0.096 mm.

Biology: Reference. Vercammen-Grandjean.[1030]

Preparasitic stage. Eggs laid contained developed miracidia, covered with ciliae,
except for apical end. Two flame cells present. Intermediate hosts. *Taphius stanleyi, T.
pfeifferi* (Planorbidae).

Intramolluscan stage. Body of redia 0.98—1.5 by 0.29—0.35 mm. Ten hair-bearing
papillae and pericecal glands were around the oral opening. Pharynx 0.03 by 0.04 mm;
cecum 0.1 by 0.05 mm. Three pairs of flame cells were found, paired ampulla opening
in posterior fourth of body. Rediae contained immature cercariae, they attained their

maturity after realizing the rediae. Body of cercarial lentiform, measuring 0.33—0.54 by 0.18—0.39 mm, tail 0.37—0.61 by 0.06—0.08 mm, with a slight construction behind the terminal ampulla, and the tip of the tail was provided with a pair of hair-bearing papillae. Pharynx without tracing of appendages, esophagus inflated; ceca short, terminating at level of excretory pore. Main collecting tubes with cross-connecting tubes. Caudal tube opening laterally near the tip of the tail. Eyespots semilunar cells containing pigment. Cystogenous gland cells full of rhabdites around the whole surface of the body. Cercariae encyst on vegetation in 15 min. Cyst wall hyaline, organogenesis was completed in 1 week after encystment.

Development in definitive host. Xenopus laevis was experimentally infected by feeding on cysts. Eggs were produced by parasites in 26—30 d postinfection.

Habitat: Rectum.

Hosts: *Xenopus laevis* (Pipidae); *Conrana crassipes* (Ranidae).

Distribution: Africa — Cameroon, South Africa, Uganda, Zaire, Zimbabwe.

References: Beverley-Burton,[84] Gassmann,[337] Macnae et al.,[577] Ortlepp,[677] Pritchard,[749] Thurston,[975] Vercammen-Grandjean.[1030]

3.2.3.1.3. *GASTRODISCIDAE Stiles et Goldberger, 1910*

Diagnosis: Cladorchoidea. Body large, flattened, ventrally excavated. It is divided into an anterior, small, subcylindrical portion and to a large, discoidal posterior portion. Ventral body surface covered with papillae, in a great number and arranged equally. Acetabulum small, ventroterminal. Pharynx with pharyngeal bulb and secondary pharyngeal sacs, esophagus short with well developed bulb, ceca long, somewhat sinuous, along outer longitudinal third of body, extending to preacetabular level. Testes deeply lobed, intercecal, diagonal in middle third of body. Modified cirrus sac present, not true in structure, including pars musculosa and pars prostatica. Genital pore at middle level of esophagus. Uterine coils in dorsal intercecal field, posteriorly ventral in pretesticular area, passing between two testes. Vitelline follicules occupying nearly the whole extracecal field, sometimes penetrate into intercecal area. Intestinal parasites in Perissodactyla and Artiodactyla.

Type genus: *Gastrodiscus* Lenkart, 1877

Diagnosis. With characters of the family.

Type species: *Gastrodiscus aegyptiacus* (Cobbold, 1876.) Ralliet, 1893 (Figure 342)

Syn. *Distoma aegyptiacum* Cobbold, 1877; *Cotylogaster cochleariformis* (Diesing, 1938.), *Gastrodiscus sonsinoi* (Cobbold, 1877); *G. polymastor* Lenkart, 1880; *G. minor* Leiper, 1913; *G. equi* Le Roux, 1939.

Diagnosis: Body length 9.0—15.0 mm, body width 5.0—7.0 mm. Acetabulum 1.37—1.58 mm in diameter, Gastrodiscus type, number of muscle units, d.e.c., 45—48; d.i.c., 24—26; v.e.c., 33—36; v.i.c., 21—23; m.e.c., 18—20 (Figure 180). Pharynx with pharyngeal bulb and secondary pharyngeal sacs 0.59—0.83 mm long, Gastrodiscus type (Figure 54); esophagus 1.58—1.98 mm in length. Anterior testis 0.79—0.86 by 0.60—0.78 mm; posterior testis 0.75—0.92 by 0.68—0.78 mm. Modified cirrus sac 1.65—1.78 by 0.43—0.47 mm. Ovary 0.38—0.43 by 0.32—0.37 mm. Genital pore bifurcal. Terminal genitalium Gastrodiscus type (Figure 74). Eggs 0.172—0.197 by 0.116 mm.

Biology: References. Le Roux,[549] Looss,[558] Malek.[582]

Preparasitic stage. ① Malek. Miracidium 0.23 mm in length, ciliated and having apical gland, a pair of penetration glands, a few germinal cells and a pair of flame cells. Life span of miracidia was about 15 h in tap water at temperatures between 25 and 27°C. Intermediate hosts. *Bulinus forskalii* (Malek) *Cleopatra bulimoiɔes, C. cyclostomoides* (Looss); *Bulinus forskalii, B. senegalensis* (Le Roux).

Intramolluscan stage. ① Looss. Body of redia 0.35 mm in size. Daughter rediae

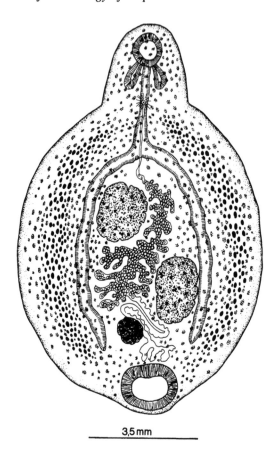

3,5 mm

FIGURE 342. *Gastrodiscus aegyptiacus* (Cobbold, 1876).
(After P. A. Maplestone, 1923.)

were found in mother rediae together with cercariae. Body of cercariae 0.50 by 0.35 mm, with eyespots. Tail shorter than body length. Pharynx 0.07 by 0.06 mm, esophagus with well developed bulb; ceca 0.17—0.20 mm. Main excretory tubes without cross-connecting tubes. Cyst 0.30 by 0.35 mm in size, cyst wall thick and resistant. Encystment occurred on any object. ② Malek. Mature redia 0.7 mm in length, birth pore situated 0.16 mm from anterior end. Rediae without collar or procrusculi. Pharynx muscular, 0.05 mm in diameter. Cercariae left rediae in immature form and developed in the tissue of the snail. Body heavily pigmented, pharynx 0.08 mm in diameter, acetabulum 0.10 by 0.15 mm, esophagus with bulb. Excretory main tubes without cross-connecting tubes. Caudal tube single. Primordia of reproductive organs were easily seen. Cercariae encyst within 0.5 to 1 h after emergence on grass blades and other materials. Adolescaria 0.33 mm in diameter.

Habitat: Large intestine.

Hosts: *Equus przewalskii* f. caballus, *E. africanus* f. asinus, *E. quagga*, *E. burchelli*, *E. grevyi*, *E. zebra*, *E. mulus* (Equidae); *Bos primigenius* f. taurus (Bovidae); *Elephas maximus Elephantidae)*; *Ceratotherium simum* (Rhinocerotidae); *Phacochoerus aethiopicus, Ph. porcus, Sus scrofa* f. domestica, *Hylochoerus meinertzhageni (Suidae)*.

Distribution: Africa — Egypt, Morocco, Chad, Gambia, Ghana, Guine, Sudan, Ethiopia, South Africa, Mozambique, Malawi, Central Africa, Niger, Mauritania, Zaire, Zambia; Asia — India, China, Vietnam; South America — Guyana; The Caribbean — Guadeloupe; Australia.

References: Applewhaise and Ruiz,[33] Chen et al.,[168] D'Huart,[214] Dollfus,[238,244] Graber,[357]

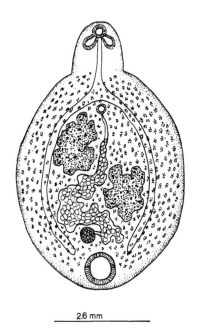

2,6 mm

FIGURE 343. *Gastrodiscus secundus*
(Looss, 1907).

Gupta and Walia,[391] Le Roux,[547] Looss,[558] Malek,[580] Mettrich,[613] Morel,[621] Ortlepp,[679] Prudhoe,[756] Seddon,[825] Thapar,[965] Tidswell,[976] Troncy et al.[993]

Gastrodiscus secundus Looss, 1907 (Figure 343)

Diagnosis: Body length 8.2—11.7 mm, body width 5.5—8.0 mm. Acetabulum 1.23 mm in diameter, Pseudodiscus type, number of muscle units, d.e.c., 46—48; d.i.c., 33—36; v.e.c., 15—17; v.i.c., 23—25; m.e.c., 14—16 (Figure 183). Pharynx with appendages 0.32—0.40 mm, Pseudodiscus type (Figure 56); esophagus 1.32—1.76 mm in length. Testes 0.18—1.20 by 0.88—0.98 mm. Cirrus sac 0.75—0.83 by 0.32—0.39 mm. Ovary 0.40—0.48 by 0.38—0.43 mm. Genital pore postbifurcal, terminal genitalium Gastrodiscus type (Figure 74). Eggs 0.119—0.191 by 0.870—0.118 mm.
Biology: References. Peter and Mudaliar,[715] Peter.[714]

Preparasitic stage. ① Peter. Eggs unsegmented when laid. Embryonic development in India took 8 to 13 d in monsoon season and 11 d in winter. Miracidia measuring 0.17 by 0.08 in contracted state. Covered with ciliated epithelial cells in four tiers, arranged by the formula 6:8:4:2. It is composed of an apical gland, one pair of penetration glands, a brain mass, a germ ball, several germ cells, and a pair of flame cells. Intermediate host: *Indoplanorbis exustus*.

Intramolluscan stage. ① Peter. Developing sporocyst was situated in the mantle tissue after 5 d of infection. Mature sporocyst elongate, saclike, 0.56 by 0.22 mm in size, containing well-formed rediae by the 13th day. ② Peter and Mudaliar. Rediae were sausage-shaped, without locomotor appendages. Rediae gave rise to daughter rediae, cercariae, or both. Young rediae measuring 0.26 by 0.08 mm, mature rediae measuring 0.74 by 0.16 mm. Birth pore was found at the first and second quarters of the body length. Pharynx 0.037 by 0.06 mm in size, cecum extending to midzone of body. Salivary glands well developed, a set of eight large cells around the pharynx. Excretory system consisted of three pairs of flame cells; it showed amphistomes construction. Posterior part of body cavity was occupied by germ tissue. ① Peter. Newly born cercariae were immature and completed their development within the snail tissue. Mature forms escaped from the intermediate host on the 55th day after infection. They were positively phototropic and swam very actively. The body was heavy and brownish, measuring 0.33—0.75 by 0.16—0.41 mm live. Tail 0.91—0.95 by 0.13—0.55 mm. Two heavily pigmented eyespots present at middle level of esophagus. Acetabulum ventroterminal, measuring 0.083 by 0.05 mm. Excretory system was characterized by the Diplocotylea type of cercariae; there were no cross-connecting tubes. The two main excretory tubes, at level of eyespots, gave off laterally a diverticulum of a short length. Caudal tube terminated as a blind dilatation, joined to the excretory system by a tubular network of the acetabulum. The rudimentary genital system was clearly defined in whole mounts. Adeloscariae hemispherical, brownish, measuring 0.33 mm in diameter. ① Peter. Development in definitive host. Ass foal was fed on 2054 cyst, the first eggs began to pass out 98 d after infection. Take was 13.6%.

Habitat: Large intestine.

Hosts: *Equus przewalskii* f. caballus, *E. africanus* f. asinus (Equidae); *Elephas maximus* (Elephantidae).

Distribution: Africa — Egypt; Asia — India, Burma.

References: Bhalerao,[88] Looss,[561] Peter,[714] Varma.[1011]

KEY TO THE SPECIES OF *GASTRODISCUS*

1. Genital pore prebifurcal, pharynx Gastrodiscus type .... *G. aegyptiacus*
   Genital pore postbifurcal, pharynx Pseudodiscus type .... *G. secundus*

3.2.3.1.4. BALANORCHIIDAE Ozaki, 1937

Diagnosis: Cladorchoidea. Body conical, almost circular in cross section. Acetabulum small, ventroterminal. Pharynx terminal, with extramural appendages; esophagus short, ceca extending to acetabulum. Testes large, entirely oval, horizontal, just before acetabulum. Ovary spherical, pretesticular. Genital pore postbifurcal, hermaphroditic pouch present, protrudible. Uterine coils intercecal, pretesticular. Vitelline follicules extending along ceca from genital pore to acetabulum. Parasitic in forestomach of ruminants.

Type genus: *Balanorchis* Fischoeder, 1901
    Syn. *Verdunia Lahille* et Joan, 1917
Diagnosis: With characters of the family.

Type species: *Balanorchis anastrophus* Fischoeder, 1901 (Figure 344)
Syn. *Verdunia tricoronata* Lahille et Joan, 1917
Diagnosis: Body length 1.7—3.5 mm, body width 0.9—1.3 mm. Acetabulum 0.38—0.53 mm, Streptocoelium type, number of muscle units, d.e.c., 12—17; d.i.c., 20—23; v.e.c., 19—29; v.i.c., 24—39; m.e.c., 10—13; (Figure 177). Anterior body end surrounded with long, fingerlike papillae. Pharynx 0.14—0.23 mm long, Balanorchis type (Figure 37); appendages 0.18—0.32 mm long; esophagus 0.50—1.25 mm in length. Right testis 0.32—0.43 mm long, left testis 0.50—0.46 mm long. Hermaphroditic pouch 0.51—0.60 by 0.23—0.32 mm, terminal genitalium Balanorchis type (Figure 80). Ovary 0.12—0.14 by 0.12—0.21 mm. Eggs 0.124—0.131 by 0.071—0.082 mm.

Biology: Not yet known.

Habitat: Rumen, reticulum.

Hosts: *Bos primigenius* f. taurus (Bovidae); *Cervus dichotomus*; *Odocoileus gymnotus* (Cervidae).

Distribution: South America — Brazil, Venezuela, Paraguay, Argentina, Uruguay.

References: Amato and Gutierras,[25] Amato et al.,[26] Caballero et al.,[151] Calzada-Verella,[155]

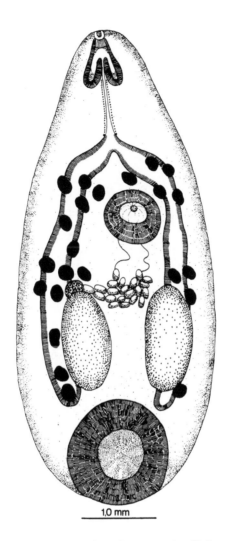

FIGURE 344. *Balanorchis anastrophus* Fischoeder, 1901. (After F. Fischoeder, 1903.)

Fischoeder,[311] Lahille and Joan,[518] Melo and Ribeiro,[603] Pinto and Almeida,[728] Roveda and Ringuelet,[792] Schiffo and Lombardero,[820] Szidat and Núñez,[942] Travassos et al.,[988] Velázquez-Moldano,[1019] Viana,[0132] Vogelsang and Rodriguez.[1038]

3.2.3.1.5. BRUMPTIIDAE (Skrjabin, 1949) Yamaguti, 1971

Diagnosis: Cladorchiidae. Body convex ventral, with or without paired auricular caudal appendages. Acetabulum ventroterminal. Pharynx terminal, with large extramural appendages; esophagus short or moderately long; ceca sinuous, terminating at acetabular field or before acetabulum. Testes horizontal or tandem, entire or lobed, situated at acetabulum or before it. Ovary spherical, median or submedian at acetabulum or before it. Genital pore bifurcal, hermaphroditic pouch present. Uterine coils winding transversely between acetabulum and genital pore. Vitelline follicles concentrated either to auricular caudal lobes or laterally along ceca. Parasitic of digestive tract of mammals.

KEY TO THE GENERA OF BRUMPTIIDAE

1. Testes horizontal, body with paired caudal appendages . . . . *Brumptia*

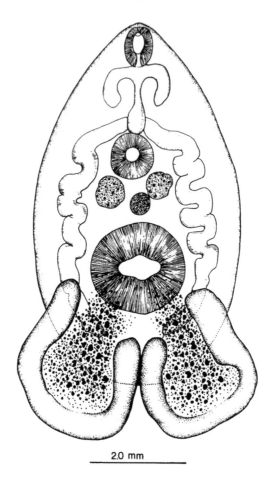

2.0 mm

FIGURE 345.   *Brumptia bicaudata* (Poirrier, 1908).
(Courtesy of H. W. Stunkard.)

Body with caudal appendages.. 2
2.   Ovary at acetabular field.. *Hawkesius*
Ovary far away from acetabulum .... *Choerocotyle*

Type genus: *Brumptia* Travassos, 1921

Diagnosis: Body large, stubby, with paired caudal appendages. Acetabulum ventroterminal, between basis of caudal lobes. Pharynx terminal, with large extramural appendages; esophagus moderately long; ceca sinuous, embracing acetabulum laterally and extending to body end. Testes horizontal, entirely spherical, median to ceca. Ovary spherical, anterodorsal to acetabulum. Genital pore bifurcal, hermaphroditic pouch present. Uterine coils mainly dorsal, between acetabulum and terminal genitalium. Vitelline follicles concentrated to caudal appendages. Parasitic in digestive tract of elephants and rhinos.

Type species: *Brumptia bicaudata* (Poirrier, 1908) Stunkard, 1926 (Figure 345)

Syn. *Amphistoma bicaudata* Poirier, 1908; *Cladorchis gigas*, MacCallum, 1917; *Brumptia gigas* Travassos, 1921. *B. bicaudata* Stunkard, 1926.

Diagnosis: With characters of the genus. Body length 12.3—16.8 mm, body width 7.6—9.3 mm. Acetabulum 2.35—3.54 mm in diameter, Brumptia type, number of muscle units, d.e.c., 48—52; d.i.e., 28—71; v.e.c., 15—17; v.i.c., 28—32; m.e.c., 10—14 (Figure 179),

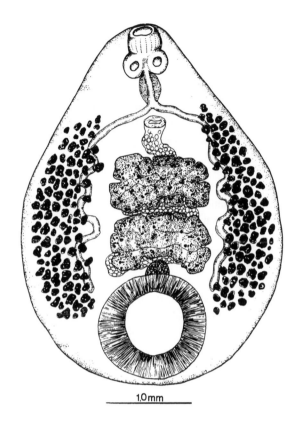

FIGURE 346. *Hawkesius hawkesi* (Cobbold, 1875).

caudal appendages 5.26—8.93 mm long. Pharynx 1.12—1.36 mm long and 1.20—1.48 mm in dorsoventral direction, Brumptia type (Figure 52); appendages 1.58—1.86 mm long; esophagus 1.68—1.73 mm long, without bulb. Right testis 3.12—3.50 by 2.89—3.11 mm, left one 2.98—3.36 by 2.86—3.24 mm. Ovary 0.86—0.98 mm in diameter. Hermaphroditic pouch 4.15—4.32 by 3.18—3.29 mm; genital pore postbifurcal, terminal genitalium Bicaudata type (Figure 81). Eggs 0.123—0.128 by 0.068—0.076 mm.

Habitat: Stomach.

Hosts: *Loxodonta africana* (Elephantidae), *Diceros bicornis* (Rhinocerotidae).

Distribution: Africa — Chad, Central Africa, Zaire, Zambia.

References: Dollfus,[239] Graber,[357] MacCallum,[573] Mettrich,[613] Poirrier,[732] Sey and Graber,[864] Strydonck,[920] Stunkard,[924] Travassos,[979] Willmott.[1066]

Hawkesius Stiles et Goldberger, 1910

Diagnosis: Brumptiidae. Body pyriform, flattened, convex dorsally. Acetabulum subterminal. Pharynx terminal with extramural appendages; esophagus short, ceca undulating, extending to acetabulum. Testes tandem, transverse oval, slightly lobed, in midzone. Ovary spherical, median anterodorsal to acetabulum. Genital pore postbifurcal, hermaphroditic pouch present. Uterine coils dorsal to testes between ovary and genital pore. Vitelline follicules lateral along ceca from bifurcation to acetabulum. Intestinal parasite of elephants and rhinos.

Type species: ***Hawkesius hawkesi*** **(Cobbold, 1875) Stiles et Goldberger, 1910 (Figure 346)**

Syn. *Amphistoma hawkesi* Cobbold, 1875; *A. ornatum* Cobbold, 1882; *Pseudodiscus (Hawkesius) hawkesi* (Cobbold, 1875) Stiles et Goldberger, 1910

Diagnosis: With characters of the genus. Body length 3.6—5.4 mm, body width 2.3—3.6 mm. Acetabulum 1.26—1.68 mm in diameter. Hawkesius type, number of muscle units, d.e.c., 42—44; d.i.c., 33—35; v.e.c., 21—24; v.i.c., 26—28; m.e.c., 10—14 (Figure 181). Pharynx 0.53—0.71 by 0.54—0.083 mm, Watsonius type (Figure 57); appendages 0.42—0.64 mm long; esophagus 1.06—1.31 mm in length. Anterior testis 0.52—1.14 by 1.23—1.52 mm, posterior testis 0.64—1.13 by 1.23—1.64 mm. Ovary 0.36—0.52 by 0.51—0.76 mm. Terminal genitalium Hawkesius type (Figure 83), hermaphroditic pouch 0.53—0.64 by 0.44—0.49 mm in immature specimens. Genital pore postbifurcal. Eggs 0.134—0.153 by 0.054—0.073 mm.

Biology: Not yet known.

Habitat: Colon.

Hosts: *Elephas maximus* (Elephantidae), *Rhinoceros unicornis* (Rhinocerotidae).

Distribution: Asia — India, Vietnam.

References: Cobbold,[177] Fukui,[334] Janchev,[447] Malik et al.,[583] Nickel,[649] Sey,[853] Srivastava and Ghosh,[911] Stiles and Goldberger,[915] Werthuysen.[1052]

Choerocotyle **Baer, 1959**

Diagnosis: Brumptiidae. Body flattened, oval, convex dorsally, with attenuated lateral edges. Acetabulum subterminal. Pharynx terminal with extramural appendages; esophagus short, ceca sinuous, extending to ovarian field. Testes tandem, strongly lobed, in midzone of body. Ovary spherical, submedian far away from acetabulum. Genital pore bifurcal, hermaphroditic pouch present. Uterine coils posterior to testes, short. Vitelline follicules extending in extracecal fields from genital pore to near acetabulum. Parasitic in intestine of African boars.

Type species: *Choerocotyle epuluensis* **Baer, 1959 (Figure 347)**

Diagnosis: With characters of the genus. Body length 15.0—18.0, body width 9.0—12.0 mm. Acetabulum 1.50—1.80 mm, Gastrodiscus type number of muscle units, d.e.c., 19—22; d.i.c., 21—25; v.e.c., 6—10; v.i.c., 19—27; m.e.c., 20—23; (Figure 180). Pharynx 1.20—1.40 mm long, Choerocotyle type (Figure 53), appendages 1.20—1.40 mm long. Genital pore bifurcal, terminal genitalium Epuluensis type (Figure 82). Eggs 0.119—0.128 by 0.055 mm.

Biology: Not yet known.

Habitat: Colon.

Hosts: *Hylochoerus meinertzhageni, Phacochoerus aethiopicus* (Suidae).

Distribution: Africa — Central Africa, Zaire.

References: Baer,[57] Graber,[360] Thal,[964] Troncy et al.[993]

3.2.3.2. PARAMPHISTOMOIDEA Stiles et Goldberger, 1910

Diagnosis: Paramphistomata. Body thick, conical, rarely flattened, ventral pouch rarely present. Acetabulum terminal, or subterminal, occasionally ventral. Pharynx terminal with or without appendages; esophagus with various length, may or may not be provided with bulb or muscular thickening, ceca simple, more or less undulating, terminating usually at or near acetabulum. Testes double, usually in middle or posterior part of body. Cirrus sac or hermaphroditic pouch absent, genital pore midventral with or without genital sucker. Terminal genitalium has various appearances. Vas deferens differentiated. Ovary posttesticular, usually spherical in outline. Uterine coils mainly intercecal, dorsal to testes. Vitelline follicules usually lateral along ceca, occasionally confluent. Excretory vesicle and Laurer's canal crossing or not crossing each other. Lymphatic system present. Parasites of alimentary tract and liver of mammals and rarely of birds.

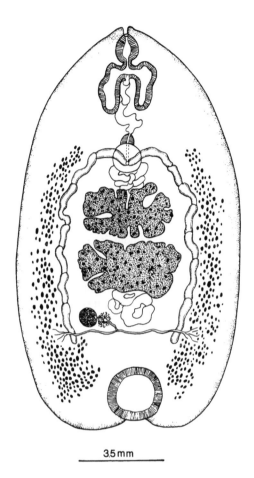

FIGURE 347. *Choerocotyle epuluensis* Baer, 1959.
(After J. G. Baer, 1959.)

KEY TO THE FAMILIES OF PARAMPHISTOMOIDEA

1. Pharyngeal appendages present, no crossing of excretory duct and Laurer's canal.....
.... Zygocotylidae
   Pharyngeal appendages absent 2
2. Ventral pouch present... Gastrothylacidae
   Ventral pouch absent... Paramphistomidae

3.2.3.2.1. ZYGOCOTYLIDAE Sey, 1988

Diagnosis: Paramphistomatoidea. Body pyriform or elongate, flattened ventrally, convex dorsally, uniform, occasionally divided. Acetabulum ventral or ventroterminal, simple or occasionally with a caudal overhanging lip on both sides. Pharynx terminal with paired or unpaired pharyngeal appendages, esophagus usually short, simple or muscular ceca extending to acetabular zone or terminating before it. Testes tandem or slightly diagonal, occasionally horizontal, entire or lobed, intercecal, rarely cecal. Ovary spherical to oval, median or submedian, posttesticular. Genital pore bifurcal pre- or postbifurcal. Uterine coils mainly dorsal to testes between ovary and bifurcation. Vitelline follicules strongly developed, along lateral sides, from esophageal zone to acetabulum. Laurer's canal opens before excretory pore. Lymphatic system present. Intestinal parasites of mammals and birds.

KEY TO THE SUBFAMILIES OF ZYGOCOTYLIDAE

1. Pharynx with primary pharyngeal sacs... 2
    Pharynx with pharyngeal bulb and secondary pharyngeal sacs.................... 4
2. Pharynx with esophagus having regular structure 3
    Pharynx with muscular esophagus .... Olveriinae
3. Pharynx with paired primary pharyngeal sacs.......................... Zygcotylinae
    Pharynx with unpaired primary pharyngeal sacs .... Stephanopharynginae
4. Testes horizontal... Pseudodiscinae
    Testes tandem.. Watsoniinae

Zygocotylinae Ward, 1917

Diagnosis: Zygocotylidae. Body elongate, oval flattened. Acetabulum ventroterminal provided with a caudal lip on each side or without it. Pharynx terminal with paired small extramural appendages, esophagus short, with or without bulb, ceca simple, extending to acetabulum. Testes tandem, lobed, in midzone of body. Ovary median, posttesticular. Genital pore just postbifurcal. Uterine coils dorsal to testes. Vitelline follicules strongly developed, lateral, from pharynx to acetabulum. Intestinal parasites of birds and mammals.
Type genus: Zygocotyle Stunkard, 1917
Diagnosis: Body elongated, flattened with caudal lips. Acetabulum ventroterminal. Pharynx terminal with paired extramural appendages, esophagus short, with bulb, ceca simple, extending to acetabular zone. Testes tandem, lobed, in middle part of body. Ovary median, intercecal. Genital pore postbifurcal. Uterine coils dorsal to testes. Vitelline follicules lateral from esophagus to acetabulum. Intestinal parasites of birds and mammals.

Type species: *Zygocotyle lunata* (Diesing, 1836) Stunkard, 1917 (Figure 348)

Syn. *Amphistoma lunatum* Diesing, 1936; *Zygocotyle ceratosa* Stunkard, 1917, *Chiorchis lunatus* Travassos, 1921.
Diagnosis: With characters of the genus. Body length 4.3—62. mm, body width 2.3—3.4 mm. Acetabulum 1.36—1.45 by 0.65—0.71 mm, Zygocotyle type, number of muscle unit, d.e.c., 60—64; d.i.c., 28—35; v.e.c., 24—27; v.i.c., 45—49; m.e.c., 10—15 (Figure 162). Pharynx 0.37—0.68 mm, Zygocotyle type (Figure 50), appendages 0.17—0.21 mm long, esophagus 0.31—0.48 mm long. Anterior testis 0.24—0.42 by 0.30—0.63, posterior testis 0.31—0.58 by 0.54—0.62 mm. Vesicula seminalis long and convoluted. Ovary 0.12—0.28 mm in diameter. Terminal genitalium Zygcotyle type (Figure 131). Eggs, 0.132—0.156 by 0.073—0.095 mm.
Biology: References. Willey,[1058—1060] Fried,[329] Fried and Nelson.[330]

Preparasitic stage. ① Willey. Eggs contained zygote, unsegmented embryo. Embryonic development occurred in 19 to 40 d under laboratory conditions. It seemed probable that hatching stimulated by light and emergence was a consequence of glandular secretion of miracidia and their body movement and cilial action. Hatched miracidia did not show phototropism or any tendency to concentrate in any one place. Body length 0.18—0.21 mm body width 0.053—0.059 mm. Covered with ciliated epidermal cells, arranged in four rows by the formula 6:8:4:2. Inner structure of miracidia composed of apical gland, two large flame cells, one on each side, several germ balls, and germinal cells. Penetration glands were not observed. Life span of miracidia up to 7 h; they usually infect intermediate host in the first 2 h. Intermediate host. ① Willey. *Helisoma antrosum*, a common planorbid snail, easy to infect with miracidia of *Z. lunata*. In laboratory, infectional rate may vary from 10 to 55%. ② Fried and Nelson. *Helisoma trivolvis*.

Intramolluscan stage. ① Willey. Mature sporocysts were found between 22—28 d postinfection. Beginning of the 25th day after infection mother rediae were present. Rediae

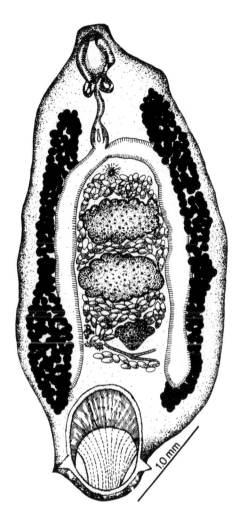

FIGURE 348. *Zygocotyle lunata* (Diesing, 1836).
(After L. Travassos, 1934.)

sausage-shaped, without locomotion appendages. Body length 0.486—0.986 mm long. Pharynx 0.032—0.040 mm in diameter, short gut and birth pore as well as 12 flame cells present. Rediae contained a few daughter rediae and cercariae in various stages of development. *Cercariae poconensis*, was the cercariae of *Z. lunata* (Willey).[1058-1060] Cercariae emerge from the rediae in an immature condition, and they complete their development in the digestive gland of the intermediate host. Body of mature cercariae 0.64—1.12 mm long, heavily pigmented. Cystogenous cells covered all the available space of the dorsal surface. A pair of eyespots were present at level of esophagus. Acetabulum spherical 0.21 mm in diameter. Pharynx 0.064 mm in diameter, with appendages. Esophagus short, with bulb; ceca terminating before acetabulum. Tail was almost twice as long as the body. The large main excretory ducts arise and run anteriorly, median to ceca. Along its course, each main duct has a number of lateral and median diverticula. The caudal tube joins the excretory bladder dorsally and passes into the tail, and neither forking nor openings can be found in the tail of fully mature cercariae. Heavily stained mass of cells constitute the rudimentary reproductive organs. Daily cercarial production of a snail is about 10—15 cercariae on the average. Emerged cercariae encysted soon on vegetation or other objects, cysts were dome-shaped with thick walls, with an average of 0.289 mm in diameter, life span of adolescariae was not longer than 4 months.

Development in definitive hosts. Laboratory-raised rats, duck, sheep proved to be successive to this fluke; attempts to infect pigeons and rabbits ended in negative results. Development in ducks was more rapid than in mammals. The usual site of this fluke is the ceca both in birds and mammals. In rats they reach their maturity in 46—61 d, in ducks 41 or 42 d after infection. Life span was more than 2 years in both ducks and rats. Worms gradually increased during this time and measured 9.10 by 4.65 mm after 7—11 d postinfection.

Habitat: Cecum.

Hosts: *Aves ipecutisi, A. melanotus, A. focata, A. platyrhynchos* f. domestica, *A. erythror-hyncha, A. subripes, A. crecca, A. acuta, Poecilone bahamensis; Aix sponsa; Anser anser* f. domesticus, *Cairina moschata* f. domestica, *Bucephala clangula, Amazonetta brasiliensis, Lophodites cucullatus, Marila americana, Nettion caroliensis* (Anatidae); *Capella gallinago* (Scolopacidae); *Recurvirostra americana, Himantopus wilsonii* (Recurvirostridae); *Gallus bankiva* f. domesticus, *Meleagris gallopavo intermedia, Meleagris gallopavo* f. domestica (Phasianidae); *Podiceps ruficollis* (Podicipedidae), Mammalia: *Alces alces, Cervus dichotomus, Odocoileus* sp. (Cervidae); *Bos primigenius* f. taurus (Bovidae); *Rattus norvegicus* f. domestica (exp.), *Mus musculus* f. domestica (exp.) (Muridae).

Distribution: North America — U.S., Canada; Central America — Panama, Mexico; South America — Brazil; Asia — U.S.S.R., India; Africa — Zimbabwe and Malagasy.

References: Bain and Threlfall,[60] Bhatnagar et al.,[92] Caballero,[144] Dollfus,[244] Duarte,[246] Fernandez and Travassos,[307] Fischoeder,[311] Khuan Shen-i,[484] Kinsella and Forrester,[486] Leonov et al.,[542] Mettrich,[612] Pinto and Almeida,[728] Price,[741] Richard and Daynes,[777] Samuel et al.,[814] Self and Bouchard,[827] Sey,[837] Stock and Barrett,[916] Stunkard,[921] Threlfall,[974] Travassos.[984]

Wardius Barker et East, 1915

Diagnosis: Zygocotylinae. Body cylindrical. Acetabulum ventroterminal. Pharynx terminal with extramural appendages, pharynx short without bulb, ceca sinuous, extending to anterior part of acetabulum. Testes tandem, indented in middle third of body. Ovary spherical, about midway between posterior testis and acetabulum. Genital pore bifurcal. Uterine coils winding posterior and dorsal testes. Vitelline follicules lateral along ceca from pharynx to acetabulum. Intestinal parasite of rodents.

Type species: *Wardius zibethicus* Barker et East, 1915 (Figure 349)

Syn. *Pseudodiscus (Watsonius) zibethicus* Fukui, 1929, *Watsonius zibethicus* Sprehn, 1952

Diagnosis: With character of the genus. Body length 4.0—13.0 mm, body width 1.0—4.4 mm, acetabulum 1.11—2.79 by 1.11—2.29 mm in size, Wardius type, number of muscle units, d.e.c., 80—85; d.i.c., 68—73; v.e.c., 48—52; v.i.c., 93—98; m.e.c., 15—17; (Figure 161). Pharynx 0.43—0.99 mm long, Zygocotype type (Figure 50), appendages 0.52—1.13 mm long, esophagus 0.72—0.86 mm long. Interior testis 0.49—1.73 mm, *posterior testis* 0.49—2.29 mm. Vesicula seminalis long and convoluted. Ovary 0.28—0.34 mm in diameter. Terminal genitalium Zygocotyle type (Figure 131).

Biology: References. Smith,[893] Murrell.[638]

Preparasitic stage. ① Smith. Incubation time of eggs during winter season in Michigan was 45 to 50 d. Eggs deposited in 2—4 cells of cleavage. At room temperature, the embryo developed within 6 d. Mature miracidia measuring 0.267 by 0.072 mm, epithelial cells arranged according to the formula 6:8:4:2. Apical gland, excretory system (consists of two flame cells) germinal masses in posterior body half were present. Life span is about 15 h. Intermediate host: *Helisoma antrosa* (Murrell[638]).

Intramolluscan stage. ② Murrell. Body of redia sausage-shaped, 0.55—0.75 by

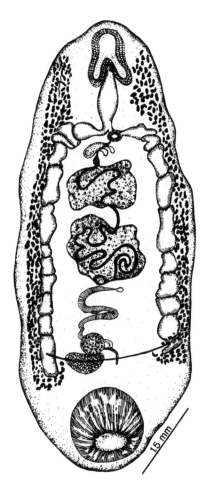

FIGURE 349. *Wardius zibethicus* Barker
et East, 1915. (After F. D. Barker and
A. M. East, 1915.)

0.11—0.28 mm. Mouth surrounded by 12 papillae, pharynx 0.031—0.049 mm in diameter, esophagus 0.02 mm long, ceca 0.10—0.12 mm long, saccate. Three pairs of flame cells present. Birth pore ventral, near posterior end of ceca. Immature cercariae escape from redia and complete their development in the snail tissue. Cercariae heavily pigmented, body measuring 0.35—0.49 by 0.20—0.29 mm. Tail 0.65—0.88 by 0.071—0.091 mm. Acetabulum ventroterminal, 0.131—0.175 in diameter. Cystogenous granules, embedded parenchyma of the dorsal side, densely packed. Two conical eyespots were located in the dorsal surface, at level of esophagus. Pharynx 0.068—0.095 mm in diameter, with two appendages, esophagus one fifth of body length, ceca terminating at level of excretory bladder. Primordia of genital organs seen in the form of heavily stained cell masses. The main excretory ducts arise at posterolateral side of bladder, extending anteriorly, giving off several blind branches. Caudal tube extending almost full tail length, without forming primary excretory pore in the tail. Cyst of adolescariae 1.24—1.26 mm in diameter, very dark. Green vegetation was preferred as a place for encystment.

Development in definitive host. Gerbils, albino rats, young racoon, and ducks did not prove to be successive definitive hosts. Incomplete development of adult stage was attained in white mice, hamster, mouse deer, and guinea pigs. Lymphatic system was observed in immature worms.

Habitat: Cecum.

Host: *Ondatra zibethica* (Cricetidae).

Distribution: North America — U.S., Canada.

References: Allen,[22] Anderson and Beaudoin,[28] Ball,[68] Barker,[70] Knight,[491] MacKinnon and Burt,[576] McKenzie and Welch,[601] Rausch,[770] Rice and Heck.[776]

Choerocotyloides Prudhoe, Yeh et Khalil, 1964

Diagnosis: Zygocotylinae. Body fleshy, flattened, somewhat fusiform in outline, ridges ventral surface with papillae. Acetabulum ventroterminal, moderately large. Pharynx terminal with extramural appendages, esophagus long, without bulb; ceca terminating in front of acetabulum. Testes tandem, lobed in middle third of body. Ovary oval, slightly lobed submedian, about midway between posterior testis and acetabulum. Genital pore prebifurcal, pars musculosa lined with ciliated epithelia and projects in form of a papilla into much dilated portion (ciliated chamber). Uterine coils composed of several transverse slings and confined to intercecal field anterior to ovary. Vitelline follicules lateral to ceca, extending from esophagus to acetabulum. Parasite of cecum of ruminants.

Type species: ***Choerocotyloides onotragi* Prudhoe, Yeh et Khalil, 1964 (Figure 350)**

Diagnosis: With characters of the genus. Body length 13.5—16.0 mm, body width 5.0—5.6 mm. Acetabulum 2.0—2.6 mm in diameter, Choerocotyloides type, number of muscle units, d.e.c., 21—24, d.i.c., 14—16, v.e.c., 10—13, v.i.c., 13—17, (Figure 137). Pharynx terminal, 1.0—1.32 by 0.9—1.1 mm, Microrchis type (Figure 43), appendages 0.30—0.36 mm long, esophagus 2.0—3.5 mm long. Terminal genitalium Gracile type (Figure 114). Testes 1.3—1.5 by 1.6—2.2 mm in size. Ovary 0.55—0.73 by 0.55—0.69 mm. Eggs, 0.112—0.120 by 0.06—0.06 mm.

Biology: Not yet known.

Habitat: Cecum.

Host: *Onotragus leche* (Bovidae).

Distribution: Africa — Zambia.

Reference: Prudhoe et al.[758]

Olveriinae Srivastava, Maurya et Prasad, 1980

Diagnosis: Zygocotylidae. Body elongate pyriform. Acetabulum terminal, moderately large. Pharynx terminal with extramural appendages, esophagus divided into a strongly muscular anterior and a short nonmuscular posterior portion, ceca long, coiled with some anteroposterior loops, terminating near ovary. Testes diagonal, more or less rounded, in posterior part of body. Ovary submedian, preacetabular. Genital pore at about bifurcation. Uterus ascending on the right, forwarding to left, and turning again to the right side. Vitelline follicules along lateral sides, extending from about genital opening to cecl end. Parasitic in forestomach of ruminants.*

Type genus. *Olveria* Thapar and Singha, 1945

Diagnosis: With character of the subfamily.

Type species. *Olveria indica* Thapar et Singha, 1945 (Figure 351)

Syn. *O. thapari* Bali et Fotedar, 1974

Diagnosis: Body length 6.1—7.0 mm, body width 2.3—3.3 mm. Acetabulum 1.30 by 0.73 mm in size, Streptocoelium type, number of muscle units, d.e.c., 24—26; d.i.c., 24—39; v.e.c., 20—26; v.i.c., 29—39; m.e.c., 2—7 (Figure 177). Pharynx 0.35—0.42 mm, Olveria type (Figure 45), appendages 0.28—0.34 mm long. Esophagus 3.10—3.90 mm long; anterior testis 1.20 by 0.95 mm, posterior testis 1.33 by 1.14 mm. Ovary 0.40—0.43 mm in diameter. Terminal genitalium Olveria type (Figure 90). Eggs 0.068—0.075 by 0.085—0.119 mm.

---

* *Surreshiella* Srivastava, Maurya et Prasad, 1980 was not included due to its incomplete description.

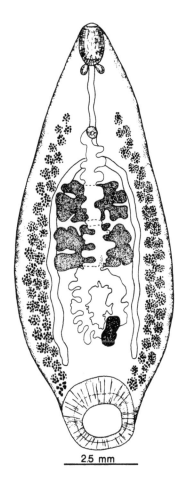

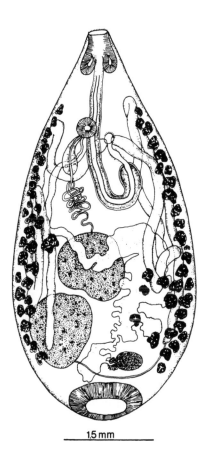

FIGURE 350. *Choerocotyloides ono-tragi* Prudhoe, Yeh et Khalil, 1964. (Courtesy of S. Prudhoe et al. and by permission of the *J. Helminthol.*, 1964, London.)

FIGURE 351. *Olveria indica* Thapar et Singha, 1945. (After G. S. Thapar and B. B. Singha, 1945.)

Biology: References. Tandon,[948] Thapar.[967]

Preparasitic stage. ① Tandon. Eggs operculated, 0.12—0.14 by 0.068—0.085 mm. Embryonic development took 15 d. Fully developed miracidium occupied the eggshell, mucoid plug present. Miracidia emerged usually in the morning. Emerged miracidia were active and preferred to move near the surface of the water. Miracidium oval or ovoid, measuring 0.126 by 0.036 mm. Epithelial cells arranged in four tiers, giving a formula 6:8:4:2. Eyespots absent contrary to Thapar's observation. Miracidia composed of a cephalic gland, two pairs of penetration glands, a brain mass, a pair of flame cells with ducts, and the germinal cells. ② Thapar: Eggs oval, 0.13 by 0.015 mm. Miracidia develop in 8—10 d in temperature 32.2 to 34.4°C. Body of miracidia emerged 0.14—0.22 by 0.054—0.064 mm, covererd with epithelial cells arranged in four tiers: 6:8:4:2. Miracidia composed of cephalis gland, penetration glands, a pair of flame cells and germinal masses. Germinal cells migrated from the subepithelial layer to the central cavity and form germinal balls in the middle and posterior third of body. Intermediate host: *Gyraulus convexiusculus.*

Intramolluscan stage. ① Thapar. On the fourth day of the infection the sporocyst was found in the mantle cavity. Its body measured 0.17 by 0.18 mm and contained germinal cells. Sporocyst of about 7-d age, besides germinal cell masses, contained young

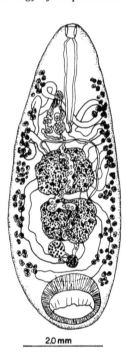

2.0 mm

FIGURE 352. *Olveria bosi*
Tandon, 1951. (After R. S. Tandon, 1951.)

rediae, later developing rediae with easily visible redial organs. Rediae were elongated, 0.20—1.0 mm long, contained germ balls of different stages. Mature rediae measures 1.0 mm in length. It has a small pharynx, followed by a saclike cecum. The excretory system consists of three pairs of flame cells. There were no locomotion appendages. Birth pore situated at the first quarter of the body. Mature rediae contained well-differentiated but immature cercaria. Body of cercaria heavily pigmented, pyriform in size, measuring 0.41 by 0.21 mm, tail 0.36 by 0.05 mm long. Pigmentation began among the eyespots, radiating over a large area on both sides. Eyespots situated at level of esophagus. Acetabulum subterminal, 0.09 mm in diameter. Pharynx 0.06 by 0.50 mm with appendages, esophagus 0.11 mm long; ceca terminating before acetabulum. Main excretory tubes with cross-connecting tubes at about middle of body. Caudal tube bifurcated and opened laterally near the tail tip. Emerged cercariae encysted on vegetation on the wall of container or near the water surface.

Development in definitive host: One buffalo calf was fed on adolescariae encysted on lettuce leaves. Autopsy after 92 d revealed immature specimens of *O. indica*.
Habitat: Rumen.
Hosts: *Bubalus arnee* f. bubalis, *Bos primigenius* f. taurus, *Ovis ammon* f. aries (Bovidae).
Reference: Bali et al.,[67] Bali and Fotedar,[66] Thapar,[965] Thapar and Singha.[968]

Oliveria bosi Tandon, 1951 (Figure 352)

Diagnosis: Body length 7.6—9.0 mm, body width 2.8—3.0 mm. Acetabulum 1.15—1.89 by 1.50—2.33 mm. Pharynx with two dorsolateral sacs. Esophagus consists of a muscular portion, measuring 2.17—2.25 mm long, and nonmuscular portion, measuring 0.92—0.95 mm. Anterior testis 1.26—1.47 by 1.55—1.83 mm, posterior testis 1.17—1.94 by 1.39—1.61 mm. Ovary 0.27—0.36 by 0.46—0.49 mm. Eggs 0.110—0.117 by 0.065—0.073 mm. Biology: Not yet known.

Habitat: Rumen.
Host: *Bubalus arnae* f. bubalis (Bovidae).
Distribution: Asia — India.
References: Tandon,[946] Thapar.[965]

KEY TO THE SPECIES OF *OLVERIA*

1. Cecal extremities turn inward and overlap each other in front of acetabulum . . . .
. . . . *O. bosi*
   Ceca without inturning and overlapping . . . . *O. indica*

Stephanopharynginae Stiles et Goldberger, 1910

Diagnosis: Zygocotylidae. Body conical or plump with rounded extremities. Acetabulum ventroterminal. Pharynx subterminal with an unpaired extramural appendage, esophagus short without bulb. Testes tandem intercecal, lobed. Ovary submedian anterodorsal to acetabulum. Genital pore bifurcal. Uterine coil dorsal to testes, between ovary and genital pore. Vitelline follicules lateral, along ceca, from bifurcation to acetabular zone. Parasitic in rumen of ruminants.

Type genus: *Stephanopharynx* Fischoeder, 1901
Diagnosis: With characters of the subfamily.

Type species: *Stephanopharynx compactus* Fischoeder, 1901 (Figure 353)

Syn. *S. secundus* Stunkard, 1929, *S. coilos* Dollfus, 1963
Diagnosis: With characters of the subfamily. Body length 3.4—9.6 mm, body width 1.6—4.3 mm. Acetabulum 1.14—2.36 mm in diameter, Stephanopharynx type, number of muscle units, d.e.c., 38—42; d.i.c., 45—53; v.e.c., 35—42; v.i.c., 38—46; m.e.c., 18—14 (Figure 177). Pharynx 0.48—1.21 mm long in dorsoventral direction, Stephanopharynx type (Figure 51); appendage 1.18—3.64 mm long, esophagus 1.28—1.39 mm. Anterior testis 0.28—0.81 by 0.71—2.14 mm, posterior testis 0.31—0.82 by 0.81—1.76 mm in dorsoventral direction. Pars musculosa thick walled, well developed. Ovary 0.12—0.47 by 0.21—0.67 mm. Terminal genitalium Stephanopharynx type (Figure 97). Eggs 0.121—0.138 by 0.061—0.072 mm.

Biology: Not completely elucidated. Dinnik[228] came to the conclusion, on grounds of examining Le Roux's amphistome material, that *Bulinus forskalii* acts as intermediate host of this species.

Habitat: Rumen.
Hosts: *Bos primigenius* f. taurus, *Kobus kob, K. leche, K. varondi; K. defassa, Redunca arundium, R. redunca, Hippotragus equinus, Bubalus arnee* f. bubalis, *Connochaetes taurinus, Syncerus caffer, Gazella rufifrons* (Bovidae).
Distribution: Africa — Uganda, Zambia, South Africa, Tanzania, Chad, Zaire, Cameroon, Nigeria, Angola, Central Africa, and Mauritius Island.
References: Dinnik,[227,228] Dollfus,[244] Eduardo,[278] Ezzat,[295] Fischoeder,[311] Graber,[357,359,360] Graber et al.,[362] Gretillat,[369] Maplestone,[591] Ortlepp,[678] Sey and Graber,[864] Strydonck,[920] Stunkard.[925]

Pseudodiscinae Näsmark, 1937

Diagnosis: Zygocotylidae. Body conical, slightly concave ventrally. Acetabulum ventral or ventroterminal. Pharynx terminal with extramural appendages, esophagus short with bulb, ceca extending to acetabular zone. Testes horizontal, cecal in part, in middle part of body. Ovary inter- or posttesticular. Genital pore bifurcal or far behind it. Uterine coils winding posterior, intercecal. Vitelline follicules lateral along ceca from esophagus to acetabulum. Intestinal parasites of mammals.

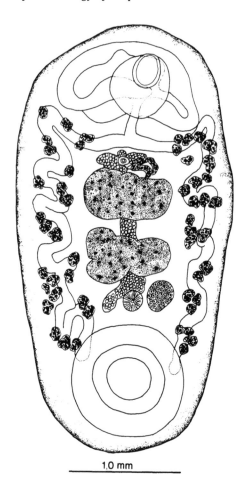

1,0 mm

FIGURE 353.   *Stephanopharynx compactus* Fischoeder, 1901. (By permission of Dr. W. Junk Publishers, 1986, Dordrecht.)

KEY TO THE GENERA OF PSEUDODISCINAE

1.   Acetabulum large, ventroterminal, papillated .... *Macropotrema*
     Acetabulum small, ventral, without papillae .... *Pseudodiscus*

Type genus: *Pseudodiscus* Sonsino, 1895
   Diagnosis: Pseudodiscinae. Body ovoid or oval. Acetabulum small, ventral. Pharynx terminal with extramural appendages, esophagus short, without bulb, ceca slightly winding, extending to acetabular zone. Testes horizontal in middle part of body. Ovary spherical, submedian. Genital pore at level of anterior part of testes. Uterine coils passing between two testes. Vitelline follicules well developed, lateral from esophagus to acetabulum. Parasitic in large intestine of horses, occasionally elephants and cattle.

Type species: *Pseudodiscus collinsi* (Cobbold, 1875) Stiles et Goldberger, 1910 (Figure 354)
   Syn. *Amphistoma collinsi* (Cobbold, 1875, *A. collinsi* Sonsino, 1895, *P. stenleyi* (Cobbold, 1875).
   Diagnosis: With characters of the genus. Body length 3.7—9.3, body width 2.2—5.1

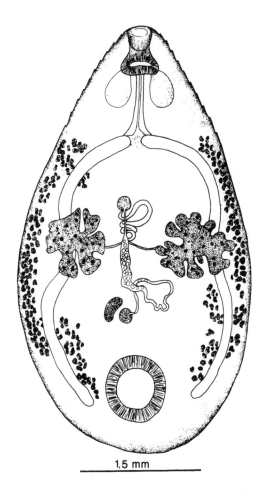

FIGURE 354. *Pseudodiscus collinsi* (Cobbold, 1875).
(Courtesy of P. Rai.)

mm. Acetabulum 1.48—1.87 mm in diameter, Pseudodiscus type, number of muscle units, d.e.c., 46—49; d.i.c., 32—36; v.e.c., 14—17; v.i.c., 23—26; m.e.c., 14—16 (Figure 183). Pharynx 0.85—1.05 mm in diameter, Pseudodiscus type (Figure 56) appendages 0.64—0.93 mm. Testes 0.74—2.30 mm by 0.63—1.67 mm. Pars musculosa relatively well developed. Ovary 0.25—0.70 mm in diameter. Terminal genitalium Pseudodiscus type (Figure 93). Eggs 0.143—0.172 by 0.092—0.101 mm.
Biology: Reference. Peter and Srivastava.[719]

Intramolluscan stage. Rediae dissected from intermediate host, (*Indoplanorbis exustus*), sausage-shaped, 0.47—1.04 by 0.17—0.29 mm. Locomotor appendages absent, birth pore at about anterior fourth of body. Pharynx 0.03—0.046 by 0.041—0.055 mm, cecum extending to level beyond birth pore. Salivary glands and three pairs of flame cells present. The rediae gave rise to cercariae or daughter rediae. Emergence of cercariae is stimulated by diffuse sunlight. On the contrary, of the other amphistome cercariae, except for *Cercaria poconensis*, the daily number of cercaria is very low, five to eight per day. It is Diplocotylea type, heavily pigmented on the dorsal side. Body dimension of alive cercariae varied 0.32—0.77 by 0.21—0.42; tail 0.62—1.14 mm long. Eyespots on dorsal surface, at level of esophagus each side. Acetabulum 0.15—0.16 mm. Pharynx terminal 0.074 by 0.083 mm in size, appendages 0.085 mm long, esophagus long, ceca terminating in front of acetabulum. Main excretory tubes running anteriorly with two lateral diverticules. The anterior continuation of the main tubes turns backward and joins the network of excretory tubes present in the substance of acetabulum. There are no cross-connectional tubes between main excretory tubes. Caudal tube takes its origin from the acetabular network, terminating blindly without lateral openings. Primordia of the reproductive system, seen on stained preparations, appeared to be strongly stained cell masses. Cercariae encysted in a period of 1/2 h. Any available substratum seems to be suitable for encystment. Adolescariae were dome shaped, measuring on an average 0.356 mm in diameter.

Development in definitive host. A 1-year-old donkey foal was infected with 233 adolescariae aged 10—35 d. First eggs were found in the feces 90 d postinfection. Later the donkey was killed, and a total of 65 flukes were collected from the large intestine (49 from cecum, 16 from colon).

Habitat: Large intestine.

Hosts: *Equus przewalskii* f. caballus, *E. africanus* f. asinus, (Equidae); *Bos primigenius* f. taurus (Bovidae), *Elephas maximus* (Elephantidae).

Distribution: Asia — India.

References: Cobbold,[177] Gupta and Walia,[391] Rai,[762] Rai and Srivastava,[763] Thapar.[965]

Macrorpotrema Blair, Beveridge et Speare, 1979

Diagnosis: Pseudodiscinae. Body conical. Acetabulum large, ventroterminal, papillated. Pharynx terminal, with extramural appendages, pharynx short, with muscular bulb, ceca simple, extending to acetabulum. Testes large, horizontal, lateral, cecal, just in front of acetabulum. Ovary submedian, intercecal. Genital pore bifurcal. Uterine coils winding, intercecal, filling up field between cecal arch and acetabulum. Vitelline follicules lateral, along ceca, extending from esophagus to cecal extremities. Parasitic in ceca of marsupials.

Type species: *Macropotrema pertinax* **Blair, Beveridge et Speare, 1979. (Figure 355)**

Diagnosis: With characters of the genus. Body length 4.0—5.0 mm, body width 2.0—2.5 mm. Acetabulum 1.90 mm in diameter, covered with numerous small spherical projections, Pertinax type, number of muscle units, d.e.c., 20—29; d.i.c., 28—39; v.e.c., 20—22; v.i.c., 31—35; m.e.c., 12—14 (Figure 174). Pharynx 0.47 by 0.50 in dorsoventral direction, Macropotrema type (Figure 55), appendages 0.55—0.62 mm, esophagus 0.5 mm long. Testes 0.85—0.93 by 0.73—0.89 in dorsoventral direction. Pars musculosa weakly developed. Ovary 0.22 by 0.32 mm. Terminal genitalium Pertinax type (Figure 125). Eggs 0.133—0.144 by 0.077—0.083 mm.

Biology: Not yet known. Speare[903] found that eggs have a thickening at antipolar end and operculum at polar end. Eggs laid contained actively moving miracidia.

Habitat: Cecum.

Host: *Macropus agilis* (Macropodidae).

Distribution: Papua New Guinea.

References: Blair et al.,[98] Owen[685] (personal communication); Speare.[903]

Watsoniinae Näsmark, 1937

Diagnosis: Zygocotylidae. Body pyriform or oval, more or less flattened, rarely convex dorsally; uniform or divided into anterior conical and posterior discoidal portions. Ventral surface sometimes papillated. Acetabulum well or moderately developed, ventral or ventroterminal. Pharynx terminal, with appendages, esophagus short, with bulb, ceca straight, terminating at level of acetabulum or before it. Testes tandem, or slightly diagonal, lobed or indented, intercecal. Ovary posttesticular, near to acetabulum. Pars prostatica either with prostatic cells or ciliated chamber. Genital pore bifurcal or prebifurcal. Uterine coils intercecal, winding posterior and dorsal to testes. Vitelline follicules in lateral fields, extending from bifurcal or testicular zone to acetabulum. Intestinal parasite of mammals.

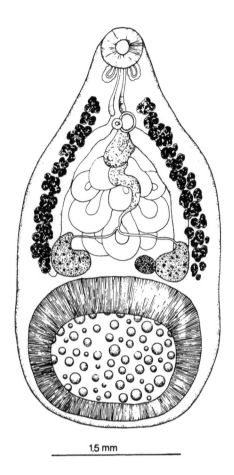

1,5 mm

FIGURE 355. *Macropotrema pertinax* Blair, Beveridge et Speare, 1979. (Courtesy of D. Blair.)

KEY TO THE GENERA OF WATSONIINAE

1. Body conical, flattened, divided into two portions................................. 2
   Body pyriform or oval not divided into two portions 3
2. Body divided into an anterior conical and posterior discoidal parts ....
   .... *Gastrodiscoides*
   Body divided into (anterior) large, flat, ventrally papillated and (posterior) small, spherical parts.. *Homalogaster*
3. Body pyriform and concave ventrally..................................... *Watsonius*
   Body oval and not convex ventrally .... *Skrjabinocladorchis*

Type genus: *Watsonius* Stiles et Goldberger, 1910

Diagnosis: Watsoniinae. Body pyriform, flattened ventrally, convex dorsally lateral margins of body tending inward. Papillar ridges on posterior part of ventral surface present. Acetabulum ventral or ventroterminal. Pharynx terminal, with extramural appendages, esophagus short, ceca terminating at acetabular zone or more anteriorly. Testes lobed, tandem, intercecal, in middle part of body. Ovary preacetabular, submedian, pars prostatica with prostate cells or ciliated chamber. Genital pore bifurcal or prebifurcal. Uterine coils winding posterior and dorsal to testes. Vitelline follicules lateral, extending from bifurcation to acetabulum. Intestinal parasite of Primates.

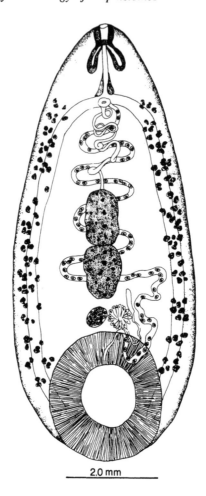

2.0 mm

FIGURE 356. *Watsonius watsoni* (Conyngham, 1964). (After A. E. Shipley, 1905, modified.)

Type species: *Watsonius watsoni* (Conyngham, 1904) Stiles et Goldberger, 1910 (Figure 356)

Syn. *Amphistoma watsoni* Conyngham, 1904, *Cladorchis watsoni*, Shipley, 1905; *Gastrodiscus watsoni* Verdun, 1905; *Paramphistomum watsoni* Manson, 1908; *Pseudodiscus watsoni* Fukui, 1929.

Diagnosis: Body length 8.0—10.0 mm, body width 4.0—5.0 mm. Acetabulum ventroterminal, 1.23 mm in diameter, Watsonius type, number of muscle units, d.e.c., 38—40; d.i.c., 36—38; v.e.c., 16—18; v.i.c., 34—36; m.e.c., 22—24 (Figure 184). Pharynx 1.23—1.86 mm long together with appendages; Watsonius type (Figure 57), esophagus 1.1 mm long. Testes slightly lobed 1.23—1.57 by 0.84—0.93 mm in size. Pars musculosa weak with several convolutions. Ovary 0.21 mm in diameter. Genital pore postbifurcal. Terminal genitalium Leydeni type (Figure 117). Eggs 0.120—0.130 by 0.075—0.080 mm.

Biology: Not yet known.

Habitat: Small and large intestine.

Hosts: *Cercopithecus callitrichus*, *C. sabaeus*, *Macaca irus*, *Mandrillus sphinx*, *Papio papio* (Cercopithecidae), *Homo sapien* (Hominidae).

Distribution: Africa — Nigeria, Guinea, Zambia; Asia — Japan (Flukes were collected from *Macaca irus*, which has Southeast-Asian distribution).

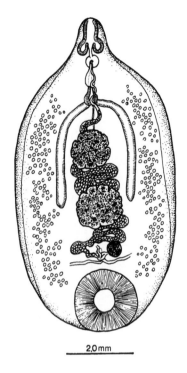

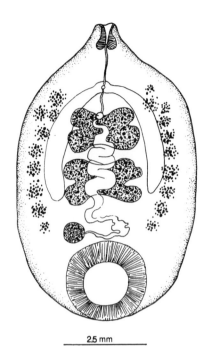

2,0 mm

2,5 mm

FIGURE 357. *Watsonius noci* (Barrois, 1908). (After R. Ph. Dollfus, 1963.)

FIGURE 358. *Watsonius macaci* Kobayashi, 1915. (After H. Kobayashi, 1915.)

References: Buckley,[135] Canyngham,[157] Deschiens,[210] Dollfus,[239] Leiper,[532] Pick and Deschiens,[725] Sey,[852] Shipley,[877] Stiles and Goldberger.[915]

Watsonius noci (Barrois, 1908) Sey, 1984 (Figure 357)

Syn. Chiorchis noci Barrois, 1908; C. (Prochiorchis) noci (Barrois, 1908) Dollfus, 1963.

Diagnosis: Body length 8.5—10.0 mm, body width 3.5—5.0 mm. Lateral margins of body turn inward. Acetabulum 1.83—2.14 mm in diameter, Pseudodiscus type, number of muscle units, d.e.c., 40—42; d.i.c., 20—22; v.e.c., 12—15; v.i.c., 10—14; m.e.c., 26—28 (Figure 183). Pharynx with appendages 1.14—1.36 mm long. Gastrodiscus type (Figure 54); esophagus 1.23—1.47 mm long. Testes 1.12—1.36 by 0.84—1.26 mm in size. Ciliated chamber present, pars musculosa weakly developed. Ovary 0.55 by 0.36 mm. Genital pore terminal. Terminal genitalium Noci type (Figure 89). Eggs 0.136—0.147 by 0.057—0.070 mm.

Biology: Not yet known.

Habitat: Cecum.

Host: *Macaca mulatta* (Cercopithecidae).

Distribution: Asia — Vietnam, Cambodia.

References: Barrois,[73] Dollfus,[243] Sey.[852]

Watsonius macaci Kobayashi, 1915 (Figure 358)

Diagnosis: Body length 8.0—11.3 mm, body width 5.0—6.5 mm. Acetabulum 2.16 mm in diameter. Pharynx 0.75 mm long, appendages 0.42—0.57 mm in diameter; esophagus 1.68 mm long, bulb present. Testes 1.54—1.86 main diameter, with 4 or 5 lobuli. Ovary 0.53 mm in diameter. Genital pore prebifurcal. Eggs 0.120 by 0.060 mm.

Biology: Not yet known.

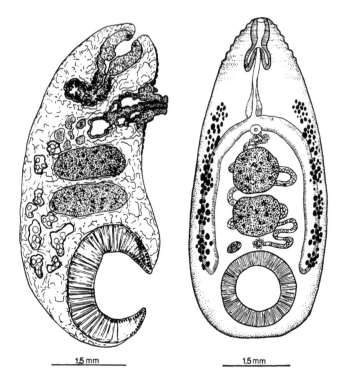

FIGURE 359. *Watsonius deschiensi* Pick, 1951 (median sagittal section).

FIGURE 360. *Watsonius papillatus* Sey, 1984. (By permission of the *Acta Zool.*, 1984, Budapest.)

Habitat: Cecum.
Host: *Macaca mulatta* (Cercopithecidae).
Distribution: Asia (The host has South and Southeast Asian distribution, the autopsy was done in Japan and Czechoslovakia.)
References: Kobayashi,[492] Zajíček and Valenta.[1089]

Watsonius deschiensi Pick, 1951 (Figure 359)

Diagnosis: Body length 5.0—8.0 mm, body width 2.0—4.0 mm. Acetabulum 1.86—2.17 mm in diameter, Watsonius type, number of muscle units, d.e.c., 30—32; d.i.c., 37—39; v.e.c., 11—13; v.i.c., 31—33; m.e.c., 12—14 (Figure 184). Pharynx with appendages 0.83—1.26 mm long, Gastrodiscus type (Figure 54); esophagus 0.73—1.12 mm long. Testes 0.61—0.92 by 0.93—1.12 mm in size. Ciliated chamber present, pars musculosa poorly developed. Ovary 0.32 by 0.43 mm in diameter. Genital pore bifurcal. Terminal genitalium Deschiensi type (Figure 107). Eggs 0.116—0.134 by 0.060—0.073 mm.
Biology: Not yet known.
Habitat: Cecum.
Hosts: *Mandrillus sphinx, Papio papio* (Cercopithecidae).
Distribution: Africa — Guinea.
References: Pick,[723,724] Pick and Deschiens,[725] Sey.[852]

Watsonius papillatus Sey, 1984 (Figure 360)

Diagnosis: Body length 6.5—7.0 mm, body width 2.0—2.2 mm. Lateral margins of body turn inward. Acetabulum 1.36—1.53 mm in diameter, Watsonius type, number of muscle units, d.e.c., 41—42; d.i.c., 33-35; v.e.c., 21—24; v.i.c., 31—35; m.e.c., 12—14 (Figure

184). Pharynx with appendages, 0.37—0.43 mm long, Pseudodiscus type (Figure 56); esophagus 0.76—0.93 mm long with bulb. Anterior testis 0.50—0.52 mm, posterior testis 0.62—0.64 mm in diameter. Pars musculosa moderately developed. Ovary 0.40—0.45 by 0.30—0.35 mm. Genital pore bifurcal. Terminal genitalium Leydeni type (Figure 117). Eggs 0.125—0.130 by 0.075—0.080 mm.

Biology: Not yet studied.

Habitat: Not given.

Host: Not given.

Distribution: Not given.

Reference: Sey.[852]

KEY TO THE SPECIES OF *WATSONIUS*

1. Testes indented or slightly lobed.. 2
   Tested strongly lobed, with 4 or 5 lobuli .... *W. macaci*
2. Pharynx of Watsonius type .... *W. watsoni*
   Pharynx of Pseudodiscus type .... *W. papillatus*
   Pharynx of Gastrodiscus type, terminal genitalium of Noci-type.............. *W. noci*
   Pharynx Gastrodiscus type, terminal genitalium of Deschiensi type..... *W. deschiensi*

Homalogaster Poirier, 1883

Diagnosis: Watsoniinae. Body divided into a large flat, ventrally papillated anterior and a small, subcylindrical posterior portion. Acetabulum terminal or ventroterminal. Pharynx terminal with extramural appendages; esophagus short with bulb; ceca somewhat sinuous, extending to acetabulum. Testes tandem, lobed, at anterior intercecal field. Ciliated chamber present, pars musculosa simple. Ovary submedian, intercecal, immediately preacetabular. Genital pore prebifurcal, terminal genitalium protrudable. Uterine coils winding posterior dorsal to testes. Vitelline follicules lateral to ceca, extending from bifurcation to acetabulum. Parasitic in large intestine, rarely in bile duct of ruminants.

Type species: *Homalogaster paloniae* Poirier, 1883 (Figures 361 and 362)

Diagnosis: With characters of the genus. Body length 8.0—15.5 mm, body width 4.5—7.5 mm. Acetabulum 2.12—3.53 mm in diameter, Homalogaster type, number of muscle units, d.e.c., 30—32; d.i.c., 31—34; v.e.c., 15—18; v.i.c., 38—41; m.e.c. (Figure 182). Pharynx with appendages 1.13—1.46 mm long, Watsonius type (Figure 57); esophagus 1.64—1.83 mm long. Testes 0.76—0.93 by 1.26—1.43 mm. Ovary 0.35—0.40 mm in diameter. Terminal genitalium Homalogaster type (Figure 88). Eggs 0.126—0.133 by 0.062—0.084 mm.

Biology: References. Buckley,[134] Chinone and Itagaki.[170]

Intramolluscan stage. Intermediate hosts: *Planorbis* (= *Indoplanorbis*) *exustus* (Buckley.[134] *Polypylis* (= *Segmentina*) *hemisphaerula* (Chinone and Itagaki).[170] ① Buckley. Rediae lack locomotor appendages. Birth pore situated at about 0.15 mm from the anterior extremity. Pharynx 0.05 mm in diameter; cecum about a quarter length of the body. Emergence of cercariae was stimulated by sunlight. Body of cercariae 0.55 by 0.45 mm. Tail was about twice the length of the body. The body was heavily pigmented, profusely supplied with cystogenous cells, paired eyespots at level of middle part of esophagus. Acetabulum 0.13 by 0.15 mm. Pharynx 0.08 by 0.07 mm, appendages 0.05 mm long; esophagus was about as long as the ceca. Main excretory tubes winding, without cross-connecting tubes and branching. Caudal tube opens from the acetabular network, it expanded posteriorly to an elongated vesicle. External pores were not observed. Cyst hemispherical 0.44 mm in diameter. ② Chinone and Itagaki. Cercariae emitted by the

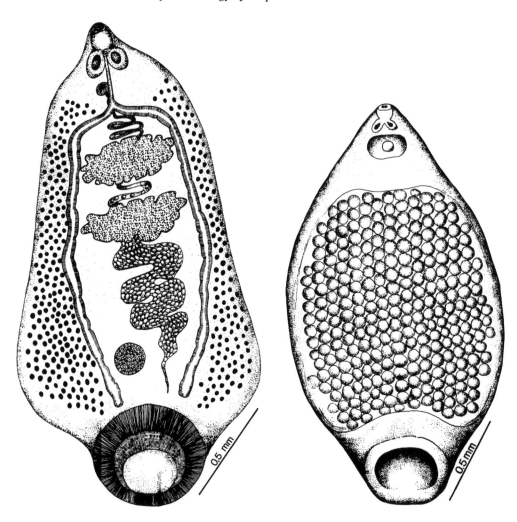

FIGURE 361.   *Homalogaster paloniae* Poirier, 1883.

FIGURE 362.   *Homalogaster paloniae* Poirier, 1883 (ventral surface with papillae).

intermediate hosts were collected from the area where cattle were infected with *H. paloniae*. Percentage infestation of naturally infected snails was 4.3%. Length of body of cercariae 0.27—0.36 mm by 0.12—0.18 mm in size. Tail 0.35—0.54 mm long. Paired eyespots with lens at level of esophagus. Acetabulum 0.068—0.108 by 0.086—0.103 mm. Pharynx with appendages 0.10 by 0.04 mm; esophagus 0.47—0.62 mm long; ceca 0.154—0.206 mm long. The main excretory tubes have cross-connecting tubes. Caudal tube opened from bladder, running posterior and expanded to elongated vesicle, without *lateral* opening. Primordia of genital organs easily seen.

Development in definitive host. ① Buckley. Two 2-month-old calves were infected with 62-d-old adolescariae and a pig with 1-month-old adolescariae. Both experimental infections proved to be negative. ② Chinone and Itagaki. Calves and goats were infected with adolescariae; eggs began to pass in the feces of calves in 57 to 94 d postinfection. Goats were sacrificed 44 d after infection; immature worms were recovered only. The take, in case of goats, was 3.2%.

Habitat: Cecum.

Hosts: *Bos gaurus, Bos primigenius* f. taurus, *Bubalus arnee* f. bubalis, *Antilope cervicapra, Capra aegagrus* f. hircus *Kobus leche, Synerus caffer, Ovis ammon* f. aries (Bovidae);

Muntiacus muntijak, Rusa timorensis, Cervus unicolor (Cervidae); *Elephas maximus* (Elephantidae).

Distribution: Asia — India, Japan, Cambodia, Malaysia, Philippines, Pakistan, Vietnam, Java, China, Indonesia, Burma, Thailand, Papua New Guinea; Africa — (locality of host was not given).

References: Chatterji,[164] Chen et al.,[168] Chinone and Itagaki,[170] Chinone et al.,[171] Euzeby,[292] Fischoeder,[311] Fukui,[334] Huq and Rahman,[428] Lancaster,[520] Leon and Juplo,[540] Le Roux,[548] Muchlis,[625] Mukherjee,[631] Owen,[685] Railliet,[764] Sakamoto et al.,[805] Segal et al.,[826] Sey,[844,853] Shanta,[871] Thapar,[965] Tubangui,[997] Wilmott and Pester,[1068] Wu et al.[1070]

Gastrodiscoides Leiper, 1913

Diagnosis: Watsoniinae. Body conical, tapering anteriorly, posterior part of body discoidal, excavated ventrally, ventral disk with ridges coalesced papillae. Acetabulum small, ventroterminal. Pharynx subterminal with extramural appendages; esophagus short, ceca simple, extending to acetabulum. Testes tandem, intercecal lobed in middle third of body. Ciliated chamber present, pars musculosa weak. Ovary submedian near posterior testis. Genital pore prebifurcal. Uterine coils intercecal, mainly dorsal to testes. Vitelline follicules in discoidal part of body, along ceca, extending from anterior testis to acetabulum. Parasitic in large intestine of mammals.

Type species: *Gastrodiscus hominis* **(Lewis et McConnel, 1876) Leiper, 1913 (Figure 363)**

Syn. *Amphistoma hominis* Lewis et McConnel, 1876, *Gastrodiscus hominis*, Fischoeder, 1902, *Gastrodiscus hominis var. suis* Warma, 1954.

Diagnosis: With characters of the genus. Body length 8.0—14.0 mm, body width 5.0—8.0 mm. Acetabulum 1.83—2.16 mm in diameter, Watsonius type number of muscle units, d.e.c, 10—27; d.i.c., 25—29; v.e.c., 12—14; m.e.c., 12—14 (Figure 184). Pharynx with appendages 1.23—1.56 mm long, Watsonius type (Figure 57). Esophagus 0.45—0.49 mm, with bulb. Anterior testis 1.26—1.43 by 1.83—2.16, posterior testis 2.32—2.54 by 3.12—3.23 mm. Ovary 0.43—0.46 mm in diameter. Terminal genitalium Parvipapillatum type (Figure 92). Eggs 0.123—0.164 by 0.062—0.074 mm.

Biology: References. Dutt and Srivastava.[256,257]

Preparasitic stage. ① Dutt and Srivastava. Eggs deposited contained embryo in an early stage of segmentation. Antiopercular end bears mintue thickening. Under laboratory conditions (24 to 33°C) the miracidium was fully mature on the 9th day. Hatching continued through 7 d. Time required for the hatching varies, according to the data of different authors (Buckley,[134] 16 or 17 d; Ahluwalia,[14] 15 d). Dimensions of hatched living miracidia were 0.16—0.14 by 0.035—0.049 mm. Body covered with 20 ciliated epithelial cells in four tiers 6:8:4:2. Inner structure of miracidia composed of a single apical gland, two pairs of penetration glands, one of which lies dorsal and the other, ventral to apical gland and large brain mass, a pair of flame cells (whose pores open between the third and fourth tiers of epithelial cells), germinal ball, and germinal cells. Buckley[134] reported one penetration gland, while Ahluwalia[14] observed two pairs but lateral in position. Intermediate host: *Helicorbis coenosus*, percentage of infestation varied between 5 to 50%.

Intramolluscan stage. Rediae sausage-shaped, lack collar and locomotor appendages. Body length 0.14—0.74 by 0.45—0.14 mm. Pharynx 0.022—0.046 mm, saccular cecum 0.042—0.164 mm long, with 8 pairs of unicellular glands. Rediae contain germinal balls and germinal cells, daughter rediae, or young cercariae. Birth pore near the level of ceca. Cercariae emerge from rediae in immature state and continue their development in the snail tissue. Prepatent period, depending on the temperature, varies from 28 to 152 d. Daily output of cercariae varies from 1 to 34, and the total number of cercariae obtained

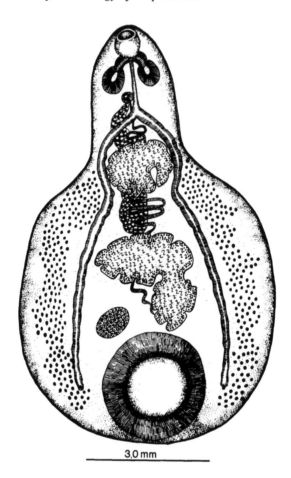

3.0 mm

FIGURE 363.   *Gastrodiscoides hominis* (Lewis et McConnel, 1876).

from individual snails ranges from 7 to 238. Mature cercariae are light brown and very active, measuring 0.43—0.86 by 0.19—0.71 mm; tail 0.46—0.92 by 0.056—0.094 mm, provided with a pair of eyespots. Acetabulum 0.067—0.14 by 0.067—0.115 mm. Pharynx with appendages, measuring 0.52—0.94 mm length. Esophagus 0.094—0.174 mm long; ceca 0.16—0.20 mm long, terminating before acetabulum. Main excretory tubes have cross-connections, caudal tube slightly dilated at distal end, without lateral openings. Rudimentary genital organs easily seen as cell-groups. Encystation takes place on different objects in the vicinity of the infected snails. Cyst measures 0.20—0.27 mm in diameter, brownish in color.

Development in definitive host. ① Dutt and Srivastava. Piglets aged between 2 weeks and 6 months were infected with adolescariae (2—28 days old) of different numbers. Percentage infestation varied between 27—43%. Eggs in uterus of flukes were recovered first, on the 74th day after infection in the autopsied swines.

Habitat: Colon.

Hosts: *Homo sapiens* (Hominidae); *Macaca irus, M. fascicularis, M. mulatta* (Cercopithecidae); *Pango pygmaeus* (Pongidae); *Arvicola terrestris, Ondata zibethica* (Cricetidae); *Myocastor coypus* (Myocastoridae; *Rattus rattus, Rattus argentiventer, Bandicota indica, B. savidei*, (Muridae); *Sus scrofa* f. domestica (Suidae); *Tragulus napu* (Tragulidae).

Distribution: Eurasia — Burma, China, Cambodia, India, Indonesia, Malaysia, Philippines, Vietnam, U.S.S.R., Thailand; South America — Guyana (intr.).

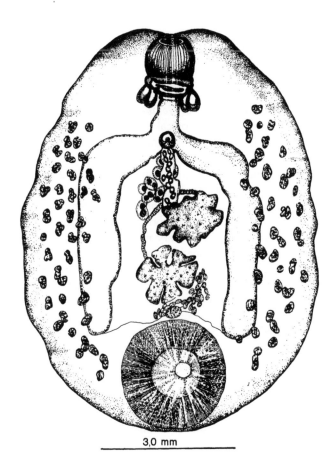

3,0 mm

FIGURE 364. *Skrjabinocladorchis jubilaricum* Tchertkova 1959. (Courtesy of A. N. Tchertkova.)

References: Ahluwalia,[14] Asadov,[41] Badanin,[53] Buckley,[134] Chatterji,[164] Chen et al.,[168] Dubinin,[247] Dwivedi and Chauhan,[261] Fox and Hall,[320] Impand et al.,[430] Jesus and Waramontri,[448] Khalil,[479] Leiper,[532] Lewis and McConnel,[550] Pester and Keymer,[712] Petrochenko and Romanovskiĭ,[722] Sadykhov,[804] Sandground,[815] Segal et al.,[826] Sey,[854] Sultanov,[930] Tchertkova,[961] Varma,[1010,1011] Zablotskiĭ,[1084] Zwicker and Carlton.[1100]

Skrjabinocladorchis Tchertkova, 1959

Diagnosis: Watsoniinae. Body oval, dorsoventrally flattened. Acetabulum ventroterminal. Pharynx terminal with appendages, esophagus short, with muscular bulb; ceca slightly sinuous, extending to acetabulum. Testes slightly oblique, lobed, intercecal, in middle third of body. Ovary median, midway between posterior testis and acetabulum. Genital pore bifurcal. Uterine coils in median field, dorsal to testis. Vitelline folliculoes lateral, along ceca, extending from bifurcation to acetabular zone. Intestinal parasite of Primates.

Type species: *Skrjabinocladorchis jubilaricum* Tchertkova, 1959 (Figure 364)

Diagnosis: With characters of the genus. Body length 8.2—9.4 mm, body width 4.9—6.5 mm. Acetabulum 2.46—2.66 mm in diameter. Pharynx 0.96—1.02 by 1.02—1.13 mm, Pseudodiscus type, (Figure 56) appendages 0.47—0.59 mm in diameter. Esophagus 0.98—1.53 mm long. Anterior testis 1.32—1.42 by 0.96—1.06, posterior testis 1.10—1.39 by 0.77—1.02 mm. Ovary 0.34—0.40 by 0.26—0.36 mm. Eggs 0.106—0.118 by 0.069—0.073 mm.

Biology: Not yet known.

Habitat: Intestine.

Host: *Pan troglodytes* (Pongidae).

Distribution: Not given (Moscow Zoo).

References: Sey,[852] Tchertkova.[961]

3.2.3.2.2. GASTROTHYLACIDAE Stiles et Goldberger, 1910

Diagnosis: Paramphistomoidea. Body conical to acorn-shaped, sometimes much elongated; ventral pouch present, its opening at anterior part of body, behind mouth orifice. Acetabulum terminal, relatively small. Pharynx terminal, without appendages, esophagus short, without or rarely with muscular bulb, ceca straight or undulating, extending to acetabular zone or middle part of body. Testes horizontal, or median (one anterodorsal to the other), postcecal or intercecal, irregular in outline. Vas deferens, seminal vesicle, pars musculosa, pars prostatica, and ductus ejaculatorius present at about median field. Vas deferens and uterus may cross each other. Ovary median or submedian, inter- or posttesticular. Ventral pouch having different shape in cross section, genital pore opens into ventral pouch, rarely outside. Uterine coils confined to dorsal median field or run laterally. Vitelline follicules in lateral or ventrolateral fields. Lymphatic system present, excretory pore behind Laurer's canal. Parasitic in forestomach of mammals.

KEY TO THE GENERA OF GASTROTHYLACIDAE*

1. Ventral pouch extensive, extending to posterior region of body 2
    Ventral pouch limited, extending to middle of body .... *Velasquezotrema*
2. Uterus and vas deferens cross over near middle .... *Gastrothylax*
    Uterus and vas deferens in dorsal median line along their length 3
3. Testes horizontal, one on each side of median line .... *Carmyeris*
    Testes not horizontal, median... *Fischoederius*

Type genus: *Gastrothylax* Poirier, 1883

Diagnosis: Gastrothylacidae. Body conical, elongate, slightly flattened ventrally, convex dorsally. Ventral pouch triangular. Acetabulum terminal, relatively small. Pharynx terminal, without appendages, esophagus small without bulb, ceca simple terminating at acetabulum or in front of it. Genital pore opens into ventral pouch. Testes horizontal, lobed, situated at posterior portion of body. Ovary posttesticular. Uterus crossing from one side to other near middle. Vitelline follicules lateral and dorsal sides from pharynx to acetabulum in form of loosely packed sets. Parasites in rumen of ruminants.

Type species: *Gastrothylax crumenifer* (Creplin, 1847) Otto, 1896 (Figure 365)

Syn. *Amphistoma crumenifer* Creplin, 1883; *Gastrothylax crumenifer* Poirier, 1883; *G. orientalis* Bambroo, 1970

Diagnosis: Body length 11.0—18.0 mm, body width 5.0—8.0 mm. Acetabulum 1.63 by 2.38 mm in size, Gastrothylax type, number of muscle units, d.e.c., 100—150; d.i.c., 102—110; v.e.c., 110—116; v.i.c., 120—124; m.e.c., 10—14 (Figure 170). Ventral pouch usually triangular in cross section with dorsally directed apex. Pharynx Paramphistomum type (Figure 67); esophagus 1.12—1.53 mm. Testes 1.14—1.23 by 0.62—0.74 mm. Pars musculosa moderately developed. Ovary 0.24—0.30 mm in diameter. Terminal genitalium opens into ventral pouch, Gracile type (Figure 114). Eggs 0.103—0.158 by 0.065—0.07 mm.

Biology: References. Tandon,[949] Peter and Srivastava,[720] Nikitin.[650,651]

Preparasitic stage. ① Tandon. Eggs oval, operculated and bore a small projection

---

* *Duttiella* Srivastava, Prasad and Maurya, 1980 was not included due to its incomplete description.

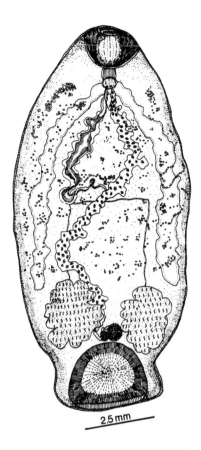

2.5mm

FIGURE 365. *Gastrothylax crumenifer*
(Creplin, 1847). (After R. S. Tandon, 1957.)

at antiopercular end. Freshly laid eggs contained zygote in various developmental stages. In August, development completed within 9 d in India. A mucoid plug present, hatching simulated by natural or artificial light, but emergence was observed even in complete darkness. Shape of miracidia pyriform, body dimensions of fixed specimens 0.10—0.14 by 0.033—0.035 mm. Body covered with ciliated 20 epidermal cells, arranged in four tiers by the formula 6:8:4:2. Miracidia composed of apical gland ("primitive gut"), two pairs of penentration glands, a pair of flame cells, irregular brain mass, germinal cells and germinal ball. Life span of miracidia 4 to 6 h. Intermediate host: *Gyraulus convexiusculus* (Tandon, Peter and Srivastava); *Gyraulus albus, Planorbis planorbis, Hippeutis complanatus, Armiger crista* (Nikitin).

Intramolluscan larval stage. ① Tandon. Miracidia entered the body of snail through the foot and mantle, and epithelial cells were shed. Entered miracidia prefer mantle wall and cavity and transformed to sporocyst within the following 5 to 7 h. Mature sporocyst measured 0.37 by 0.125 mm, with some developing rediae and germinal balls. Rediae were liberated through the terminal opening of the sporocyst. Rediae began to shed by 10th or 11th day and situated in the mantle cavity of the snail. Mature rediae measuring 0.56—1.03 by 0.17—0.23 mm. Pharynx 0.041 mm long; cecum reached a little beyond the middle of the body. Birth pore was about 0.078 mm away from the anterior end of body. Three pairs of flame cells were present. Cercariae escaped through the birth pore in immature form. Mature cercariae measuring 0.55—0.71 by 0.32—0.44 mm. Tail 0.55—0.63 mm long. Pigment granules spread on both dorsal and ventral body surfaces. Brown cystogenous cells were numerous below the cuticle (= integument) and gelatinous substance for cyst formation produced by them. A pair of eyespots situated just behind the pharynx. Acetabulum 0.05 by 0.068 mm. Pharynx 0.095—0.115 by 0.07—0.075 mm; esophagus 0.09—0.16 mm long, ceca terminating in front of acetabulum. Main excretory tubes with cross-connecting tubes, caudal tube expanded near the tip of the tail, without lateral openings. Encystment took place on the surface of the objects found in the vicinity of the infected snails. Cyst measuring 0.355 mm in diameter. ② Nikitin. Eggs laid contained embryo composed of some cells. Embryonic development followed 12 d at 28°C. Miracidia 0.176 by 0.056 mm, covered with ciliated epithelial cells arranged in four tiers. Inner structure of miracidia composed of apical gland (''cecum''), penetration glands, brain mass, a pair of flame cells, germinal ball, and germinal cells. Longevity of miracidia was about 13 h. Penetrated miracidia transformed into sporocyst in some days. Mature sporocysts were found in the 17th day or so, in summer temperature of Moscow district, it measured 0.28—0.35 by 0.19—0.23 mm, having some developing redia. Mature rediae measured 0.52 by 0.16 mm. Pharynx 0.039—0.043 mm in diameter. Redia contains 9—14 germinal balls. First cercariae were found in the snail by the 37th day after infection. They emerged in immature form; their development continued in the snail tissue. Mature cercariae began to discard in the 66 d after infestation. Body of cercariae 0.50—0.60 by 0.14—0.19 mm, tail 0.58—0.65 mm long, belonging to Pigmentata group. Main excretory tubes with cross-connecting tubes, caudal tube with opening at tip of the tail. Encystment took place on vegetation or other objects, 2—3 mm under the water surface. Adolescariae 0.33—0.38 mm in diameter. ③ Peter and Srivastava. Mature rediae 0.5 by 0.2 mm in size, without locomotor appendages. Diameter of pharynx was about 1/14 of body length. Three pairs of flame cells existed. Birth pore at anterior third of body. Size of body of cercariae 0.25—0.56 by 0.14—0.27 mm. Tail 0.28—0.63 mm in length. Eyespots at level of esophagus. Acetabulum 0.072 by 0.132 mm. Pharynx 0.11 by 0.062 mm, esophagus 1/5 of body; ceca at posterior half of body. Main excretory tubes with cross-connecting tubes. Caudal tube expanded near to the tip, without lateral openings. Primordia of genital organs easily seen in the form of cell masses. Freshly encyst adolescariae 0.27 mm in diameter on an average. *Cercaria chungati*, described by Peter and Srivastava[717] from naturally infected snail, *Gyraulus convexiusculus* proved to be the cercaria of *Gastrothylax crumenifer*.

Development in definitive host. ① Tandon. One goat kid was infected with 26-day-old adolescariae. The kid was sacrificed after 9 months of infestation, and one mature and three immature flukes were recovered. Infestation percentage was 7.14%. ② Peter and Srivastava. Prepatent period was 118 d in the kid and 114 d in the buffalo calf, respectively. Percentage of infestation was 9.8% in kid and 10.4% in buffalo calf. ③ Nikitin. Prepatent period was more than 3 months in calf. Infestation percentage was 28.4 to 32.6%.

Habitat: Rumen.

Hosts: *Bos primigenius* f. taurus, *Ovis ammon* f. aries, *Capra aegagrus* f. hircus, *Bubalus arnee* f. bubalis, *Syncerus caffer*, *Kobus leche*, *Tragelaphus spekei*, *Antilope carvicapra*, *Buselaphus tragocamelus*, *Hyelaphus porcinus* (Bovidae); *Rangifer tarandus*, *Cervus unicolor* (Cervidae).

Distribution: Asia — India, U.S.S.R., Vietnam, Cambodia, Philippines, Afghanistan, Iraq, China, Sri Lanka, Taiwan, Malaysia, Pakistan; Africa — Zambia, South Africa.

References: Agrawal,[10] Agrawal and Ahluwalia,[13] Bali,[64] Bhalerao,[90] Chen et al.,[168] Crusz,[191] Dawes,[205] Erbolatov,[286] Fukui,[334] Gupta and Dutta,[390] Kotrlá et al.,[501] Kurtpinar and Latif,[515] Le Roux,[545] Nikitin,[651,655] Ortlepp,[678] Rybaltovskiĭ,[799] Segal et al.,[826] Seneviratna,[828] Sey,[844,845,849,854] Sey and Eslami,[866] Shagraev and Zhaltsanova,[869] Sultanov et al.,[930] Thapar,[965] Tubangui,[997] Velitchko,[1022] Yusuf and Chaudhry.[1081]

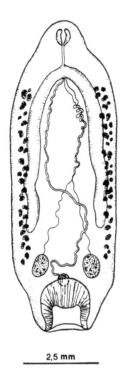

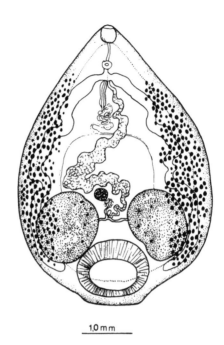

FIGURE 366. *Gastrothylax compressus* Brandes, 1898. (After G. Brandes, 1898, modified.)

FIGURE 367. *Gastrothylax globoformis* Wang, 1977. (After P. Q. Wang, 1977.)

Gastrothylax compressus **Brandes, 1898 (Figure 366)**

Syn. *G. glandiformis* Yamaguti, 1939, *G. indicus*, Dutt, 1978, *Gastrothylax zhonghuaensis* Wang, 1979

Diagnosis: Body length 4.5—15.0 mm, body width 3.5—5.2 mm. Acetabulum 1.23—1.36 mm in diameter, Carmyerius type, number of muscle units, d.e.c., 42—45; d.i.c., 35—38; v.e.c., 28—31; v.i.c., 29—34; m.e.c., 7—9 (Figure 164). Ventral pouch usually triangular with dorsally directed apex. Pharynx 0.82—0.85 by 0.68—0.72 mm, Paramphistomum type (Figure 67); esophagus 0.46—0.77 mm in length, without bulb. Testes lobed 0.42—0.15 mm in diameter. Pars musculosa weakly developed. Ovary 0.28—0.39 mm in diameter. Terminal genitalium within ventral pouch, Gracile type (Figure 114). Eggs 0.115—0.150 by 0.060—0.081 mm.

Biology: Not yet known.

Habitat: Rumen.

Hosts: *Bos primigenius* f. taurus, *Bubalus arnee* f. bubalis, *Ovis ammon* f. aries (Bovidae).

Distribution: Asia — India, Japan.

References: Brandes,[115] Dutt,[255] Sey,[850] Yamaguti.[1072]

Gastrothylax globoformis **Wang, 1977 (Figure 367)**

Syn. *Gastrothylax zhonghuaensis* Wang, 1979.

Diagnosis: Body length 5.2—7.0 mm, body width 3.4—3.8 mm. Acetabulum 0.90—1.28 by 1.36—1.86 mm. Pharynx 0.35—0.48 by 0.35—0.43 mm, esophagus 0.32—0.64 mm long. Testes 1.28—1.60 by 1.16—1.28 mm. Ovary 0.32—0.48 by 0.32—0.46 mm. Terminal genitalium within ventral pouch. Eggs 0.105—0.122 by 0.063—0.070 mm.

Biology: Not yet known.

Habitat: Rumen.

Host: *Bubalus arnee* f. bubalis (Bovidae).
Distribution: Asia — China.
Reference: Wang.[1049]

KEY TO THE SPECIES OF *GASTROTHYLAX*

1. Ceca extending to anterior part of testes or before them 2
    Ceca extending to posterior part of testes .... *G. globoformis*
2. Testes irregular in shape ceca extend to anterior part of testes .... *G. crumenifer*
    Testes regular in shape, ceca extend far before testes .... *G. compressus*

Carmyerius Stiles et Goldberger, 1910

Syn. *Wellmanius* Stiles et Goldberger, 1910
Diagnosis: Gastrothylacidae. Body conical, straight or curved, circular in cross section. Acetabulum terminal. Ventral pouch assumes different forms in appearance. Pharynx without appendages, esophagus short without or occasionally with bulb; ceca in lateral fields, long or short, sinuous or not. Testes symmetrical, lobed, preacetabular. Ovary intertesticular, prevesicular. Genital pore within ventral pouch or rarely without it. Uterine coils to dorsal median field along its length. Vitelline follicules in lateral fields, ventral to ceca extending from bifurcation to acetabular zone. Parasitic in forestomach of mammals.

Type species: *Carmyerius gregarius* **(Looss, 1896) Stiles et Goldberger, 1910 (Figure 368)**
Diagnosis: Body length 6.5—10.2 mm, body width 1.5—4.2 mm. Acetabulum 1.37—1.62 mm in diameter, Carmyerius type, number of muscle units, d.e.c., 38—40; d.i.c., 45—52; v.e.c., 32—36; v.i.c., 31—34; m.e.c., 7—9, (Figure 164). Ventral pouch usually triangular with a ramifying dorsally directed apex. Pharynx 0.58—0.61 mm long Gregarius type (Figure 64), esophagus 0.74—1.26 mm. Testes lobed 0.83—1.16 by 0.52—0.78 mm. Pars musculosa moderately developed. Ovary 0.73 by 0.58 mm in diameter. Terminal genitalium within ventral pouch, Gregarius type (Figure 87). Eggs 0.114—0.135 by 0.080—0.085 mm.
Biology: References. Looss.[558]

Eggs laid contained embryo composed of 3 to 5 cells. Embryonic development took place 12 to 14 d on 22°C. Miracidia ("free embryo") covered with ciliated epidermal cells, measuring 0.3 mm in length. Miracidia composed of cephalic gland ("rudimental intestine") excretory system and germ masses. Snails *Physa alexandrina* (= *Bulinus contortus*) and *Ph. micropleura* (= *Bulinus forskalii*) did not prove to be successive intermediate hosts.
Habitat: Rumen.
Hosts: *Bos primigenius* f. taurus, *Bubalus arnee* f. bubalis, *Syncerus caffer*, *Antilope cervicapra*, *Kobus megaceros*, *Tragelaphus scriptus*, *T. sylvaticus* (Bovidae).
Distribution: Africa — Central Africa, Chad, Cameroon, Nigeria, Zaire, Egypt, South Africa, Sudan; Asia — India, Philippines.
References: Dollfus,[244] Ezzat,[295] Graber,[357] Graber et al.,[362] Looss,[558] Myers et al.,[640] Patnaik,[704] Schillhorn et al.,[821] Sey,[838,840,850] Tubangui.[997]

Carmyerius spatiosus (Brandes, 1898) Stiles et Goldberger, 1910. (Figure 369)

Syn. *Gastrothylax spatiosus* Brandes, 1898.
Diagnosis: Body length 9.2—12.1, body width 2.8—3.1 mm. Acetabulum 1.38—1.43 by 1.12—1.23 mm, Gastrothylax type, number of muscle units, d.e.c., 65—68, d.i.c., 42—46, v.e.c., 45—49, 44—48, m.e.c., 7—9 (Figure 170). Ventral pouch either circular or triangular with blunt angles. Pharynx 0.65—0.72 by 0.58—0.62, Paramphistomum type

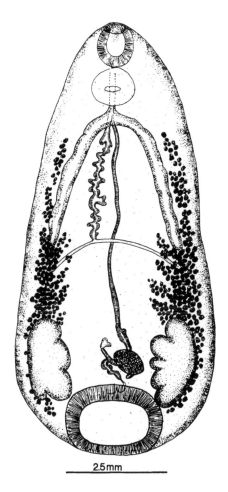

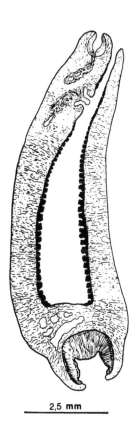

2.5mm

2,5 mm

FIGURE 368. *Carmyerius gregarius* Looss, 1896. (After A. Looss, 1896.)

FIGURE 369. *Carmyerius spatiosus* (Brandes, 1898) (median sagittal section).

(Figure 67), esophagus 0.36—0.37 mm long. Pars musculosa weakly developed. Ovary 0.21—0.28 mm in diameter. Terminal genitalium within ventral pouch, Gracile type (Figure 114). Eggs 0.115—0.125 by 0.060—0.068 mm.

Biology: Not known yet.

Habitat: Rumen.

Hosts: *Bos primigenius* f. taurus, *Bubalus arnee* f. bubalis, *Ovis ammon* f. aries, *Syncerus caffer, Redunca redunca, Alcelaphus lichteinsteini, R. arundinum, A. buselaphus, Hippotragus equinus, H. beckeri, Kobus defassa, K. varondi, K. leche, K. kob, K. thomasi, Tragelaphus spekei, T. scriptus, Damaliscus korrigum, Capra aegagrus* f. hircus, *Taurotragus derbianus* (Bovidae); *Muntiacus muntjak* (Cervidae).

Distribution: Africa — Zambia, Chad, Central Africa, South Africa Sudan, Egypt, Mozambique, Zimbabwe, Tanzania, Liberia, Cameroon, Senegal, Kenya; Asia — Vietnam, Mongolia, India, Malaysia and Arabia.

References: Baer,[54] Brandes,[115] Cruz E. Silva,[195] Dawes,[205] Dinnik,[227] Ezzat,[295] Graber,[357,360] Graber et al.,[362] Graber and Thal,[365] Gretillat,[372] Leiper,[530] Le Roux,[546] Maplestone,[591] Ortlepp,[678] Pačenovsky et al.,[688] Pike and Condy,[727] Sey,[850,853] Strydonck,[920] Szidat.[939]

Carmyerius mancupatus (Fischoeder, 1901) Stiles et Goldberger, 1910 (Figure 370)

Syn. *Gastrothylax mancupatus* (Fischoeder, 1901 *Carmyerius minutus*, (Fischoeder, 1901),

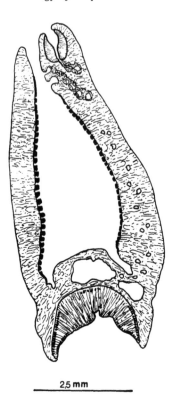

2,5 mm

FIGURE 370. *Carmyerius mancupatus* (Fischoederius, 1901) (median
sagittal section).

Wellmanius wellmani Stiles et Goldberger, 1910; *Carmyerius dollfusi* Golvan Chabaud et
Gretillat, 1957.

Diagnosis: Body length 5.2—11.3, body width 2.3—4.5 mm. Acetabulum 1.83—1.97
mm in diameter, Gastrothylax type, number of muscle units, d.e.c., 64—78; d.i.c., 45—48;
v.e.c., 53—56; v.i.c., 52—58; m.e.c., 6—8 (Figure 170). Ventral pouch usually triangular
with apex ventrally directed but circular or pentagonal forms also occurred. Pharynx 0.23—0.32
mm in diameter, Paramphistomum type (Figure 67). Esophagus 0.28—0.31 mm long. Testes
1.66—1.93 by 1.28—1.4 mm. Pars musculosa moderately developed. Ovary 0.15—0.43
mm in diameter. Terminal genitalium Mancupatus type, (Figure 120). Eggs 0.128—0.149
by 0.067—0.076 mm.
Biology: References. Gretillat,[370] (as C. dollfusi); Dinnik.[228]

Preparasitic stage. ① Gretillat. Eggs laid contained embryo with 3 to 5 cells. Embryonic development took place 15 to 18 d at temperatures of 29—30°C. Longevity of
miracidium at this temperature 12 to 18 h. Intermediate host *Bulinus mariei* (Gretillat),
Anisus natalensis (Dinnik).

Intramolluscan stage. ① Gretillat. Mature rediae containing young cercariae, measuring 0.80—0.90 mm in length and 0.18—0.20 mm in width. Pharynx 0.025—0.030
mm in diameter, saccular cecum 0.05—0.06 mm in diameter. Rediae contained germ
cells and embryo balls. Cercariae passed through birth pore in immature form, and their
development contained in the snail tissue. Mature cercariae measured 0.35 mm in size,
tail 0.70 mm long. Main excretory tubes without cross-connecting tubes. Emerged cercariae encyst on vegetation. Adolescariae 0.25 mm in diameter.

Development in definitive host. ① Gretillat. Calf infected with adolescariae and the
first eggs were passed 76 d after infestation.

1,0 mm

FIGURE 371. *Carmyerius synethes* (Fischoeder, 1901). (After X. Y. Wang, 1979.)

Habitat: Rumen.

Hosts: *Bos primigenius* f. taurus, *Capra aegagrus* f. hircus *Syncerus caffer, Kobus defassa, K. leche, Hippotragus equinus, Redunca redunca, Taurotragus oryx, Tragelophus scriptus* (Bovidae).

Distribution: Africa — Zaire, Cameroon, Malagasy, Kenya, Tanzania, Central Africa, Guinea; Asia — Vietnam.

References: Dinnik,[228] Dinnik and Dinnik,[227] Fischoeder,[309,311] Gretillat,[374] Prudhoe,[756] Railliet,[764] Sey.[845,850]

Carmyerius synethes (Fischoeder, 1901) Stiles et Goldberger, 1910 (Figure 371)

Syn. *Gastrothylax synethes* Fischoeder, 1901.

Diagnosis: Body length 7.2—15.3 mm, body width 3.2—5.5 mm. Acetabulum 1.16—1.32 mm in diameter, Gastrothylax type; number of muscle units, d.e.c., 63—68; d.i.c., 34—38; d.e.c., 58—62; v.i.c., 43—48; m.e.c., 7—9 (Figure 170). Ventral pouch triangular, with apex ventrally directed. Pharynx 0.58—0.62 by 0.42—0.45 mm, Paramphistomum type (Figure 67), esophagus 0.67—0.83 mm in length. Testes 1.26—1.53 by 0.76—0.93 mm. Pars musculosa weakly developed. Ovary 0.30—0.34 by 0.29—0.32 mm. Terminal genitalium within ventral pouch, Synethes type (Figure 98). Eggs 0.100—0.125 by 0.060—0.065 mm.

Biology: Not yet known.

Habitat: Rumen.

Hosts: *Bos primigenius* f. taurus, *Capra aegagrus* f. hircus, *Bubalus arnee* f. bubalis (Bovidae); *Cervus unicolor* (Cervidae).

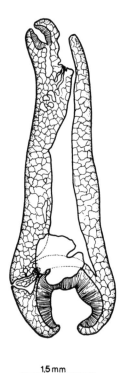 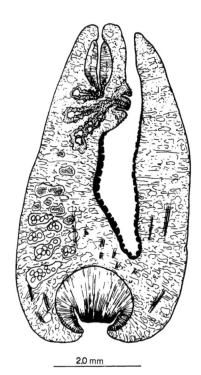

1,5 mm

2,0 mm

FIGURE 372.   *Carmyerius wen-yoni* (Leiper, 1908) (median sag-ittal section). (After R. T. Leiper, 1908.)

FIGURE 373.   *Carmyerius cruciformis* (Leiper, 1910) (median sagittal section).

Distribution: Asia — Borneo, Malaysia, Philippines, Sri Lanka, Vietnam, China.
References: Chen et al.,[168] Eduardo,[264] Eduardo and Manuel,[279] Fischoeder,[311] Schad et al.,[819] Sey,[850,853] Wang.[1050]

Carmyerius wenyoni (Leiper, 1908) Fukui, 1923. (Figure 372)

Syn. *Gastrothylax wenyoni* Leiper, 1908.

Diagnosis: Body length 6.3 mm, body width 2.1 mm. Acetabulum 1.26 mm in diameter, Carmyerius type, number of muscle units, d.e.c., 43; d.i.c., 41; v.e.c., 48; v.i.c., 41; m.e.c., 9—12 (Figure 164). Ventral pouch with apex dorsally directed divided at its tip by a ridge. Pharynx 0.53 mm in diameter. Paramphistomum type, (Figure 67), esophagus 0.68—0.73 mm. Testes 1.12 by 0.26 mm. Ovary 0.23 mm in diameter. Terminal genitalium within ventral pouch, probably Endopapillatus type (Figure 86). Eggs 0.130—0.150 by 0.050—0.075 mm. Excretory duct and Laurer's canal open with a joint orifice.
Biology: Not yet known.
Habitat: Forestomach.
Host: *Kobus megaceros* (Bovidae).
Distribution: Africa — Sudan.
References: Leiper,[530] Sey.[850]

Carmyerius cruciformis (Leiper, 1910) Fukui, 1923 (Figure 373)

Syn. *Gastrothylax cruciformis* Leiper, 1910.

Diagnosis: Body length 4.4—8.0, body width 1.5—2.2 mm. Acetabulum 0.95—0.12 by 0.85—0.91 mm. Carmyerius type, number of muscle units, d.e.c., 22—25; d.i.c., 19—22;

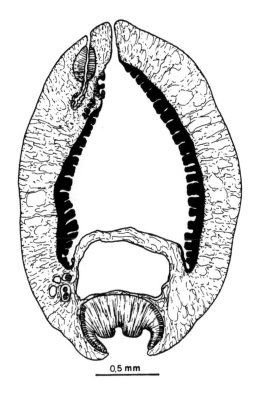

0,5 mm

FIGURE 374. *Carmyerius bubalis* Innes, 1912 (median sagittal section).

v.e.c., 20—22; m.e.c., 7—9 (Figure 164). Ventral pouch triangular, quadrangular, or pentagonal. Pharynx 0.68—0.71 by 0.36—0.42 mm, Paramphistomum type (Figure 67). Esophagus 0.26 mm long. Testes 0.14—0.18 mm in diameter. Pars musculosa weakly developed. Ovary 0.28—0.32 mm in diameter. Terminal genitalium Cruciformis type (Figure 84). Eggs 0.123—0.148 by 0.070—0.075 mm.
Biology: Not yet known.
Habitat: Stomach.
Host: *Hippopotamus amphibius* (Hippopotamidae).
Distribution: Africa — Uganda, Dahomey, Kenya, Chad.
References: Joyeux and Baer,[457] Leiper,[531] Maplestone,[591] Sey,[846,850] Sey and Graber.[864]

Carmyerius bubalis (Innes 1912) Stunkard, 1925 (Figure 374)

Syn. *Gastrothylax bubalis* Innes, 1912, *Carmyerius gretillati* Strydonck, 1970.
Diagnosis: Body length 3.2—12.5 mm, body width 1.5—4.2 mm. Acetabulum 0.82—0.85 by 0.45—0.50 Carmyerius type, number of muscle units, d.e.c., 38—40; d.i.c., 23—28; v.e.c., 39—42; v.i.c., 29—32; m.e.c., 7—9 (Figure 164). Ventral pouch triangular, with ventrally directed apex. Pharynx 0.28—0.32 by 0.25—0.32 mm, Calicophoron type (Figure 58), esophagus 2.26—4.84 mm. Testes 0.43—1.16 by 0.34—1.35 mm. Pars musculosa weakly developed. Ovary 0.15—0.62 mm in diameter. Terminal genitalium within ventral pouch, Bubalis type (Figure 101). Eggs 0.116—0.124 by 0.060—0.080 mm.
Biology: Not yet known.
Habitat: Rumen.
Hosts: *Alcelaphus* sp. *Boocercus eurycerus* (Bovidae).
Distribution: Asia — China; Africa — Zaire, Zimbabwe.
References: Chen et al.,[168] Innes,[432] Sey,[850] Strydonck,[920] Stunkard.[923]

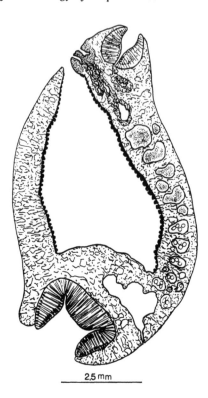

2,5 mm

FIGURE 375. *Carmyerius exoporus* Maplestone, 1923 (median sagittal section).

Carmyerius exoporus Maplestone, 1923. (Figure 375)

Diagnosis: Body length 5.3—11.5, body width 2.6—3.8 mm. Acetabulum 1.54—1.63 mm in diameter, Carmyerius type, number of muscle units, d.e.c., 45—48; d.i.c., 30—34; v.e.c., 43—45; v.i.c., 34—38; m.e.c., 7—9 (Figure 164). Ventral pouch hexagonal. Pharynx 0.68—0.71 by 0.85—0.91 mm, Paramphistomum type (Figure 67), esophagus 0.63 mm long. Testes 1.32—1.56 by 0.73—1.21 mm. Ovary minal genitalium without ventral pouch, Microbothrium type (Figure 122). Eggs 0.112—0.134 by 0.062—0.071 mm.
Biology: References. Dinnik and Dinnik,[232] Dinnik.[228]

Preparasitic stage. Newly laid eggs contained embryo composed of several cells, miracidia developed in about 15 d. Hatched miracidium measured 0.16 by 0.025 mm. Covered with 20 ciliated epidermal cells arranged in four tiers: 6:8:4:2. Three glands were present at the anterior extremity; moreover, a pair of flame cells and germinal cells were found in miracidia. Intermediate host *Anisus natalensis*.

Intramolluscan stage. Penetrated miracidia gradually developed sporocysts localized in tissue, surrounded intestine. By 10—12 d postinfection, sporocysts reach their maturity. Mature sporocyst measuring 0.18—0.33 by 0.10—0.19 mm. Germinal cells developed into germinal balls; the number of embryos developed varied from 6 to 11 per sporocyst. Miracidium gives rise to one sporocyst. Developing rediae ruptured the wall of sporocyst, escaped from it, and were situated in the adjacent tissue. Infection of snails with a single miracidium ranging 3 to 9. Mature rediae measured 0.35—0.65 by 0.10—0.14 mm. Pharynx 0.035—0.044 mm long, saccular cecum extends to level of the birth pore. There were three pairs of flame cells present; germinal mass, cells, and embryo balls situated in the middle part of body. First-generation rediae give birth to daughter rediae and cercariae. Cercariae emerge from rediae in immature form and continue their development in the tissue of the snail. Mature cercariae begin to emerge from the infected snail

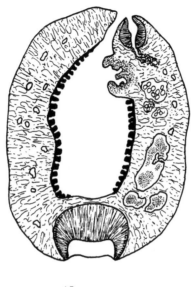

1,5 mm

FIGURE 376. *Carmyerius graberi* Gretillat, 1960 (median sagittal section).

31—33 d postinfection, measuring 0.25—0.30 mm in length, tail 0.50—0.70 mm.
Cercarial productivity about 20/d. Cercariae encysted in about 1/2 h, diameter of cyst
0.21—0.26 mm.

Development in definitive host. A 5-month-old calf was infected with 928 adolescariae.
First eggs appeared in feces 100 d later.

Habitat: Rumen.

Hosts: *Bos primigenius* f. taurus, *Ovis ammon* f. aries, *Syncerus caffer, Tragelaphus spekei,
Damaliscus korrigum, Hippotragus equinus, Kobus defassa, Onotragus leche, Redunca
redunca* (Bovidae).

Distribution: Africa — Zaire, Malawi, Kenya, Tanzania, Chad, Central Africa.

References: Dinnik and Dinnik,[227] Graber,[360] Graber et al.,[362] Maplestone,[591] Prudhoe,[756]
Sey,[850] Sey and Graber.[864]

Carmyerius graberi Gretillat, 1960. (Figure 376)

Diagnosis: Body length 3.0—5.0 mm, body width 2.5—3.2 mm. Acetabulum 1.12—1.23
mm in diameter, Carmyerius type, number of muscle units, d.e.c., 35—37; d.i.c., 35—38;
v.e.c., 43—48; v.i.c., 44—48; m.e.c., 7—9 (Figure 164). Ventral pouch circular. Pharynx
0.43—0.52 mm in diameter, Paramphistomum type (Figure 67), esophagus 0.30—0.32 mm
long. Testes 1.53—1.76 by 0.73—0.94 mm. Ovary 0.46—0.53 mm in diameter. Pars
musculosa weakly developed. Terminal genitalium within ventral pouch, Synethus type
(Figure 98). Eggs 0.125—0.130 by 0.070—0.075 mm.

Biology: Not yet known.

Habitat: Rumen.

Hosts: *Bos primigenius* f. taurus, *Ovis ammon* f. aries, *Syncerus caffer, Kobus defassa, K.
thomasi, K. kob, K. ellipsiprymnus, Alcelaphus buselaphus, Gazella rufifrons, Redunca
redunca, Tragelaphus scriptus* (Bovidae).

Distribution: Africa — Chad, Central Africa, Cameroon.

References: Graber,[357,360] Graber et al.,[362] Gretillat,[370,371,373] Sey,[850] Sey and Graber.[864]

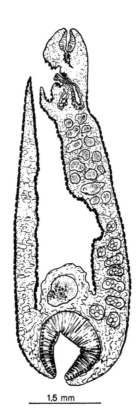

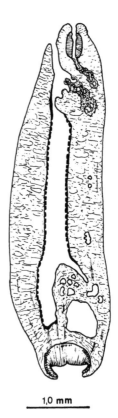

FIGURE 377.   *Carmyerius en-*
dopapillatus Dollfus, 1962 (median sagittal section).

FIGURE 378.   *Carmyerius par-*
vipapillatus Gretillat, 1962 (median sagittal section).

Carmyerius endopapillatus **Dollfus, 1962 (Figure 377)**

Syn. *Carmyerius papillatus* Gretillat, 1962

Diagnosis: Body length 4.2—9.6 mm, body width 2.5—5.2 mm. Acetabulum 0.98—1.2 mm in diameter, Carmyerius type, number of muscle units, d.e.c., 36—39; d.i.c., 26—29; v.e.c., 34—38; v.i.c., 19—33; m.e.c., 7—9 (Figure 164). Ventral pouch circular, pentagonal, triangular (in the latter with ventrally directed apex). Pharynx 0.38—0.42 by 0.28—0.31 mm Calicophoron type (Figure 58), esophagus 0.23—1.26 mm long, without bulb. Testes 1.13—1.78 by 1.26—1.37 mm. Pars musculosa moderately developed. Ovary 0.33—0.46 by 0.24—0.43 mm. Terminal genitalium within ventral pouch, Endopapillatus type (Figure 86). Eggs 0.123—0.146 by 0.0665—0.0791 mm.

Biology: Not yet known.

Habitat: Rumen.

Hosts: *Bos primigenius* f. taurus, *Hippotragus equinus, Kobus defassa, K. kob, K. thomsoni, Redunca redunca, Syncerus caffer, Tragelaphus scriptus* (Bovidae); *Hippopotamus amphibius* (Hippopotamidae).

Distribution: Africa — Zaire, Chad, Central Africa.

References: Dollfus,[242] Graber,[357] Graber et al.,[362] Gretillat,[372] Sey,[850] Strydonck,[920] Thal.[964]

Carmyerius parvipapillatus **Gretillat, 1962 (Figure 378)**

Syn. *C. gregarius congolensis* Dollfus, 1963

Diagnosis: Body length 3.7—7.2 mm, body width 2.5—2.8 m. Acetabulum 1.24—1.36 by 0.83—0.92 mm. Carmyerius type number of muscular units, d.e.c., 40—48; d.i.c.,

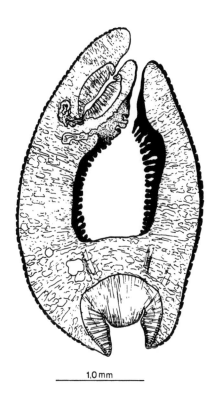

FIGURE 379. *Carmyerius schoutedeni*
Gretillat, 1964 (median sagittal section).

40—43; v.e.c., 39—41; v.i.c., 47—53; m.e.c., 7—9 (Figure 164). Ventral pouch usually quadrangular. Pharynx 0.38—0.43 by 0.30—0.32 mm, Paramphistomum type (Figure 67), esophagus 0.45—0.75 mm long. Testes 1.26—1.32 by 0.78—0.93 mm. Pars musculosa moderately developed, ovary 0.46—0.53 by 0.51—0.58 mm. Terminal genitalium Parvipapillatum type (Figure 92). Eggs 0.120—0.125 by 0.068—0.075 mm.
Biology: Not yet known.
Habitat: Rumen.
Hosts: *Bos primigenius* f. taurus, *Capra aegagrus* f. hircus (exp.) *Ovis ammon* f. aries (exp.), *Damaliscus karrigum, Hippotragus beckeri, Kobus defassa, K. ellipsiprymnus, K. kob, Tragelaphus scriptus* (Bovidae).
Distribution: Africa — Cameroon, Chad, Zaire, Guinea, Kenya, Niger, Zambia, Central Africa.
References: Dinnik,[228] Graber,[357] Graber et al.,[362] Gretillat,[372] Sey.[850]

Carmyerius schoutedeni Gretillat, 1964 (Figure 379)

Diagnosis: Body length 3.2—4.1, body width 2.2—2.5 mm. Acetabulum 1.63—1.82 mm in diameter, Carmyerius type number of muscle units, d.e.c., 32—36; d.i.c., 40—45; v.e.c., 34—38; v.i.c., 32—36; m.e.c., 7—9 (Figure 164). Ventral pouch triangular with apex ventrally directed. Pharynx 0.72—0.83 mm long, Paramphistomum type (Figure 67), esophagus 0.28—0.32 mm long. Testes 0.61—0.78 by 0.25—0.37 mm. Pars musculosa weakly developed. Ovary 0.22—0.25 mm. Terminal genitalium within ventral pouch, Schoutedeni type (Figure 95). Eggs 0.106—0.113 by 0.055—0.065 mm.
Biology: Not yet known.
Habitat: Rumen, stomach.
Hosts: *Syncerus caffer, Bubalus arnee* f. bubalis, *Cephalophus nigrifrons, Boocercus eu-*

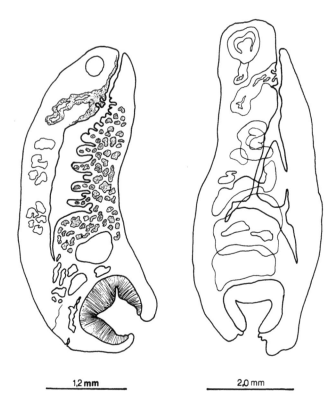

1,2 mm   2,0 mm

FIGURE 380. *Carmyerius mul-*   FIGURE 381. *Carmyerius di-*
tivitellarius Strydonck, 1970   *plopharyngialis* (Strydonck, 1970)
(median sagittal section). (Cour-   (median sagittal section). (Courtesy of D. Strydonck.)   tesy of D. Strydonck.)

ryceros, Kobus defassa, K. kob, Redunca redunca (Bovidae); *Hippopotamus amphibius*
(Hippopotamidae).
Distribution: Africa — Central Africa, Chad, Zaire, Senegal.
References: Diaw et al.,[217] Graber,[360] Gretillat,[373,374] Sey,[850] Strydonck.[920]

Carmyerius multivitellarius Strydonck, 1970 (Figure 380)

Diagnosis: Body length 4.5—5.2 mm, body width 1.6—1.9 mm. Acetabulum 1.04 by
1.10 mm, Gastrothylax type, number of muscle units, d.e.c., 72—79; d.i.c., 59—63; v.e.c.,
80—87; v.i.c., 64—65; m.e.c., 7—9 (Figure 170). Ventral pouch characterized by large
tissue invagination. Pharynx 0.49 by 0.42, Paramphistomum type (Figure 67), esophagus
1.30—1.90 mm long. Testes 1.22—1.28 by 0.98—0.20 mm. Ovary 0.22 by 0.51 mm. Pars
musculosa moderately developed, convoluted. Terminal genitalium Mancupatus type (Figure
120). Eggs 0.118—0.125 by 0.061—0.074.
Biology: Not yet known.
Habitat: Forestomach.
Hosts: *Gazella thomsoni* (Bovidae).
Distribution: Africa — Zaire.
References: Sey,[850] Strydonck.[920]

Carmyerius diplopharyngialis Strydonck, 1970 (Figure 381)

Diagnosis: Body length 7.4 mm, body width 2.7 mm. Acetabulum 1.36 mm in diameter,
Carmyerius type, number of muscle units, d.e.c., 27—35; d.i.c., 30—33; v.e.c., 29—30;

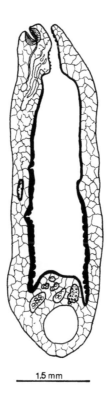

1.5 mm

FIGURE 382. *Carmyerius chabaudi* Strydonck, 1970 (median sagittal section). (Courtesy of D. Strydonck.)

v.i.c., 30—34; m.e.c., 7—9 (Figure 164). Ventral pouch was a slit in appearance. Pharynx 0.84—0.85 mm, type not designated, esophagus 0.59 mm long. Testes 1.65 by 1.98 and 1.74 by 2.26 mm. Ovary 0.40 by 0.78 mm. Pars musculosa well developed. Terminal genitalium within ventral pouch, Microbothrium-type (Figure 122). Eggs 0.110—0.120 by 0.060—0.074 mm.
Biology: Not yet known.
Habitat: Rumen.
Host: *Bovidae*.
Distribution: Zaire.
References: Sey,[850] Strydonck.[920]

Carmyerius chabaudi **Strydonck, 1970. (Figure 382)**

Diagnosis: Body length 4.5—8.8 mm, body width 1.7—2.1 mm. Acetabulum 0.82—1.21 mm in diameter, Carmyerius type, number of muscle units, d.e.c., 32—40; d.i.c., 30—36; v.e.c., 31—36; v.i.c., 30—34; m.e.c., 7—9 (Figure 164). Ventral pouch polygonal in appearance. Pharynx 0.36—0.46 in diameter, esophagus 0.53—0.63 mm long. Testes 0.72—1.26 by 0.63—1.04 mm. Pars musculosa moderately developed. Ovary 0.16—0.22 by 0.22—0.33 mm. Terminal genitalium within ventral pouch, Elongatus type (Figure 85). Eggs 0.100—0.110 by 0.110 by 0.055—0.065 mm.
Biology: Not yet known.
Habitat: Rumen.
Host: *Gazella thomsoni* (Bovidae).
Distribution: Africa-Congo.
References: Sey,[850] Strydonck.[920]

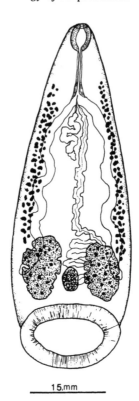

15.mm

FIGURE 383. *Carmyerius bulbosus* Sey, 1985. (By permission of the *Acta Zool.*, 1985, Budapest.)

Carmyerius bulbosus Sey, 1985 (Figure 383)

Diagnosis: Body length 6.5—7.2 mm, body width 2.5—2.8 mm. Acetabulum 1.26—1.53 mm in diameter, Gastrothylax type, number of muscle units, d.e.c., 56—60 mm; d.i.c., 43—45; v.e.c., 58—62; v.i.c., 47—49; m.e.c., 5—7 (Figure 170). Ventral pouch quadrangal, with dorsally directed apex. Pharynx 0.52—0.78 by 0.63—0.77 mm, Paramphistomum type, (Figure 67), esophagus 0.38—0.45 mm long. Right testis 0.91—1.2 by 0.70—0.75, left testis 0.85—1.1 by 0.81—0.95 mm. Pars musculosa weakly developed. Ovary 0.51—0.60 by 0.29—0.32 m. Terminal genitalium within ventral pouch, Gracile type (Figure 114). Eggs 0.118—0.126 by 0.500 mm.

Biology: Not yet known.

Habitat: Rumen.

Hosts: *Bubalus arnee* f. bubalis (Bovidae); *Cervus unicolor* (Cervidae).

Distribution: Asia — Vietnam.

Reference: Sey.[854]

KEY TO THE SPECIES OF *CARMYERIUS*

1. Terminal genitalium without tegumental papillae 2
   Terminal genitalium with tegumental papillae...................................... 5
2. Terminal genitalium inside ventral pouch 3
   Terminal genitalium outside ventral pouch............................. *C. exoporus*
3. Vitellaria normally developed... 4
   Vitellaria strongly developed....................................... *C. multivitellarius*

4. Terminal genitalium Gracile type .... *C. spatiosus*
   Terminal genitalium Microbothrium type .... *C. diplopharingialis*
   Terminal genitalium Bubalis type....................................... *C. bubalis*
   Terminal genitalium Mancupatus type .... *C. mancupatus*
5. Ceca terminating at middle part of body... 6
   Ceca terminating at level of testes 7
6. Terminal genitalium Gregarius type.................................... *C. gregarius*
   Terminal genitalium Elongatus type.................................... *C. chabaudi*
7. Esophagus without bulb... 8
   Esophagus with bulb.. *C. bulbosus*
8. Excretory duct and Laurer's canal open separately................................. 9
   Excretory duct and Laurer's canal unite before opening .... *C. wenyoni*
9. Terminal genitalium Synethes type, pars prostatica 1.2—1.4 mm long ... *C. synethes*
   Terminal genitalium Synethes type, pars prostatica 0.5—0.6 mm long .... *C. graberi*
   Terminal genitalium Parvipapillatus type......................... *C. parvipapillatus*
   Terminal genitalium Endopapillatus type......................... *C. endopapillatus*
   Terminal genitalium Cruciformis type .... *C. cruciformis*
   Terminal genitalium Schoutedeni type .... *C. schoutedeni*

Fischoederius Stiles et Goldberger, 1910.

Diagnosis: Gastrothylacidae. Body conical or elongate, circular or rounded in cross section. Acetabulum terminal. Pharynx terminal without appendages, esophagus short, without bulb; ceca more or less sinuous, in dorsal or lateral fields, extending to midzone of body or to acetabular region. Testes median line, one testis anterodorsal to other, in front of acetabulum. Ovary submedian. Uterine coils winding posteriorly but straight anteriorly in median line. Vitelline follicules in ventrolateral fields extending from behind genital pore to testicular zone. Parasitic in forestomach of mammals.

Type species: *Fischoederius elongatus* (Poirier, 1883) Stiles et Goldberger, 1910 (Figure 384)

Syn. *Gastrothylax elongatus* Poirrier, 1883, *Fischoederius fischoederi* Stiles et Goldberger, 1910, *F. ceylonensis* Stiles et Goldberger, 1910, *F. siamensis* Stiles et Goldberger, 1910.

Diagnosis: Body length 3.7—20.8 mm, body width 1.8—3.7 mm. Acetabulum 1.13—1.26 mm in diameter, Fischoederius type, number of muscle units, d.e.c., 43—45 + 10—12; d.i.c., 39—42; v.e.c., 50—55; v.i.c., 59—66; m.e.c., 18—20 (Figure 169). Ventral pouch usually triangular with apex ventrally directed. Pharynx 0.63—0.86 by 0.04—0.67 mm, Paramphistomum type (Figure 67), esophagus 0.63—0.84 mm. Testes 0.82—1.26 mm in diameter. Pars musculosa moderately developed. Ovary 0.22—0.31 by 0.27—0.42 mm. Terminal genitalium within ventral pouch, Elongatus type (Figure 85). Eggs 0.125—0.132 by 0.067—0.072 mm.

Biology: References. Sewell,[830] Rao and Ayyar,[767] Tandon,[951] Peter and Srivastava,[718] Mukherjee,[630] Agrawal and Pande.[12]

Preparasitic stage. ① Mukherjee. Eggs with thickening at antiopercular end. Freshly deposited eggs contained zygote of 8 to 16 cell stage. Embryonic development completed 7 to 8 d at room temperature of summer months in India. Mass hatching began on 8th day. Body length 0.12—0.20 by 0.044—0.088 mm. Covered with 20 ciliated epithelial cells arranged in four rows by the formula 6:8:4:2. Miracidia composed of an apical gland, two pairs of penetration glands, a pair of flame cells, germinal cells, and germinal balls. Intermediate hosts: *Lymnaea acuminata*, *L. luteola*, *Gyraulus euphraticus*.

Intramolluscan stage. ① Sewell. Rediae elongate in shape, without locomotor ap-

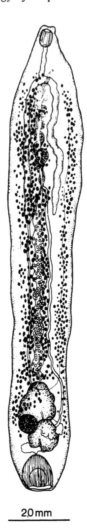

2.0 mm

FIGURE 384. *Fischoederius
elongatus* (Poirier, 1883). (After
S. Yamaguti, 1939.)

pendages. Body size 0.80—0.10 mm. Pharynx 0.053 mm in diameter, cecum saccular, extending beyond birth pore and situated at anterior quarter of body, consists of three pairs of flame cells. Rediae contained daughter rediae and daughter cercariae alike. Cercaria of *F. elongatus* identical with *Cercaria indicae XXIX*. Sewell, 1922. Cercariae realized from rediae in an immature form, and they continued their development in snail tissue. Body 0.26—0.69 by 0.35—0.15 mm. Tail 0.75 mm long. A pair of pigmented eyespots situated near to anterior extremity. The whole body surface is covered with rodlike cytogenous cells. Acetabulum 0.087—0.12 mm in diameter. Pharynx 0.069 mm in diameter, esophagus short, ceca extending to postequatorial part of body. Main excretory tubes with cross-connecting tubes. Caudal tube somewhat dilated at posterior tip, with lateral openings. Primordia of genital organs consist of masses of small, round granular cells. ② Rao and Ayyar. Well developed rediae measure 0.8 by 0.105 mm. Pharynx 0.053 mm in diameter, saccular cecum one fourth of body, at anterior surrounded by a mass of pyriform granular cells. Three pairs of flame cells present. ③ Peter and Srivastava. *Cercariae indicae XXIX* Sewell, 1922 and *Cercaria chungathi* Peter and Srivastava 1961 belong to this species. Pigmentata type. Body 0.25—0.77 by 0.15—0.30 mm, tail 0.37—0.70 mm long. Eyespots at level of bifurcation. Acetabulum 0.098—0.148 mm long. Pharynx 0.09 by 0.06 mm, esophagus short, ceca terminate at posterior half of body length. Main excretory tubes with cross-connecting tubes. Primordia of genital organs were seen as groups of cells. Caudal tube expands near to tip of tail, without lateral openings. Adolescariae 0.33 mm in diameter. ④ Tandon. Mature cercaria 0.85—0.95 mm, tail 0.74—0.81 mm long. Acetabulum 0.01—0.07 by 0.13—0.14 mm. Pharynx 0.10—0.20 by 0.06—0.09 mm, esophagus short, ceca ending little beyond the half of body length. Primordia of genital organs were rounded masses of cells.

Development in definitive host. ① Rao and Ayyar: Cyst were fed to a calf, and 8 weeks later some fairly mature flukes were recovered. ② Mukherjee: Goats and a cow were infected by adolescariae with a total of 140 and 120 cysts, respectively. Worms were not found in goats when they were sacrificed 101 and 121 d after infection. A total of 500 cysts were given to a clean cow and first amphistome eggs were found in feces 126 d postinfection. ③ Agrawal and Pande. Guinea pig, rabbit, albino rat, kid, lamb, kitten, and piglet were infected with cysts. Maturing flukes were only found in the rumen of lamb, slaughtered 126 d after infection.

Habitat: Rumen.

Host: *Anoa depressicornis, Bos primigenius* f. taurus, *Bubalus arnee* f. bubalis, *Capra aegagrus* f. hircus, *Ovis ammon* f. aries, *Bos gaurus* (Bovidae); *Cervus unicolor, Muntjacus muntjak, Rangifer tarandus* f. domestica, *Axis axis, Cervus canadensis, Capreolus capreolus, Alces alces* (Cervidae).

Distribution: Asia — Malaysia, Taiwan, Celebes, Thailand, India, Vietnam, Cambodia, Borneo, Philippines, China, Sri Lanka, U.S.S.R., Japan, Korea; Africa — Egypt, Java, Papua New Guinea.

References: Ashizawa et al.,[47] Bali,[64] Boer,[101] Chen,[167] Chen et al.,[168] Chinone,[169] Chu,[176] Crusz,[191] Dawes,[205] Eduardo and Manuel,[279] Gohar,[352] Hsu,[426] Kadenatsii,[459] Lee and Lowe,[527] Mukherjee,[631] Oshmarin and Demshin,[681] Owen,[685] Paudi et al.,[707] Schad et al.,[819] Segal et al.,[826] Seneviratna,[828] Sey,[846,850,853] Shagraev and Zhaltsanova,[869] Shant,[871] Thapar,[965] Vaidyanathan,[1008] Wu et al.,[1070] Yamaguti,[1071] Zhaltsanova.[1096]

Fischoederius cobboldi (Poirier, 1883) Stiles et Goldberger, 1910. (Figures 385 and 386)

Syn. *Gastrothylax cobboldi* Poirier, 1883.

Diagnosis: Body length 8.2—12.3 mm, body width 5.3—7.6 mm. Acetabulum 1.7—2.1 mm in diameter, Gastrothylax type, number of muscle units, d.e.c., 62—69; d.i.c., 32—36; v.e.c., 58—60; v.i.c., 42—47; m.e.c., 7—9 (Figure 170). Pharynx 0.67—0.85 mm long, Paramphistomum type (Figure 67), esophagus 0.87—1.32 mm. Ventral pouch usually triangular with apex directed ventrally. Testes 0.84—1.26 by 0.63—0.98 mm. Pars musculosa moderately developed. Ovary 0.33—0.39 mm in diameter. Terminal genitalium Microbothrium type (Figure 122). Eggs 0.110—0.120 by 0.060—0.065 mm.

Biology: Not yet known.

Habitat: Rumen.

Hosts: *Bos primigenius* f. taurus, *Bubalus arnee* f. bubalis, *Boselaphus tragocamelus, Capra aegagrus* f. hircus, *Ovis ammon* f. aries (Bovidae).

Distribution: Asia — China, Japan, Malaysia, Sri Lanka, India, Cambodia, Philippines, Afghanistan, Korea, Thailand, North Borneo, Vietnam, China.

References: Bali et al.,[67] Chen et al.,[168] Chu,[176] Crusz,[191] Eduardo and Manuel,[279] Fischoeder,[311] Fukui,[334] Gupta and Dutta,[390] Kotrlá et al.,[501] Lee and Lowe,[527] Patnaik and Acharjyo,[705] Schad et al.,[819] Segal et al.,[826] Seneviratna,[828] Sey,[846,850,854] Shanta,[871] Thapar.[965]

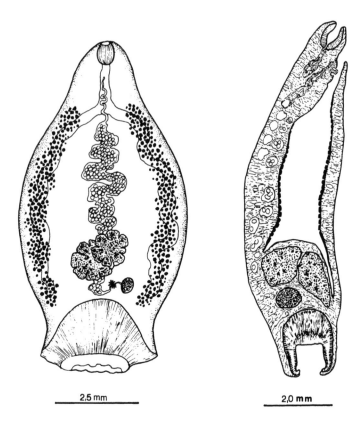

2,5 mm   2,0 mm

FIGURE 385. *Fischoederius cobboldi* (Poirrier, 1883).

FIGURE 386. *Fischoederius cobboldi* (Poirier, 1883) (median sagittal section).

Fischoederius japonicus (Fukui, 1922) Yamaguti, 1939 (Figure 387)

Syn. *Fischoederius siamensis* var. *japonica* Fukui, 1922; *Gastrothylax elongatus* var. *japonica* Fukui, 1922.

Diagnosis: Body length 3.6—8.2 mm, body width 1.8—3.6 m. Acetabulum 1.13—1.73 mm in diameter, Carmyerius type, number of muscle units, d.e.c., 33—35; d.i.c., 41—43; v.e.c., 33—37; v.i.c., 33—37; m.e.c., 7—9 (Figure 164). Pharynx 0.37—0.67 by 0.43—0.58 mm, Paramphistomum type (Figure 67); esophagus 0.36—0.51 mm in length. Ventral pouch triangular, apex directed ventrally. Testes 0.78—1.12 by 0.81—0.93 mm. Pars musculosa weakly developed. Ovary 0.25—0.36 by 0.27—0.43 mm. Terminal genitalium Papillogenitalis type (Figure 91). Eggs 0.127—0.153 by 0.076—0.096 mm.

Biology: Not yet known.

Habitat: Rumen.

Hosts: *Bos primigenius* f. taurus, *Bubalus arnee* f. bubalis (Bovidae); *Cervus unicolor*, *Muntiacus muntjak* (Cervidae).

Distribution: Asia — Japan, Vietnam, China.

References: Chen et al.,[168] Sey,[853,854] Yamaguti.[1072]

Fischoederius skrjabini Kadenatsii, 1962 (Figure 388)

Diagnosis: Body length 20.3—25.4 mm, body width 3.8—4.9 mm. Acetabulum 0.28—1.74 by 1.93—2.20 mm. Pharynx 0.73—0.82 by 0.82—0.92 mm, esophagus 0.82—0.92 mm long. Anterior testis 1.93—3.14 by 1.84—2.76, posterior testis 2.02—3.3 by 1.63—3.31 mm. Pars musculosa weakly developed. Ovary 0.82—0.92 by 0.64 mm. Eggs 0.138—0.152 by 0.072—0.076 mm.

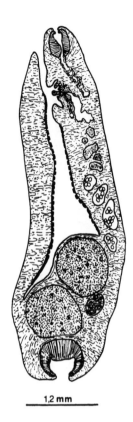

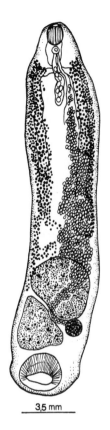

FIGURE 387. *Fischoederius japonicus* (Fukui, 1922) (median sagittal section).

FIGURE 388. *Fischoederius skrjabini* Kadenatsii, 1962. (After A. N. Kadenatsii, 1962.)

Biology: Not yet known.
Habitat: Rumen.
Hosts: *Alces alces, Capreolus capreolus, Cervus canadensis, Rangifer tarandus* (Cervidae).
Distribution: Asia — U.S.S.R.
References: Kadenatsii,[458,459] Shagraev and Zhaltsanova.[869]

Fischoederius ovatus Wang, 1977 (Figure 389)

Syn. *Fischoederius ovis* Zhang et Yang, 1986
Diagnosis: Body length 6.8—8.8 mm, body width 3.1—3.8 mm. Acetabulum 0.80—1.20 by 1.20—1.44 mm. Pharynx 0.32—0.40 by 0.48—0.56 mm, esophagus 0.32—0.48 mm long. Anterior testis 1.60—1.90 by 1.70—2.00 mm, posterior testis 1.60—2.00 by 1.80—2.20 mm. Ovary 0.40—0.50 mm in diameter. Eggs 0.126—0.130 by 0.080—0.090 mm.
Biology: Not yet known.
Habitat: Rumen.
Hosts: *Bos primigenius* f. taurus, *Bubalus arnee* f. bubalis (Bovidae).
Distribution: Asia — China.
Reference: Wang.[1049]

Fischoederius compressus Wang, 1979 (Figure 390)

Diagnosis: Body length 3.7—5.8, body width 1.0—1.5 mm. Acetabulum 0.87—1.08 by 0.71—1.05 mm. Pharynx 0.28—0.38 by 0.22—0.28, esophagus 0.31—0.38 mm long. Anterior testis 0.35—0.61 by 0.26—0.43 mm, posterior testis 0.35—0.69 by 0.43—0.49 mm. Ovary 0.15—0.17 by 0.21—0.27 mm. Eggs 0.108—0.133 by 0.056—0.073 mm.

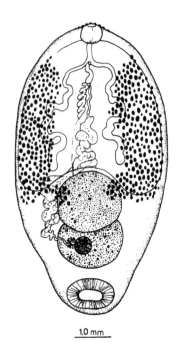

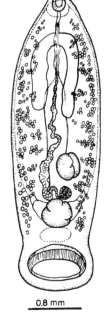

1,0 mm

0,8 mm

FIGURE 389. *Fischoederius ovatus* Wang, 1977. (After P. Q. Wang, 1977.)

FIGURE 390. *Fischoederius compressus* Wang, 1979. (After X. Y. Wang, 1979.)

Biology: Not yet known.
Habitat: Rumen.
Host: *Hidropotes inermis (Cervidae)*.
Distribution: Asia — China.
Reference: Wang.[1050]

KEY TO THE SPECIES OF *FISCHOEDERIUS*

1. Ceca in lateral fields 2
   Ceca in dorsal field 4
2. Ceca extending to posterior testis 3
   Ceca extending to middle part of body .... *F. skrjabini*
3. Testes irregular in shape .... *F. cobboldi*
   Testes entire in shape .... *F. ovatus*
4. Ceca pre-equatorial 5
   Ceca postequatorial .... *F. japonicus*
5. Testes overlap each other .... *F. elongatus*
   Testes separated by uterine coil.. *F. compressus*

Velasquesotrema Eduardo et Javellana, 1987

Diagnosis: Gastrothylacidae. Body subcylindrical, nearly round in cross section. Ventral pouch limited in extent, up to middle of body. Acetabulum subterminal. Pharynx terminal without appendages, esophagus moderately long, without bulb; ceca in lateral sides of body, sinuous; extending to ovarian zone. Testes obliquely tandem, lobed, at the beginning of the posterior body half; seminal vesicle thin walled; pars musculosa thick walled. Ovary post-testicular, spherical. Uterine coils in median line of body. Vitelline follicules in lateral fields from genital pore to anterior end of anterior testis. Parasitic in forestomach of ruminants.

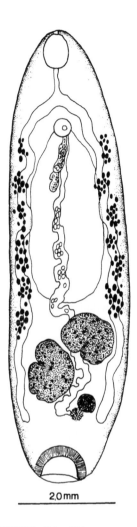

2.0 mm

FIGURE 391. *Velasquezotrema
brevisaccus* (Eduardo, 1981).
(Courtesy of S. L. Eduardo.)

Type species: *Velasquezotrema brevisaccus* **Eduardo and Javellana, 1987 (Figure 391)**
Syn. *Fischoederius brevisaccus* Eduardo, 1981.

Diagnosis: With characters of the genus. Body length 3.0—8.7 mm, body width 0.9—1.6
mm in dorsoventral direction. Acetabulum 0.70—0.80 mm in diameter in dorsoventral
direction, Gastrothylax type, number of muscle units, d.e.c., 42—54; d.i.c., 23—36; v.e.c.,
44—50; v.i.c., 22—26; m.e.c., 4—5 (Figure 170). Pharynx 0.4—0.5 mm long, Calicophoron type (Figure 58), esophagus 0.30—0.80 mm long. Anterior testis 0.62—0.69 mm
long; 0.26—0.35 mm in width in dorsoventral direction; posterior testis 0.58—0.66 mm
long; 0.32—0.37 mm in width in dorsoventral direction. Pars musculosa moderately developed. Terminal genitalium within ventral pouch, Brevisaccus type (Figure 100). Eggs
0.120 by 0.070 mm.

Biology: Not yet studied.

Habitat: Forestomach.

Host: *Bos primigenius* f. taurus, *Bubalus arnee* f. bubalis (Bovidae).

Distribution: Asia — Philippines.

References: Eduardo and Javellana.[281]

3.2.3.2.3. PARAMPHISTOMIDAE Fischoeder, 1901

Diagnosis: Body moderately large or small, conical, oval, pyriform, elliptical, rarely strongly flattened, and with conspicuous, hooplike thickening. Acetabulum ventral, terminal or ventroterminal. Ventral pouch absent. Pharynx terminal, proportionately developed, rarely enormous, without appendages; esophagus short with or without bulb or muscular thickening; ceca usully simple or undulating or sinuous, along lateral sides, extending to acetabular zone. Testes tandem, diagonal or horizontal, in middle or posterior part of body, lobed, globular or rounded in outline. Ovary median or submedian, posterior or posterodorsal to hind testis. Uterine coils winding forward dorsal, occasionally lateral to testes. Genital pore near to bifurcation, terminal genitalium well differentiated. Vitelline follicles extending in lateral fields, commencing at varying levels, occasionally confluent medially. Laurer's canal and excretory duct cross or do not cross each other. Lymphatic system present. Parasitic in forestomach of mammals (mainly undulates).

KEY TO THE SUBFAMILY OF PARAMPHISTOMIDAE

1. Excretory duct and Laurer's canal cross each other .... Paramphistominae
    Excretory duct and Laurer's canal without crossing................... Orthocoeliinae

3.2.3.2.3.1. ORTHOCOELIINAE Price et McIntosh, 1953

Diagnosis: Paramphistomidae. Body conical or oval, occasionally strongly flattened. Acetabulum terminal or ventroterminal. Pharynx terminal, well developed, and with differentiated musculature, esophagus with or without muscular thickening, ceca simple or winding, extending to acetabular zone. Testes tandem or horizontal, in middle or posterior zone of body, entire or irregular in outline. Ovary median or submedian, near or at acetabulum. Genital pore at about bifurcation. Uterine coils winding, dorsal to testes. Vitelline follicules in lateral fields, commencing at different levels. Parasitic in forestomach and stomach of mammals.

KEY TO THE GENERA OF ORTHOCOELIINAE

1. Beaker-shaped retractile vestibule present at anterior end .... *Palamphistomum*
    Retractile vestibule absent 2
2. Testes horizontal.. 3
    Testes tandem or slightly diagonal.. 4
3. Acetabulum with muscular plug .... *Gemellicotyle*
    Acetabulum without muscular plug .... *Bilatorchis*
4. Body conical, not flattened 5
    Body strongly flattened dorsoventrally.............................. *Platyamphistoma*
5. Pharynx normally developed... 6
    Pharynx enormously developed .... *Macropharynx**
6. Genital sucker present 7
    Genital sucker absent 8
7. Genital papilla poorly musculated..................................... *Leiperocotyle*
    Genital papilla strongly muscular .... *Sellsitrema*
8. Pars musculosa weakly developed 9
    Pars musculosa well developed .... *Orthocoelium*
9. Terminal genitalium not enormously developed................................ 10
    Terminal genitalium enormous....................................... *Gigantoatrium*

* Lacking information on the position of Laurer's canal and of excretory duct, this subfamily was assigned tentatively to this.

10. Body without looplike thickening 11
Body with looplike thickening .... *Glyptamphistoma*
11. Body conical, not concave ventrally 12
Body flattened, concave ventrally.. *Buxifrons*
12. Vitellaria well developed, consisting of densely packed follicules extending from pharynx to acetabulum .... *Paramphistomoides*
Vitellaria moderately developed, consisting of loosely packed follicules, extending from genital opening to acetabulum .... *Pseudoparamphistoma*
Vitellaria weakly developed, limited in number, extending from bifurcation to acetabulum.. *Nilocotyle*

Type genus: *Orthocoelium* (Stiles et Goldberger, 1910) Price et McIntosh, 1953
Syn. *Ceylonocotyle* Näsmark, 1937; *Cochinocotyle* Gupta et Gupta, 1970, *Chenocoelium* Wang, 1966.
Diagnosis: Orthocoeliinae. Body conical to elongated. Acetabulum ventroterminal. Pharynx terminal, esophagus with or without muscular thickening, ceca simple, occasionally winding, extending to acetabular zone. Testes tandem, preovarian, entire or lobed. Pars musculosa well developed and convoluted; pars prostatica shorter or longer. Ovary anterodorsal to acetabulum. Genital pore pre- or postbifurcal. Uterine coils winding, posterior and dorsal to testes. Vitelline follicules lateral, extending along entire length of ceca. Lymphatic system present. Parasitic in forestomach of mammals.

Type species: ***Orthocoelium orthocoelium*** **(Fischoeder, 1901) Price et McIntosh, 1953**
(Figure 392)
Syn. *Paramphistomum orthocoelium* (Fischoeder, 1901) Price et McIntosh, 1953 (Figure 392) *Paramphistomum orthocoelium* Fischoeder, 1901; *P. spinicephalus* Tandon, 1955; *O. saccocoelium* Sey, 1980; *P. chinensis* Hsu, 1935; *Chenocoelium orthocoelium* of Chen et al. 1985.
Diagnosis: Body length 5.3—7.8 mm, body width 1.5—3.3 mm. Acetabulum 0.93—1.53 mm in dorsoventral direction, Streptocoelium type, number of muscle units, d.e.c., 28—32; d.i.c., 29—42; v.e.c., 25—32; v.i.c., 33-42; m.e.c., 5—6 (Figure 177). Pharynx 0.78—1.42 mm long, Orthocoelium type (Figure 66), esophagus 1.23—1.42 mm long. Anterior testis 0.38—0.59 mm long, 0.82—1.18 mm in dorsoventral direction, posterior testis 0.52—0.62 mm long, 0.79—0.97 mm in dorsoventral direction. Pars musculosa relatively well developed. Ovary 0.21—0.38 mm in diameter. Terminal genitalium Orthocoelium type (Figure 124). Genital pore prebifurcal. Eggs 0.104—0.163 by 0.061—0.07 mm.
Biology: Not yet known.
Habitat: Rumen and reticulum.
Hosts: *Bubalus arnee* f. bubalis, *Capra aegagrus* f. hircus, *Bos primigenius* f. taurus, *Ovis ammon* f. aries (Bovidae); swamp deer (Cervidae).
Distribution: Asia — India, Sri Lanka, Vietnam, Malaysia, Thailand, Cambodia, China, Bangladesh, Philippines, Japan, Korea, Burma.
References: Ashizawa et al.,[46] Balasubramaniam et al.,[63] Chatterji,[164] Chen,[167] Chu,[176] Crusz,[191] Eduardo,[276] Fischoeder,[311] Khan,[483] Lee and Lowe,[527] Maxwell,[597] Näsmark,[647] Segal et al.,[826] Sey,[846] Shanta,[871] Tandon,[947] Thapar.[965]

Orthocoelium streptocoelium **(Fischoeder, 1901) Yamaguti, 1971 (Figure 393)**
Syn. *Paramphistomum streptocoelium* Fischoeder, 1901
Diagnosis: Body length 4.8—9.7 mm, body width 1.1—2.8 mm. Acetabulum 0.92—1.56 mm in diameter, Streptocoelium type, number of muscle units, d.e.c., 17—22; d.i.c., 22—29; v.e.c., 16—21; v.i.c., 20—35; m.e.c., 5—9 (Figure 177). Pharynx 0.38—1.12

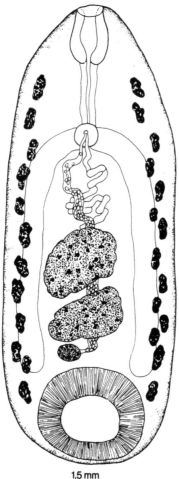

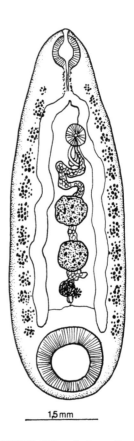

FIGURE 392. *Orthocoelium orthocoelium* (Fischoeder, 1901). (By permission of Dr. W. Junk Publishers, 1985, Dordrecht.)

FIGURE 393. *Orthocoelium streptocoelium* (Fischoeder, 1901). (After F. Fischoeder, 1903.)

mm long, Calicophoron type (Figure 58), esophagus 0.48—1.26 mm long, with muscular sphincter. Anterior testis 0.49—0.96 mm long and 0.81—0.93 mm in dorsoventral direction, posterior testis 0.28—0.69 mm long, and 0.48—0.86 mm in dorsoventral direction. Pars musculosa well developed. Ovary 0.22—0.61 mm in diameter. Terminal genitalium Streptocoelium type (Figure 129), genital pore postbifurcal. Eggs 0.105—0.120 by 0.060—0.069 mm. Biology: Reference. Durie.[253]

Preparasitic stage. Durie. Eggs were operculated with a small projection at the antiopercular end. Freshly deposited eggs were in early, four-cell stage. Embryonic development took place within 16 d, mass hatching occurred on the 17th day in an incubation at 27°C. Miracidia measure 0.20—0.30 mm in length, covered with epidermal cells, arranged by the formula 6:8:4:2. Miracidia composed of a single, apical gland ("primitive gut"), two pairs of penetration glands, a brain mass, a pair of apical glands, germinal cells, and germinal balls. Miracidiae were active for 4 to 5 h in the presence of the susceptive intermediate host. Intermediate host *Glyptanisus gilberti* (Planorbidae).

Intramolluscan stage. Durie. During the period of the first 5 to 8 d, a rapid formation of rediae can be observed in sporocyst. Rediae liberated by rupture of the anterior body wall. Free rediae were found in the snail tissue 10 or 11 d after infection. Mature rediae measuring 0.50—0.90 by 0.09—0.10 mm, pharynx 0.027 by 0.040 mm in size, esophagus 0.030 mm, gut 0.10 by 0.050 mm with salivary glands. There were three pairs of flame cells. Birth pore situated about 0.17 mm from the anterior end. Cercariae were released in immature condition, and their development continued in the snail tissue. First cercariae emerged from rediae 21 d after infection at 27°C. Mature cercariae measured 0.33—0.38 by 0.18—0.21 mm. Tail 0.45 mm long. Eyespots placed anterodorsally and lateral to esophagus. Acetabulum ventroterminal, 0.10 mm in diameter. Pharynx 0.065 by 0.045 mm, esophagus with muscular sphincter, ceca terminating at posterior third of body. Main excretory tubes with cross-connecting tubes. The reproductive system was represented by primordia, consisting of masses of cells. Caudal tube was expanded at the tip of the tail, with dorsal and ventral ducts, openings on the exterior. Adolescariae measure 0.22 mm in diameter.

Development in definitive host. Durie. Adolescariae were fed on cattle and sheep, the flukes reached their maturity within 56 d and 48 d, respectively.

Habitat: Rumen.

Host: *Bubalus arnee* f. bubalis, *Bos primigenius* f. taurus, *Capra aegagrus* f. hircus, *Kobos thomasi*, *K. defassa*, *Tragelophus scriptus*, *Ovis ammon* f. aries (Bovidae); *Cervus unicolor*, (Cervidae).

Distribution: Asia — Sri Lanka, Malaysia (North Borneo), India, Japan, China; Africa — Zaire, Australia, New Zealand; The Caribbean — Cuba.

References. Bali et al.,[67] Chen et al.,[168] Chinone,[169] Crusz,[191] Durie,[253] Eduardo,[276] Fischoeder,[311] Gupta and Nakhasi,[396] Hovorka et al.,[425] Keith and Keith,[475] Kotrlá and Prokopič,[499] Lee and Lowe,[527] Näsmark,[647] Presidente,[738] Prokopič and Kotrlá,[752] Schad et al.,[819] Sey,[846] Strydonck,[920] Tandon and Sharma.[959]

Orthocoelium dicranocoelium (Fischoeder, 1901) Yamaguti, 1971 (Figure 394)

Syn. *Paramphistomum dicranocoelium* Fischoeder, 1901; *Ceylonocotyle tamilensis* Gupta et Bakhshi in Gupta and Nakhasi, 1977.

Diagnosis: Body length 4.2—7.3 mm, body width 1.2—2.4 mm. Acetabulum 0.68—1.22 mm in diameter, Streptocoelium type, number of muscle units, d.e.c., 12—17; d.i.c., 20—27; v.e.c., 13—18; v.i.c., 20—27; m.e.c., 6—9 (Figure 177). Pharynx 0.43—0.96 mm long, Dicranocoelium type (Figure 60), esophagus 0.48—1.26 mm long with muscular thickening. Anterior testis 0.86—1.38 mm long and 0.68—1.12 mm in dorsoventral direction, posterior testis 0.98—1.38 mm long and 0.68—1.13 mm in dorsoventral direction. Ovary 0.33—0.45 mm in diameter. Terminal genitalium Gracile type (Figure 114), genital pore postbifurcal, pars musculosa well developed. Eggs 1.120—0.140 by 0.081—0.092 mm.

Biology: Not yet known.

Habitat: Rumen.

Hosts: *Bos primigenius* f. taurus, *Bubalus arnee* f. bubalis, *Capra aegagrus* f. hircus, *Ovis ammon* f. aries (Bovidae).

Distribution: Asia — Sri Lanka, Philippines, India, Indonesia, Vietnam, China.

References: Chen et al.,[168] Crusz,[191] Eduardo,[276] Fischoeder,[311] Gupta and Nakhasi,[396] Seneviratna,[828] Sey,[846,854] Tandon and Sharma.[959]

Orthocoelium scoliocoelium (Fischoeder, 1904) Yamaguti, 1971 (Figure 395)

Syn. *Ceylonocotyle cheni* Wang, 1966; *C. longicoelium* Wang, 1977 *Paramphistomum scoliocoelium*, Fischoeder, 1904; *P. shipleyi* Stiles et Goldberger, 1910; *Cotylophoron ovatum* Harshey, 1934

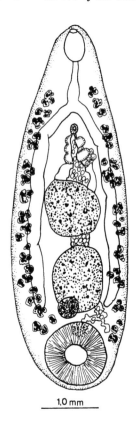

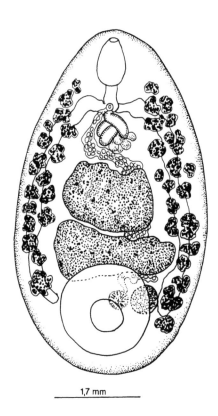

1,0 mm   1,7 mm

FIGURE 394. *Orthocoelium di-cranocoelium* (Fischoeder, 1901). (By permission of Dr. W. Junk Publishers, 1985, Dordrecht.)

FIGURE 395. *Orthocoelium scoliocoelium* (Fischoeder, 1904). (By permission of Dr. W. Junk Publishers, 1985, Dordrecht.)

Diagnosis: Body length 2.8—5.1 mm, body width 0.8—2.3 mm. Acetabulum 0.49—1.14 mm in diameter, Streptocoelium type, number of muscle units, d.e.c., 14—19; d.i.c., 18—27; v.e.c., 12—16; v.i.c., 21—29; m.e.c., 8—12 (Figure 177). Pharynx 0.34—0.69 mm long, Dicranocoelium type (Figure 60), esophagus 0.38—0.98 mm long. Anterior testis 0.53—1.18 mm long and 0.48—1.54 mm in dorsoventral direction, posterior testis 0.58—0.98 mm long and 1.28—1.83 mm in dorsoventral direction. Pars musculosa well developed and convoluted. Ovary 0.22—0.48 by 0.23—0.46 mm. Terminal genitalium Scoliocoelium type (Figure 127), genital pore bifurcal. Eggs 0.120—0.150 by 0.067—0.086 mm.

Biology: References. Dinnik,[224] Jain and Srivastava,[444] Mukherjee.[635]

Preparasitic stage. ① Jain and Srivastava. Eggs oval, operculated with a small thickening at the antiopercular end. Freshly laid eggs contained zygote having 4- to 16-celled stage. Embryonic development at room temperature (34.4 to 43.2°C) in India took place within 10 to 13 d. Hatching was stimulated by sunlight or artificial light alike. Living miracidia measure 0.13—0.18 by 0.04—0.06 mm. Body covered with ciliated epithelial cells, arranged by the formula 6:8:4:2. There were 26—28 sensory papillae on the terebratorium and 12 more along the body surface. Inner structure of miracidia consists of a single cephalic gland ("primitive gut"), two pairs of penetration glands, a pair of flame cells, a brain mass, and germinal cells. Intermediate host: *Bulimus pulchellus* ② Mukherjee. Eggs yellowish brown, operculated with shell thickening at antiopercular end. Zygote unsegmented. Under controlled temperatures (32—35°C) embryonic development took place in 9 to 13 d. Body of fixed miracidia 0.10—0.13 by 0.05—0.06 mm, covered with epidermal cells, arranged by the formula 6:8:4:2. There were 31 or 32 sensory papillae on the terebratorium and 14 along body surface. Two pairs of penetration glands, one apical gland, a pair of flame cells, one brain mass, and germinal cells were present. Intermediate host: *Digoniostoma* (= *Bulimus) pulchellus*. ③ Dinnik. *Anisus natalensis*.

Intramolluscan stage. ① Jain and Srivastava. Young sporocyst measures 0.109 by 0.058 mm. The body cavity contains germ cells and germ balls. Excretory system had two flame cells and their ducts. Mature sporocyst 0.16—0.17 by 0.052—0.064 mm in size. They contain developing rediae. Rediae measure 0.57—0.70 by 0.15—0.23 mm in size. Pharynx 0.03 mm in diameter, cecum 0.10 mm long, salivary glands present. There are three pairs of flame cells. The germ balls fill up the posterior part of body, the mature rediae contained about four numbers of developing cercariae. The birth pore was situated about 0.15 mm from the anterior body. Cercariae discarded in immature form and continued their development in the snail tissue. Mature cercariae of *O. scoliocoelium* belonged to the "Pigmentata" group. Body dimensions 0.52 by 0.27, tail measures 0.36 mm in length. A pair of eyespots were situated on the lateral surface; the whole body of cercariae covered with dark-brown patches of pigments. The cytogenous cells with a number of cystogenous rods were distributed unevenly under tegument. Acetabulum ventroterminal, 0.088—0.96 mm in diameter. Pharynx 0.096 by 0.066 mm in size, esophagus 0.133 mm long, with muscular thickening, ceca extending to the middle of body length. Main excretory tubes with cross-connecting tubes, the latter formed a loop with the main excretory tubes of either side through which passes the corresponding cecum. Caudal tube expanded at the tip of the tail, without lateral openings. Features of *Cercaria bulimusi* Peter and Srivastava, 1955 agreed well with those of cercariae of *O. scoliocoelium*. Shedding of cercariae was usually stimulated by sunlight, daily production was about 5 to 9 cercariae. The emerged cercariae are ready to encyst on vegetation, adolescariae measure 0.4 mm in diameter. ② Mukherjee. Sporocyst 0.17 by 0.053 mm in normal condition. It has a pair of flame cells, sporocysts usually become mature in 11 d or so. Redia without locomotor appendages. Mature rediae measure 0.45—0.84 by 0.19—0.29 mm, pharynx 0.03—0.05 by 0.02—0.05 mm, cecum 0.063 mm long. Three pairs of flame cells present. Birth pore was situated about 0.13—0.16 mm from anterior body end. Cercaria emerged from rediae in immature form, and they continued their development in the snail tissue. Body dimensions of mature cercariae in extended condition 0.54—0.72 by 0.15—0.32 mm, tail 0.34—0.68 mm long. A pair of eyespots were present on the lateral surface of the body, and the whole body was covered with evenly distributed brown pigment; the round cystogenous cells were densely packed and contain small cystogenous rods. Acetabulum ventroterminal, 0.076—0.099 by 0.072—0.096 mm in diameter. Pharynx 0.076—0.11 by 0.038—0.065 mm in size, esophagus 0.11—0.21 mm in length, with muscular thickening ceca terminating anterior to acetabulum. Main excretory tubes with cross-connecting tubes, which formed a loop with the main excretory tubes and the loops may or may not enclose the ceca. Caudal tube expands somewhat terminally, having no lateral openings. Primordia of genital organs were represented by masses of cells. Cercaria of *O. scoliocoelium* belong to the Pigmentata group and *Cercaria bulimusi* Peter and Srivastava 1955 closely resemble this cercaria. Mature cercariae emerged from the infected snails 31—33 days postinfection, and their daily production was about 40 per snail. Emerged cercariae were ready to encyst on vegetation, adolescariae 0.28—0.35 mm in diameter.

Development in definitive host. ① Jain and Srivastava. Kid was infected with adolescariae 36—62 d old, the minimum time for parasites to become mature was 125 d. Percentage infestation was 56.4%. ② Mukherjee. Goats were infected with adolescariae of 20 to 58 d old; prepatent period was 173 d. Percentage of infection was 12.9%.
Habitat: Rumen and reticulum.

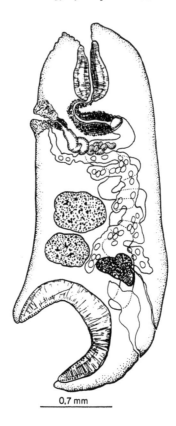

0,7 mm

FIGURE 396. *Orthocoelium par-vipapillatum* (Stiles et Goldberger, 1910) (median sagittal section). (By permission of Dr. W. Junk Publishers, 1985, Dordrecht.)

Hosts: *Bos primigenius* f. taurus, *Bubalus arnee* f. bubalis, *Ovis ammon* f. aries, *Capra aegagrus* f. hircus, (Bovidae); *Muntiacus muntjak* (Cervidae).

Distribution: Asia — India, Vietnam, Japan, Malaysia (North Borneo), Indonesia, China, Philippines, Pakistan; Africa — Kenya, Chad, Central Africa.

References: Bali and Fotedar,[65] Bali et al.,[67] Chellappa,[165] Chen et al.,[168] Dinnik,[224] Eduardo,[276] Fischoeder,[311] Gretillat,[374] Gupta,[387] Lee,[525] Lee and Lowe,[527] Näsmark,[647] Oshmarin and Demshin,[681] Shad et al.,[819] Sey,[846,854] Subbarao et al.,[928] Tandon and Sharma,[959] Yamaguti.[1072,1074]

Orthocoelium parvipapillatum **(Stiles et Goldberger, 1910) Eduardo, 1980 (Figure 396)**
Syn. *Paramphistomum parvipapillatum* Stiles et Goldberger, 1910

Diagnosis: Body length 2.9—3.9 mm, body width 1.0—2.3 mm. Acetabulum 0.50—0.91 mm, Streptocoelium type, number of muscle units, d.e.c., 15—19; d.i.c., 21—27; v.e.c., 12—19; v.i.c., 22—27; m.e.c., 3—9 (Figure 177). Pharynx 0.48—0.54 mm long, Dicranocoelium type (Figure 60), esophagus 0.65—0.72 mm long. Anterior testis 0.34—0.68 mm long and 0.77—0.45 mm in dorsoventral direction, posterior testis 0.29—1.09 long and 0.15—0.18 mm in dorsoventral direction. Pars musculosa moderately developed. Ovary 0.15—0.36 by 0.20—0.30 mm. Terminal genitalium Parvipapillatum type (Figure 92) genital pore postbifurcal. Eggs 0.135—0.141 by 0.653—0.072 mm.

Biology: Not yet known.

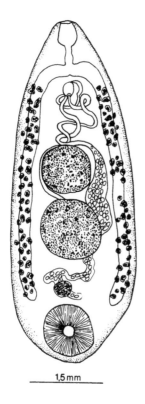

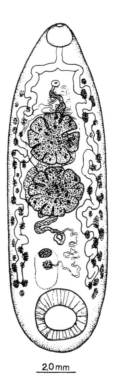

1,5 mm

2,0 mm

FIGURE 397. *Orthocoelium dawesi* (Gupta, 1958). (After N. K. Gupta and U. Nakhasi, 1977.)

FIGURE 398. *Orthocoelium parastreptocoelium* (Wang, 1959). (After P. Q. Wang, 1959.)

Habitat: Rumen, reticulum.
Hosts: *Bubalus arnee* f. bubalis, *Ovis ammon* f. aries, *Bos primigenius* f. taurus. (Bovidae).
Distribution: Asia — Thailand, India, Indonesia.
References: Eduardo,[276] Stiles et Goldberger.[915]

Orthocoelium dawesi (Gupta, 1958) Yamaguti, 1971 (Figure 397)

Diagnosis: Body length 2.7—4.3 mm, body width 1.6—2.6 mm. Acetabulum 0.35—0.82 mm in dorsoventral direction, Streptocoelium type, number of muscular units, d.e.c., 13—16; d.i.c., 18—27; v.e.c., 13—16; v.i.c., 21—25; m.e.c., 6—9; (Figure 177). Pharynx 0.47—0.85 mm long, Dicranocoelium type (Figure 60), esophagus 0.38—0.43 mm long. Anterior testis 0.32—0.48 mm long, 1.10—1.58 mm in dorsoventral direction, posterior testis 0.42—0.68 mm long, 0.82—1.28 mm in dorsoventral direction. Pars musculosa strongly developed, thick walled and convoluted. Ovary 0.20—0.51 mm in diameter. Terminal genitalium Dawesi type (Figure 106). Genital pore postbifurcal. Eggs 0.098—0.143 by 0.058—0.073 mm.
Biology: Not yet known.
Habitat: Rumen.
Host: *Bos primigenius* f. taurus, *Ovis ammon* f. aries, *Capra aegagrus* f. hircus (Bovidae).
Distribution: Asia — India.
References: Eduardo,[275] Gupta,[385] Gupta and Nakhasi.[396]

Orthocoelium parastreptocoelium (Wang, 1959) Sey, 1989 (Figure 398)

Syn. *Ceylonocotyle parastreptocoelium* Wang, 1959.

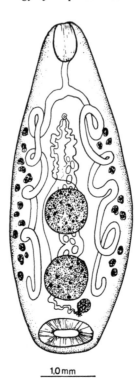

1.0 mm

FIGURE 399. *Orthocoelium sinuocoelium* (Wang, 1959). (After P. Q. Wang, 1959.)

Diagnosis: Body length 13.8—18.5 mm, body width 4.0—5.0 mm. Acetabulum 2.08—2.70 by 2.08—2.92 mm. Pharynx 0.88-1.60 by 1.20—1.60 mm, esophagus 0.96—1.20 mm long. Anterior testis 1.92—3.82 by 1.92—3.46 mm, posterior testis 1.92—3.46 mm by 1.78—3.46 mm. Ovary 0.48—0.64 by 0.64—0.78 mm. Genital pore postbifurcal. Eggs 0.125—0.136 by 0.065—0.072 mm.
Biology: Not yet known.
Habitat: Rumen.
Host: *Bos primigenius* f. taurus, *Bubalus arnae* f. bubalis (Bovidae).
Distribution: Asia — China.
Reference: Wang.[1046]

Orthocoelium sinuocoelium (Wang, 1959) (Sey, 1989) (Figure 399)

Syn. *Ceylonocotyle sinuocoelium* Wang, 1959.
Diagnosis: Body length 3.6—6.0 mm, body width 1.6—1.8 mm. Acetabulum terminal, 0.50—0.64 by 0.80—0.88 mm in size. Pharynx measuring 0.64—0.72 by 0.48—0.52 mm, esophagus 0.26—0.40 mm in length. Ceca form four convolutions in lateral fields. Anterior testis 0.64—0.82 by 0.64—0.72 m, posterior testis 0.22—0.80 by 0.64—0.66 mm. Ovary 0.16—0.18 by 0.16—0.19 mm in size. Genital pore postbifurcal. Eggs 0.140—0.146 by 0.068—0.072 mm.
Biology: Not yet known.
Habitat: Forestomach.
Host: *Bos primigenius* f. taurus (Bovidae).
Distribution: Asia — China.
Reference: Wang.[1046]

 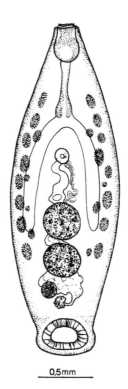

1,5 mm   0,5mm

FIGURE 400. *Orthocoelium gi-gantopharynx* Schad, Kuntz, Anteson, et Webster, 1964. (By permission of Dr. W. Junk Publishers, 1985, Dordrecht.)

FIGURE 401. *Orthocoelium brevicaeca* (Wang, 1966). (After X. Y. Wang, 1966.)

Orthocoelium gigantopharynx (Schad, Kuntz, Anteson et Webster, 1964) Eduardo, 1985 (Figure 400)

Syn. *Ceylonocotyle hsui*, Wang, 1966; *C. chinensis*, Want, 1959, *Paramphistomum gotoi* of Dawes, 1936; of Tandon, 1955; and of Lee and Lowe, 1971.

Diagnosis: Body length 4.6—5.2 mm, body width 2.1—2.7 mm. Acetabulum 0.78—1.56 mm, Streptocoelium type, number of muscle units, d.e.c., 1.16; d.i.c., 2.14; v.i.c., 29—31; v.e.c., 26—29; v.i.c., 19—25; m.e.c., 8—10, (Figure 177). Pharynx 1.32—1.76 mm long, Gigantopharynx type (Figure 63), esophagus 0.67—1.32 mm long, with poorly developed sphincter. Anterior testis 0.39—1.41 mm long and 0.68—1.32 mm in dorsoventral direction. Pars musculosa moderately developed. Ovary 0.18—0.52 by 0.21—0.32 mm. Terminal genitalium Microbothrium type (Figure 122), genital pore bifurcal. Eggs 0.107—0.118 by 0.573—0.063 mm.

Biology: Not yet known.

Habitat: Rumen.

Hosts: *Babalus arnee* f. bubalis, *Antilope cervicapra* (Bovidae).

Distribution: Asia — Malaysia (North Borneo), Philippines, India.

References: Eduardo,[276] Patnaik,[703] Schad et al.,[819] Sey.[843,846]

Orthocoelium brevicaeca (Wang, 1966) Sey, 1989 (Figure 401)

Syn. *Ceylonocotyle brevicaeca* Wang, 1966

Diagnosis: Body length 2.5—2.9 by 0.89—1.1 mm, body width 0.89—1.1 mm. Ace-

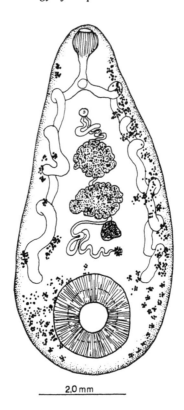

FIGURE 402. *Orthocoelium bovini*
(Gupta et Gupta, 1970). (After N. K.
Gupta and P. Gupta, 1970.)

tabulum 0.31—0.38 by 0.31—0.43 mm. Pharynx 0.24—0.34 by 0.26—0.34 mm, esophagus
0.52—0.64 mm long, bulb present, ceca short extending to anterior part of posterior testis.
Anterior testis 0.38—0.47 by 0.35—0.47 mm, posterior testis 0.38—0.43 by 0.35—0.44
mm. Ovary 0.14—0.17 in diameter. Genital pore postbifurcal. Eggs 0.095—0.130 by
0.059—0.081 mm.
Biology: Not yet known.
Habitat: Rumen.
Host: *Bos primigenius* f. taurus (Bovidae).
Distribution: Asia — China.
Reference: Wang.[1047]

Orthocoelium bovini (Gupta et Gupta, 1970) Eduardo, 1985 (Figure 402)

Syn. *Cochinocotyle bovini* Gupta et Gupta, 1970
Diagnosis: Body length 5.5—9.54 mm, body width 2.3—4.1 mm. Acetabulum 1.05—2.30
mm in diameter, Paramphistomum type, number of muscular units, d.e.c., 1,14; d.e.c.,
2,12; d.i.c., 31—32; v.e.c., 28—29; v.i.c., 21—23 (Figure 173). Pharynx 0.72—0.85 mm
long, Orthocoelium type (Figure 66) (Paramphistomum by Gupta and Gupta), esophagus
0.48—0.68 mm long, without muscular thickening. Anterior testis 0.62—0.81 by 0.90 mm,
posterior testis 0.33—0.76 by 1.2—1.3 mm. Pars musculosa strongly muscular and convoluted. Ovary triangular 0.29—0.52 mm in diameter. Terminal genitalium Microbothrium
type (Figure 122), genital pore postbifurcal. There were no eggs.
Biology: Not yet known.
Habitat: Forestomach.

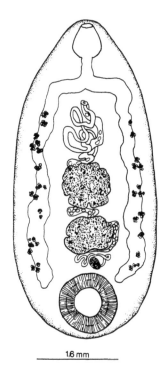

FIGURE 403. *Orthocoelium narayanai* (Gupta et Gupta, 1972). (After N. K. Gupta and P. Gupta, 1972.)

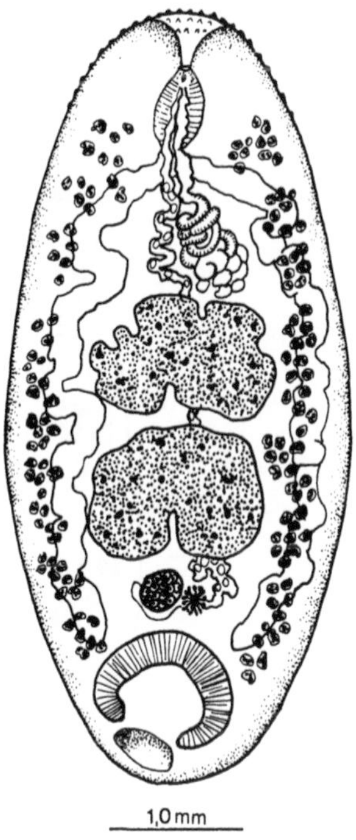

FIGURE 404. *Orthocoelium indonesiense* Eduardo, 1980. (By permission of Dr. W. Junk Publishers, 1980, Dordrecht.)

Host: *Bos primigenius* f. taurus (Bovidae).
Distribution: Asia — India.
Reference: Gupta and Gupta.[392]

Orthocoelium narayanai (Gupta et Gupta, 1972) Eduardo, 1985 (Figure 403)

Syn. *Ceylonocotyle narayanai* Gupta et Gupta, 1972
Diagnosis: Body length 4.9—6.9 mm, body width 2.2 mm in dorsoventral direction. Acetabulum 0.72—0.90 mm in diameter, Calicophoron type, number of muscle units, d.e.c., 14—16; d.i.c., 32—34; v.e.c., 12—14; v.i.c., 32—34; m.e.c., 10—12, (Figure 165). Pharynx 0.54—0.75 mm long, Calicophoron type (Figure 58), esophagus 0.27—0.70 mm long. Anterior testis 0.60—0.63 by 0.70—0.72 mm, posterior testis 0.56—0.72 by 0.63—0.72 mm. Pars musculosa well developed. Ovary 0.18 by 0.09 mm. Terminal genitalium probably Liorchis type (Figure 118). Genital pore postbifurcal. Eggs 0.110—0.130 by 0.056—0.060 mm.
Biology: Not yet known.
Habitat: Forestomach.
Hosts: *Bos primigenius* f. taurus, *Bubalus arnee* f. bubalis (Bovidae).
Distribution: Asia — India.
References: Gupta and Gupta,[394] Gupta and Nakhasi.[396]

Orthocoelium indonesiense Eduardo, 1980 (Figure 404)

Diagnosis: Body length 4.9—7.0, body width 1.0—2.0 mm. Acetabulum 0.64—0.89 mm in diameter, Gastrothylax (= Carmyerius type) number of muscle units, d.e.c., 33—37; d.i.c., 20—30; v.e.c., 32—44; v.i.c., 22—28; m.e.c., 6—10, (Figure 164). Pharynx

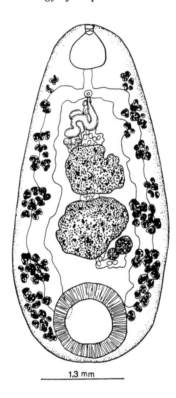

1,3 mm

FIGURE 405. *Orthocoelium dinniki*
Eduardo, 1985. (By permission of Dr.
W. Junk Publishers, 1985, Dordrecht.)

0.50—0.68 mm long, Paramphistomum type; (Figure 67), esophagus 0.20—0.41 mm long.
Anterior testis 0.70—1.53 by 0.73—1.43 mm, posterior testis 0.76—1.09 by 1.00—1.18
mm. Pars musculosa well developed, convoluted. Ovary 0.20—0.35 by 0.45—0.49 mm.
Terminal genitalium Papillogenitalis (Figure 91). Genital pore prebifurcal. Eggs 0.120—0.137
by 0.060—0.070 mm.
Biology: Not yet known.
Habitat: Rumen.
Hosts: *Bos primigenius* f. taurus, *Ovis* sp. (Bovidae).
Distribution: Asia — Indonesia.
Reference: Eduardo.[268]

Orthocoelium dinniki Eduardo, 1985 (Figure 405)
   Diagnosis: Body length 4.2—5.6 mm, body width 1.6—2.1 mm. Acetabulum 1.12—1.30
mm in diameter, Streptocoelium type, number of muscle units, d.e.c., 13—21; d.i.c.,
27—32; v.e.c., 9—17; v.i.c., 20—32; m.e.c., 3—11 (Figure 177). Pharynx 0.43—0.76
mm long, Calicophoron type (Figure 58), esophagus 0.30—0.59 mm long. Anterior testis
0.48—0.93 mm long and 0.82—1.63 mm in dorsoventral direction. Pars musculosa moderately developed, convoluted. Ovary 0.18—0.47 mm in diameter. Terminal genitalium
Papillogenitalis type (Figure 91), genital pore bifurcal. Eggs. 0.120—0.160 by 0.720—0.860
mm.
Biology: Not yet known.
Habitat: Rumen.
Hosts: *Bubalus arnee* f. bubalis, *Bos primigenius* f. taurus, *Capr aegagrus* f. hircus (Bovidae);
   Cervus unicolor (Cervidae).

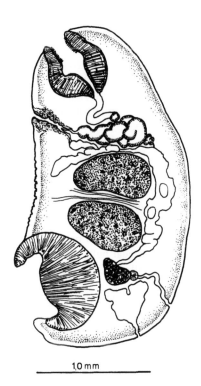

FIGURE 406. *Orthocoelium serpenti-caecum* Eduardo et Peralta, 1987 (median sagittal section). (By permission of Dr. W. Junk Publishers, 1987, Dordrecht.)

Distribution: Asia — China, Philippines, Japan, India, Vietnam.
References: Eduardo,[276] Raina et al.,[765] Sey.[854]

Orthocoelium serpenticaecum Eduardo et Peralta, 1987 (Figure 406)

Diagnosis: Body length 2.4—3.1 mm, body width 1.0—1.4 mm in dorsoventral direction. Acetabulum 0.59—0.90 mm in diameter, Streptocoelium type, number of muscle units, d.e.c., 9—20; d.i.c., 12—23; v.e.c., 18—28; v.i.c., 15-24; m.e.c., 7—25 (Figure 177). Pharynx 0.48—0.56 mm long, Serpenticaecum type (Figure 69) esophagus 0.22—0.48 mm long, without muscular thickening. Anterior testis 0.29—0.70 mm long and 0.44—0.86 mm in dorsoventral direction; posterior testis 0.25—0.68 mm long, and 0.46—0.88 in dorsoventral direction. Pars musculosa well developed. Ovary 0.26—0.32 by 0.15—0.22 mm. Terminal genitalium Serpenticaecum type (Figure 96), genital pore bifurcal. Eggs 0.090—0.130 mm.
Biology: Not yet known.
Habitat: Rumen.
Host: *Bubalus arnee* f. bubalis (Bovidae).
Distribution: Asia — Philippines.
Reference: Eduardo and Peralta.[280]

KEY TO THE SPECIES OF *ORTHOCOELIUM**
(after Eduardo, 1985, modified)

1. Testes entire in outline.. 2

---

* *O. naesmarki* (Mukherjee, 1963) Yamaguti 1971 is considered species inquirenda.

Testes deeply lobed . . . . *O. parastreptocoelium*
2. Ceca extending to acetabular zone 3
    Ceca extending to posterior testis . . . . *O. brevicaeca*
3. Ceca straight or slightly convoluted along their full length 4
    Ceca strongly convoluted, tortuous in lateral fields 13
4. Ceca cross dorsomedian line in their posterior part, forward opposite lateral side; internal
    surface of pharynx with long papillae . . . . *O. gigantopharynx*
    Ceca do not cross dorsomedian line, internal surface of pharynx lacks long
    papillae 5
5. Anterior pharyngeal sphincter present 6
    Anterior pharyngeal sphincter absent 7
6. Esophagus long, with a posterior sphincter, ceca straight . . . . *O. orthocoelium*
    Esophagus short, lacks a posterior sphincter, ceca form coils . . . . *O. bovini*
7. Lip sphincter of pharynx and esophageal bulb present 8
    Lip sphincter of pharynx and esophageal bulb absent 9
8. Genital sphincter and sphincter papillae present, radial musculature fairly strong . . . . . .
    . . . . *O. scoliocoelium*
    Genital sphincter present, sphincter papilla absent, radial musculature fairly strong . . .
    . . . . *O. dawesi*
    Genital sphincter and sphincter papilla absent, radial musculature weak 10
9. Genital fold with tegumental papillae . . . . *O. parvipapillatum*
    Genital fold lacks tegumental papillae . . . . *O. dicranocoelium*
10. Posterior esophageal sphincter present 11
    Posterior esophageal sphincter absent. 12
11. Genital sphincter and sphincter papilla well developed, the latter extensive and connected
    to the former; genital fold lacks tegumental papillae . . . . *O. streptocoelium*
    Genital sphincter and sphincter papilla not well developed, not extensive, and not
    connected to each other; genital fold with tegumental papilla . . . . *O. dinniki*
12. Pars prostatica long; genital sphincter and sphincter papilla present; genital fold with
    tegumental papilla; excretory vesicle deep in the dorsoventral direction . . . .
    . . . . *O. indoesiense*
    Pars prostatica short; genital sphincter present; sphincter papilla absent, genital fold
    lacks tegumental papillae; excretory vesicle not deep in dorsoventral direction . . . .
    . . . . *O. narayanai*
13. Genital opening bifurcal; terminal genitalium with sphincter . . . .
    . . . . *O. serpenticaecum*
    Genital opening postbifurcal; terminal genitalium without sphincter . . . .
    . . . . *O. sinuocoelium*

Leiperocotyle **Eduardo, 1980**
    Diagnosis: Orthocoeliinae. Body oval to elongate. Acetabulum ventroterminal. Pharynx
subterminal, esophagus short, with or without muscular thickening, ceca undulating, extending to acetabular zone. Testes tandem, transverse oval, irregular in outline. Ovary
anterodorsal to acetabulum. Genital pore postbifurcal, genital sucker present. Uterine coils
dorsal to testes. Pars musculosa well, pars prostatica weakly developed. Vitelline follicules
lateral to testes, extending from pharynx to acetabular zone. Parasitic of rumen of ruminants.

Type species: *Leiperocotyle okapi* **(Leiper, 1935) Eduardo, 1980 (Figure 407)**
    Syn. *Cotylophoron okapi* Leiper, 1935
    Diagnosis: Body length 2.9—5.0 mm, body width 1.3—2.1 mm. Acetabulum 1.00—1.24
mm in diameter, Cotylophoron type, number of muscle units, d.e.c., 15—18; d.i.c., 30—43;
v.e.c., 11—16; v.i.c., 31—42; m.e.c., 19—20; (Figure 166). Pharynx 0.37—0.53 mm

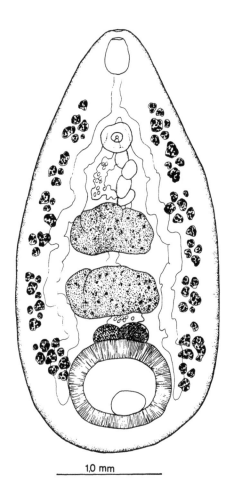

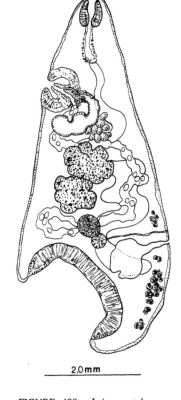

FIGURE 407. *Leiperocotyle okapi* (Leiper, 1935). (By permission of Dr. W. Junk Publishers, 1980, Dordrecht.)

FIGURE 408. *Leiperocotyle congolensis* (Baer, 1936) (median sagittal section). (By permission of Dr. W. Junk Publishers, 1985, Dordrecht.)

long, Calicophoron type (Figure 58), esophagus 0.44—0.66 mm long. Anterior testis 0.41—0.71 by 0.50—1.06, posterior testis 0.41—0.68 by 0.93—1.14 mm. Pars musculosa well developed. Ovary measures 0.21—0.30 by 0.21—0.35 mm. Terminal genitalium Cotylophoron type (Figure 105). Genital pore postbifurcal. Eggs 0.091—0.122 by 0.050—0.067 mm.

Biology: Not yet known.

Habitat: Digestive tube.

Host: *Okapia johnstoni* (Giraffidae).

Distribution: Africa — Zaire (dissected in London).

References: Baer,[55,56] Eduardo,[268] Leiper.[533]

Leiperocotyle congolense (Baer, 1936) Eduardo, 1980 (Figure 408)

Syn. *Cotylophoron congolense* Baer, 1936

Diagnosis: Body length 6.0—8.5 mm, body width 2.0—3.5 mm. Acetabulum 1.00—2.64 mm in diameter, Cotylophoron type, number of muscle units, d.e.c., 14—18; d.i.c., 40—51; v.e.c., 10—15; v.i.c., 48—61; m.e.c., 12—18 (Figure 166). Pharynx 0.80—0.92 mm long, Calicophoron type (Figure 58), esophagus 0.65—0.76 mm with sphincter. Anterior testis

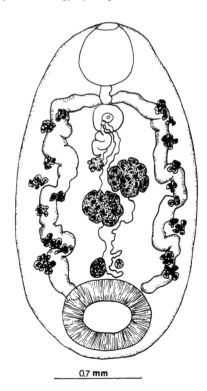

FIGURE 409.   *Leiperocotyle gretillati* Eduardo, 1985. (By permission of Dr. W. Junk Publishers, 1985, Dordrecht.)

0.63—0.86 m long and 0.92—1.13 mm in dorsoventral direction, posterior testis 0.82—1.03 mm long and 1.02—1.33 mm in dorsoventral direction. Ovary 0.42—0.48 by 0.49—0.50 mm. Pars musculosa well developed, convoluted. Terminal genitalium Macrosphinctris type (Figure 119). Genital pore postbifurcal. Eggs 0.130—0.140 by 0.058—0.061 mm.
Biology: Not yet known.
Habitat: Stomach.
Host: *Okapia johnstoni* (Giraffidae).
Distribution: Africa — Zaire.
References: Baer,[55] Eduardo.[268]

Leiperocotyle gretillati Eduardo, 1985 (Figure 409)

Syn. *Ceylonocotyle scoliocoelium var. benoiti* Gretillat, 1966
Diagnosis: Body length 2.3—2.9 mm, body width 1.4—1.5 mm. Acetabulum 0.75—0.96 mm in diameter, Cotylophoron type, number of muscle units, d.e.c., 10—16; d.i.c., 30—41; v.e.c., 11—14; v.i.c., 35—46; m.e.c., 7—18 (Figure 166). Pharynx 0.57—0.65 mm long, Calicophoron type (Figure 58), esophagus 0.33—0.45 mm long, with muscular thickening. Anterior testis 0.20—0.25 mm long, and 0.24—0.27 mm in dorsoventral direction, posterior testis 0.20—0.23 mm long and 0.27—0.30 mm in dorsoventral direction. Ovary 0.09—0.13 by 0.09—0.14 mm. Terminal genitalium Cotylophoron type (Figure 105). Genital pore bifurcal. Pars musculosa moderately developed and convoluted. Size of eggs were not given.
Biology: Not yet known.
Habitat: Intestine.
Host: *Syncerus caffer* (Bovidae).
Distribution: Africa — Zaire, Central Africa.
References: Eduardo,[276] Gretillat.[374]

KEY TO THE SPECIES OF *LEIPEROCOTYLE*
(after Eduardo, 1985)

1. Esophagus with muscular thickening...................................... *L. okapi*
   Esophagus without muscular thickening 2
2. Vitelline follicules confluent dorsomedially, at their posterior limit, esophageal sphincter present .... *L. congolensis*
   Vitellaria not confluent dorsomedially at their posterior limit, esophageal sphincter absent.. *L. gretillati*

Paramphistomoides Yamaguti, 1958
Diagnosis: Orthocoeliinae. Body elongate, elliptical. Acetabulum ventroterminal. Pharynx terminal, well developed, esophagus short, without muscular thickening, ceca simple, terminating in front of acetabulum. Testes tandem, lobed, in middle third of body. Ovary rounded, slightly submedian. Genital pore postbifurcal. Uterine coils winding, dorsal to testes. Vitelline follicules small, densely packed, extending from pharynx to acetabulum, confluent posteriorly. Parasitic in rumen of ruminants.

Type species: *Paramphistomoides maplestonei* (Bhalerao, 1937) Yamaguti, 1958 (Figure 410)
Syn. *Paramphistomum pseudocuonum* Wang, 1979, *Paramphistomum maplestonei* Bhalerao, 1937
Diagnosis: With characters of the genus. Body length 4.0—4.6 mm, body width 1.1—1.2 mm. Acetabulum 0.67 mm in diameter. Pharynx 0.46—0.57 by 0.40—0.43; esophagus 0.20—0.22 mm long. Testes 0.63—0.73 by 0.63—0.77 mm. Ovary 0.23—0.32 mm in diameter. Eggs 0.115—0.119 by 0.050—0.052 mm.
Biology: Not yet known.
Habitat: Small and large intestine.
Host: *Hyelaphus porcinus* (Bovidae).
Distribution: Asia — India.
References: Bhalerao,[90] Yamaguti.[1076]

Platyamphistoma Yamaguti, 1958
Diagnosis: Orthocoeliinae. Body more or less elliptical in outline, strongly flattened, convex on both surfaces and with a median notch behind acetabulum. Acetabulum prominent, ventral, opening surrounded by circular body fold. Pharynx small, prominent; esophagus with bulb, ceca undulating, extending to acetabular zone. Testes tandem, irregular in outline, in middle third of body. Ovary median, midway between acetabulum and posterior testis. Uterine coils in midline dorsal to testes. Genital pore postbifurcal, genital sucker present. Vitelline follicules well developed, extending along ceca from bifurcation to ovarian zone. Parasitic in stomach of hippopotamus.

Type species: *Platyamphistoma polycladiforme* (Näsmark, 1937) Yamaguti, 1958 (Figure 411)
Syn. *Nilocotyle polycladiforme* Näsmark, 1937
Diagnosis: With character of the genus. Body length 2.8—3.8 mm, body width 3.1—3.4 mm. Acetabulum 0.31—0.43 mm in diameter, Nilocotyle type, number of muscle units, d.e.c., 22—26; d.i.c., 22—26; v.e.c., 20—23; v.i.c., 22—25; m.e.c., 6—8 (Figure 172). Pharynx 0.25—0.28 by 0.18—0.23 mm, Paramphistomum type (Figure 67), esophagus 0.25—0.28 mm. Testes 0.41—0.53 by 0.28—0.32 mm in size. Pars musculosa weakly developed. Ovary 0.18—0.23 by 0.14—0.17 mm. Terminal genitalium Sellsi type (Figure 128). Eggs 0.117—0.119 by 0.068—0.071 mm.

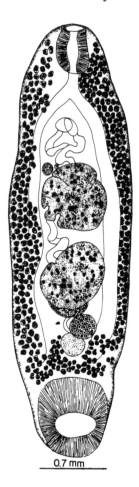

FIGURE 410. *Paramphisto-moides maplestonei* (Bhalerao, 1937). (After G. S. Bhalerao, 1937.)

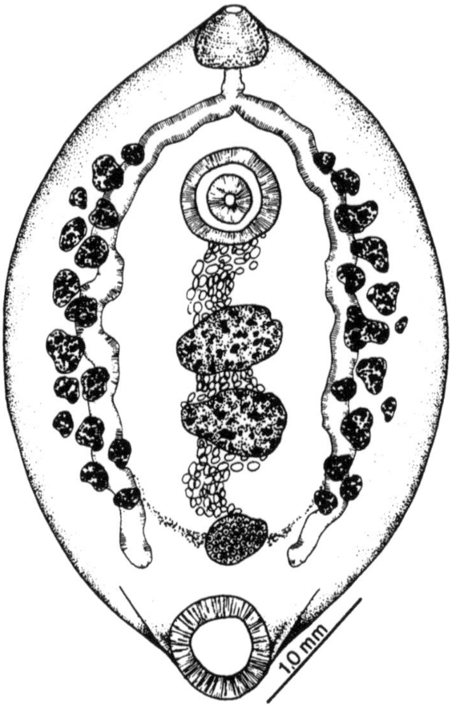

FIGURE 411. *Platyamphistoma polycladiforme* (Näs-mark, 1937).

Biology: Not yet known.
Habitat: Stomach.
Host: *Hippopotamus amphibius* (Hippopotamidae).
Distribution: Africa — Sudan, Chad.
References: Dollfus,[239] Näsmark,[647] Sey and Graber.[864]

Bilatorchis Eduardo, 1980

Diagnosis: Orthocoeliinae. Body elongate to oval. Acetabulum ventroterminal. Pharynx subterminal, esophagus short, without muscular thickening, ceca with dorsoventral bends, extending to anterior border of testes. Testes horizontal, postcecal, in posterior third of body. Pars musculosa well developed. Ovary intertesticular, anterodorsal to acetabulum. Genital pore bifurcal. Uterine coils in median line from ovary to cecal arch. Vitelline follicules lateral along ceca extending from bifurcation to testicular zone. Parasitic in rumen of ruminants.

Type species: *Bilatorchis papillogenitalis* Eduardo, 1980 (Figure 412)

Diagnosis: With characters of the genus. Body length 1.5—4.1 mm, body width 0.7—1.6 mm. Acetabulum 0.61—0.79 mm in diameter dorsoventrally, Nilocotyle type, number of

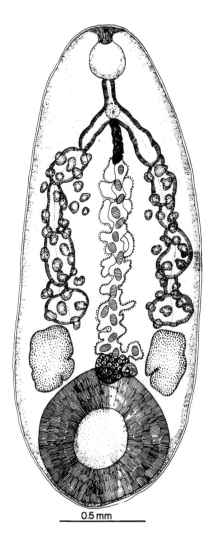

0.5 mm

FIGURE 412. *Bilatorchis papillogenitalis*
Eduardo, 1980. (By permission of Dr. W.
Junk Publishers, 1980, Dordrecht.)

muscle units, d.e.c., 30—34; d.i.c., 30—32; v.e.c., 30—32; v.i.c., 31—35; m.e.c., 4—8
(Figure 172). Pharynx 0.10—0.26 mm long, Calicophoron type, (Figure 58), esophagus
0.20—0.29 mm long. Testes 0.29—0.56 anteroposteriorly and 0.66—1.17 mm dorsoven-
trally. Ovary 0.08—0.38 mm anteroposteriorly, 0.17—0.20 mm dorsoventrally. Terminal
genitalium Papillogenitalis type (Figure 91). Eggs 0.105—0.129 by 0.059—0.067 mm.
Biology: Not yet known.
Habitat: Rumen.
Host: *Kobus leche* (Bovidae).
Distribution: Africa — Zambia.
Reference: Eduardo.[269]

Buxifrons Näsmark, 1937

Diagnosis: Orthocoeliinae. Body conical to oval flattened, convex dorsally and concave
ventrally, with strongly developed, longitudinal body musculature. Acetabulum terminal,
muscular. Pharynx terminal or subterminal, esophagus short, without bulb, ceca winding,

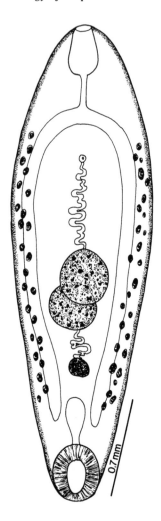

FIGURE 413. *Buxifrons buxifrons*
(Leiper, 1910).

terminating before acetabulum. Testes tandem or diagonal, entire in outline, in middle third of body. Ovary anterior to acetabulum. A few uterine coils, genital pore postbifurcal. Vitelline follicules lateral to testes, a few in number. Parasitic in stomach of hippopotamus.

Type species: *Buxifrons buxifrons* **(Leiper, 1910) Näsmark, 1937 (Figure 413)**
   Syn. *Paramphistomum buxifrons* Leiper, 1910; *Buxifrons maxima* Näsmark, 1937
   Diagnosis: Body length 7.5—8.0, body width 4.0—4.5 mm. Acetabulum 0.97 mm in diameter, Buxifrons type, number of muscle units, d.e.c., 43—48; d.i.c., 53—55; v.e.c., 46—48; v.i.c., 52—54 (Figure 163). Pharynx 0.32—0.36 by 0.28—0.34 mm, Paramphistomum type (Figure 67), esophagus 0.48—0.53 mm long. Anterior testis 0.49—0.55 by 0.39—0.43, posterior testis 0.76—0.82 by 0.43—0.49 mm. Ovary 0.29—0.32 mm in diameter. Terminal genitalium Buxifrons type (Figure 102). Eggs 0.115—0.135 by 0.065—0.070 mm.
Biology: Not yet known.
Habitat: Stomach.
 Hosts: *Bos primigenius* f. taurus (Bovidae); *Hippopotamus amphibius* (Hippopotamidae), *Loxodonta africana* (Elephantidae).

Distribution: Africa — Zaire, Uganda, Chad, Central Africa.

References: Dollfus,[244] Gretillat,[374] Leiper,[531] Näsmark,[647] Sey and Graber.[864]

Notes: When Näsmark,[647] described the species *B. maxima* the main, specific characters were designated the large size of body and the diverse acetabulum, body length ratio to the species of *B. buxifrons*. Gretillat[374] described this species on samples derived from cattle, and the mature species of *B. buxifrons* were larger; thus, Näsmark's specific characters did not seem to be stable. Hence, *B. buxifrons* was regarded to be a contracted form of *B. maxima*, and the latter species was suppressed to a junior synonym of the former.

Gigantoatrium (Yamaguti, 1958) spelling amended

Diagnosis: Orthocoeliinae. Body elongate; conical to elliptical. Acetabulum subterminal. Pharynx terminal, esophagus short with muscular thickening. Testes tandem, entire in outline, in posterior half of body. Ovary near posterior testis. Uterine coils dorsal to testes. Terminal genitalium large but not suckerlike, postbifurcal, genital pore postbifurcal. Parasitic in stomach of hippopotamus.

Type species: ***Gigantoatrium gigantoatrium* (Näsmark, 1937) Yamaguti, 1958 (Figure 414)**

Syn. *Nilocotyle gigantoatrium* Näsmark, 1937

Diagnosis: With characters of the genus. Body length 3.0 mm on an average, body width 1.1 mm on an average. Acetabulum 0.57 mm in diameter, Nilocotyle type, number of muscle units, d.e.c., 22—26; d.i.c., 22—26; v.e.c., 20—23; v.i.c., 22—25; m.e.c., 5—7 (Figure 172). Pharynx 0.38 mm long, Paramphistomum type (Figure 67), esophagus 0.20 mm long. Testes 0.40 mm in length, 0.40 mm in dorsoventral direction. Terminal genitalium Gigantoatrium type (Figure 112). Eggs 0.140 by 0.060 mm.

Biology: Not yet known.

Habitat: Stomach.

Host: *Hippopotamus amphibius* (Hippopotamidae).

Distribution: Africa — Sudan.

Reference: Näsmark.[647]

Nilocotyle Näsmark, 1937

Diagnosis: Orthocoeliinae. Body oval or subcylindrical, circular in cross section. Acetabulum terminal or subterminal. Pharynx terminal, esophagus with or without muscular thickening. Testes tandem, oval or irregular in outline. Ovary between posterior testis and excretory vesicle. Uterine coils winding mainly dorsal to testes. Genital pore prebifurcal, postbifurcal, or bifurcal, genital sucker absent. Vitelline follicules in lateral sides, from diverse level anteriorly to acetabular zone. Parasitic in stomach of hippopotamus.

Type species: *Nilocotyle pygmaea* Näsmark, 1937 (Figure 415)

Diagnosis: Body length 1.2—1.5 mm, body width 1.1—1.2 mm. Acetabulum 0.47—0.52 mm in diameter, Nilocotyle type, number of muscle units, d.e.c., 21—24; d.i.c., 23—26; v.e.c., 19—24; v.i.c., 20—23; m.e.c., 7—9 (Figure 172). Pharynx 0.28—0.33 mm long, Dicranocoelium type (Figure 60), esophagus 0.25—0.31 mm long, with muscular thickening. Testes 0.14—0.18 long and 0.38—0.42 mm in dorsoventral direction. Pars musculosa weakly developed. Ovary 0.18—0.20 mm in diameter. Terminal genitalium Buxifrons type (Figure 102). Genital pore prebifurcal. Eggs 0.080—0.124 by 0.054—0.063 mm.

Biology: Not yet known.

Habitat: Stomach.

Host: *Hippopotamus amphibius* (Hippopotamidae).

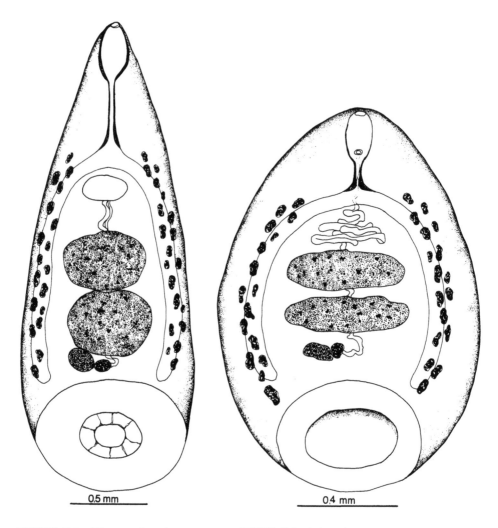

FIGURE 414.   *Gigantoatrium gigantoa-trium* (Näsmark, 1937).

FIGURE 415.   *Nilocotyle pygmaea* Näsmark, 1937.

Distribution: Africa — Sudan, Chad.
References: Näsmark,[647] Sey and Graber.[864]

Nilocotyle minuta (Leiper, 1910) Näsmark, 1937 (Figure 416)

Syn. *Paramphistomum minutum* Leiper, 1910

Diagnosis: Body length 2.1—2.8 mm, dorsoventral dimension 0.9—1.1 mm. Acetabulum 0.45—0.49 mm in diameter, Nilocotyle type, number of muscle units, d.e.c., 17—19; d.i.c., 22—24; v.e.c., 24—26; v.i.c., 23—25; m.e.c., 8—10 (Figure 172). Pharynx 0.36—0.49 mm long, Paramphistomum type (Figure 67); esophagus 0.19—0.23 mm long, with muscular thickening. Anterior testis 0.13—0.15 mm long, 0.31—0.40 mm in dorsoventral dimension, posterior testis 0.12—0.15 mm long and 0.35—0.45 mm in dorsoventral dimension. Ovary 0.12—0.14 mm in diameter. Terminal genitalium Minutum type (Figure 123). Genital pore prebifurcal. Eggs 0.114-0.126 by 0.600—0.750 mm.
Biology: Not yet known.
Habitat: Stomach.
Host: *Hippopotamus amphibius* (Hippopotamidae).
Distribution: Africa — Uganda, Sudan, Chad, Dahomey.

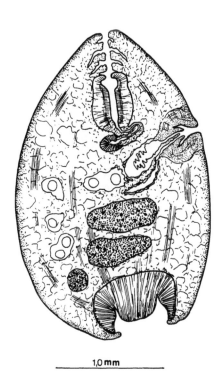

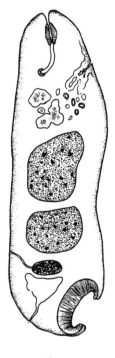

FIGURE 416. *Nilocotyle minuta* (Leiper, 2910) (median sagittal section).

FIGURE 417. *Nilocotyle wagandi* (Leiper, 1910) (median sagittal section). (After R. T. Leiper, 1910.)

References: Bwangamoi,[140] Dinnik et al.,[234] Leiper,[531] Näsmark,[647] Sey,[846] Sey and Graber.[864]

Nilocotyle wagandi (Leiper, 1910) Näsmark, 1937 (Figure 417)

Syn. *Paramphistomum wagandi* Leiper, 1910

Diagnosis: Body length 2.9—3.5 mm, body width 0.9—1.2 mm in dorsoventral direction. Acetabulum 0.50—0.78 mm in diameter, Nilocotyle type, number of muscle units, d.e.c., 20—22; d.i.c., 30—31; v.e.c., 21—22; v.i.c., 35—36; m.e.c., 6—9 (Figure 172). Pharynx 0.33 mm long, Paramphistomum type (Figure 67); esophagus 0.20 mm long without muscular thickening. Testes 0.45 mm in length, and 0.62 mm in dorsoventral direction. Terminal genitalium Wagandi type (Figure 130). There were no eggs.

Biology: Not yet known.

Habitat: Stomach.

Host: *Hippopotamus amphibius* (Hippopotamidae).

Distribution: Africa — Uganda, Sudan, Chad.

References: Bwangamoi,[140] Dinnik et al.,[234] Graber,[357] Leiper,[531] Näsmark.[647]

Nilocotyle circularis Näsmark, 1937 (Figure 418)

Diagnosis: Body length 1.2—1.6 mm, body width in dorsoventral direction 0.6—0.9 mm. Acetabulum 0.38—0.46 mm in diameter. Nilocotyle type, number of muscle units, d.e.c., 20—23; d.i.c., 21—24; v.e.c., 20—23; v.i.c., 19—24; m.e.c., 8—11 (Figure 172). Pharynx 0.18—0.25 mm long, Paramphistomum type (Figure 67), esophagus 0.25—0.28 mm long, with muscular thickening. Testes 0.23—0.25 mm in length, 0.34—0.38 mm in

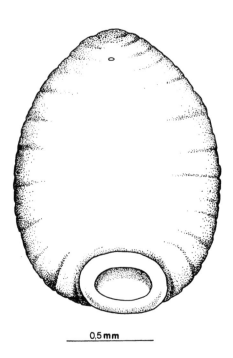

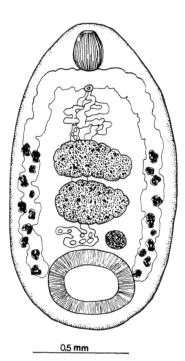

0,5 mm

0,5 mm

FIGURE 418.   *Nilocotyle circularis* Näsmark, 1937.

FIGURE 419.   *Nilocotyle hippopotami* Näsmark, 1937.

dorsoventral direction. Pars musculosa weakly developed. Terminal genitalium Hippopotami type (Figure 115). Genital pore bifurcal. Eggs 0.123—0.132 by 0.050—0.080 mm.
Biology: Not yet known.
Habitat: Stomach.
Host: *Hippopotamus amphibius* (Hippopotamidae).
Distribution: Africa — Sudan.
Reference: Näsmark.[647]

Nilocotyle hippopotami Näsmark, 1937 (Figure 419)

Diagnosis: Body length 1.3—1.6 mm, body width in dorsoventral direction 0.5—0.9 mm. Acetabulum 0.32—0.46 in diameter, Nilocotyle type, number of muscle units, d.e.c., 26—28; d.i.c., 20—24; v.e.c., 24—26; v.i.c., 16—18; m.e.c., 7—10 (Figure 172). Pharynx 0.18—0.23 mm long, Paramphistomum type (Figure 67), esophagus 0.10—0.15 mm long, without muscular thickening. Anterior testis 0.50—1.10 by 0.40—0.93, posterior testis 0.62—0.94 by 0.26—0.32 mm. Pars musculosa weakly developed. Ovary 0.32—0.35 by 0.23—0.28 mm. Terminal genitalium Hippopotami type (Figure 115). Genital pore prebifurcal. Eggs 0.130—0.160 by 0.068—0.075 mm.
Biology: Not yet known.
Habitat: Stomach.
Host: *Hippopotamus amphibius* (Hippopotamidae).
Distribution: Africa — Sudan, Ethiopia, Chad.
References: Graber,[357] Näsmark,[647] Sey and Graber.[864]

Nilocotyle leiperi Näsmark, 1937

Diagnosis: Body length 2.4 mm, body width 0.9 mm in dorsoventral direction. Acetabulum 0.47 mm in diameter, Nilocotyle type, number of muscle units, d.e.c., 24—27; d.i.c.,

19—23; v.e.c., 22—25; v.i.c., 14—17; m.e.c., 9—10 (Figure 172). Pharynx 0.29 mm long, Paramphistomum type (Figure 67); esophagus 0.20 mm long, with muscular thickening. Testes 0.15 mm long, and 0.23 mm in dorsoventral direction. Terminal genitalium Minutum type (Figure 123). There were no eggs.

Biology: Not yet known.

Habitat: Stomach.

Host: *Hippopotamus amphibius* (Hippopotamidae).

Distribution: Africa — Sudan.

Reference: Näsmark.[647]

Nilocotyle microatrium Näsmark, 1937

Diagnosis: Body length 2.2 mm, body width in dorsoventral direction 0.9 mm. Acetabulum 0.46 mm in diameter, Nilocotyle type, number of muscle units, d.e.c., 25—27; d.i.c., 20—22; v.e.c., 21—24; v.i.c., 15—17; m.e.c., 7—10 (Figure 172). Pharynx 0.35 mm long, Dicranocoelium type (Figure 60), esophagus 0.16 mm long with muscular thickening. Testes undeveloped. Terminal genitalium Microatrium type (Figure 121). There were no eggs.

Biology: Not yet known.

Habitat: Stomach.

Host: *Hippopotamus amphibius* (Hippopotamidae).

Distribution: Africa — Sudan, Dahomey.

References: Näsmark,[647] Sey.[846]

Nilocotyle praesphinctris Näsmark, 1937 (Figure 420)

Diagnosis: Body length 3.8—4.1, dorsoventral thickness 1.5—1.7 mm. Acetabulum 0.85—0.94 mm in diameter, Nilocotyle type, number of muscle units, d.e.c., 17—20; d.i.c., 20—24; v.e.c., 16—19; v.i.c., 22—25; m.e.c., 7—10 (Figure 172). Pharynx 0.51—0.52 mm long, Dicranocoelium type (Figure 60); esophagus 0.25—0.31 mm in length with muscular thickening. Length of testes dorsoventrally 0.55—1.02 mm, width 0.72—0.35 mm. Pars musculosa weakly developed. Ovary 0.28—0.31 mm in diameter. Terminal genitalium Minutum type (Figure 123). Genital pore postbifurcal. Eggs 0.138—0.151 by 0.063—0.095 mm.

Biology: Not yet known.

Habitat: Stomach.

Host: *Hippopotamus amphibius* (Hippopotamidae).

Distribution: Africa — Sudan, South Africa, Chad.

References: McCulley et al.,[600] Näsmark,[647] Sey and Graber,[864] Swart.[936]

Nilocotyle hepaticae Swart, 1961 (Figure 421)

Diagnosis: Body length 2.3—2.6 mm, body width 1.1—1.2 mm. Acetabulum 0.70—0.60 mm anteroposteriorly, Nilocotyle type, number of muscle units, d.e.c., 17; d.i.c., 24; v.e.c., 17; v.i.c., 19; m.e.c., 7—10 (Figure 172). Pharynx subterminal, 0.24—0.26 mm long, Dicranocoelium type (Figure 60), esophagus 0.69 mm in length with muscular thickening. Testes 0.26 mm long and 0.73 mm in dorsoventral direction. Pars musculosa well developed, convoluted. Ovary 0.21 by 0.14 mm in diameter. Terminal genitalium Minutum type (Figure 123). Genital pore bifurcal. Eggs 0.136 by 0.062 mm.

Biology: Not yet known.

Habitat: Liver.

Host: *Hippopotamus amphibius* (Hippopotamidae).

Distribution: Africa — South Africa.

Reference: Swart.[935]

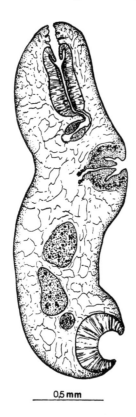

0,5 mm

0,5 mm

FIGURE 420.   *Nilocotyle praes-phinctris* Näsmark, 1937 (median sagittal section). (After P. J. Swart, 1961.)

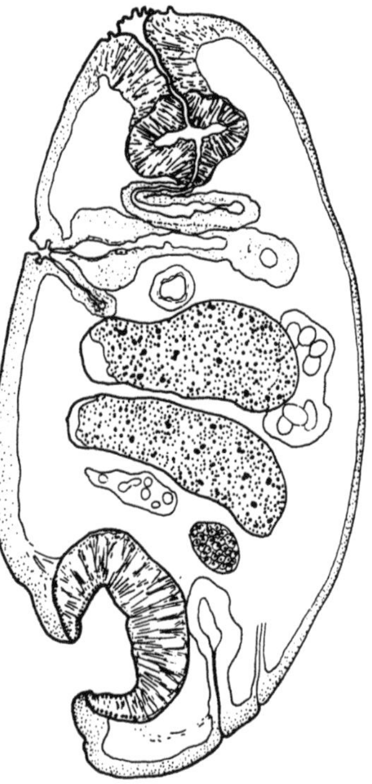

FIGURE 421.   *Nilocotyle hepaticae* Swart, 1961 (median sagittal section). (By permission of the *Onderstepoort J. Vet. Res.*, 1961, Onderstepoort.)

Nilocotyle duplicisphinctris Sey et Graber, 1980 (Figure 422)

Diagnosis: Body length 2.8—4.0 mm, body width 0.8—1.7 mm. Acetabulum 0.80—0.89 by 0.24—0.36 mm. Nilocotyle type, number of muscle units, d.e.c., 12—15; d.i.c., 45—52; v.e.c., 10—14; v.i.c., 42—46; m.e.c., 10—14 (Figure 172). Pharynx 0.43—0.68 mm long, Dicranocoelium type (Figure 60), esophagus 0.40—0.76 mm long with muscular thickening. Anterior testis 0.20—0.48 by 0.32—0.78, posterior testis 0.25—0.40 by 0.30—0.50 mm. Pars musculosa well developed. Ovary 0.26—0.28 by 0.18—0.20 mm. Terminal genitalium Duplicisphinctris type (Figure 108). Genital pore bifurcal. Eggs 0.152—0.163 by 0.062—0.084 mm.

Biology: Not yet known.

Habitat: Stomach.

Host: *Hippopotamus amphibius* (Hippopotamidae).

Distribution: Africa — Ethiopia, Dahomey.

References: Sey,[846] Sey and Graber.[865]

KEY TO THE SPECIES OF *NILOCOTYLE*

1. Pharynx Dicranocoelium type... 2
    Pharynx Paramphistomum type 3
2. Terminal genitalium Buxifrons type....................................... *N. pygmaea*
    Terminal genitalium Microatrium type................................. *N. microatrium*

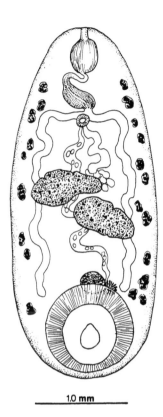

FIGURE 422. *Nilocotyle duplicis-phinctris* Sey et Graber, 1980. (By permission of the *J. Helminthol.*, 1980, London.)

Terminal genitalium Minutum type .... *N. praesphinctris*
Terminal genitalium Ichikawai type..................................... *N. hepaticae*
Terminal genitalium Duplicisphinctris type .... *N. duplicisphinctris*
3. Terminal genitalium Hippopotami type 4
Terminal genitalium Minutum type 5
Terminal genitalium Wagandi type .... *N. wagandi*
4. Esophagus with strongly developed musculature....................... *N. circularis*
Esophagus without musculature..................................... *N. hippopotami*
5. Pharynx relative large, ratio of length of pharynx and body length 1:6.0 .. *N. minuta*
Pharynx relative small, ratio of length of pharynx and body length 1:8.5... *N. leiperi*

Sellsitrema (Yamaguti, 1958) Eduardo, 1980

Diagnosis: Orthocoeliinae. Body conical or flattened dorsoventrally. Acetabulum terminal. Pharynx subterminal, esophagus short, without muscular thickening, ceca simple extending to acetabular zone. Testes tandem, oval in shape in posterior half of body. Ovary in midway between posterior testes and excretory vesicle. Genital pore postbifurcal, genital sucker present. Uterine coils dorsal to testes. Vitelline follicules lateral, extending from pharynx to acetabular zone. Parasitic in stomach of hippopotamus.

Type species: *Sellstrema sellsi* (Leiper, 1910) Eduardo, 1980 (Figure 423)

Syn. *Paramphistomum sellsi* Leiper, 1910
Diagnosis: With characters of the genus. Body length 4.0—6.0 mm, dorsoventral thickness

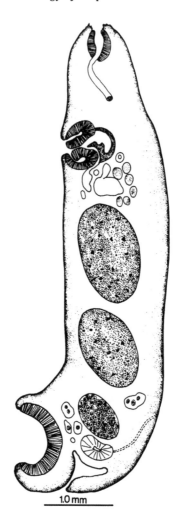

FIGURE 423. *Sellsitrema sellsi* (Leiper, 1910) (median sagittal section). (After R. T. Leiper, 1910.)

0.8—1.2 mm. Acetabulum 0.53 mm in diameter, probably Nilocotyle type. Pharynx 0.35 mm long, probably Paramphistomum type; esophagus 0.5 mm long without muscular thickening. Testes 1.0 mm in length, and 0.7 mm in width. Ovary 0.5 mm in diameter. Terminal genitalium Sellsi type (Figure 128). There were no eggs.

Biology: Not yet known.

Habitat: Stomach.

Host: *Hippopotamus amphibius* (Hippopotamidae).

Distribution: Africa — Uganda.

References: Bwangamoi,[140] Dinnik et al.,[234] Leiper,[531] Näsmark.[647]

Pseudoparamphistoma **Yamaguti, 1958**

Diagnosis: Orthocoeliinae. Body elongated, tapering anteriorly. Acetabulum ventroterminal. Pharynx terminal, esophagus short, with a small sphincter, ceca straight, terminating at posterior testicular zone. Testes tandem, lobed, in posterior half of body. Ovary submedian, midway between posterior testis and acetabulum. Genital pore postbifurcal. Uterine coils winding forward lateral to testes. Vitelline follicules relatively large, loosely packed, ex-

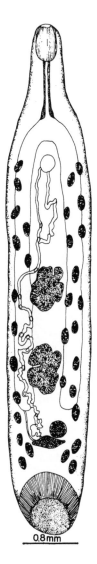

FIGURE 424. *Pseudoparamphistoma cuonum* (Bhalerao, 1937). (After G. D. Bhalerao, 1937.)

tending from genital pore to acetabulum. It was found in the stomach of Indian wild dogs and buffalo.

Type species: ***Pseudoparamphistoma cuonum* (Bhalerao, 1937) Yamaguti, 1971 (Figure 424)**

Syn. *Paramphistomum cuonum* Bhalerao, 1937

Diagnosis: With characters of the genus. Body length 3.3—5.7 mm, body width 0.6—0.9 mm. Acetabulum 0.47—0.55 mm in diameter. Pharynx 0.32—0.43 by 0.30—0.32 mm; esophagus 0.46—0.58 mm long. Testes 0.47—0.69 by 0.31—0.34 mm in size. Ovary 0.13 by 0.23 mm. Eggs 0.120—0.123 by 0.054—0.065 mm.

Biology: Not yet known.

Habitat: Stomach, small intestine.

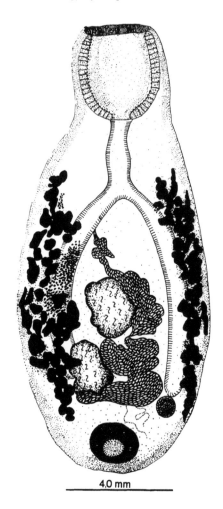

4,0 mm

FIGURE 425.   *Macropharynx sudanensis* Näsmark, 1937. (After K. E. Näsmark, 1937.)

Hosts: *Cuon dukhunensis* (Canidae); *Bubalus arnee* f. *bubalis* (Bovidae).
Distribution: Asia — India.
References: Bhalerao,[90] Nama.[644]

Macropharynx **Näsmark, 1937**
    Diagnosis: Orthocoeliinae. Body conical. Acetabulum subterminal, relatively small. Pharynx terminal, enormous, occupying almost entire breadth of anterior extremity; esophagus without muscular thickening; ceca simple, extending to ovarian zone. Testes slightly diagonal, lobed, situated in posterior half of body. Ovary submedian, preacetabular. Genital opening postbifurcal. Uterine coils winding posteriorly, ventral to testes. Vitelline follicules in lateral sides, extending from esophageal zone to acetabulum. Parasitic in stomach of hippopotamus.

Type species: *Macropharynx sudanensis* Näsmark, 1937 (Figure 425)
    Diagnosis: Body length 16.6 mm, body width 7.1 mm. Acetabulum 1.92 mm in diameter. Pharynx 3.04 mm in length, esophagus 1.20 mm long, without muscular thickening. Anterior testis 2.08 by 1.92 mm, posterior testis 1.60 by 2.80 mm. Eggs 0.148 by 0.070 mm. Biology: Not yet known.

Habitat: Stomach.
Host: *Hippopotamus amphibius* (Hippopotamidae).
Distribution: Africa — Sudan.
Reference: Näsmark.[647]

Palamphistomum Srivastava et Tripathi, 1980

Diagnosis: Orthocoeliinae. Body subcylindrical, slightly curved ventrally, anterior body end with dome-shaped papillae and transverse striations. Acetabulum subterminal. At anterior end, a beaker-shaped, highly retractile vestibule present, its wall is supported by muscular rays arranged in a frill. Pharynx well developed, esophagus with long bulb, ceca simple extending to acetabular zone. Testes tandem, entire or lobed in posterior half of body. Ovary median or submedian, preacetabular. Genital pore bifurcal or just postbifurcal. Lymphatic system consists of a pair of longitudinal tubes. Parasitic in rumen and reticulum of domestic ruminants.

Type species: *Palamphistomum dutti* Srivastava et Tripathi, 1980 (Figure 426)

Diagnosis: Body length 5.1—6.5 mm, body width 1.5—2.0 mm. Acetabulum 0.67—0.69 by 0.90—0.97 mm. Pharynx muscular, esophagus 0.51—0.58 mm long. Anterior testis 0.67—0.70 by 0.72—0.84, posterior testis 0.63—0.69 by 0.70—0.80 mm. Ovary 0.14—0.20 by 0.27 mm. Pars musculosa convoluted and strongly developed. Genital pore postbifurcal. Eggs 0.126—0.178 by 0.072—0.098 mm.

Biology: Not completely elucidated. Tripathi and Srivastava[992] found that *Gyraulus convexiusculus* served as intermediate host both under laboratory circumstances and in nature.

Habitat: Rumen and reticulum.

Hosts: *Bubalus arnee* f. bubalis, *Ovis ammon* f. aries, *Capra aegagrus* f. hircus (Bovidae).

Distribution: Asia — India.

References: Srivastava and Tripathi,[909] Tripathi and Srivastava.[992]

Palamphistomum lobatum Srivastava and Tripathi, 1980 (Figure 427)

Diagnosis: Body length 4.7—6.5 mm, body width 1.8—2.1 mm. Acetabulum 0.67—0.69 by 0.90—0.97 mm. Pharynx 0.58—0.65 by 0.47—0.53; esophagus long, with bulb. Anterior testis 0.77-1.22 by 0.49—0.94 mm, posterior testis 0.75—0.14 by 0.63—1.01 mm. Ovary 0.33—0.58 by 0.17—0.33 mm in size. Genital pore just bifurcal. Eggs 0.091—0.100 by 0.04—0.06 mm.

Biology: Not yet elucidated completely. Tripathi and Srivastava[991] reported that *Gyraulus convexiusculus* served as intermediate host both in the laboratory and in nature.

Habitat: Rumen and reticulum.

Hosts: *Bubalus arnee* f. bubalis, *Ovis ammon* f. aries, *Capra aegagrus* f. hircus (Bovidae).

Distribution: Asia — India.

References: Srivastava and Tripathi,[909] Tripathi and Srivastava.[992]

KEY TO THE SPECIES OF *PALAMPHISTOMUM*

1. Testes spherical entire in outline, vitelline follicules commence from bifurcation......
   ....*P. dutti*
   Testes lobed, vitelline follicules commence just behind genital pore...... *P. lobatum*

Gemellicotyle Prudhoe, 1975

Diagnosis: Orthocoeliinae. Body conical. Acetabulum terminal and large, provided with a central adhesive organ. Pharynx terminal, esophagus simple, without muscular thickening; ceca extending to acetabular zone. Testes horizontal, preacetabular, irregular in outline.

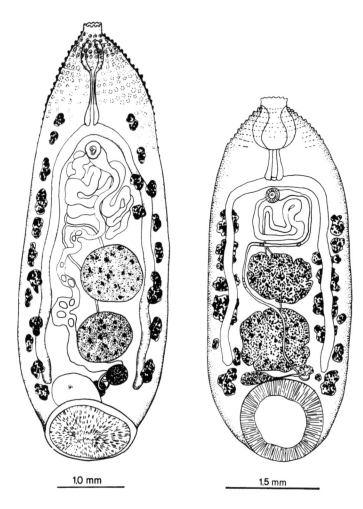

1,0 mm   1,5 mm

FIGURE 426. *Palamphistomum*
dutti Tripathi et Srivastava, 1980.
(After H. D. Srivastava and H. N.
Tripathi, 1987.)

FIGURE 427. *Palamphistomum*
lobatum Tripathi et Srivastava, 1980.
(After H. D. Srivastava and H. N.
Tripathi, 1987.)

Ovary median, anterodorsal to acetabulum. Genital pore bifurcal. Uterine coils winding
between testes and cecal arch. Vitelline follicules along ventral surface of ceca. Parasitic in
stomach of wallaby.

Type species: *Gemellicotyle wallabicola* **Prudhoe, 1975 (Figure 428)**
    Diagnosis: With characters of the genus. Body length 10.0—14.0 mm, body width 7.0—8.0
mm. Acetabulum 4.00—5.50 mm in diameter, Gemellicotyle type, number of muscle units,
d.e.c., 20; d.i.c., 60; v.e.c., 12; v.i.c., 45 (Figure 144). Pharynx 2.00 mm in diameter,
Gemellicotyle type (Figure 62). Testes 3.20 by 2.10 mm in size. Ovary about 0.60 mm in
diameter. Terminal genitalium bifurcal, Gemellicotyle type (Figure 111). Eggs 0.145—0.160
by 0.082—0.009 mm, containing embryo of morula stage (Speare).[903]
Biology: Not yet known.
Host: *Macropus agilis* (Macropodidae).
Distribution: Papua New Guinea, Australia.
References: Prudhoe,[757] Speare,[903] Speare et al.[904]

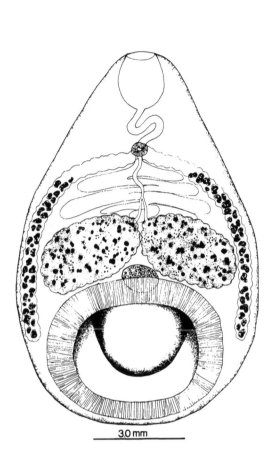

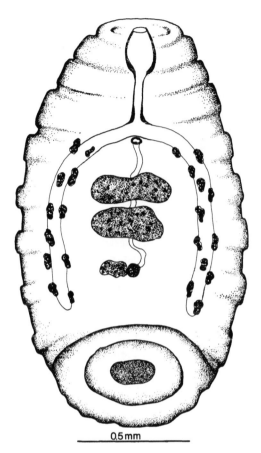

FIGURE 428. *Gemellicotyle wallabicola* Prudhoe, 1975.

FIGURE 429. *Glyptamphistoma paradoxum* (Näsmark, 1937).

Glyptamphistoma Yamaguti, 1958

Diagnosis: Orthocoeliinae. Body small, approximately barrel-shaped, with a number of parallel hoops. Acetabulum terminal. Pharynx subterminal, esophagus with weakly developed musculature, ceca undulating, terminating in front of acetabulum. Testes tandem, transverse oval, entire, in middle third of body. Ovary oval, irregular in outline slightly submedian, between posterior testis and acetabulum. Genital pore postbifurcal. Uterus weakly developed, dorsal to testes. Vitelline follicules not numerous, loosely packed, extending along ceca. Parasitic in stomach of hippopotamus.

Type species: *Glyptamphistoma paradoxum* (**Näsmark, 1937**) **Yamaguti, 1958 (Figure 429)**

Syn. *Nilocotyle paradoxum* Näsmark, 1937

Diagnosis: With characters of the genus. Body length 1.5—1.9 mm, body width 1.0—1.2 mm. Acetabulum 0.41—0.62 mm in diameter, Nilocotyle type, number of muscle units, d.e.c., 19—21; d.i.c., 23—25; v.e.c., 16—18; v.i.c., 20—22; m.e.c., 9—12 (Figure 172). Pharynx terminal, 0.20—0.25 mm long, Paramphistomum type (Figure 67); esophagus 0.21—0.27 mm long. Testes 0.21—0.28 by 0.41—0.48 mm in size. Ovary 0.13—0.17 by 0.07—0.09 mm. Terminal genitalium Epiclitum type (Figure 109). Eggs 0.062—0.068 by 0.028—0.042 mm.

Biology: Not yet known.

Habitat: Stomach.
Host: *Hippopotamus amphibius* (Hippopotamidae).
Distribution: Africa — Sudan, Chad.
References: Graber,[357] Näsmark,[647] Sey and Graber.[864]

3.2.3.2.3.2. PARAMPHISTOMINAE Fischoeder, 1901

Diagnosis: Paramphistomidae. Body pyriform, conical, oval, subcylindrical, oval or round in cross section. Acetabulum terminal or ventroterminal. Pharynx strongly developed without appendages, esophagus short without or occasionally with muscular thickening; ceca usually extending to acetabular zone. Testes tandem or diagonal lobed or spherical in outline. Pars musculosa strongly or poorly developed. Ovary median or submedian posterior or posterodorsal to posterior testis. Genital pore bifurcal or postbifurcal, terminal genitalium with well developed musculature. Uterine coils winding forward dorsal to testes. Vitelline follicules extending in lateral fields and may or may not be confluent posteriorly. Laurer's canal and excretory vesicle cross each other. Lymphatic system present. Parasitic in forestomach and bile duct of mammals.

KEY TO THE GENERA OF PARAMPHISTOMINAE

1. Body conical, medium or large in size; acetabulum medium or large in size 2
   Body oval, with small acetabulum .... *Ugandocotyle*
2. Genital sucker present .... *Cotylophoron*
   Genital sucker absent 3
3. Pars musculosa strongly developed 4
   Pars musculosa weakly developed 5
4. Pharynx Calicophoron type .... *Calicophoron*
   Pharynx Explanatum type... *Gigantocotyle*
5. Acetabulum Explanatum type... *Explanatum*
   Acetabulum Paramphistomum or Pisum type .... *Paramphistomum*

Type genus: *Paramphistomum*, Fischoeder, 1901

Syn. *Liorchis* Velitchko, 1966; *Srivastavaia* Singh, 1970

Diagnosis: Paramphistomidae. Body conical, round in cross section. Acetabulum ventro-terminal, moderate in size. Pharynx terminal, without appendages; esophagus without muscular thickening; ceca in lateral fields, usually straight, occasionally sinuous. Testes: rounded or lobed, tandem or slightly diagonal. Pars musculosa weakly developed. Ovary posttesticular. Genital pore bifurcal, postbifurcal occasionally prebifurcal. Uterine coils winding forward, dorsal to testes and ventral to male ducts. Vitelline follicles in lateral fields, may or may not be confluent dorsomedially. Parasitic in forestomach of ruminants.

Type species: *Paramphistomum cervi* (Zeder, 1790) Fischoeder, 1901 (Figure 430)

Syn. *Festucaria cervi* Zeder, 1790; *Fasciola cervi* Schrank, 1790; *Fasciola elaphi* Gmelin, 1791; *Monostoma elaphi* (Zeder, 1800); *Monostoma conicum* Zeder, 1803; *Amphistoma conicum* (Rudolphi, 1809); *Amphistoma cervi* Stiles et Hassad, 1900; *Paramphistomum cervi* (Zeder, 1790) of Maplestone; 1923 in part; of Travassos, 1934 in part; *P. cervi* (Schrank, 1790) of Dawes, 1936 in part; *P. (Paramphistomum) cervi* (Schrank, 1790) of Fukui, 1929 in part.

Diagnosis: Body length 6.2—14.3 mm, body width 2.2—4.3 mm. Acetabulum 1.26—2.68 mm in diameter, Paramphistomum type, number of muscle units, d.e.c.1, 12—15; d.e.c.2, d.i.c., 38—49; v.e.c., 15—23; v.i.c., 39—54; m.e.c., 7—10 (Figure 173). Pharynx 0.87—1.28 mm long, Liorchis type (Figure 65), esophagus *without muscular thickening*. Anterior testis 0.62—2.20 mm long, and 1.30—1.80 in dorsoventral direction; posterior

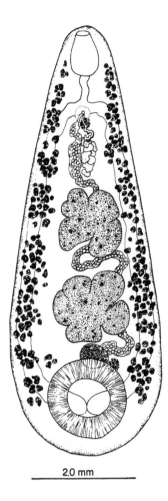

2.0 mm

FIGURE 430. *Paramphistomum cervi*
(Zeder, 1970).

testis 1.12—1.76 mm long and 2.73—3.68 mm in dorsoventral direction. Pars musculosa short and weakly developed. Ovary 0.32—0.79 by 0.29—0.62 mm. Terminal genitalium Gracile type (Figure 114). Genital pore bifurcal. Eggs 0.116—0.189 by 0.052—0.064 mm.

Biology: Before listing the papers dealing with the description of the life history of this species, it should be mentioned that a great number of papers have been published on the occurrence of *P. cervi* in various parts of the world. These papers can be divided into two parts as regards their scope. The first group comprises papers containing nomination of this species only, without further information. Since *P. cervi* and some of the related species are characterized by minute histomorphological features, it is almost impossible to ascertain the proper systematic status of species involved in these papers. The second group of papers either deals with this species explicitly or include descriptions or drawings from which the status of the species can be decided more or less accurately. These papers were reviewed by Sey,[848] and he found that *P. cervi* is a species having Palaearctic distribution. Distributional data, referring to the occurrence of this species outside the Palaearctic region, are probably incorrect, and they include other than this speices of the local amphistome fauna. Therefore, the presentation of the life-cycle pattern was based on the following references which probably dealt with this species: Nöller and Schmid,[660] Szidat,[940] Kryukova,[512] Pogoreliĭ and Mereminskiĭ,[731] Bortnovskiĭ,[108] Kraneburg and Hasslinger,[506] Kraneburg,[505] Kraneburg and Boch,[507] Odening et al.,[669,670] Schmid et al.,[822] Sey.[848] To avoid repetition of the same larval stages, described by various authors, the presentation of the following life-cycle pattern is a summary of the papers listed.

Preparasitic stage. Eggs are greenish yellow, covered by colorless shell. Their shape may be pyriform, oval, or elliptical. The antiopercular pole bears a minute spine. Variations of egg size could be observed not only among the eggs laid by a worm population, but also among those of separate specimens. No direct correlation appears to exist between the body dimension of the fluke and that of the eggs. Newly laid eggs contain zygote situated near the opercular end of the egg, and during the subsequent divisions it moves toward the center of it and becomes surrounded by vitelline cells. During incubation at 27°C no significant change can be observed in the first 4 or 5 d except for the growth of the embryo. On the 5th or 6th day, the terebratorium, the single apical, as well as the four penetration glands appear. On the 7th day the flame cells, found by the joining of the second and third epidermal cell rows, begin their activity. The formation of the germinal tissue also takes place during this period. With the growth of the embryo, the vitelline cells gradually decrease, and their place is occupied by two large vacuoles. Mucoid plug was not observable. On the 8th and 10th days, the embryo reaches a size of 0.110-0.145 mm, and the embryo (miracidium) is ready for hatching. Under controlled temperature (27°C) hatching of miracidia usually began on the 8th day, and it continued on the following 2—4 d. After the optimal period of hatchability (10th day), the life span of the embryos was much shorter than that of *C. daubneyi* (Sey).[842] The eggs maintain vitality under low temperature (4—6°C) for 5—6 months, suggesting the possibility of hibernation in the temperature belt. The hatched and free-swimming miracidium is fusiform without eye-spots, covered with cilia. The measurements of the living miracidia: 0.175—0.200 by 0.040—0.050 mm. The body is covered with flat, epidermal cells in four transverse rows according to the formula 6:8:4:2. At the anterior end of the miracidium lies the cilium-free terebratorium with argentophilic structures. They are situated along three axes, $A_1 = 5$, $A_2 = 14$, $A_3 = 10$; a further 12 similar structures can be found along the body, 10 of them between the first and the second epidermal cell rows and two between the second and the third ones. The inner miracidial structure consists of the apical gland, the penetration glands, the nervous system, the excretory ducts and the germinal tissue, similar to that of other miracidia of rumen flukes. The life span of miracidia kept in tap water at room temperature was 10—12 h but their virulence was limited to the first 4 h only. Intermediate hosts: *Anisus leucostomus, A. septemgiratus, A. vortex, Argimer crista, A. inermis, Bathyomphalus contortus, Choanophalus anophalus, Gyraulus albus, G. gredleri, G. ehrenbergi, Hippeutis complanatus, Planorbis carinatus, P. planorbis,* and *Segmentina nitida.*

Intramolluscan larval stage. After penetration, the invading miracidium continues its development in the snail tissue; undergoing noticeable changes: shedding of epidermal plates, losing some internal structures (apical papilla, apical, and penentration glands) and breaking down of embryo balls into separate germinal cells. The sporocyst shows a marked increase in size; e.g., the 4-d-old specimens measure 0.16—0.17 by 0.14—0.15 mm; it reaches maturity in about 10—15 d. Mature sporocysts are located in the body cavity of the snail, along the digestive tract; it is covered by a thin, transparent envelope and contains some fully developed rediae and numerous embryo balls. One pair of flame cells was found in the sporocyst; neither the process of liberation of the rediae nor the opening of the sporocyst's surface was observed.

The first free rediae emerged on the 13th to 15th day after infection. Young rediae measure 0.15—0.23 mm in length and 0.075—0.180 mm in width. During their further development they grow markedly and appear to reach their maximum when liberation of cercariae begins. The first mature rediae were perceived on the 15th day after liberation, and their measurements were 0.70—1.10 by 0.20—0.25 mm. In the inner structure, a digestive system, a nervous system, and excretory system, and a germinal tissue can be distinguished. The digestive system includes a mouth, a pharynx, an esophagus, and an unpaired gut. The mouth is a minute opening followed by the muscular pharynx, measuring 0.03—0.05 by 0.03—0.05 mm. The esophagus is short, 0.15—0.20 mm in length, surrounded by the so-called salivary glands and leading to the gut. The latter measures 0.07—0.12 mm in length and 0.05—0.08 mm in width. The center of the nervous system is situated at the level of the esophagus, and it is very similar to that of the miracidium. The excretory system comprises three pairs of flame cells and their ducts which unite at about the middle of the body, and they open to the outside through a small common bladder. The germinal system is situated in the second half of the young rediae consisting of the germinal epithelium, germinal cells, and embryo balls. The latter develop gradually into cercariae. The birth pore becomes prominent only later, when young cercariae begin to accumulate in the first part of the body. The distance from the anterior body end is 0.30—0.32 mm. In about 50 d the signs of aging can be discerned (dirty gray and dark spots are found, body cavity is empty) and these rediae are gradually dying off.

Cercariae at birth are poorly developed; the first free cercariae are recovered on the 30th to 37th day after infection, measuring 0.25—0.37 mm with a tail appendage of 0.10—0.12 mm in length and 0.75—0.18 mm in width. The eyespots can already be seen when cercariae are within the redia. The first mature cercariae develop between 45 and 55 d after infection. They are dark brown, quickly swimming organisms, with a body measuring 0.30—0.34 by 0.20—0.32 mm and tail 0.40—0.50 by 0.06—0.07 mm. The body is covered with integument, having cystogenous cells and rods. These rods make the body opaque; the scattered pigment granules lend a brown color to the body. A pair of eyes are located on the dorsal surface of the mature cercaria; they are conical in shape. The acetabulum is situated at the posterior end on the ventral body surface, measuring 0.09—0.11 by 0.09—0.11 mm in living specimens. The inner structure of the cercariae consists of the digestive system (pharynx, esophagus, gut), excretory system (flame cells, ascending and descending excretory ducts, caudal excretory tube, and bladder), nervous system, and primordia of the reproductive system. The number of the flame cells is not exactly detectable due to the opaque of the body. Sensory papillae can be observed along the tail; 25 pairs are found in the region above the excretory pores and two pairs beyond them. Larval stages of *P. cervi* hibernate in the intermediate host (*Planorbis planorbis*); in an endemic area of Hungary (Gemenc) 11.9% of *P. planorbis*, born in the previous year and dissected at the end of April, 1975, contained fully developed cercariae.[1101]

Shortly after emergence, the cercariae encyst on the vegetation. The adolescariae are spherical in appearance, measuring 0.18—0.25 mm in diameter. Out of the environmental factors, strong illumination stimulates the emergence. Under low temperature (4—8°C) the life span of adolescariae lasts for 2—3 months.

Development in definitive host. In order to determine the prepatent period of *P. cervi*, a 7-month-old roe deer was infected with 2000, 2-week-old adolescariae. Fecal samples were regularly controlled, and the first eggs were found on the 85th day after infection. The percentage take was 42.3% (Sey).[848] The prepatent period was 96—130 d in cattle (Nikitin,[654] Gluzman,[347] Gluzman and Artemenko,[350] Mereminksiĭ et al.,[607] Klesov and Mereminskiĭ,[489] Kraneburg and Boch,[507]), 96—107 d in sheep (Gluzman,[347] Gluzman and Artemenko,[350] Kraneburg and Boch,[507]), and 82—96 d in roe deer (Kraneburg and Boch,[507] Sey).[848] The life span of *P. cervi* was estimated to be 4 years in cattle (Klesov and Mereminkskiĭ).[489]

Habitat: Rumen and reticulum.

Hosts: *Bos primigenius* f. taurus, *Bos mutus* f. grunniens, *Bison bonasus*, *Capra aegagrus* f. hircus, *Ovis ammon* f. aries, *O. musimon* (Bovidae); *Capreolus capreolus*, *Cervus nippon*, *C. canadensis*, *C. elaphus*, *Rangifer tarandus*, *Alces alces*, *Dama dama* (Cervidae).

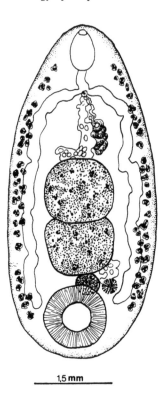

1,5 mm

FIGURE 431. *Paramphistomum
liorchis* Fischoeder, 1901. (By permission of Dr. W. Junk Publishers,
1982, Dordrecht.)

Distribution: Europe and Asia — Austria, Germany, Italy, Poland, Yugoslavia, U.S.S.R.,
Spain, Portugal, Turkey, Czechoslovakia, Albania, Rumania, Bulgaria, China, England,
Iraq, Iran, Ireland, Hungary, Sweden, Japan, France, Mongolia, Norway, Holland, Den-
mark, Finland, Pakistan; North America — U.S., Canada; South America — Brazil.

References: Afzal et al.,[4] Aleksandrova,[18] Aleksandrowska et al.,[19] Arru and Deiana,[36]
Asadov,[42,43] Ashizawa et al.,[46] Audi,[48] Babič,[52] Bobkova,[99] Boch et al.,[100] Brenes,[120]
Brumpt,[128] Burgu,[136] Chen et al.,[168] Craig and Davies,[187] Deiana,[208] Dodbiba,[237] Edgar,[263]
Eduardo,[272] Erhardová and Kotrlý,[288] Eslami and Faizy,[291] Euzeby,[293] Evranova,[294] Ezzat,[295]
Fischoeder,[311] Gagarin,[335] Georgiev,[339] Gluzman,[347] Gluzman and Artemenko,[350] Gotts-
chalk,[354] Graubmann et al.,[367] Gubanov,[379] Guilhon and Priouzeau,[382] Halvorsen and Wis-
sler,[405] Helle,[410] Henriksen and Nansen,[412] Hsu,[426] Iqbal,[433] Ivashkin,[438] Kadenatsii,[459]
Kelly,[476] Klesov and Mereminskiĭ,[489] Kotrlá and Kotrlý,[503] Kraneburg and Boch,[507] Kutzer
and Hinaidy,[516] Lankester et al.,[522] Leitão,[535] Merdivenci,[604] Mereminskiĭ et al.,[607] Mosk-
vin,[622] Motl and Pav,[624] Näsmark,[647] Nikitin,[654] Nilsson,[656] Odening et al.,[669] Olteanu and
Lungu,[674] Oshmarin and Oparin,[680] Ovcharenko,[683] Pav et al.,[708] Podlesniĭ,[730] Prokopič and
Kotrlá,[752] Quiroz and Ochoa,[760] Rinses,[779] Rockett et al.,[781] Romashov,[788] Schmid et al.,[822]
Schoon,[823] Sey,[836,841,845,858] Sey and Eslami,[866] Sivsteva,[888] Swales,[933] Szidat,[940] Threlfall,[973]
Toshchev,[978] Tudor and Anton,[1000] Vujič,[1042] Wieczorowski,[1055] Wu et al.,[1070] Yusuf and
Chaudhry,[1081] Zadura,[1085] Zadura and Niec,[1086] Zdun,[1091] Zdzitowiecki et al.,[1092] Zgardin
and Frukhtman,[1093] Zhalstanova.[1094]

Paramphistomum liorchis Fischoeder, 1901 (Figure 431)

Diagnosis: Body length 3.2—9.5 mm, body width 0.7—1.8 mm in dorsoventral direction.
Diameter of acetabulum 0.87—1.86 mm in dorsoventral direction, Paramphistomum type,

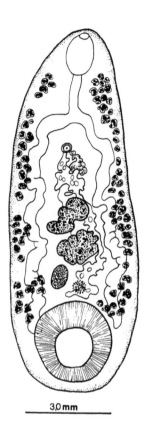

FIGURE 432. *Paramphistomum
gracile* Fischoeder, 1901. (By
permission of Dr. W. Junk Publishers, 1982, Dordrecht.)

number of muscle units, d.e.c.1, 9—16; d.e.c.2, 19—29; d.i.c., 21—42; v.e.c., 13—18;
m.e.c., 8—19 (Figure 173). Pharynx 0.48—1.22 mm long, Liorchis type (Figure 65),
esophagus 0.25—0.56 mm long, without muscular thickening. Anterior testis 0.68—1.42
mm long and 0.52—1.64 mm in dorsoventral direction; posterior testis 0.49—1.34 mm long
and 0.41—1.39 mm in dorsoventral direction. Ovary 0.19—0.31 by 0.10—0.25 mm. Pars
musculosa short and weakly developed. Genital pore bifurcal. Terminal genitalium Liorchis
type (Figure 118). Eggs 0.121—0.132 by 0.061—0.065 mm.

Biology: Not yet completely elucidated. Snider and Lankester[896] found that *Helisoma trivolvis*
and *H. complanatus* were natural intermediate hosts of this fluke.

Habitat: Rumen.

Hosts: *Bos primigenius* f. taurus (Bovidae), *Mozama simplicicornis, M. campestris, M.
mexicanus, M. rufus, M. dichotomus, M. namuby, M. americana, Alces alces, Odocoileus
besoaricus, O. virginianus* (Cervidae).

Distribution: North America — Canada, U.S.; South America — Brazil, Guinea.

References: Becklung,[77] Eduardo,[272] Fischoeder,[311] Prestwood et al.,[739,740] Price,[745] Sey,[846]
Snider and Lankester.[896]

Paramphistomum gracile Fischoeder, 1901 (Figure 432)

Syn. *Pramphistomum indicum* Stiles and Goldberger, 1910 in part, *P. bombayiensis* Gupta
et Verma in Gupta and Nakhasi, 1977

Diagnosis: Body length 5.4—10.8 mm, body width 1.5—2.9 mm in dorsoventral direc-

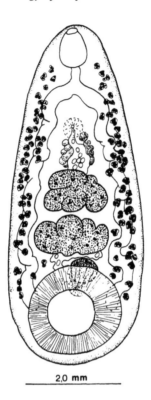

2,0 mm

FIGURE 433. *Paramphistomum
epiclitum* Fischoeder, 1904. (By
permission of Dr. W. Junk Publishers, 1982, Dordrecht.)

tion. Acetabulum 1.25—1.76 mm in diameter in dorsoventral direction, Paramphistomum
type, number of muscle units, d.e.c.1, 13—21; d.e.c.2, 23—44; d.i.c., 32—43; v.e.c.,
13—19; v.i.c., 40—53; m.e.c., 9—23 (Figure 173). Pharynx 0.56—0.98 mm long, Calicophoron type (Figure 58); esophagus 0.46—1.37 mm long, without muscular thickening.
Anterior testis 0.74—0.98 mm long and 0.72—0.86 mm in dorsoventral direction; posterior
testis 0.48—1.26 mm long and 0.41—1.16 mm in dorsoventral direction. Pars musculosa
short and weakly developed. Ovary 0.31—0.42 by 0.28—0.31 mm. Terminal genitalium
Gracile type (Figure 114). Genital pore postbifurcal. Eggs 0.119—0.128 by 0.065—0.073
mm.

Biology: Not yet known.

Habitat: Rumen.

Hosts: *Bos primigenius* f. taurus, *Bubalus arnee* f. bubalis, *Boselaphus tragocamelus* (Bovidae).

Distribution: Asia — Sri Lanka, India, Japan, Mongolia, Iran, China, Thailand.

References: Chen et al.,[168] Eduardo,[272] Fischoeder,[311] Gupta and Nakhasi,[396] Mukherjee,[632]
Näsmark,[647] Pačenovský et al.,[688] Patnaik and Acharjyo,[705] Sey,[843] Sey and Eslami.[866]

Paramphistomum epiclitum Fischoeder, 1904 (Figure 433)

Syns. *Paramphistomum indicum* Stiles et Goldberger, 1910; in part, *P. thapari* Price et
McIntosh, 1953; *P. malayi* Lee et Lowe, 1971; *Cotylophoron indicum* Stiles et Goldberger,
1910; *C. madrasense* Gupta, 1958; *C. chauhani* Gupta et Gupta, 1972; *Srivastavaia indica*
Singh, 1970.

Diagnosis: Body length 5.1—8.7 mm, body width 0.7—2.1 mm in dorsoventral direction.

Acetabulum 0.98—1.73 mm in dorsoventral direction, Paramphistomum type, number of muscle units, d.e.c.1, 8—27; d.e.c.2, 28—41; d.i.c., 28—53; v.e.c., 10—21; v.i.c., 29—63; m.e.c., 18—25 (Figure 173). Pharynx 0.68—1.09 mm long, Calicophoron type (Figure 58), esophagus 0.35—0.98 mm, without muscular thickening. Anterior testis 0.48—0.98 mm long and 0.68—1.39 mm in dorsoventral direction; posterior testis 0.78—1.36 mm long and 1.26—1.54 mm in dorsoventral direction, pars musculosa short, weakly developed. Ovary 0.21—0.48 by 0.25—0.32 mm. Terminal genitalium Epiclitum type (Figure 109). Genital pore postbifurcal. Eggs 0.141—0.153 by 0.079—0.081 mm.

Biology: References. Mukherjee,[633] under the name: *Cotylophoron indicum*, Singh,[884] under the name: *Srivastavaia indica*.

Preparasitic stage. ① Mukherjee. Freshly laid eggs unsegmented. Embryonic development completed 12 d at room temperature in May (India). Newly hatched miracidia measured 0.121—0.191 by 0.065—0.086 mm. Ciliated epidermal cells were situated in four rows, arranged by the formula 6:8:4:2. The inner structure of miracidia is composed of apical gland (gut), a pair of penetration glands, nerve mass, a pair of flame cells, germ cells, and 54 sensory papillae. Intermediate host: *Indoplanorbis exustus* (both in Mukherjee and Singh).

Intramolluscan larval stage. ① Mukherjee. Miracidia, after penetration into the snail tissue, lost inner structure and transformed into sporocyst. Mature sporocyst 0.102 by 0.066 mm. The body cavity enclosed the germ balls and the developing and mature rediae as well. First rediae emerged on the 6th or 7th day after infection. First generation rediae measured 0.834—0.956 by 0.191—0.208 mm. ② Singh. Mother rediae measured 1.60 by 0.15 mm. Pharynx 0.042 mm in diameter, cecum about 0.19 mm long. Birth pore present about 0.22 mm from the anterior end. Excretory system consisted of five pairs of flame cells. The rediae contained 3 to 5 developing cercariae and up to 16 embryos or germ balls. ① Mukherjee. Cercaria of *P. epiclitum* proved to be the same as of *Cercariae indicae XXVI*. Sewell.[830] Adolescariae dome-shaped structure measuring 0.261—0.313 mm in diameter. ② Singh. Cercariae long and 0.35—0.51 mm. Acetabulum 0.11—0.15 mm in diameter. Pharynx 0.052—0.096 mm in diameter. Length of tail 0.76—0.89 mm long. A pair of eyespots present, dark brown pigmentation forms a sort of network. Main excretory tubes with cross-connecting tubes, caudal tubes with expansion near its tip and opened laterally with two orifices. Genital primordia consisted of irregular masses of cells. Adolescariae measure 0.25—0.26 mm in diameter.

Development in definitive host. ① Mukherjee. One goat was infected with 451 cysts and first eggs were detectable in 4 months and 29 d after infestation. At post-mortem examination 53 trematodes (take 11.72%) were recovered. ② Singh. Mature flukes were found 82 d after infection in lamb and 141 d in kid.

Habitat: Rumen.

Hosts: *Bos gaurus, Bos primigenius* f. taurus, *Bubalus arnee* f. bubalis, *Capra aegagrus* f. hircus, *Hippotragus niger, Ovis ammon* f. aries (Bovidae).

Distribution: Asia — Vietnam, Afghanistan, India, Malaysia, Burma, Pakistan, Bangladesh.

References: Bali,[64] Bhalerao,[90] Eduardo,[272] Fischoeder,[312] Gupta,[386] Gupta and Gupta,[395] Kotrlá et al.,[501] Lee and Lowe,[527] Sey,[843,854] Zahedi et al.[1087]

Paramphistomum gotoi Fukui, 1922 (Figure 434)

Diagnosis: Body length 4.7—7.8 mm, body width 1.3—2.2 mm in dorsoventral direction. Acetabulum 1.12—1.87 mm in dorsoventral direction, Paramphistomum type, number of muscle units, d.e.c.1, 8—12; d.e.c.2, 22—35; d.i.c., 28—43; v.i.c., 40—43; v.e.c., 8—13; m.e.c., 13—20 (Figure 173). Pharynx 0.78—0.89 mm long, Liorchis type (Figure 65); esophagus 0.68—0.98 mm long, without muscular thickening. Anterior testis 0.69—1.11 mm long, 0.87—0.92 mm in dorsoventral direction; posterior testis 0.43—0.92 mm long,

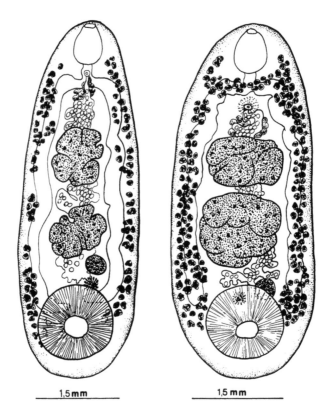

1,5 mm   1,5 mm

FIGURE 434. *Paramphistomum*
gotoi Fukui, 1922. (By permission of Dr. W. Junk Publishers,
1982, Dordrecht.)

FIGURE 435. *Paramphistomum*
ichikawai Fukui, 1922. (By permission of Dr. W. Junk Publishers, 1982,
Dordrecht.)

0.42—0.98 mm in dorsoventral direction. Pars musculosa short and weakly developed.
Ovary 0.22—0.29 by 0.31—0.37 mm. Genital pore prebifurcal, terminal genitalium Gracile
type (Figure 114). Eggs 0.132—0.145 by 0.067—0.072 mm.
Biology: Not completely known. Stepanov[914] examined embryonic development of eggs of
    this species. Embryonic development depends on temperature. Miracidia develop in 12
    to 13 d at 28 to 29°C; the minimal temperature which is necessary to commence embryonic
    development is 10°C. Under natural conditions[1102] embryogenesis completes in 40 to
    50 d.
Habitat: Rumen.
Hosts: *Antilope cervicapra, Bos primigenius* f.taurus, *Bubalus arnee* f. bubalis, *Ovis ammon*
    f. aries (Bovidae).
Distribution: Eurasia — Japan, Korea, U.S.S.R., Mongolia, Rumania, Malaysia, Philip-
    pines, Iran, Vietnam, China; Africa — Egypt, Syria.
References: Chen et al.,[168] Chinone,[169] Chu,[176] Davydova-Velitchko,[203] Eduardo,[272] El-
    Moukdad,[285] Fukui,[332] Lee,[525] Lee and Lowe,[527] Näsmark,[647] Patnaik,[703] Sey,[840,841,845,853]
    Sey and Eslami,[866] Shanta,[871] Sivtseva,[888] Thapar,[965] Yamaguti.[1072]

Paramphistomum ichikawai Fukui, 1922 (Figure 435)

Syn. *Cotylophoron vigisi* Davydova, 1963.
    Diagnosis: Body length 4.5—9.3 mm, body width 1.6—3.1 mm in dorsoventral direction.
Diameter of acetabulum 0.89—1.5 mm in dorsoventral direction, Pisum type, number of
muscle units, d.e.c.1, 20—23; d.e.c.2, 7—16; d.i.c., 33-60; v.e.c., 16—28; v.i.c., 32—61;

m.e.c., 12—18 (Figure 175). Pharynx 0.52—0.98 mm long, Calicophoron type (Figure 58); esophagus 0.31—0.52 mm long without muscular thickening. Anterior testis 0.41—1.11 mm long and 0.68—1.69 mm in dorsoventral direction; posterior testis 0.32—1.86 mm long and 1.12—1.98 mm in dorsoventral direction. Pars musculosa short and weakly developed. Ovary 0.31—0.58 by 0.42—0.59 mm. Terminal genitalium Ichikawai type (Figure 116). Genital pore postbifurcal. Eggs 0.121—0.128 by 0.059—0.072 mm.

Biology: References: Durie,[253] Kisilev,[488] Katkov,[469] Gluzman,[348] Dvoryadkhin and Besprozvannykh.[258]

Preparasitic stage. ① Durie. Deposited eggs were usually segmented in the four-cell stage. Embryonic development completed within 12 d at 27°C. Sizes and structures of miracidia were similar to those of *Orthocoelium streptocoelium*. Intermediate host: *Segnitilia alphena* (= *Helicorbis australiensis*). ② Kisilev. Freshly laid eggs contained embryo with four-cell cleavage stage. Length of incubation depended on the temperature applied. The lower temperature, necessary for commencing of embryonic development, was 7°C; at temperature of 30 to 32°C embryonic development lasted 6 d. Longevity of miracidia 8 h at room temperature. Intermediate hosts: *Helicorbis sujfunensis, Gyraulus filiaris*. ③ Gluzman. Embryogenesis completed 8 d after incubation of eggs at temperatures between 27 and 28°C. Body of miracidia covered with epidermal cells, arranged by the formula 6:8:4:2, measure 0.142—0.198 by 0.030—0.068 mm. Inner structure composed of a pair of flame cells, a single apical gland, two penetration glands, nerve mass, germ cells, and germ ball. Several sensory structures were found at the terebratorium and along the body of the miracidium. Intermediate host: *Segmentina nitida*. ④ Dvoryadkhin and Besprozvannykh. Freshly laid eggs unsegmented. Embryogenesis lasted 11 to 12 d at 27°C. Size of body of miracidia 0.264—0.283 by 0.033—0.039 mm. Inner structure composed of two pairs of penetration glands, a single apical gland, a pair of flame cells, nerve mass, and germ cells. Intermediate hosts: *Anisus centrifugus, A. subfiliaris, A. minusculus, Polypilis semiglobosa, Helicorbis sujfunensis*. ⑤ Katkov. Deposited eggs contained embryo in early cleavage stage. Miracidia developed in 12 d at temperature of 26°C. Body of miracidia measured 0.17—0.24 by 0.033—0.055 mm. Body covered by epidermal cells situated in four tiers, arranged by the formula 6:8:4:2. Inner structure of miracidia composed of two pairs of penetration cells, a cephalic gland, nerve mass, two flame cells, and germ cells. There were 14 sensory formations along the body of miracidia.

Intramolluscan larval stage. ① Durie. Sporocyst firmly established in the tissue of the mantle cavity. They reached their maturity, and the first rediae were realized 8 d after infection. Rediae were similar in structure to that of *Orthocoelium streptocoelium*. First cercariae liberated from the rediae via birth pore in an immature condition, 15 d after infection. Cercariae measured 0.280—0.336 by 0.196—0.224 mm. Main excretory tubes had cross-connecting tubes. A lateral diverticulum was given off from the main duct in the region of the eyespots. The rate of output of cercariae sporadic: high outputs were followed by periods of low secretion. Liberated cercariae encysted in 30 min, adolescariae measured 0.226 mm in diameter. ② Kisilev. After penetration of miracidia to snail, the first mature sporocyst was found 8 to 17 d later. Mature sporocyst measured 0.651 by 0.294 mm. Young rediae located in the tissue of the mantle and the hepatopancreas; mature rediae 1.008 by 0.252 mm in size. First cercariae liberated from rediae between the 15th and 55th d after snail exposure to miracidia. Mature cercaria measured 0.336 by 0.231 mm in fixed state. Encysted cercariae 0.189—0.231 mm in diameter. ③ Gluzman. Sporocyst and redia elongate oval in shape. Mature rediae 1.020 mm long and 0.250 mm in width. First cercariae were found on the 16th d in the rediae. Mature cercariae 0.246—0.320 by 0.092—0.144 mm, tail 0.458—0.670 by 0.05—0.08 mm. Measurements of adolescariae 0.185—0.225 mm in diameter. ④ Dvoryadkhin and Besprozvannykh. Under laboratory conditions percentage of infestation varied 7.1—7.5% (*Anisus centrifugus*), 5.5—6.0% (*A. subfilialis*), 9.4—12.0% (*A. minusculus*), 71.6—73.3% (*H. sujfunensi*), and 30.0—35.0% (*P. semiglobosa*). Mature sporocyst was found in the infected snail on the 10th day, measuring 0.55 by 0.25 mm. Mature mother rediae developed by the 22nd day after infestation, measuring 0.59—0.60 by 0.21—0.23 mm. Each redia contained 3 to 5 daughter rediae and germ balls up to 15. Cercariae liberated from rediae in an immature form. First cercariae were secreted on the 48th day after infestation. Body measured: 0.24—0.30 by 0.21—0.23 mm, tail: 0.52 mm in length. Free-swimming cercariae encysted in 3 to 5 h, adolescariae 0.205 mm in diameter. Under natural conditions of Primore and Priamure, duration of larval stages of *P. ichikawai* was 2.5 to 4 months or more.

Development in definitive host. ① Durie. A lamb was fed 1000 cysts, and the first eggs were recovered 49 to 51 d later. On post-mortem 456 flukes were found in the alimentary tract. ② Kisilev. Prepatent period was 42—51 d in sheep. ③ Gluzman. Artificially infected cattle and sheep excreted the first eggs in feces on 40 to 41 d after infestation.

Habitat: Rumen.

Hosts: *Bos primigenius* f. taurus, *Bubalus arnee* f. bubalis, *Capra aegagrus* f. hircus, *Ovis ammon* f. aries (Bovidae); *Cervus elaphus, C. nippon* (Cervidae).

Distribution: Eurasia — Japan, China, Yugoslavia, U.S.S.R., Bulgaria, Hungary, Germany, Poland, Czechoslovakia, Vietnam, South Korea, France, Rumania, Australia, Taiwan; India — questionable.

References: Babič,[52] Boray,[106] Bryden,[130] Chen et al.,[168] Davydova,[202] Davydova-Velitchko,[204] Durie,[252] Dvoryadkhin and Besprozvannykh,[258] Eduardo,[272] Erbolatov,[286] Euzeby,[293] Fukui,[334] Kisilev,[488] Kotrlá and Chroust,[502] Kotrlá and Kotrlý,[503] Näsmark,[647] Odening et al.,[669] Pačenovský et al.,[689] Sey,[836,841,845,853,858] Vasil'ev et al.,[1014] Velitchko,[1022,1025] Vujič,[1042] Zdzitowiecki et al.,[1092] Zhaltsanova.[1094]

Paramphistomum leydeni Näsmark, 1937 (Figure 436)

Syn. *Paramphistomum scotiae* Willmott, 1950; *P. julimarinorum* Velasquez-Maldonado, 1976; *P. nicabrasilosum* Velasquez-Maldonado, 1976; *P. procapri* Wang, 1979; *Cotylophoron skrjabini* Mitskevich, 1958 in part.

Diagnosis: Body length 3.8—6.2 mm, body width 1.9—3.2 mm in dorsoventral direction. Diameter of acetabulum 1.12—1.76 mm in dorsoventral direction, Paramphistomum type, number of muscle units, d.e.c.1, 8—13; d.e.c.2, 30—36; d.i.c., 33—41; v.e.c., 14—22; v.i.c., 41—49; m.e.c., 12—23 (Figure 173). Pharynx 0.49—0.82 mm long, Liorchis type (Figure 65), esophagus 0.48—0.75 mm long, without muscular thickening. Anterior testis 0.42—0.96 mm long and 0.98—1.59 mm in dorsoventral direction; posterior testis 0.51—0.89 mm long and 0.79—1.87 mm in dorsoventral direction. Pars musculosa short and weakly developed. Ovary 0.25—0.52 by 0.21—0.72 mm. Terminal genitalium Leydeni type (Figure 117). Genital pore bifurcal. Eggs 0.142—0.153 by 0.071—0.079 mm.

Biology: References: Nikitin,[654] Gluzman,[346-349] Gluzman and Artemenko,[350] Zhaltsanova,[1097] Katkov.[471] (Under the name *Liorchis scotiae*.)

Preparasitic stage. ① Nikitin. Newly laid eggs contained embryo in early cleavage stage. Embryonic development took place 9 d at 28°C. Freely swimming miracidia 0.146—0.215 by 0.039—0.099 mm in size. Longevity of miracidia a period of 21 h. Intermediate hosts: *Planorbis planoribis, Gyraulus albus, Hippeutis complanatus, Armiger crista.* ② Gluzman. Freshly deposited eggs contained embryo in early (2 to 8) cleavage stage. Embryogenesis completed during 8 d at temperatures of 27—28°C. Body of miracidia covered epidermal cells, arranged by the formula 6:8:4:2. Inner structure of miracidia composed of two pairs of penetration glands, a single apical gland, two flame cells, nerve mass, germ cells, and germ balls. Life span of miracidia: 7 h at temperatures

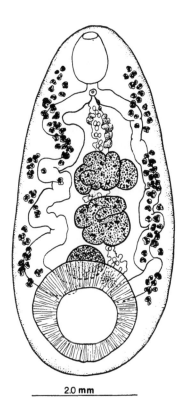

2.0 mm

FIGURE 436. *Paramphistomum leydeni* Näsmark, 1937. (By permission of
Dr. W. Junk Publishers, 1982, Dordrecht.)

18—21°C. Intermediate hosts: *Planorbis planorbis, Anisus vortex, A. spirorbis, A. contortus, A. septemgyratus, A. leucostoma, Segmentina nitida.* ③ Zhaltsanova. Deposited
eggs contained embryos in early cleavage stage. Embryonic development completed 9 to
10 d at 28°C. Miracidia measuring 0.132—0.172 by 0.024—0.043 mm. Intermediate
hosts: *Gyraulus gredleri, Segmentina nitida.*

Intramolluscan stage. ① Nikitin. Sporocyst reached maturity 22 to 23 d after infestation; measuring 0.480—0.0615 by 0.195—0.214 mm. Sporocyst at this stage contained
9 to 11 developing rediae. First rediae liberated from sporocyst on the 23rd day after
exposure of snail to miracidium. Mature redia measures 0.405—0.645 by 0.135—0.169
mm, containing 5 to 8 developing cercariae. First cercariae liberated from redia on the
61st day after infestation, and they were secreted by snails on the 83rd day after infestation.
Mature cercariae measure 0.28—0.45 by 0.12—0.16 mm; tail 0.51—0.64 mm long. Free-
swimming cercariae encysted in period of some minutes to some or more hours. Encysted
cercariae measure 0.227—0.279 mm in diameter. ② Gluzman. Sporocysts measure 0.510
by 0.628 mm in size on the 11th to 13th day after infestation, containing developing
rediae. First rediae liberated from sporocyst on the 13th day, and they measured
0.370—0.610 by 0.120—0.153 mm on the 18th day after infection. Cercariae were
secreted by rediae from the 29th day. Mature cercariae 0.290—0.425 by 0.108—0.159
mm in size; tail 0.475—0.694 mm in length. Free-swimming cercariae encysted soon on
appropriate objects; encysted cercariae 0.221—0.276 mm in diameter. Life span of adolescariae 13 months at temperatures between 6 and 8°C. ③ Katkov. Developmental process
of larval stages in *Planorbis planorbis* composed of the formation of sporocyst from

miracidia after penetration to the snail. Mother rediae evolved from the germinal cells, and they migrated to the hepatopancreas. Mother rediae gave rise to daughter rediae and cercariae, daughter rediae again granddaughter rediae and cercariae and so on. The change of generations in such a way continued up to the death of the snail. Under favorable conditions (23 to 29°C) cercariae mainly developed from rediae, but at low temperatures (5 to 6°C) rediae developed in greater number than cercariae did. Starvation delayed the tempo of development of larvae and reduced the number of cercariae excreted; but the state of larvae became normal in 2 months after finishing starvation. Infected snail survived desiccation on the surface up to 1.5 months; and larval development continued and mature cercariae gradually accumulated in the snail. After placing these snails into water, mass production of cercariae was experienced. The production of larvae and the number of hatched cercariae, under optimal conditions, were directly proportional to the temperature, to the number of the penetrated miracidia, to the duration of infection, and to the size of the snails. Emergence of cercariae was stimulated by bright light. ④ Zhaltsanova. Mature sporocysts were found on the 17th to 19th day and the first rediae liberated on the 18th day after infection. First cercariae were born in an immature stage and emerged on the 42nd day after infection. Mature cercariae measuring 0.42—0.54 by 0.14—0.20 mm; tail 0.60—0.66 by 0.06—0.072 mm. Adolescariae 0.334—0.365 mm in diameter.

Development in definitive host. ① Zhaltsanova. Prepatent period in cattle varied between 94—96 d. ② Gluzman and Artemenko. Calves were infected with 12,000—20,000 adolescariae, and the first eggs were found in the feces 97—114 d later. ③ Nikitin. Flukes reached their maturity between 104—127 d in experimentally infected cattle and sheep. Life span of adults: supposedly several years. ④ Gluzman. Prepatent period 96—97 d, longevity of this fluke 1 year in sheep and 4 years in cattle.

Habitat: Rumen and reticulum.

Hosts: *Bos primigenius* f. taurus, *Bison bonasus*, *Bubalus arnee* f. bubalis, *Ovis ammon* f. aries, *Procapra picticaudata* (Bovidae); *Alces alces*, *Capreolus capreolus*, *Cervus elaphus*, *C. nippon*, *Rangifer tarandus* (Cervidae).

Distribution: Eurasia — U.S.S.R., Sweden, Yugoslavia, France, Hungary, Germany, England, Ireland, Norway, Turkey, Austria, China; South America — Brazil, Venezuela.

References: Asadov,[42] Asadov et al.,[44] Babič,[52] Chen et al.,[168] Eduardo,[272] Erbolatov,[286] Gluzman,[346] Gluzman and Artemenko,[350] Graber and Thal,[365] Katkov et al.,[472] Mitskevitch,[617] Nasimov,[645] Näsmark,[647] Nekrasov and Smirnov,[648] Nikitin,[651,652] Ruziev,[798] Salimov et al.,[807] Sey,[836,858] Velázquez-Moldano,[1019] Velitchko,[1021,1022,1025] Willmott.[1063]

Paramphistomum hibernae Willmott, 1950 (Figure 437)

Diagnosis: Body length 4.0—10.8, body width 1.5—3.9 mm in dorsoventral direction. Diameter of acetabulum 1.32—2.28 mm in dorsoventral direction, Paramphistomum type, number of muscle units, d.e.c.1, 8—21; d.e.c.2, 27—32; d.i.c., 29—43; v.e.c., 10—21; v.i.c., 43—51; m.e.c., 3—9 (Figure 173). Pharynx 0.53—0.92 mm long, Liorchis type (Figure 65); esophagus 0.41—0.92 mm long, without muscular thickening. Anterior testis 0.53—0.86 mm long and 0.57—0.92 mm in dorsoventral direction; posterior testis 0.59—0.75 mm long and 0.72—1.13 mm in dorsoventral direction. Pars musculosa short and weakly developed. Ovary 0.21—0.69 by 0.49—0.63 mm. Terminal genitalium Leydeni type (Figure 117). Genital pore prebifurcal. Eggs 0.138—0.149 by 0.069—0.076 mm.

Biology: References: Willmott,[1065] Willmott and Pester,[1067] Katkov.[468]

Preparasitic stage. ① Willmott. Freshly laid eggs contained embryo in early cleavage stage. Embryonic development took place within 14—20 d at room temperature. Hatching was stimulated by light. Freely swimming miracidia measured 0.259 by 0.035 mm. Body covered by ciliated epithelial cells, situated in four tiers, arranged by the formula 6:8:4:2. Inner structure of miracidia composed of two pairs of penetration glands, a single apical

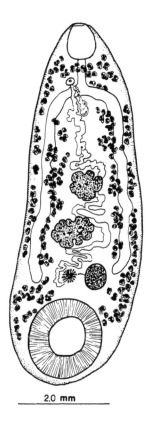

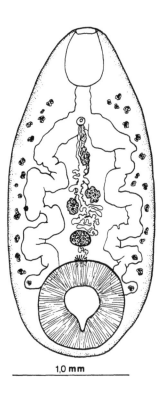

2,0 mm | 1,0 mm

FIGURE 437. *Paramphistomum hiberniae* Willmott, 1950. (By permission of Dr. W. Junk Publishers, 1982, Dordrecht.)

FIGURE 438. *Paramphistomum cephalophi* Eduardo, 1982. (By permission of Dr. W. Junk Publishers, 1982, Dordrecht.)

gland, nervous system, a pair of flame cells, and germinal tissue (germ cells). Longevity of miracidia in laboratory temperature (20 to 22°C) was 6 to 8 h. Intermediate host: *Planorbis leucostoma* (Willmott and Pester).[1067] ② Katkov. Deposited eggs contained centrally located embryo. Miracidia developed in 10 to 12 d at temperatures between 26 and 28°C. Newly hatched miracidia measured 0.172—0.275 by 0.034—0.077 mm. Body covered with ciliated epithelial cells, arranged by the formula 6:8:4:2. There were 14 sensory structures along the surface of the body. Inner structure of miracidia: apical gland, penetration glands, nervous mass, flame cells, and germinal cells were mentioned. Life span of miracidia 6 to 16 h.

Habitat: Rumen.

Hosts: *Bos primigenius* f. taurus, *Ovis ammon* f. aries (Bovidae).

Distribution: Eurasia — England, Ireland, U.S.S.R.

References: Eduardo,[272] Nikitin,[653] Salimov et al.,[807] Toktouchikova,[977] Velitchko,[1021,1022] Willmott.[1063]

Paramphistomum cephalophi Eduardo, 1982 (Figure 438)

Diagnosis: Body length 2.5—2.9 mm, body width 0.7—1.2 mm in dorsoventral direction. Acetabulum 0.70—0.78 mm in dorsoventral direction, Paramphistomum type, number of muscle units, d.e.c.1, 13—15; d.e.c.2, 20—26; d.i.c., 30—35; v.e.c., 8—15; v.i.c., 32—37; m.e.c., 17—20 (Figure 173). Pharynx 0.49—0.49 mm long, Cephalophi type (Figure 59); esophagus 0.22—0.38 mm long. Anterior testis 0.10—0.13 mm long and 0.14—0.15 mm

in dorsoventral direction; posterior testis 0.12—0.14 mm long and 0.09—0.13 mm in dorsoventral direction. Pars musculosa short and coiled. Genital pore bifurcal, terminal genitalium Gracile type (Figure 114). Eggs absent.
Biology: Not yet known.
Habitat: Small intestine.
Host: *Cephalophus nigrifrons* (Bovidae).
Distribution: Africa — Tanzania.
Reference: Eduardo.[272]

KEY TO THE SPECIES OF *PARAMPHISTOMUM*
(after Eduardo, 1982, with slight modification)

1. Pharynx with sphincter of the Cephalophi type; acetabular rim with posterior notch...
.... *P. cephaloni*
Pharynx without anterior sphincter; acetabular rim without posterior notch 2
2. Vitellaria confluent dorsomedially in their anterior limit 3
Vitellaria not confluent dorsomedially 4
3. Pharynx of the Liorchis type; terminal genitalium of the Leydeni type .. *P. hiberniae*
Pharynx of the Calicophoron type; terminal genitalium of the Ichikawai type ....
.... *P. ichikawai*
4. Testes lobed 5
Testes not lobed .... *P. liorchis*
5. Pharynx of the Liorchis type 6
Pharynx of the Calicophoron type 8
6. Blind cecal ends usually meet dorsomedially; terminal genitalium of the Gracile
type .... *P. gotoi*
Blind cecal ends do not meet dorsomedially 7
7. Terminal genitalium of the Gracile type; tegumental papillae in some rows around the
anterior extremity, not extending to the genital pore .... *P. cervi*
Terminal genitalium of the Leydeni type, tegumental papillae in several rows around
enterior extremity, extending to genital pore .... *P. leydeni*
8. Terminal genitalium of the Epiclitum type, tegumental papillae extensive, present around
oral opening and on anterior ventral half of body .... *P. epiclitum*
Terminal genitalium of the Gracile type, tegumental papillae present only around oral
opening .... *P. gracile*

Ugandocotyle Näsmark, 1937
Diagnosis: Paramphistominae. Body oval, bulky. Acetabulum small, terminal. Pharynx subterminal, large, esophagus short, without muscular thickening; ceca undulating, extending to acetabular zone. Testes tandem, ventral in position, lobed. Ovary median, anterodorsal to acetabulum. Genital pore prebifurcal. Pars musculosa weakly developed. Uterine coils well developed, dorsal to testes. Vitelline follicules numerous, lateral to ceca, extending from pharynx to acetabulum. Parasitic in stomach and intestine of hippopotamus.

Type species: *Ugandocotyle pisum* (Leiper, 1910) Näsmark, 1937 (Figure 439)
Syn. *Paramphistomum pisum* Leiper, 1910
Diagnosis: With characters of the genus. Body length 5.3—6.8 mm, body width 3.2—4.3 mm. Acetabulum 1.58—1.87 mm in diameter, Pisum type, number of muscle units, d.e.c.1, 23—24; d.e.c.2, 13—21; d.i.c., 43—46; v.e.c., 18—19; v.i.c., 60—63 (Figure 175). Pharynx 1.26—1.58 mm long, Pisum type (Figure 68), esophagus 0.28—0.33 mm long. Testes 0.63—1.49 mm long and 0.76—1.6 mm in dorsoventral direction. Ovary 0.41—0.48

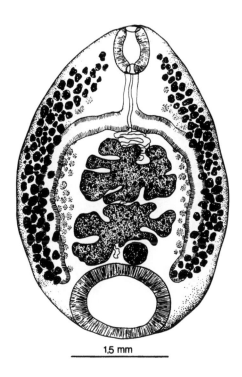

FIGURE 439. *Ugandocotyle pisum* (Leiper, 1910).

mm long and 0.75—0.93 mm in dorsoventral direction. Terminal genitalium Pisum type (Figure 126). Eggs 0.150—0.163 by 0.070—0.090 mm.

Biology: Not yet known.

Habitat: Stomach and intestine.

Host: *Hippopotamus amphibius* (Hippopotamidae).

Distribution: Africa — Uganda, Sudan, Chad.

References: Leiper,[531] Maplestone,[591] Näsmark,[647] Sey and Graber.[864]

Explanatum Fukui, 1922

Diagnosis: Paramphistominae. Body conical, curved ventrally, small to large in size. Acetabulum terminal or ventroterminal, enormous. Pharynx terminal without appendages; esophagus short without muscular thickening; ceca in lateral sides, simple or sinuous, extending to acetabulum. Testes tandem or slightly diagonal in middle third of body. Ovary submedian, posttesticular. Pars musculosa short, weakly developed. Uterine coils well developed, posterior to testes. Vitelline follicules in lateral fields, extending from bifurcation to acetabulum. Genital pore bifurcal or prebifurcal. Parasitic in bile duct and liver of ruminants.

Type species: *Explanatum explanatum* (Creplin 1847) Fukui, 1929 (Figure 440)

Syn. *Paraphistomum fraternum* Stiles et Goldberger, 1910; *P. siamense* Stiles et Goldberger, 1910; *Amphistomum explanatum* Creplin, 1847.

Diagnosis: Body length 6.9—13.6 mm, body width 3.2—6.3 mm in dorsoventral direction. Acetabulum enormous 3.26—4.63 mm in diameter, Explanatum type, number of muscle units, d.e.c.1, 9—14; d.e.c.2, 18—31; d.i.c., 38—49; v.e.c., 10—13; v.i.c., 28—46; m.e.c., 4—8 (Figure 168). Pharynx 0.68—0.92 mm long, Explanatum type (Figure 61);

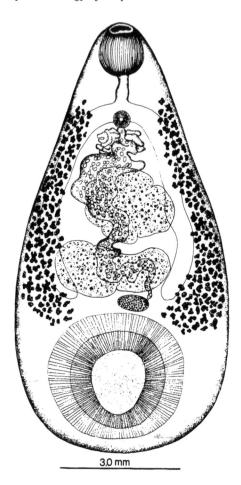

3,0 mm

FIGURE 440.   *Explanatum explanatum* (Creplin,
1847). (After R. P. Mukherjee, 1960.)

esophagus 0.76—0.98 mm long. Anterior testis 1.02—1.63 mm long and 1.26—2.73 mm
in dorsoventral direction, posterior testis 1.22—1.76 mm long and 1.98—2.74 mm in
dorsoventral direction. Ovary 0.35—0.63 by 0.42—0.54 mm. Pars musculosa short and
weakly developed. Terminal genitalium Explanatum type (Figure 110). Genital pore bifurcal.
Eggs 0.117—0.131 by 0.072—0.081 mm.
Biology: References: Ozaki,[687] Singh.[883]

Preparasitic stage. ① Ozaki. Deposited eggs contained zygote in four-cell stage.
Duration of time required for embryonic development 8 to 15 d at summer temperature
(23—32°C). It was found that stirring stimulated hatching even if eggs were kept in a
dark room. Measurements of fixed miracidia varied 0.11—0.19 by 0.027—0.050 mm.
Body covered with ciliated epidermal cells arranged in four rows by the formula 6:8:4:2.
Inner structure of miracidia composed of one pair of penetration glands, one apical gland,
brain mass, a pair of flame cells, and germ cells. ② Singh. Freshly laid eggs yellowish-
green with a simple propagatory cell central in position. Miracidia hatch in about 9 to 16
d, depending on temperature (room temperature and 27.7°C). Body measurements 0.15
by 0.067 mm. Ciliated epidermal cells were present, arranged in four rows, by the formula
6:8:4:2. A single pair of penetration glands, one apical gland, a pair of flame cells, and
germ balls, composed of several large cells. Intermediate host: *Gyraulus convexiusculus.*

Intramolluscan larval stage. ① Singh. The 4-d-old sporocyst measured 0.22 by 0.13

mm. Number of germ balls varies between 2 and 11. Sporocyst seemed to be fully developed by the 6th day. An opening along the sporocyst body, caused by the escaped rediae, was observed. First rediae were found on the 12th day. Mature rediae measured 0.99—1.05 by 0.19—0.21 mm. Three pairs of flame cells, muscular pharynx, saccular cecum with salivary glands and birth pore, near anterior end of body, were present. Liberated cercariae were usually immature, their development continued in the snail tissue. Body measurements: 0.49 by 0.34 mm. Tail 0.62 mm long. Acetabulum 0.155 mm in diameter. Body heavily pigmented. Two eyespots were present in the anterior part on the dorsal side. Pharynx 0.062 mm in diameter, esophagus elongated, ceca extending to acetabulum. Main excretory vesicles with cross-connecting tubes, caudal tube somewhat expanded at the tip of the tail, lateral openings present. Cercariae emitted from the snail hosts on 27th day after infection. Encystment took place within 40 min on an average. Cyst 0.20—0.23 mm in diameter.

   Development in definitive host. ① Singh. Goat kid was infected orally, flukes reached maturity in less than 16 weeks after infection with adolescariae.

Habitat: Liver, gallbladder, bile ducts.

Hosts: *Bos primigenius* f. taurus, *Hippotragus equinus, Kobus leche, Redunca arondinum, Alcelaphus* sp., *Capra aegagrus* f. hircus, *Ovis ammon* f. aries, *Damaliscus albifrons, Bubalus arnee* f. bubalis, *Bos javanicus* f. domestica (Bovidae); *Cervus unicolor* (Cervidae).

Distribution: Africa — Angola, Malawi, Zambia, South Africa, Guinea, Senegal, Zaire; Asia — Malaysia, Indonesia (Celebes), China, India, Philippines, Japan, Iraq, Afghanistan, Vietnam, Cambodia, Korea, Sri Lanka, Burma, Pakistan, Thailand; The Caribbean — Cuba.

References: Adiwinata,[3] Agrawal,[10] Ashizawa et al.,[46] Aziz et al.,[49] Chatterji,[164] Chen et al.,[168] Crusz,[191] Dubois,[250] Eduardo,[274] Fukui,[334] Gupta,[389] Hovorka et al.,[425] Kadhim et al.,[460] Khan,[483] Kim,[485] Kotrlá et al.,[501] Leiper,[534] Le Roux,[545] Maplestone,[591] Morel,[621] Näsmark,[647] Ortlepp,[678] Presidente,[738] Rahman,[761] Rao and Acharjyo,[768] Sadana et al.,[803] Segal et al.,[826] Sey,[846,853] Tendeiro,[962] Thapar,[965] Yamaguti,[1075] Yusuf and Chaudhry.[1081]

Explanatum bathycotyle (Fischoeder, 1901) Yamaguti, 1958 (Figure 441)

   Syn. *Paramphistomum bathycotyle* Fischoeder, 1901

   Diagnosis: Body length 8.9—14.6 mm, body width 3.2—4.8 mm. Acetabulum enormous, 2.57—4.32 mm in diameter, Explanatum type, number of muscle units, d.e.c.1, 9—14; d.e.c.2, 23—29; d.i.c., 28—39; v.e.c., 8—13; v.i.c., 30—45; m.e.c., 9—13 (Figure 168). Pharynx 0.62—0.98 mm long, Explanatum type (Figure 61); esophagus 0.32—1.23 mm long. Anterior testis 0.83—1.29 mm long, and 1.36—1.89 mm in dorsoventral direction, posterior testis 0.89—1.38 mm long, and 1.38—1.91 mm in dorsoventral direction. Pars musculosa weakly developed, short. Ovary 0.42—0.59 by 0.41—0.63 mm. Terminal genitalium Gracile type (Figure 114). Genital pore bifurcal. Eggs 0.114—0.131 by 0.071—0.079 mm.

Biology: Not completely elucidated. Ghafoor[342] found that the juvenile flukes reach their predilection place via the bile duct and not transperitoneally.

Habitat: Liver, bile ducts, gallbladder.

Hosts: *Axis axis* (Cervidae); *Bos primigenius* f. taurus, *Bubalus arnee* f. bubalis, *Capra aegagrus* f. hircus, *Bos javanicus* f. (Bovidae)

Distribution: Asia — Sri Lanka, Malaysia, India, Vietnam, Philippines, Burma, China, Indonesia.

References: Chen et al.,[168] Crusz,[191,192] Crusz and Nagaliyadde,[194] Eduardo,[274] Fischoeder,[311] Ghafoor,[342] Gupta,[384] Gupta et al.,[399] Kraneveld and Douwes,[508] Lee and Lowe,[527] Maxwell,[597] Näsmark,[647] Railliet,[764] Shanta.[871]

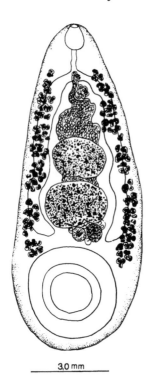

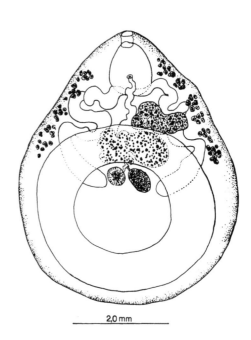

FIGURE 441. *Explanatum bathycotyle* (Fischoeder, 1901). (By permission of Dr. W. Junk Publishers, 1984, Dordrecht.)

FIGURE 442. *Explanatum anisocotyle* (Faust, 1920). (By permission of Dr. W. Junk Publishers, 1984, Dordrecht.)

Explanatum anisocotyle (Faust, 1920) Yamaguti, 1958 (Figure 442)

Syn. *Paramphistomum anisocotyle* Faust, 1920

Diagnosis: Body length 4.6—6.4 mm, body width 3.3—4.1 mm. Acetabulum enormous, 3.12—3.52 mm in diameter, Explanatum type, number of muscle units, d.e.c.1, 7—13; d.e.c.2, 18—27; v.e.c., 8—13; v.i.c., 29—39; m.e.c., 4—8 (Figure 168). Pharynx 0.69—0.96 mm long, Explanatum type (Figure 61); esophagus 0.48—0.56 mm long, without muscular thickening. Anterior testis 0.51—1.50 mm long, and 1.03—1.34 mm in dorsoventral direction; posterior testis 0.56—0.85 mm long, and 1.02—1.49 mm in dorsoventral direction. Pars musculosa short, weakly developed. Ovary 0.43—0.56 by 0.45—0.50 mm. Terminal genitalium Explanatum type (Figure 110). Genital pore prebifurcal. Eggs 0.135—0.160 by 0.070—0.085 mm.

Biology: Not yet known.

Habitat: Liver and bile ducts.

Hosts: *Bubalus arnee* f. bubalis, *Bos primigenius* f. taurus, *Capra aegagrus* f. hircus (Bovidae).

Distribution: Asia — India, Philippines, Malaysia.

References: Eduardo,[274] Faust,[302] Näsmark,[647] Tubangui.[996,997]

KEY TO THE SPECIES OF *EXPLANATUM*

(After Eduardo, 1984 slightly modified)

1. Terminal genitalium Explanatum type 2
    Terminal genitalium Gracile type .... *E. bathycotyle*

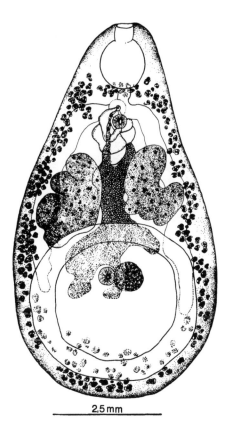

FIGURE 443. *Gigantocotyle gigantocotyle*
(Brandes in Otto, 1896). (By permission of Dr. W.
Junk Publishers, 1984, Dordrecht.)

2.   Pharynx and acetabulum very close together.......................... *E. anisocotyle*
   Pharynx and acetabulum far apart; body large, ceca almost straight ... *E. explanatum*

Gigantocotyle Näsmark, 1937

Diagnosis: Paramphistominae. Body conical, curved ventrally. Acetabulum enormous, ventroterminal. Pharynx subterminal, without appendages; esophagus short with or without muscular thickening; ceca straight or sinuous, extending to acetabular zone. Testes tandem, slightly diagonal or obliquely horizontal. Pars musculosa well developed. Ovary submedian dorsal to acetabulum. Genital pore bifurcal. Uterine coils posterior to testes. Vitelline follicules lateral extending from pharynx to acetabular zone, may or may not be confluent anteriorly and posteriorly. Terminal genitalium bifurcal. Parasitic in stomach and intestine of Artiodactyla.

Type species: *Gigantocotyle gigantocotyle* (Brandes in Otto, 1896) Näsmark, 1937 (Figure 443)

Syn. *Amphistomum gigantocotyle* Brandes, 1896.

Diagnosis: Body length 6.9—10.3 mm, body width 4.0—5.5 mm. Acetabulum 3.96—6.00 in diameter, Gigantocotyle type, number of muscle units, d.e.c.1, 8—19; d.e.c.2, 34—48; d.i.c., 38—52; v.e.c.1, 13—15; v.e.c.2, 8—11; v.i.c., 42—51; m.e.c., 8—13 (Figure 171). Pharynx 1.15—1.53 m long, Explanatum type (Figure 61); esophagus 1.28—1.62 mm long. Right testis 1.50—0.83 mm long and 1.20—2.09 mm in dorsoventral direction, left testis 1.90—2.03 mm long and 1.30—1.86 mm in dorsoventral direction. Terminal genitalium Gigantocotyle type (Figure 113). Eggs 0.100—0.122 by 0.66-0.80 mm.

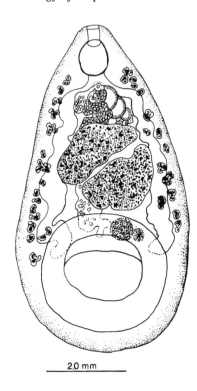

2.0 mm

FIGURE 444. *Gigantocotyle formosana* (Fukui, 1929). (By permission of Dr. W. Junk Publishers, 1984, Dordrecht.)

Biology: Not yet known.
Habitat: Stomach.
Host: *Hippopotamus amphibius* (Hippopotamidae).
Distribution: Africa — Angola, Zaire, Uganda, South Africa, Sudan, Zambia.
References: Dollfus,[244] Eduardo,[274] Leiper,[531] Mettam,[611] Näsmark,[647] Otto,[682] Sey,[846] Strong and Shattock,[918] Swart.[935]

Gigantocotyle formosana (Fukui, 1929) Näsmark, 1937 (Figure 444)

Syn. *Paramphistomum fraternum* Stiles et Goldberger, 1910

Diagnosis: Body length 6.3—9.1 mm, body width 2.9—4.6 mm. Acetabulum 2.89—3.62 mm in diameter, Symmeri type, number of muscle units, d.e.c.1, 12—21; d.e.c.2, 5—9; d.i.c., 27—34; v.e.c., 11—18; v.i.c., 34—42; m.e.c., 8—10 (Figure 178). Pharynx 0.83—1.12 mm long, Explanatum type (Figure 61); esophagus 0.78—0.91 mm long. Anterior testis 0.87—1.13 mm long and 1.73—2.14 mm in dorsoventral direction, posterior testis 0.74—1.14 mm long and 1.73—1.92 mm in dorsoventral direction. Pars musculosa well developed. Ovary 0.41—0.52 by 0.38—0.71 mm. Terminal genitalium Microbothrium type (Figure 122). Eggs 0.980—0.124 by 0.049—0.076 mm.
Biology: Not yet known.
Habitat: Rumen and abomasum.
Hosts: *Bos primigenius* f. taurus, *Bubalus arnee* f. bubalis, *Kobus leche*, *Redunca arondinum* (Bovidae).
Distribution: Asia — Japan, Vietnam, Taiwan, India, Philippines, China.
References: Chen et al.,[168] Dróżdż and Malczewski,[245] Eduardo,[274] Fukui,[334] Le Roux,[545] Näsmark,[647] Ortlepp,[678] Sey.[853]

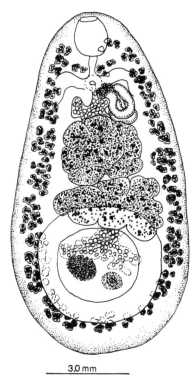

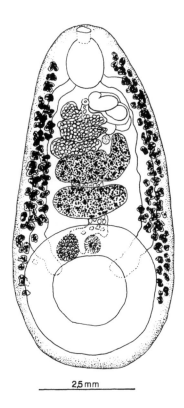

FIGURE 445. *Gigantocotyle symmeri* (Fukui, 1929). (By permission of Dr. W. Junk Publishers, 1984, Dordrecht.)

FIGURE 446. *Gigantocotyle duplicitestorum* Näsmark, 1937. (By permission of Dr. W. Junk Publishers, 1984, Dordrecht.)

Gigantocotyle symmeri Näsmark, 1937 (Figure 445)

Syn. *Gigantocotyle lerouxi* Yeh, 1957

Diagnosis: Body length 9.8—12.8 mm, body width 5.1—5.9 mm. Acetabulum 3.98—4.73 mm, Symmeri type, number of muscle units, d.e.c.1, 20—23; d.e.c.2, 4—7; d.i.c., 29—39; v.e.c., 12—16; v.i.c., 34—46; m.e.c., 6—9 (Figure 178). Pharynx 1.13—1.64 mm long, Explanatum type (Figure 61); esophagus 0.18—1.32 mm long. Anterior testis 1.00—1.73 mm long and 2.13—2.36 mm in dorsoventral direction, posterior testis 1.02—2.16 mm long and 2.16—2.73 mm in dorsoventral direction. Pars musculosa well developed. Ovary 0.68—1.13 by 0.78—0.98 mm. Terminal genitalium Gigantocotyle type (Figure 113). Eggs 0.098—0.128 by 0.048—0.078 mm.

Biology: Not yet known.

Habitat: Rumen, abomasum.

Hosts: *Bos primigenius* f. taurus, *Alcelaphus buselaphus, Hypotragus equinus, Kobus leche, Tragelaphus spekei, Syncerus caffer* (Bovidae).

Distribution: Africa — Sudan, Chad, Cameroon, Zambia, Zimbabwe, Central Africa, Botswana, South Africa.

References: Eduardo,[274] Graber,[357] Graber and Thal,[365] Näsmark,[647] Sey and Graber,[864] Yeh.[1079]

Gigantocotyle duplicitestorum Näsmark, 1937 (Figure 446)

Diagnosis: Body length 8.4—9.5 mm, body width 3.2—4.6 mm. Acetabulum 3.00—4.09 mm in diameter, Duplicitestorum type, number of muscle units, d.e.c.1, 13—18; d.e.c.2, 21—24; d.i.c., 31—42; v.e.c., 8—14; v.i.c., 31—37; m.e.c., 10—12 (Figure 167). Pharynx

1.05—1.52 mm long, Explanatum type (Figure 61), esophagus 0.56—0.98 mm long. Anterior testis 0.70—0.89 mm long and 1.38—2.38 mm in dorsoventral direction, posterior testis 0.66—0.82 mm long and 1.09—2.08 mm in dorsoventral direction. Pars musculosa well developed. Ovary 0.56—0.62 by 0.47—0.51 mm. Terminal genitalium Microbothrium type (Figure 122). Eggs 0.140—0.153 by 0.075—0.078 mm.

Biology: Not yet known.

Habitat: Stomach, small intestine.

Host: *Hippopotamus amphibius* (Hippopotamidae).

Distribution: Africa — Uganda, Sudan, Ethiopia, South Africa, Zambia.

References: Dinnik et al.,[234] Eduardo,[274] Graber et al.,[366] Leiper,[531] Näsmark,[647] Swart.[935]

KEY TO THE SPECIES OF *GIGANTOCOTYLE**
(After Eduardo, 1984 slightly modified)

1. Vitelline follicules confluent anteriorly and posteriorly, terminal genitalium Gigantocotyle type 2
    Vitelline follicules not confluent dorsomedially, terminal genitalium Microbothrium type 3
2. Acetabulum Symmeri type .... *G. symeri*
    Acetabulum Gigantocotyle type .... *G. gigantocotyle*
3. Acetabulum Duplicitestorum type .... *G. duplicitestorum*
    Acetabulum Symmeri type .... *G. formosana*

Cotylophoron Stiles et Goldberger, 1910

Diagnosis: Paramphistominae. Body conical or oval. Acetabulum ventroterminal, moderate in size. Pharynx terminal, without appendages, esophagus short with or without muscular thickening; ceca in lateral fields, form dorsoventral bands, blind ends directed dorsally. Testes tandem, diagonal, or horizontal, strongly lobed. Pars musculosa well developed. Ovary submedian, just before or dorsal to acetabulum. Genital pore bifurcal or postbifurcal, genital sucker present. Uterine coils winding, dorsal to testes. Vitelline follicules in lateral fields, extending from esophageal zone to acetabulum, occasionally confluent posteriorly. Parasitic in forestomach of ruminants.

Type species: *Cotylophoron cotylophorum* **(Fischoeder, 1901) Stiles et Goldberger, 1910 (Figure 447)**

Syn. *Paramphistomum cotylophorum* Fischoeder, 1901 *Cotylophoron guongdongense* Wang, 1979

Diagnosis: Body length 4.1—8.2 mm, body width 2.8—3.9 mm. Acetabulum 1.21—2.68 mm in diameter, Cotylophoron type, number of muscle units, d.e.c., 8—14; d.i.c., 34—50; v.e.c., 8—15; v.i.c., 42—53; m.e.c., 8—13 (Figure 166). Pharynx 0.59—1.26 mm long, Calicophoron type (Figure 58); esophagus 0.39—1.26 mm long, with muscular thickening. Anterior testis 0.59—1.23 mm long and 1.18—2.26 mm in dorsoventral direction, posterior testis 0.69—2.16 mm long and 1.98—2.23 mm in dorsoventral direction. Pars musculosa well developed, convoluted. Ovary 0.29—0.53 by 0.41—0.62 mm. Terminal genitalium Cotylophoron type (Figure 105). Genital pore bifurcal. Eggs 0.121—0.145 by 0.052—0.068 mm.

Biology: The life-cycle examinations of amphistome species under the name *Cotylophoron cotylophorum* were reported by several authors (Le Roux,[544] Krull,[510] Bennett,[79] Srivastava,[908] Varma,[1012] Zajíček[1088]). Later it was pointed out that none of this material was actually *C. cotylophoron*. Dinnik[228] found that Le Roux's material was identical with *Calicophoron microbothrium*. Bennett and Krull's specimens proved to be *Calicophoron*

---

* G. birmense (Railliet, 1924), species inquirenda.

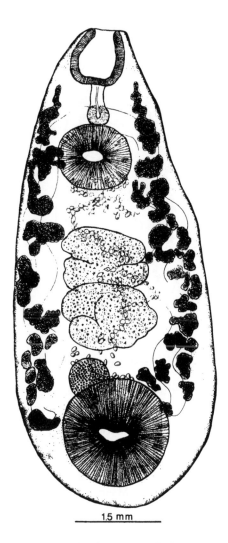

1,5 mm

FIGURE 447. *Cotylophoron cotylophorum* (Fischoeder, 1901). (After R. Ph. Dollfus, 1963.)

microbothrioides (Price,[745] Sey[836]). Examination of large amphistome collection of different hosts and parts of India (Sey,[843] Raina et al.[765]) *Cotylophoron* spp. have not been recovered; species under this name proved to be identical with species of the local fauna. Examined, the Czechoslovakian (Zajiček[1088]) "*Cotylophoron cotylophorum*" material proved to be identical with *C. fülleborni*.[1101] This species very probably has only African distribution, and its life cycle has not been studied yet.

Habitat: Rumen and reticulum.

Hosts: *Alcelaphus buselaphus, A. lichtensteini, Alcelaphus* sp. *Aepyceros melampus, Bos primigenius* f. taurus, *Bubalus arnee* f. bubalis, *Capra aegagrus* f. hircus, *Connochaetus taurinus, Cephalophus harveyi, Damaliscus lunatus, D. corrigum, Hypotragus niger, H. equinus, Kobus ellipsiprymnus, K. defassa, K. kob, K. leche, K. varondi, Neotragus pygmaeus, Oreotragus oreotragus, Ovis ammon*, f. aries, *Ourebia ourebia, Redunca arundinum, R. redunca, Raphicercus campestris, Syrcerus caffer, Sylvicapra grimmi, Taurotragus oryx, Tragelaphus angasi, T. scriptus, T. spekei, T. strepsicerus* (Bovidae); *Mosama simplicicornis* (Cervidae); *Equus burchelli* (Equidae); *Hippopotamus amphibius* (Hippopotamidae); *Loxodonta africana* (Elephantidae).

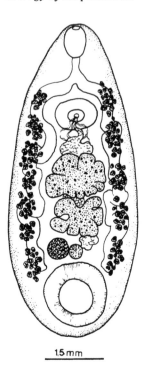

1.5 mm

FIGURE 448. *Cotylophoron jacksoni* Näsmark, 1937. (By permission of Dr. W. Junk Publishers, 1985, Dordrecht.)

Distribution: Africa — Dahomey, Kenya, South Africa, Uganda, Chad, Zambia, Malawi, Mozambique, Central Africa, Zaire, Zimbabwe, Liberia, Senegal, Guinea, Sudan, Egypt, Tanzania, Cameroon, Ethiopia, Angola, Mauritania, Algeria; Asia — Malaysia, Philippines, China, Burma, India, Pakistan, Vietnam, Cambodia; West Indies — Trinidad; South America — Venezuela.

References: Anderson,[29] Balasubramanian et al.,[63] Bali,[64] Bergeon,[83] Bwangamoi,[140] Caballero and Diaz-Ungria,[152] Cameron,[156] Chatterji,[164] Chen,[166] Chen et al.,[168] Cruz e Silva,[195] De Leon and Juplo,[209] Diaw et al.,[217] Diaz-Ungria,[219] Dinnik et al.,[234] Dollfus,[239] Eduardo,[277] Euzeby,[292] Fischoeder,[311] Fitzsimmons,[318] Graber,[356] Graber et al.,[362,366] Graber and Thal,[365] Hsu,[426] Joyeux and Baer,[457] Khan,[483] Leiper,[531] Le Roux,[544,545] Malek,[580] Manuel and Madriaga,[590] Maplestone,[591] Mettam,[610] Mettrich,[613] Morrel,[621] Nagaty,[642,643] Näsmark,[647] Ortlepp,[678] Prokopič and Kotrlá,[752] Prudhoe,[756] Quiroz et al.,[759] Roth and Dalchow,[791] Schwartz,[824] Segal et al.,[826] Serrano,[829] Sey,[844-847] Singh and Lakra,[885] Srivastava,[908] Stiles and Goldberger,[915] Strydonck,[920] Stunkard,[925] Szidat,[939] Thapar,[965] Tubangui,[998] Vercruysee,[1031] Vogelsang,[1037] Yusuf and Chaudhry.[1081]

Cotylophoron jacksoni Näsmark, 1937 (Figure 448)

Diagnosis: Body length 7.3—9.4 mm, body width 2.3—3.1 mm in dorsoventral direction. Acetabulum 1.24—2.19 mm in diameter, Cotylophoron type, number of muscle units, d.e.c., 9—14; d.i.c., 33-47; v.e.c., 31—42; m.e.c., 4—10 (Figure 166). Pharynx 0.72—1.11 mm long, Calicophoron type (Figure 58); esophagus 0.73—0.92 mm long, without muscular thickening. Anterior testis 1.12—1.43 mm long and 1.42—1.87 mm in dorsoventral direction; posterior testis 0.81—1.22 mm long and 1.43—1.58 mm in dorsoventral direction. Pars musculosa well developed, convoluted. Ovary 0.32—0.58 by 0.47—0.62 mm. Terminal

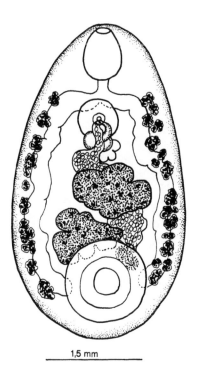

1,5 mm

FIGURE 449. *Cotylophoron fülleborni* Näsmark, 1937. (By permission of Dr. W. Junk
Publishers, 1985, Dordrecht.)

genitalium Cotylophoron type (Figure 105), genital opening postbifurcal. Uterine coils wind-
ing, dorsal to testes. Vitelline follicules in lateral fields from bifurcation to anterior margin
of acetabulum. Eggs 0.113—0.132 by 0.053—0.062 mm.
Biology: Not yet known.
Habitat: Rumen.
Hosts: *Alcelaphus buselaphus*, *Bos primigenius* f. taurus, *Hippotragus niger*, *Tragelaphus
strepsicerus* (Bovidae).
Distribution: Africa — Zambia, Uganda, Gambia, Kenya, Tanzania, Zimbabwe.
References: Eduardo,[277] Näsmark.[647]

Cotylophoron fülleborni Näsmark, 1937 (Figure 449)

Syn. *Cotylophoron indicum* of Näsmark, 1937, *C. noveboracensis* Price et McIntosh,
1953, *C. kwantungensis* Wang, 1979; *C. sinuointestinum* Wang, 1977, *C. shangkiangensis*
Wang, 1979; *W. guangdongense* Wang, 1979.

Diagnosis: Body length 3.1—5.9 mm, body width 1.1—2.6 mm. Acetabulum 1.26—1.58
mm in diameter, Cotylophoron type, number of muscle units, d.e.c., 10—21; d.i.c., 28—37;
v.e.c., 12—17; v.i.c., 29—37; m.e.c., 5—13 (Figure 166). Pharynx 0.38—0.89 mm long,
Calicophoron type (Figure 58), esophagus 0.29—0.64 mm long and 0.89—1.26 mm in
dorsoventral direction, posterior testis 0.39—0.68 mm long and 0.69—1.26 mm in dorsoventral direction. Pars musculosa well developed, convoluted. Ovary 0.19—0.36 by 0.22—0.38
mm. Genital pore postbifurcal. Terminal genitalium Cotylophoron type (Figure 105). Uterine
coils undulating, dorsal to testes. Vitelline follicules in lateral fields, along ceca, extending
from bifurcation to acetabular zone. Eggs 0.110—0.143 by 0.051—0.073 mm.
Biology: Not completely elucidated. Zajiček[1088] described the preparasitic stages of this
species under the name *Cotylophoron cotylophorum* which was identical with this species

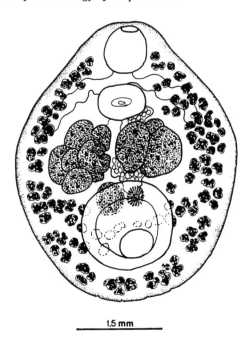

1.5 mm

FIGURE 450.   *Cotylophoron panamense* Price et
McIntosh, 1953. (By permission of Dr. W. Junk Publishers, 1985, Dordrecht.)

(see "Notes" under *Cotylophoron cotylophorum*). Eggs were operculated; when deposited
they contained embryo of advanced cleavage stage, and miracidia began to form within
3—4 d at room temperature (24—32°C). It was also found that freshly laid eggs contained
fully formed miracidia, which were ready to hatch in 2 h. Length of body 0.260—0.288
mm. Covered with ciliated epidermal cells, arranged in four rows, by the formula 6:8:4:2.
Inner structure of miracidia composed of apical gland, penetration glands, a pair of flame
cells and germinal cells. *Lymmaea cubensis*, *Physa cubensis*, and *Viviparus* sp. did not
prove to be susceptible intermediate hosts of this species.

Habitat: Rumen.

Hosts: *Aepiceros melampus*, *Bos primigenius* f. taurus, *Bubalus arnee* f. bubalis, *Boocerus
euryceros*, *Syncerus caffer* (Bovidae).

Distribution: Africa — Malawi, South Africa, Uganda, Cameroon, Senegal, Kenya, Zambia,
Zaire, Tanzania.

References: Anderson,[29] Chen et al.,[168] Dinnik et al.,[234] Eduardo,[277] Maplestone,[591] Näsmark,[647] Sey,[847] Strydonck,[920] Vassiliades.[1013]

Cotylophoron panamense Price et McIntosh, 1953 (Figure 450)

Diagnosis: Body length 3.6—5.9 mm, body width 2.7—4.0 mm. Acetabulum 1.56—2.03
mm in diameter, Cotylophoron type, number of muscle units, d.e.c., 11—15; d.i.c., 42—53;
v.e.c., 8—14; v.i.c., 48—57; m.e.c., 5—10 (Figure 166). Pharynx 0.43—0.97 mm long,
Calicophoron type (Figure 58), esophagus 0.50—0.72 mm long without muscular thickening.
Right testis 0.93—1.57 mm long and 1.20—2.86 mm in dorsoventral direction; left testis
0.82 mm in diameter. Pars musculosa well developed, convoluted. Ovary 0.26—0.53 by
0.48—0.62 mm. Terminal genitalium Cotylophoron type (Figure 105). Genital pore postbifurcal. Uterine coils posterior to testes. Vitelline follicles in lateral fields extending from
pharynx to acetabular zone, confluent posteriorly. Eggs 0.128—0.132 by 0.058—0.060
mm.

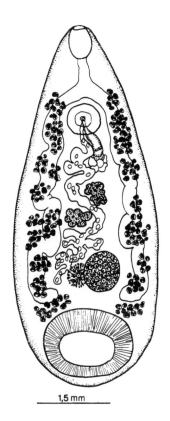

FIGURE 451. *Cotylophoron bareilliense* (Mukherjee, 1963). (By permission of Dr. W. Junk Publishers, 1985, Dordrecht.)

Biology: Not yet known.

Habitat: Rumen.

Hosts: *Bos primigenius* f. taurus, *Ovis ammon* f. aries (Bovidae).

Distribution: North America — U.S.; Central America — Panama; West Indies — Trinidad, Dominican Republic; The Caribbean — Cuba; South America — Colombia.

References: Eduardo,[277] Price and McIntosh.[748]

Cotylophoron bareilliense **(Mukherjee 1963) Mukherjee et Chauhan, 1965 (Figure 451)**

Syn. *Cotylophoron skrjabini* Mukherjee, 1963

Diagnosis: Body length 5.1—9.7 mm, body width 2.7—3.4 mm. Acetabulum 1.43—1.84 mm in diameter, Cotylophoron type, number of muscle units, d.e.c., 10—24; d.i.c., 43—58; v.e.c., 14—21; v.i.c., 41—52; m.e.c., 5—9 (Figure 166). Pharynx 0.52—0.84 mm long, Calicophoron type (Figure 58); esophagus 0.96—1.08 mm long, without muscular thickening. Anterior testis 0.21—0.50 mm long and 0.30—0.79 mm in dorsoventral direction, posterior testis 0.22-0.61 mm long and 0.27—0.74 mm in dorsoventral direction. Pars musculosa well developed, convoluted. Ovary 1.00—1.14 by 0.78—1.06 mm. Terminal genitalium Cotylophoron type (Figure 105). Genital pore postbifurcal. Uterine coils winding posterior to testes and passing between them. Vitelline follicules lateral, along ceca, extending from bifurcation to acetabulum. Eggs 0.150—0.161 by 0.061—0.081 mm.

Habitat: Rumen.

Hosts: *Bos primigenius* f. taurus, *Bubalus arnee* f. bubalis, *Capra aegagrus* f. hircus (Bovidae).

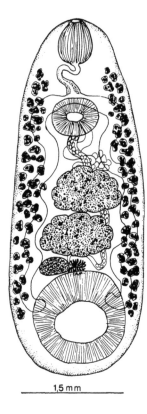

FIGURE 452. *Cotylophoron macrosphinctris* Sey et Graber, 1979. (From O. Sey and M. Graber, 1979.)

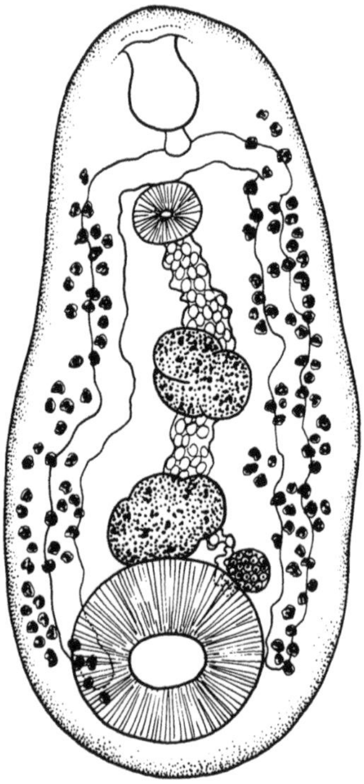

FIGURE 453. *Cotylophoron xiangjiangense* Wang, 1979. (After X. Y. Wang, 1979.)

Distribution: Asia — India, Sri Lanka, Philippines.
References: Eduardo,[277] Mukherjee.[628]

Cotylophoron macrosphinctris Sey et Graber, 1979 (Figure 452)

Diagnosis: Body length 5.5—7.8 mm, body width 1.5—3.0 mm. Acetabulum 1.20—2.05 mm in diameter, Cotylophoron type, number of muscle units, d.e.c., 8—13; d.i.c., 43—49; v.e.c., 10—16; v.i.c., 40—46; m.e.c., 10—15 (Figure 166). Pharynx 0.80—1.23 mm long, Calicophoron type (Figure 58); esophagus 1.20—1.52 mm long, with muscular thickening. Anterior testis 0.40—1.80 mm in length and 0.8—1.9 mm in width, posterior testis 0.50—1.80 mm in length and 0.70—2.10 mm in width. Pars musculosa well developed, convoluted. Ovary 0.28—0.84 by 0.20—0.48 mm. Terminal genitalium Macrosphinctris type (Figure 120). Genital pore postbifurcal. Uterine coils winding, dorsal to testes. Vitelline follicules in lateral field from esophagus to acetabular zone. Eggs 0.100—0.150 by 0.06—0.070 mm.
Biology: Not yet known.
Habitat: Rumen.
Hosts: *Alcelaphus buselaphus, Ourebia ourebia, Syncerus caffer* (Bovidae).
Distribution: Africa — Central Africa, Chad, Cameroon, Uganda, Zaire.
References: Eduardo,[277] Graber,[360] Graber and Thal,[365] Sey,[847] Sey and Graber.[864]

Cotylophoron xiangjiangense Wang, 1979 (Figure 453)

Diagnosis: Body length 3.8—5.6 mm, body width 1.9—3.0 mm. Acetabulum 1.23—1.72 by 1.23—1.58 mm, Paramphistomum type, number of muscle units, d.e.c.1, 16; d.e.c.2,

14; d.i.c., 55; v.e.c., 18; v.i.c., 63 (Figure 173). Pharynx 0.43—0.78 mm long, close to Liorchis type; esophagus 0.67 mm long with muscular thickening. Anterior testis 0.79 by 0.57 mm, posterior testis 0.76 by 0.73 mm. Pars musculosa well developed, convoluted. Ovary 0.37 by 0.40 mm. Terminal genitalium Cotylophoron type (Figure 105). Genital pore bifurcal. Eggs 0.115—0.140 by 0.063—0.080 mm.

Biology: Not yet known.
Habitat: Forestomach.
Host: *Bubalus arnee* f. bubalis (Bovidae).
Distribution: Asia — China.
References: Eduardo,[277] Wang.[1050]

KEY TO THE SPECIES OF *COTYLOPHORON*
(After Eduardo, 1985 modified)

1. Esophageal muscular thickening present 2
   Esophageal muscular thickening absent 4
2. Pharynx Calicophoron type 3
   Pharynx Liorchis type... *C. xiangjiangense*
3. Genital sucker Macrosphinctris .... *C. macrosphinctris*
   Genital sucker Cotylophoron type................................ *C. cotylophorum*
4. Vitelline follicules confluent posteriorly .... *C. panamense*
   Vitelline follicules not confluent 5
5. Testes longer than ovary... *C. fülleborni*
   Testes smaller than ovary .... *C. bareilliense*

Calicophoron Näsmark, 1937
Syn. *Bothriophoron* Stiles et Goldberger, 1910; *Lorisia* Penso, 1940.

Diagnosis: Paramphistominae. Body conical tapering anteriorly, almost round in cross section. Acetabulum ventroterminal, moderate in size. Pharynx muscular, without appendages; esophagus short, with or without muscular thickening; ceca straight or sinuous, terminating at acetabular zone. Testes tandem, diagonal, in middle or posterior half of body. Pars musculosa strongly developed, convoluted. Ovary median or submedian in front of acetabulum. Terminal genitalium well developed, without genital sucker. Genital pore bifurcal or postbifurcal. Uterine coils winding forward dorsal to testes, then ventral to male ducts. Vitelline follicules in lateral field, extending from bifurcal zone to acetabular zone, occasionally confluent anteriorly. Excretory vesicle and Laurer's canal cross each other. Parasitic in forestomach of ruminants.

Type species: *Calicophoron calicophorum* (Fischoeder, 1910) Näsmark, 1937 (Figure 454)
Syn. *Paramphistomum calicophorum* (Fischoeder, 1901); *P. crassum* Stiles et Goldberger 1910; *P. cauliorchis* Stiles et Goldberger 1910; *P. ijimai* Fukui, 1922; *P. (Cauliuchis) skrjabini* Popova, 1937; *P. erschovi* Davydova, 1959; *Calicophoron orientalis* Mukherjee, 1966; *C. wuchengense* Wang, 1979; *C. fusum* Wang et Xia 1979; *C. villum* Wang et Liu, 1977 *C. zhejiangense* Wang, 1979 *Cotylophoron skrjabini* Mitskevich, 1959 in part.

Diagnosis: Body length 7.9—15.9 mm, body width 3.2—7.8 mm. Acetabulum 1.21—3.36 mm, Calicophoron type, number of muscle units, d.e.c., 10—22, d.i.c., 34—41, v.e.c., v.i.c., 18—23; m.e.c., 8—14; (Figure 165). Pharynx 0.68—2.13 mm long, Calicophoron type (Figure 58); esophagus 0.52—1.63 mm long, without muscular thickening. Anterior testis 1.21—4.23 mm long and 1.32—4.63 mm in dorsoventral direction; posterior testis 1.16—4.36 mm long and 1.63—4.56 mm in dorsoventral direction. Pars musculosa well

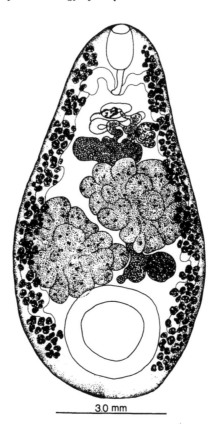

3.0 mm

FIGURE 454. *Calicophoron calicophorum*
(Fischoeder, 1901). (By permission of Dr. W.
Junk Publishers, 1983, Dordrecht.)

developed, convoluted; pars prostatica long. Ovary 0.56—1.42 by 0.78—1.53 mm. Terminal
genitalium Calicophoron type (Figure 103). Genital pore postbifurcal. Eggs 0.121—0.158
by 0.061—0.093 mm.

Biology: References. Porter,[733,734] Jonathan,[452] Durie,[254] Dvoryadkhin and Besprozvannykh.[258]

Preparasitic stage. ① Jonathan (described under the name *Calicophoron ijimai).*
Freshly deposited eggs contained zygote consisting of some cells. Embryonic development
took place within 13 d at 27°C. Newly hatched miracidium measured 0.24—0.25 by
0.033—0.039 mm. Inner structure of miracidia composed of two pairs of penetration
glands, a single apical gland, brain mass, a pair of flame cells, and germ cells. Life span
of miracidia varied between 4 and 24 h under laboratory conditions. Intermediate hosts:
① Jonathan. *Planorbis kakuika.* ② Porter. *Bulinus tropicus.* ③ Durie. *Pygmanisus pi-
larius.* ④ Dvoryadkhin and Besprozvannykh. *Helicorbis sujfenensis.*

Intramolluscan larval stage. ① Jonathan. Sporocyst on the 10th day measured 0.55
by 0.25 mm in size and contained 9 to 13 developing rediae. Rediae liberated after
rupturing of sporocyst's wall and became mature by the 27th day after emergence, measuring 0.59—0.60 by 0.21—0.23 mm. First cercariae came out from snails 59 d after
infestation. Body of cercariae 0.24—0.30 by 0.21—0.23 mm; tail 0.52 mm in length;
strongly brown in color. Adolescariae 0.20 mm in diameter. Two infected snails yielded
37 and 21 cercariae, respectively, during a 50-h period. ② Porter. Redia muscular,
measured 3.0 mm. Body of cercariae measured 0.40—0.50 by 0.30—0.40 mm. Two

pigmented eyespots and cystogenous cells are present, excretory system has two main tubes with cross-connecting tubes. Pharynx and acetabulum well marked. Genital rudiments consisted of cell masses. Cyst 0.40—0.55 mm in diameter. ③ Durie. Fully matured redia measured 0.40 by 0.11 mm and contained 13 embryos and germ balls. Excretory system consisted of three pairs of flame cell with their ducts. Cercariae dark brown, fixed specimens measured 0.26—0.29 by 0.15—0.18 mm; tail 0.49 mm in length. Acetabulum 0.077 mm in diameter. Main excretory tubes with cross-connecting tubes. Caudal tube extending to posterior end, no indication of lateral openings. Adolescariae 0.12—0.20 mm in diameter and approximately 0.12 mm in height.

Development in definitive host. ① Durie. Lambs were infected with cyst at different intervals, and it was concluded that prepatent period was about 80—96 d.

Habitat: Rumen, reticulum.

Hosts: *Alcelaphus buselaphus, A. lewel, Aepyceros melampus, Antidorcal marsupialis, Bos primigenius* f. taurus, *Bubalus arnee* f. bubalis, *Capra aegagrus* f. hircus, *Connachaetes taurinus, Damaliscus albifrons, Hippotragus equinus, H. niger, Kobus ellipsiprymnus, K. kob, K. leche, Ovis ammon* f. aries, *Redunca redunca, Taurotragus oryx, Tragelaphus angasi, T. scriptus, T. strepsiceros, Syncerus caffer* (Bovidae); *Muntiacus muntjak, Axis axis, Cervus unicolor* (Cervidae).

Distribution: Africa — Zambia, Zimbabwe, Kenya, Cameroon, Angola, Central Africa, Chad, South Africa, Zaire, Senegal, Mozambique; Asia — Vietnam, U.S.S.R., India, Philippines, Sri Lanka, Papua New Guinea, Indonesia, (Celebes), Taiwan, Korea, China, Japan, Malaysia, Solomon Islands, Thailand, Burma; The Caribbean — Cuba; and Australia, New Zealand, Mariana Islands, Fiji, New Caledonia.

References: Anderson,[29] Asadov,[43] Bali et al.,[67] Caeiro,[154] Chen et al.,[168] Chinone,[169] Crusz,[191] Cruz e Silva,[195] Davydova,[200] Durie,[254] Eduardo,[265,273] Graber,[357,360] Graber et al.,[362,363] Gretillat,[371] Gupta,[383] Hovorka et al.,[425] Hsu,[426] Jain and Kamalapur,[445] Jonathan,[452] Kadenatsii,[459] Kelly and Henderson,[477] Kotrlá and Prokopič,[499] Mukherjee,[631] Näsmark,[647] Ortlepp,[678] Owen,[685] Penso,[710] Prokopič and Kotrlá,[752] Roth and Dalchow,[791] Seneviratna,[828] Sey,[838,844,853] Swart,[934] Tandon and Sharma,[959] Varma,[1011] Vassiliades,[1013] Velitcho,[1022-1024] Wang,[1050] Yamaguti.[1075]

Calicophoron bothriophoron (Braun, 1892) Eduardo, 1983 (Figure 455)

Syn. *Amphistomum bothriophoron* Braun, 1892

Diagnosis: Body length 5.4—7.6 mm, body width 2.2—2.9 mm. Acetabulum 0.89—1.53 mm in diameter, Pisum type, number of muscle units, d.e.c.1, 8—17; d.e.c.2, 6—9; d.i.c., 8—12; v.e.c., 10—15; v.i.c., 32—41; m.e.c., 13—23, (Figure 175). Pharynx 0.51—0.73 mm long, Calicophoron type (Figure 58); esophagus 0.52—0.75 mm long without muscular thickening. Anterior testis 0.69—0.94 mm long and 0.68—1.42 mm in dorsoventral direction. Pars musculosa well developed, convoluted. Ovary 0.31—0.64 by 0.35—0.52 mm. Terminal genitalium Bothriophoron type (Figure 99). Genital pore postbifurcal. Eggs 0.131—0.158 by 0.060—0.076 mm.

Biology: Not yet known.

Habitat: Rumen and reticulum.

Hosts: *Bos primigenius* f. taurus, *Capra aegagrus* f. hircus, *Kobus defassa, Ovis ammon* f. aries, *Redunca fulvorufula, Syncerus caffer* (Bovidae).

Distribution: Africa — South Africa, Malagasy, Chad, Cameroon, Tanzania, Mauritius, Somalia, Central Africa, Kenya; India.

References: Daynes,[206] Dinnik,[227] Eduardo,[273] Fischoeder,[310] Graber,[357,359,360] Gretillat,[368] Näsmark,[647] Sey,[846] Sey and Graber,[864] Vegli.[1018]

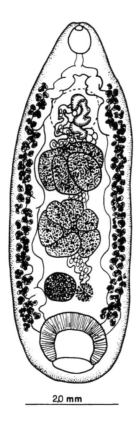

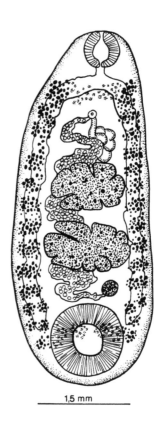

2,0 mm   1,5 mm

FIGURE 455. *Calicophoron*
bothriophoron (Braun, 1892). (By
permission of Dr. W. Junk Publishers, 1983, Dordrecht.)

FIGURE 456. *Calicophoron*
microbothrium (Fischoeder,
1901).

Calicophoron microbothrium (Fischoeder, 1910) Eduardo, 1983 (Figure 456)

Syn. *Paramphistomum microbothrium* Fischoeder, 1901

Diagnosis: Body length 3.1—13.0 mm, body width 1.4—4.3 mm. Acetabulum 1.28—2.42 mm in dorsoventral direction, Paramphistomum type, number of muscle units, d.e.c.1, 10—17; d.e.c.2, 23—39; d.i.c., 29—53; v.e.c., 10—19; v.i.c., 39—64; m.e.c., 9—15 (Figure 173). Pharynx 0.38—1.42 mm long, Calicophoron type (Figure 58); esophagus 0.38—0.82 mm long, without muscular thickening. Anterior testis 0.13—1.86 mm long 0.23—2.58 mm in dorsoventral direction; posterior testis 0.68—2.13 mm long and 1.26—2.93 mm in dorsoventral direction. Pars musculosa strongly developed and convoluted. Ovary 0.48—0.83 by 0.52—0.98 mm. Terminal genitalium Microbothrium type (Figure 122). Genital pore postbifurcal. Eggs 0.152—0.216 by 0.075—0.096 mm.

Biology: References. Dinnik and Dinnik,[229,233] Lengy,[536] Horak,[422] Prod'hon et al.[750]

Preparasitic stage. ① Dinnik and Dinnik. Freshly deposited eggs contained an embryo at an early stage of cleavage (3 to 5 cells). Embryonic development took place within 14—16 d at 26—28°C. Newly hatched miracidia measured 0.22—0.29 by 0.068—0.081 mm. Body covered with ciliated epidermal cells in four rows and arranged by the formula 6:8:4:2. Central cavity of miracidia contained germinal tissue, germinal cells, and germinal balls. Intermediate host: *Bulinus tropicus* ② Lengy. Freshly laid eggs contained either zygote or embryo in early cleavage (4—8 cells). By the 10th day inner structural elements of miracidia developed, and a mucoid plug was seen around the opercular end of eggs. Hatching began on the 17th day at 28°C. Freely swimming miracidia measured 0.20—0.30

by 0.044—0.066 mm. Body covered with ciliated cells in four rows and arranged by the formula 6:8:4:2. Internal structure of the miracidiae consisted of an apical gland, two pairs of penetration glands, a nervous system, a pair of flame cells with their ducts, and germinal tissue. Longevity of miracidia in tap water and at 24°C was as long as 17 h. Intermediate host: *Bulinus truncatus*, (Lengy),[536] *Bulinus liratus* (Prod'hon et al.).[750]

Intramolluscan stage. ① Dinnik and Dinnik. Sporocyst usually situated in the tissue of the respiratory cavity or in the body space surrounding the intestine. Young sporocyst measured 0.32 mm long, fully developed sporocyst 3.6 mm long. Sporocyst completed its development by the 25th day after exposure to miracidia. First rediae emerged 14 d after infection of the snail. Mature rediae developed by the 25th to 36th days postinfection, and they measured 0.55—0.75 by 0.17—0.22 mm. Pharynx 0.046—0.049 mm in diameter, sacculate cecum 0.15—0.22 mm long. Birth pore situated at level of posteior half of cecum. Mother rediae usually contained daughter rediae (1 to 3 or more) in the anterior part of body cavity and embryo balls (15 to 25) in its posterior part. No morphological differences were found between young rediae of the first and those of the second generation. Mother rediae died after completion of development of all embryo balls into daughter rediae. First cercariae emerged 28 to 30 d after exposure of snail to miracidia. Four redial generations were observed; in an artificially infected snail 186 small young rediae, 395 rediae with embryo balls, 19 containing daughter rediae and embryo balls, 185 rediae contained developing cercariae, 10 aged rediae with daughter rediae, and 760 developing and developed free cercariae. Cercariae liberate in an immature form and continue their development in the snail tissue. Mature cercariae measure 0.40—0.58 by 0.25 mm; tail 0.80 mm in length. Body heavily pigmented and cystogenous cells with rods present. Eyespots situated just posterior to the oral sucker. Acetabulum 0.114—0.137 mm in diameter. Pharynx 0.068—0.082 mm in diameter. Main excretory ducts with cross-connecting tubes. Caudal tube expanding near to tip of tail and has two lateral orifices. Adolescariae 0.20—0.25 mm in diameter. Most of the infected snails surveyed for about 6 months. ② Lengy. Miracidia, situated in the mantle cavity began their penetration to the snail tissue. Complete process of penetration lasted about 40 min. Penetrated miracidia retained their epidermal cells and were shedded some hours later. Mature sporocysts were observed from the 10th day onward after exposure to miracidia, the largest sporocyst measured 0.34 by 0.55 mm and contained 87 rediae and embryo balls. Rediae began their development from embryo balls forming in sporocyst. Rediae were observed both in the body space surrounding the digestive tract or within hepatopancreas, ovotestis, and mantle tissue. Rediae reached their maturity about 10 d after liberation from sporocyst. Body measured 0.72—1.13 by 0.12—0.76 mm. Pharynx 0.040—0.059 by 0.044—0.059 mm. Cecum 0.119—0.239 mm long. In close connection with the gut and pharynx, closely packed salivary glands present. There were successive rediae generations, but it was difficult to detect their exact number due to their overlapping. First cercariae usually emerge by the 23rd day after infestation. They were liberated in an immature stage, and they contained their development in the snail tissue. Mature cercariae were present in the hepatopancreas. Body measured 0.20—0.53 by 0.17—0.35 mm, tail 0.67—0.81 by 0.070—0.094 mm. The latter with a mucoid fin, divided into three sections. Body of cercariae covered with pigment granules, except for three areas in a dorsal view. Two eyespots on the dorsal side and a great number of cystogenous rods present. Acetabulum 0.090—0.138 by 0.075—0.091 mm. Pharynx 0.044—0.062 by 0.047—0.110 mm, esophagus 0.040 mm in length, ceca extending to acetabulum. Main excretory tubes with cross-connecting tubes. Fourteen pairs of flame cells were found in immature cercariae. Caudal tube expands near the tip of tail and has two openings to the opposite lateral surfaces. After release of cercariae from infected snails they commenced to encyst. Adolescariae dome-shaped, measuring 0.14—0.20 by 0.22—0.26 mm. Viability of adolescariae persisted for at least 2 months.

Development in definitive host. ① Dinnik and Dinnik. Calf was infected with 3000 adolescariae, and post-mortem examination was made on the 82nd day, and 1665 young flukes were recovered. In experimentally infected cattle, flukes reached their maturity and began to pass out eggs in about 100 d postinfection. This species reached their full size and achieved their maximum egg production after 5—9 months of development. There was no noticeable increase in the size of specimens after this. Longevity of this fluke was postulated so that it may be a rather long period, exceeding 7 years. Egg production of a calf, infected with 3360 adolescariae, increased gradually and reached its maximum (133 EPG) on the 150th day after infestation. Seven years later this experimentally infected cattle henceforward passed out eggs (80 EPG). ② Lengy. One lamb was infected with 1500 adolescariae and was sacrificed 48 d later. A total of 60 young flukes were recovered (percentage take 4%; in another lamb, 45.0%). Flukes reached their maturity in lamb 89 d after infection with adolescariae. ③ Horak. It was found that 38.0—90.0% of adolescariae excysted in the alimentary tract of domestic ruminant. Prepatent period was longest in sheep (71 d) and shortest in cattle (56 d). Overcrowding caused stunted growth, and massive infestation delayed migration of worms both in sheep and in cattle.

Habitat: Rumen and reticulum.

Hosts: *Alcelaphus lelwel, Alcelaphus buselaphus, Adenota leucotis, Aepyceros melampus, Bos primigenius* f. taurus, *Bubalus arnee* f. bubalis, *Capra aegagrus* f. hircus, *Capra nubiana, Damaliscus dorcas, D. korrigum, Gazella gazella, G. thomsoni, G. rufifrons, Hippotragus equinus, Kobus defassa, K. kob, K. leche, K. varondi, K. ellipsiprymnus, K. leucotis, Ovis ammon* f. aries, *Ourebia ourebi, Oryx gazella, Redunca arundinum, Redunca redunca, Syncerus caffer, Taurotragus oryx, Tragelaphus angasi, Capra ibex* (Bovidae); *Camelus dromedarius* (Camelidae).

Distribution: Africa — Egypt, Kenya, Zaire, Chad, Ethiopia, Angola, Zambia, South Africa, Tanzania, Cameroon, Central Africa, Zimbabwe, Sudan, Uganda, Senegal, Botswana, Mozambique, Lesotho; Europe — Italy (Sardinia), Portugal, France; Asia — Iraq, Israel, Iran, and Malagasy.

References: Arfaa,[34] Caeiro,[154] Cruz e Silva,[195] Diaw et al.,[217] Dinnik,[228] Dinnik and Dinnik,[230] Dinnik et al.,[234] Eduardo,[273] Eisa,[284] Ezzat,[295] Fitzsimmons,[319] Graber,[357,358,360] Graber et al.,[362,363] Graber and Thal,[365] Gretillat,[371] Kurtpinar and Latif,[515] Leiper,[534] Lengy,[536] Näsmark,[647] Ortlepp,[678] Prod'hon et al.,[750] Prokopič and Kotrlá,[752] Prudhoe,[756] Reinhardt,[772] Roach and Lopes,[780] Roth and Dalchow,[791] Schillhorn et al.,[821] Sey,[838,840,845,846] Sey and Arru,[863] Sey and Eslami,[866] Swart,[934] Wandera.[1045]

Calicophoron papillosum (Stiles et Goldberger, 1910) Eduardo 1983 (Figure 457)

Syn. *Paramphistomum papillosum* Stiles et Goldberger, 1910, *Calicophoron zhejiangense* Wang, 1979

Diagnosis: Body length 4.0—7.5 mm, body width 2.1—3.3 mm. Acetabulum 1.47—2.01 mm in dorsoventral direction, Calicophoron type, number of muscle units, d.e.c., 16—21; d.i.c., 31—38; v.e.c., 13—17; v.i.c., 39—47; m.e.c., 4—9 (Figure 165). Pharynx 0.80—0.87 mm long, Calicophoron type (Figure 58); esophagus 0.42—0.59 mm long, without muscular thickening. Anterior testis 0.31—1.01 mm long and 0.57—1.74 mm in dorsoventral direction, posterior testis 0.35—1.01 mm long and 0.61—2.32 in dorsoventral direction. Pars musculosa well developed, convoluted. Ovary 0.17—0.29 by 0.25—0.44 mm. Terminal genitalium Papillogenitalis type (Figure 91). Genital pore bifurcal. Eggs 0.129—0.132 by 0.056—0.063 mm.

Biology: Not yet known.

Habitat: Rumen.

Hosts: *Bos primigenius* f. taurus, *Bubalus arnee* f. bubalis, *Ovis ammon* f. aries (Bovidae).

Distribution: Asia — India, Indonesia, Bangladesh, Vietnam, Iran, Pakistan.

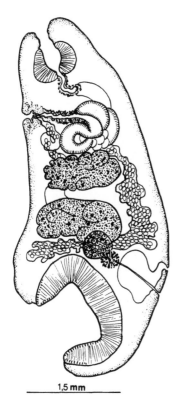 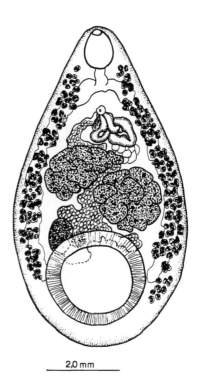

FIGURE 457. *Calicophoron papillosum* (Stiles et Goldberger, 1910) (median sagittal section). (By permission of Dr. W. Junk Publishers, 1983, Dordrecht.)

FIGURE 458. *Calicophoron raja* Näsmark, 1937. (By permission of Dr. W. Junk Publishers, 1983, Dordrecht.)

References: Eduardo,[273] Gupta,[388] Gupta and Nakhasi,[396] Khan,[483] Näsmark,[647] Sey,[853] Sey and Eslami,[866] Stiles and Goldberger,[915] Tandon and Maitra.[956]

Calicophoron papilligerum (Stiles et Goldberger, 1910) Eduardo, 1983

Syn. *Paramphistomum papilligerum* Stiles et Goldberger, 1910

Diagnosis: Body length 8.5, body width 4.3 mm. Acetabulum 1.88 mm in transverse diameter. Pharynx oval, internal surface papillated, esophagus without muscular thickening. Testes large, lobed, tandem in position. Pars musculosa well developed, convoluted. Ovary submedian, preacetabular. Genital opening postbifurcal. Eggs 0.135 by 0.067 mm.

Biology: Not yet known.

Habitat: Forestomach.

Host: *Cervus eldi* (Cervidae).

Distribution: Not known (probably India).

References: Eduardo,[273] Stiles and Goldberger.[915]

Calicophoron raja Näsmark, 1937 (Figure 458)

Diagnosis: Body length 4.1—11.2 mm, body width 3.6—5.2 mm. Diameter of acetabulum 0.98—3.16 mm in dorsoventral direction, Pisum type, number of muscle units, d.e.c.1, 9—15; d.e.c.2, 5—8; d.i.c., 42—51; v.e.c., 8—14; v.i.c., 31—40; m.e.c., 11—21 (Figure 175). Pharynx 0.39—0.98 mm long, Calicophoron type (Figure 58); esophagus 0.69—0.76 mm long, without muscular thickening. Anterior testis 1.16—1.68 mm long and 1.98—3.26

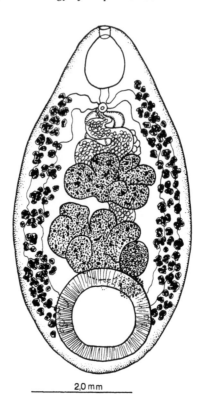

2,0 mm

FIGURE 459.   *Calicophoron clavula* (Näsmark, 1937). (By permission of Dr. W. Junk
Publishers, 1983, Dordrecht.)

mm in dorsoventral direction; posterior testis 1.28—2.13 mm long and 2.16—3.68 mm in
dorsoventral direction. Pars musculosa well developed, convoluted. Ovary 0.38—0.79 by
0.49—0.73 mm. Terminal genitalium Raja type (Figure 94). Genital pore postbifurcal. Eggs
0.132—0.165 by 0.069—0.083 mm.

Biology: Dinnik and Hammond[235] found that *Physopsis globosus* proved to be susceptible
to miracidia tested under laboratory conditions.

Habitat: Rumen, reticulum rarely.

Hosts: *Aepyceros melampus, Alcelaphus buselaphus, Bos primigenius* f. taurus, *Capra ae-*
gagrus f. hircus, *Connochaetus taurinus, Damaliscus korrigum, Gazella thomsoni, Kobus*
defassa, K. leche, K. varondi, Ovis ammon f. aries, *Oryx gazella, Redunca redunca,*
Syncerus caffer, Taurotragus oryx, Tragelaphus strepsiceros, T. scriptus (Bovidae).

Distribution: Africa — Kenya, Tanzania, Chad, South Africa, Botswana, Zambia, Namibia,
Zimbabwe, Sudan; The Caribbean — Cuba.

References: Dinnik,[227] Dinnik and Dinnik,[230] Eduardo,[273] Graber,[357] Näsmark.[647]

Calicophoron clavula (**Näsmark, 1937) Eduardo, 1983 (Figure 459)**

Syn. *Paramphistomum clavula* Näsmark, 1937

Diagnosis: Body length 4.3—8.9 mm, and 2.2—42. mm in dorsoventral direction. Acetabulum 1.48—2.67 mm in diameter, Paramphistomum type, number of muscle units,
d.e.c.1, 12—31; d.e.c.2, 19—33; d.i.c., 39—53; v.e.c., 10—19; v.i.c., 35—59; m.e.c.,
15—28 (Figure 173). Pharynx 0.68—1.43 mm in length, Calicophoron type (Figure 58);
esophagus 0.36—0.68 mm. Anterior testis 0.48—1.36 mm long and 0.96—2.58 mm in
dorsoventral direction, posterior testis 0.48—1.76 mm long and 1.02—2.36 mm in dorsov-

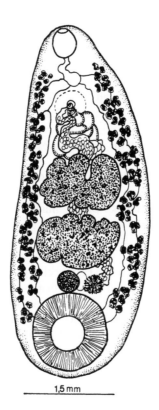

1,5 mm

FIGURE 460. *Calicophoron microbothrioides* (Price et Mc-Intosh, 1949). (By permission of Dr. W. Junk Publishers, 1983, Dordrecht.)

entral direction. Ovary 0.18—0.83 by 0.39—0.86 mm. Pars musculosa well developed, convoluted. Terminal genitalium Clavula type (Figure 104). Genital pore bifurcal. Eggs 0.173—0.164 by 0.071—0.089 mm.

Biology: Not yet completely known. Sobrero[896] found that intermediate host of this species was *Bulinus abyssinicus*. Measurements of mature cercariae fixed in alcohol: 0.41—0.46 by 0.30—0.32, tail 0.50—0.56 mm in length. Adolescariae 0.34—0.38 mm in diameter.

Habitat: Rumen.

Hosts: *Aepyceros melampus, Alcelaphus buselaphus, Bos primigenius* f. taurus, *Capra aegagrus* f. hircus, *Hippotragus equinus, H. niger, Kobus kob, K. varondi, Ourebia ourebia, Redunca redunca, Synerus caffer, K. defassa, K. leucotis* (Bovidae).

Distribution: Africa — Zaire, Chad, Somalia, Tanzania, Central Africa, Sudan, Niger, Egypt, Nigeria, Uganda.

References: Eduardo,[273] Gentile et al.,[338] Graber and Thal,[365] Güralp and Oğuz,[402] Manter and Pritchard,[589] Näsmark,[647] Prokopič and Kotrlá,[752] Sey.[845]

Calicophoron microbothrioides (Price et McIntosh, 1949) Eduardo, 1983 (Figure 460)

Syn. *Paramphistomum microbothrioides* Price and McIntosh, 1944; *Ceylonocotyle petrovi* Davydova, 1961; *Cotylophoron cotylophorum* of Bennett, 1936; *Paramphistomum microbothrium* of Chen et al. 1985.

Diagnosis: Body 3.8—7.6 mm and 1.7—2.6 mm in dorsoventral direction. Acetabulum 1.2—1.7 mm in diameter, Pisum type, number of muscle units, d.e.c.1, 13—18; d.e.c.2,

3—11; d.i.c., 31—41; v.e.c., 10—17; v.i.c., 29—43; m.e.c., 12—20 (Figure 175). Pharynx 0.40—0.72 mm long, Calicophoron type (Figure 58); esophagus 0.20—0.76 mm long, with muscular thickening. Anterior testis 0.48—0.92 mm long and 1.12—1.76 mm in dorsoventral direction, posterior testis 0.48—1.12 mm long and 1.12—1.87 mm in dorsoventral direction. Pars musculosa well developed, convoluted. Ovary 0.28—0.49 by 0.35—0.69 mm. Terminal genitalium Microbothrium type (Figure 122). Genital pore postbifurcal. Eggs 0.11—0.13 by 0.65—0.73 mm.

Biology: References. Krull,[510] (under the name *Cotylophoron cotylophorum*) Colvin,[180] Samnaliev and Vassilev,[809] Dvoryadkhin et al.[259]

Preparasitic stage. ① Bennett. Freshly laid eggs contained zygote without cleavage or in second cleavage stage. Embryonic development took place within 13 to 15 d in summer at laboratory temperatures. Living miracidia measure 0.15—0.21 mm by 0.032—0.063 mm. Body covered with ciliated epidermal cells, situated in four rows and arranged by the formula 6:8:4:2. Inner structure of miracidia composed of two pairs of penetration glands, a single cephalic gland, (primitive gut), brain mass, two pairs of flame cells with their ducts and germinal tissue (germ cells, germ balls). Longevity of miracidia 8 to 10 h at room temperature. Colvin[180] found that the life span of miracidia was longer in natural media (12.19 and 13.54 h) than in some artificial ones. ② Samnaliev and Vassilev. Miracidia hatched in a great number, incubated at 27—28°C between the 11th and 14th days. They measure 0.13—0.17 by 0.032—0.046 mm. ③ Dvoryadkhin et al.[259] Embryonic development lasted 14 d at 26 to 27°C. Longevity of miracidia 24 h.

Intermediate host: *Fossaria* (= *Lymnaea*) *parva*, *F.* (= *Lymnaea*) *modicella* (Bennett,[79] Krull)[510] *Lymnaea trucatula* (Samnaliev and Vassilev),[809] *Anisus minusculus* (Dvoryadkhin et al.)[259]

Intramolluscan stage. ① Bennett. Miracidia retained their epidermal cells after penetration. They gradually developed into sporocyst; the latter reached their maturity 10 to 15 d after penetration. Mature sporocyst 0.27—0.47 by 0.13—0.28 mm. Sporocyst contained developing rediae up to 9 in number; they liberate by rupturing of the body wall. Rediae when liberated have all the structures except for birth pore. Mature redia 0.55—1.01 by 0.12—0.22 mm. Pharynx 0.042—0.050 mm in diameter, cecum extending to anterior fifth of body. Excretory system consisted of three pairs of flame cells. Birth pore situated at about 0.16—0.18 mm from anterior body end. Mass of salivary glands well developed. Germinal tissue located in the posterior extremity of the body. Number of cercariae varied from 10 to 23. Cercariae were shed 15 d after infestation in a very immature condition. Development continued in the snail tissue. Mature cercariae escaped from the snail 32 d after infestation. Body measurements 0.30—0.49 by 0.15—0.25 mm, tail 0.58—0.70 in length. Acetabulum 0.043 mm in diameter. Pigmentation distributed over entire body surface. Eyespots located lateral to and posterior to oral sucker. Oral sucker 0.047 by 0.046 mm, esophagus with muscular thickening, ceca extending to acetabulum. Main tubes of excretory system have cross-connecting tubes. Caudal tube expanded at the tip of tail and had two (a posterior and a lateral one) short tubes reaching the surface. After escaping, cercariae remain free swimming from 10 to 60 min. Cyst 0.154 mm in diameter. Under optimal conditions cysts live for several months. ② Samnaliev and Vassilev. Mature rediae measure 0.55—1.18 by 0.14—0.32 mm. Pharynx 0.062 by 0.040—0.059 mm, esophagus 0.009—0.025 mm, cecum 0.074—0.13 mm long. Birth pore was 0.15—0.25 mm from the anterior body end. Cercariae liberated from rediae in an immature form. Mature cercariae were shed 40 to 42 d after infestation. Cercaria belongs to the Pigmentata group. Body 0.14—0.25 by 0.17—0.25 mm in size, tail 0.46—0.68 mm in length. Eyespots at level of esophagus. Acetabulum 0.050—0.057 by 0.032—0.050 mm. Adolescariae measure 0.192—0.224 by 0.174—0.223 mm. Main excretory ducts have cross-connecting tubes. Cercariae tail provided with finfold. ③ Dvoryadhkin et al. Mature

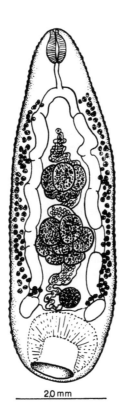

FIGURE 461. *Calicophoron
sukari* (Dinnik, 1954). (Courtesy
of N. N. Dinnik.)

sporocyst measure 0.27 by 0.14 mm and contain from 1 to 3 mother redia. Mature redia
0.59 by 0.20 mm in size. Cercariae, excreted by snails, were densely pigmented. Body
measurements 0.19 by 0.20 mm, tail 0.44 mm long. Adolescaria 0.151 mm in diameter.

Development in definitive host. ① Bennett. Calves were infected with 150 metacercariae. Maturity was reached in about $3^1/_2$ months. Habitat: Rumen.

Hosts: *Bos primigenius* f. taurus, *Bison bison*, *Bubalus arnee* f. bubalis, *Ovis ammon* f.
aries (Bovidae); *Cervus nippon*, *C. unicolor*, *Muntiacus muntjak* (Cervidae).

Distribution: North America — U.S.; The Caribbean — Cuba, Puerto Rico; Asia — U.S.S.R.,
Vietnam; Europe — Bulgaria.

References: Chen et al.,[168] Davydova,[201] Dvoryadkhin et al.,[259] Eduardo,[273] Herd and Hull,[415]
Kamburov et al.,[463] Lee et al.,[528] Nazarova,[646] Prestwood et al.,[740] Price and McIntosh,[747]
Sey,[836,845,854] Velitchko.[1020,1022]

Calicophoron sukari (Dinnik, 1954) Eduardo, 1983 (Figure 461)

Syn. *Paramphistomum sukari* Dinnik, 1954

Diagnosis: Body length 5.8—9.4 mm, body width 1.4—3.3 mm. Acetabulum 1.23—1.54
mm diameter in dorsoventral direction, Calicophoron type, number of muscle units, d.e.c.,
11—17; d.i.c., 28—39; v.i.c., 10—14; v.i.c., 26—43; m.e.c., 4—8 (Figure 165). Pharynx
0.43—0.79 mm long, Calicophoron type (Figure 58); esophagus 0.45—0.82 mm long
without muscular thickening. Anterior testis 0.53—1.28 mm long and 1.12—2.36 mm in
dorsoventral direction, posterior testis 1.36—1.52 mm long and 1.31—2.16 mm in dorsoventral direction. Pars musculosa well developed and convoluted. Ovary 0.22—0.59 by
0.41—0.69 mm. Terminal genitalium Microbothrium type (Figure 122). Genital pore postbifurcal. Eggs 0.125—0.162 by 0.068—0.089 mm.

Biology: References. Dinnik and Dinnik,[229,231] Dinnik and Hammond.[235]

Preparasitic stage. ① Dinnik. Freshly laid eggs contained embryo in early morula stage. Miracidia developed and hatched within 15—16 d at temperatures of 26—27°C. Body of miracidia measured 0.20—0.23 by 0.049—0.059 mm and was covered with ciliated epidermal cell in four rows, arranged by the formula 6:8:4:2. Inner structures of miracidia were a pair of flame cells and 20 to 30 germinal cells in the middle part of the body cavity. Intermediate host: *Biomphalaria pfeifferi.*

Intramolluscan stage. ① Dinnik. Measurements of sporocyst, examined 19 d after infection of snails, were 0.84 by 0.41 mm. First free rediae were found on the 14th day postinfection. First generation rediae measured 1.0—1.8 by 0.18—0.30 mm. Pharynx 0.036—0.050 mm in diameter and the cecum 0.12 mm long. Two successive redial generations were observed. Cercariae liberated from rediae in an immature form and continued their development in the snail tissue. Mature cercariae heavily pigmented, measuring 0.25 mm in length and 0.17 mm in width. First cercariae excreted from the infected snail 35 d after exposure to miracidia. Main excretory tubes have cross-connecting tubes. Caudal tube with a slight expansion near the tip of body's end, and they opened with two lateral openings.

Habitat: Rumen and reticulum.

Hosts: *Bos primigenius* f. taurus, *Capra aegagrus* f. hircus, *Ovis ammon* f. aries, *Syncerus caffer* (Bovidae).

Distribution: Africa—Kenya, Tanzania, Angola, Central Africa, Ethiopia, Zambia, Uganda, Zimbabwe.

References: Dinnik,[227,228] Dinnik and Dinnik,[230,231] Eduardo,[273] Graber,[358] Gretillat,[372] Roth and Dalchow,[791] Sachs and Sachs.[802]

Calicophoron phillerouxi (Dinnik, 1961) Eduardo 1983 (Figure 462)

Syn. *Paramphistomum phillerouxi* Dinnik, 1961; *P. vangrenbergeni* Strydonck, 1970; *P. togolense* Albaret, Bayssade-Dufour, Guilhon, Kulo et Picot, 1978.

Diagnosis: Body length 4.8—11.4 mm, body width 1.6—3.7 mm. Acetabulum 1.28—2.47 mm in dorsoventral direction, Paramphistomum type, number of muscle units, d.e.c.1, 12—22; d.e.c.2, 14—28; d.i.c., 35—51; v.e.c., 10—24; v.i.c., 38—50; m.e.c., 8—12 (Figure 173). Pharynx 0.62—1.13 mm long, Calicophoron type (Figure 58); esophagus 0.47—0.75 mm long without muscular thickening. Anterior testis 0.57—1.26 mm long and 0.75—2.76 mm in dorsoventral direction, posterior testis 0.98—1.2 mm long and 2.16—2.87 mm in dorsoventral direction. Pars musculosa moderately developed and convoluted. Ovary 0.48—0.59 by 0.62—0.89 mm. Terminal genitalium Microbothrium type (Figure 122). Genital pore postbifurcal. Eggs 0.128—0.149 by 0.063—0.079 mm.

Biology: References. Dinnik,[225] Dinnik and Hammond,[235] Albaret et al.[16]

Preparasitic stage. ① Dinnik. Freshly laid eggs contained embryo in early cleavage stage. Miracidia developed and hatched in 15 d at 26°C. Body of miracidia measured 0.20—0.23 mm in length and was covered with ciliated epidermal cells, situated in four tiers, and arranged by the formula 6:8:4:2. Inner structure of miracidia composed of a pair of flame cells; germinal tissue, situated in the middle part of body cavity. The special structure of the miracidia was characterized by the two granulated masses of protoplasm extending from the posterior extremity along the dorsal and ventral sides of body cavity. Floating, large, round cells were seen in these protoplasmic bodies. The significance of these cells has not yet been known. Intermediate hosts: *Bulinus forskalii, B. cercinus, B. senegalensis.* ② Albaret et al. Miracidia fixed measured 0.20 by 0.040 mm. Body covered with 20 ciliated epidermal cells, arranged by the formula 6:8:4:2. On the apical extremity of the terebratorium there were argentophilic structures situated along three axes (T$_1$: one central and two lateral; T$_2$: two large and 5 to 6 smaller; T$_3$: 6 smaller formations). Inner

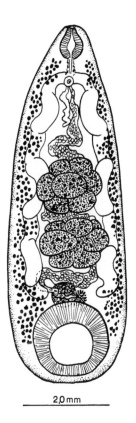

FIGURE 462. *Calicophoron phillerouxi* (Dinnik, 1961). (Courtesy of N. N. Dinnik.)

2.0 mm

structure of miracidia composed of penetration glands, apical gland, two flame cells with their ducts and nerve mass which has trunks from the central mass to anteriorly, laterally, and posteriorly. Intermediate host: *Bulinus (B.) forskalii*.

Intramolluscan larval stage. ① Dinnik. Sporocysts measured 0.56 by 0.45 mm, by the 16th day of development, contained 16 to 23 young rediae and several embryo balls. First rediae emerged from the 12th day onward. Mature rediae measured 0.72—0.91 mm long, included several developing cercariae embryos and embryo balls. Pharynx 0.043—0.046 by 0.033—0.043 mm. Cercariae liberated from rediae in an immature form. They reached their maturity in the snail tissue. Mature cercariae heavily pigmented, with two eyespots and about 0.30 mm length. Tail was about twice the length of body. Main excretory tubes have cross-connecting tubes. First cercariae were shed 42 d after exposure of snail to miracidium. Cercarial production varied between 150 and 459 cercariae daily. Encysted cercariae 0.17 to 0.19 mm in diameter. ② Albaret et al. Sporocyst after 10 to 14 d infestation was V-shaped, measured 1.40—2.4 mm. Redia, after 22 d infestation measured 0.39—0.77 mm. Pharynx 0.040 mm in diameter, cecum 1/5 to 1/6 of the total body length. First cercariae emerged on the 26th day after infestation. Body 0.40—0.45 mm in diameter, tail longer than body length. Pharynx 0.050 mm in diameter, acetabulum 0.10 mm in diameter. Main excretory tubes with cross-connecting tubes, caudal one with apical expansion and with two lateral openings. Free-swimming cercariae encysted soon, and cysts were similar to those of other species of the genus. Along the body surface and the tail a great number of superficial argentophilic structures were observed, arranged in a special system and consideration (Table 1).

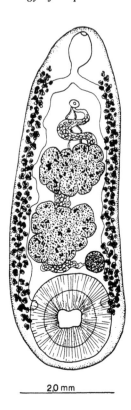

2.0 mm

FIGURE 463. *Calicophoron daubneyi* (Dinnik, 1962). (By permission of the *Acta Vet. Acad. Sci. Hung.*, 1974, Budapest.)

Habitat: Rumen.

Hosts: *Aepyceros melampus, Bos primigenius* f. taurus, *Damaliscus korrigum, Hippotragus equinus, H. niger, Kobus kob, K. defassa, K. wardoni, Ovis ammon* f. aries, *Redunca redunca, R. arundinum, Syncerus caffer, Tragelaphus strepsiceros* (Bovidae).

Distribution: Africa — Zimbabwe, Chad, Tanzania, Niger, South Africa, Central Africa, Zambia, Zaire, Togo, Uganda, Mauritius, Sierra Leone, Kenya, Senegal.

References: Albaret et al.,[16] Diaw et al.,[217] Dinnik,[225] Dinnik et al.,[234] Eduardo,[273] Graber,[360] Graber and Thal,[365] Roth and Dalchow,[791] Sey,[846] Strydonck.[920]

Calicophoron daubneyi (Dinnik, 1962) Eduardo, 1983 (Figure 463)

Syn. *Paramphistomum daubneyi* Dinnik, 1962

Diagnosis: Body length, 3.12—10.86 mm, body width 1.8—2.9 mm. Acetabulum 1.89—2.87 mm, in diameter, Paramphistomum type, number of muscle units, d.e.c.1, 13—19; d.e.c.2, 19—26; d.i.c., 29—54; v.e.c., 12—17; v.i.c., 37—45; m.e.c., 8—14 (Figure 173). Pharynx 0.82—1.28 mm long, Calicophoron type (Figure 58); esophagus 0.42—0.73 mm long. Anterior testis 1.38—2.97 mm long and 2.46—2.97 mm in dorsoventral direction, posterior testis 0.98—1.23 mm long and 2.48—2.93 mm in dorsoventral direction. Pars musculosa well developed, convoluted. Ovary 0.53—0.62 by 0.52—0.85 mm. Terminal genitalium Microbothrium type (Figure 122). Genital pore postbifurcal. Eggs 0.129—0.142 by 0.063—0.078 mm.

Biology: References. Dinnik,[226] Sey.[842]

Preparasitic stage. ① Dinnik. Eggs laid contained embryo with one- to four-cell stage. Embryonic development took place within 15 d when eggs were kept at a temperature of 26°C. Miracidium 0.20 mm long, covered with ciliated epidermal cells, situated in four rows, and arranged by the formula 6:8:4:2. In the body cavity germinal cells present. Intermediate host: *Lymnaea truncatula.* ② Sey. Freshly laid eggs contained embryo in an early cleavage stage. Embryonic development lasted 3 to 10 d at 27°C. Eggs with fully developed miracidia had the so-called mucoid plug. Freely swimming miracidia measured 0.15—0.22 by 0.034—0.055 mm, coverred with ciliated epidermal cells, arranged by the formula 6:8:4:2. Inner structure of miracidia consisted of a single cephalic gland, brain mass, a pair of flame cells, two pairs of penetration glands, and germinal tissue. Longevity of miracidia kept in tap water at room tempeature was 9—10 h. Intermediate hosts: *Lymnaea truncatula* and *L. peregra* (with very low susceptibility).

Intramolluscan larval stage. ① Dinnik. After penetration into snails germ cells develop into germinal balls, and later into rediae in sporocysts. Sporocysts measure 0.37 by 0.23 mm in size and contain 25—35 embryos or developing rediae. Mature rediae 0.55—0.84 mm long, contain up to six developing cercariae and several embryo balls and embryo cells. Cercariae are shed in immature condition and continue their development in the snail tissue. First cercariae emerged on the 60th day after infection. These cercariae are heavily pigmented and with two eyespots. ② Sey. After penetration of miracidium into snail it developed into sporocyst. Mature sporocyst measured 0.60—0.70 by 0.20—0.26 mm. Rediae emerged from sporocyst on the 9th day after infection. Mature redia measures 0.45—0.68 by 0.18—0.25 mm. Pharynx 0.030—0.034 mm in diameter, gut 0.080—0.095 mm in length, salivary glands present. Three pairs of flame cells also present. Germinal cavity of redia filled with embryos of different stages: 8—10 developing cercariae, 17—32 embryo balls and several germinal cells. Birth pore exists at anterior fifth of anterior body end. First cercariae liberated on the 16th day after infection. Mature cecariae 0.28—0.40 by 0.16—0.24 mm in size, tail 0.19—0.30 mm in length. Body heavily pigmented and covered with cystogenous cells and rods. There are two pairs of eyespots on the dorsal surface. Acetabulum 0.080—0.125 mm in diameter. Pharynx 0.40—0.075 mm in diameter, esophagus 0.034 mm long, ceca extending to acetabular zone. Main excretory tubes have cross-connecting tubes and caudal tube terminating at caudal tip with two lateral openings. Primorida of reproductive organs seen as groups of cells. Adolescariae 0.20 mm in diameter.

Development in definitive host. ① Sey. Kid was infected with 1500 2-week-old adolescariae. First eggs were found in feces 12 and 13 weeks after infection. Post-mortem examination revealed 536 flukes (take 35.7%).

Habitat: Rumen.

Hosts: *Bos primigenius* f. taurus, *Bubalus arnee* f. bubalis, *Capra aegagrus* f. hircus, *Ovis ammon* f. aries, *Oryx beisa, Ovis ammon, Oryx gazella* (Bovidae).

Distribution: Europe — Hungary, Bulgaria, Germany, France, Greece, Italy, Turkey, Rumania, Yugoslavia, Czechoslovakia, Albania; Africa — Kenya, Ethiopia, Algeria, Somalia.

References: Daynes and Graber,[207] Dinnik,[226] Eduardo,[273] Graber et al.,[362] Erhardová,[287] Graber,[358] Graber and Thal,[365] Kamburov and Osikovski,[464] Kosaroff and Mihailova,[498] Kotrlá et al.,[500] Kotrlá nd Kotrlý,[503] Lepojev and Cvetković,[543] Moskvin,[623] Pačenovský et al.,[690] Pavlović,[709] Sey,[836,840,841,845] Sey and Visnyakov,[862] Sey and Arru.[863]

Calicophoron sukumum (Dinnik, 1964) Eduardo, 1983 (Figure 464)

Syn. *Paramphistomum sukumum* Dinnik, 1964

Diagnosis: Body length 7.1—9.3 mm, body width 2.5—3.7 mm. Acetabulum 1.52—2.43 mm in dorsoventral direction, Pisum type, number of muscle units, d.e.c.1, 16—22; d.e.c.2, 4—10; d.i.c., 34—52; v.e.c., 13—21; v.i.c., 44—63; m.e.c., 16—24 (Figure 175). Pharynx 0.78—1.12 mm long, Calicophoron type (Figure 58); esophagus 0.58—0.64 mm long, without muscular thickening. Anterior testis 1.24—1.48 mm long and 1.21—2.16 mm in

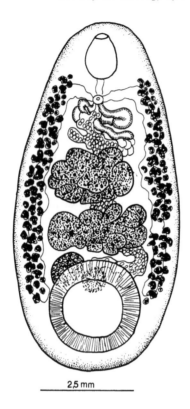 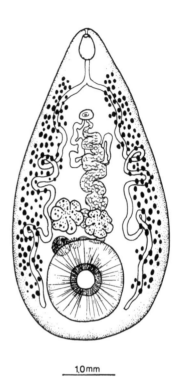

FIGURE 464. *Calicophoron sukumum* (Dinnik, 1964). (By permission of Dr. W. Junk Publishers, 1983, Dordrecht.)

FIGURE 465. *Calicophoron wuchengensis,* Wang, 1979. (After X. Y. Wang, 1979.)

dorsoventral direction, posterior testis 1.02—3.18 mm long and 1.87—2.76 mm in dorsoventral direction. Pars musculosa well developed and convoluted. Ovary 0.48—0.76 by 0.59—0.98 mm. Terminal genitalium Microbothrium type (Figure 122). Genital pore bifurcal. Eggs 0.139—0.154 by 0.058—0.079 mm.

Biology: Not yet known.

Habitat: Rumen.

Hosts: *Bos primigenius* f. taurus, *Connochaetes taurinus, Damaliscus korrigum, Hippotragus equinus, Kobus defassa, K. leche, Redunca arrundinum, Syncerus caffer, Taurotragus oryx* (Bovidae).

Distribution: Africa — Tanzania, Zambia, Zimbabwe.

References: Dinnik,[227] Eduardo,[273] Roth and Dalchow,[791] Sachs and Sachs.[802]

Calicophoron wuchengense Wang, 1979 (Figure 465)

Diagnosis: Body length 5.1—7.0 mm, body width 2.4—3.6 mm. Acetabulum 1.64—1.92 by 1.76—2.08 mm, Calicophoron type (Figure 165). Pharynx 0.64—0.80 by 0.64—0.78 mm, Calicophoron type (Figure 58); esophagus 0.45 mm long, ceca strongly convoluted, ending behind acetabulum. Testes horizontal, right testis 0.48—0.72 by 0.40—0.72 mm, left testis 0.48—0.64 by 0.40—0.80 mm. Pars musculosa strongly developed and convoluted. Ovary 0.22—0.40 by 0.27—0.40 mm. Terminal genitalium Calicophoron type (Figure 103). Genital pore postbifurcal. Eggs 0.089—0.099 by 0.056—0.077 mm.

Biology: Not yet known.

Habitat: Forestomach.

Host: *Bubalus arnee* f. bubalis (Bovidae).

Distribution: Asia — China.
Reference: Wang.[1050]

KEY TO THE SPECIES OF *CALICOPHORON**
(after Eduardo, 1983)

1. Testes horizontal and relatively small .... *C. wuchengense*
   Testes not horizontal and relatively big 2
2. True ventral atrium present 3
   True ventral atrium absent 4
3. Papillae present on wall of ventral atrium.......................... *C. papilligerum*
   Papillae absent on wall of ventral atrium .... *C. bothriophoron*
4. Terminal genitalium of Calicophoron type .... *C. calicophorum*
   Terminal genitalium of Raja type .... *C. raja*
   Terminal genitalium of Clavula type...................................... *C. clavula*
   Terminal genitalium of Papillogenitalis type .... *C. papillosum*
   Terminal genitalium of Microbothrium type 5
5. Esophageal bulb present .... *C. microbothrioides*
   Esophageal bulb absent 6
6. Blind cecal ends directed dorsally 7
   Blind cecal ends directed ventrally... 8
7. Cecal ends meet medially, bigger terminal genitalium .... *C. microbothrium*
   Cecal ends do not meet medially, smaller terminal genitalium........ *C. phillerouxi*
8. Posterior part of internal surface of esophagus lined by ciliated epithelium, vitellaria
   not confluent dorsomedially... *C. sukari*
   Posterior part of internal surface of esophagus not lined by ciliated epithelium; vitellaria
   not confluent dorsomedially... 9
9. Acetabulum of Paramphistomum type; circular elevated area around genital pore
   absent... *C. daubneyi*
   Acetabulum of Pisum type; circular elevated area around genital pore present.........
   ....*C. sukumum*

3.3. PHYLOGENETIC RELATIONSHIPS AMONG AMPHISTOMES

The conception of the phylogenetic systematics, formulated by Henning,[411] renders classification of organisms possible on the basis of their observable traits. At the present state of the amphistome investigations, the holomorphological (macro- and histomorphological and semaphoront traits), ecological (intermediate and definitive hosts), and distributional data offer useful characters in the analysis of phylogenetic relationships.

Whatever character states are taken into account, the most valuable seem to be those which (1) show an altering series of character states within the taxon, (2) show alterations of the characters that are related to one another, and (3) are not adaptive in nature. In the phylogenetic analysis the weighting of the character states at hand has primary importance. Examinations of the evolutionary history of the amphistomes show that the evolutionary events have affected the aspects of their way of life which were connected with the reproduction, excretion, feeding, and ecological segregation (habitat segregation and formation of body) of the species. According to these functions, the character states can be divided into five categories, and in each, primary, secondary, and tertiary ones were differentiated. The reason for the qualification of these character states is to assess the dissimilar information content they bear. The phylogenetic analyses of the amphistomes were based on 17 primary,

---

* *Lorisia cardonae* Penso, 1940 spec. inquirenda.

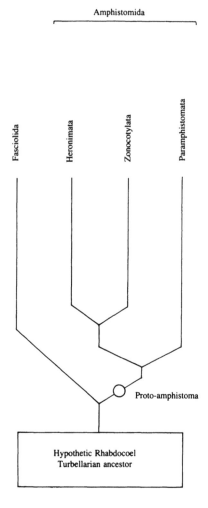

FIGURE 466. Phylogenetic relationships among higher taxa of Amphistomida.

16 secondary, and 31 tertiary character states.[856,859] The special traits of the primary characters (apomorphic) substantiate the monophiletic status of the taxon in question (in-group states); their generalized traits (plesiomorphic), however, connect the taxon studied to another taxon having the same taxic level (out-group states). The secondary character states indicate the transformation series among the studied taxa. The tertiary character states are rather the manifestations of the specific states of the progressive developmental lines. The cladograms, which depict the phylogenetic relationships of certain groups of the amphistomes, are based on the synapomorphic states, borne by the secondary and tertiary character traits.

By analyzing the character states it is possible to make a distinction between the characters of general significance (plesiomorph) and those of special significance, (apomorph) which is the first step in establishing phylogenetic relationships. By combining all traits at hand in such a manner, the cladograms, based on these weighted characters, will prove to be the best representation of phylogenetic relationships. The Amphistomida, a monophyletic group of the digenetic trematodes, shows closer similarity to Fasciolida than to that of any other groups of the digenetic trematodes. The out-group traits which contact these two taxa are the large size of the operculated eggs and the freely encysted cercariae (Figure 466). The examinations of the character states of trematodes, now regarded to be the monophiletic group of amphistomes, show three taxa having the same taxic level (suborder) which can be differentiated: Heronimata Skrjabin et Schultz, 1937; Zonocotylata Sey, 1988; and Paramphistomata Szidat, 1936. Their phylogenetic relationships are depicted in the cladogram (Figure 466).

Heronimata is a monotypic taxon; its taxonomic rank is determined by the monostome plan of body and by the adults' special morphological structure as well as by their ancient type of reproduction.

Zonocotylata is a reduced taxon, including one genera and its two species. It has morphological characters of its own (reproductive system, fixative apparatus, etc.) justifying its taxonomic position in the systematics of the Amphistomata in the present sense.

The most successful evolutionary line of amphistomes is the taxon Paramphistomata whose representatives are parasitic in each grade of the vertebrate definitive hosts. In the following pages examples to demonstrate the phylogenetic relationships in certain groups of amphistomes of various taxic levels will be introduced.

The first dichotomy within the suborder Paramphistomata is determined by the presence or absence of the cirrus sac. This state designates the superfamily Cladorchoidea Skrjabin, 1949, the ancient branch of the suborder and the Paramphistomoidea Stiles et Goldberger, 1910, the derived sister group of Paramphistomata. An analysis of the morphological similarities of the species included by the taxon Cladorchoidea, five monophyletic groups, with equal taxonomic level have been separated: Cladorchiidae Skrjabin, 1949; Diplodiscidae Skrjabin, 1949; Gastrodiscidae Stiles et Goldberger, 1949, Balanorchiidae Ozaki, 1957; Brumptiidae Skrjabin, 1949. The sister group of Paramphistomoidea includes the taxa Zygocotylidae Sey, 1988; Gastrothylacidae Stiles et Goldberger, 1910; and Paramphistomidae Fischoeder, 1901. These taxa are defined by clearly cut morphological and/or biological characters, and their phylogenetic relationships are depicted in the cladogram (Figure 467).

Bearing the phylogenetic value of the structure of the reproductive system in mind, we proposed a new systematic scheme for the Cladorchiidae[859] in which the position of the testes was considered as the main guiding principle in establishing of the transformation series: tandem, diagonal, horizontal. Accordingly, the taxon Cladorchiidae was divided into 20 subfamilies. The splitting of the family into a relatively great number of subfamilies is in connection with the ancestry of the species included. At the same time, the morphological similarities did not give a solid basis for elevating any one of the subfamily series to family rank. The phylogenetic relationships of the subfamilies are shown in the cladogram (Figure 468).

The taxonomic structure of the derived developmental line, Paramphistomoidea formularized by Sey,[859] contains three families (Zygocotylidae, Gastrothylacidae, and Paramphistomidae) including some subfamilies, several genera, and species. The phyletic relationships of the taxa representing dissimilar levels are indicated in the cladogram (Figure 469). Finally, the cladogram (Figure 470) illustrates the phylogenetic relationships of the species, the taxa Paramphistominae.

Besides examinations of the phylogenetic relationships, based on holomorphological traits, the cytogenetic analysis can give further information to recognize tendencies in the evolutionary history of the amphistomes.

The specific constancy of the the number of chromosomes and their morphological traits has held out promises of the more precise identification of closely related species, and the pattern and process of chromosomal changes often provide clues to phylogenetic relationships. Data which refer to amphistomes are summarized in Table 10.

Before analyzing data as tabulated, some of their weak points should be pointed out. Of the amphistomes a relatively small number of species has been studied karyologically; they are 12.2% of the total sum of species and were confined mainly to amphistomes parasitic in mammals. No information is available on the karyotype of fish amphistomes, and data

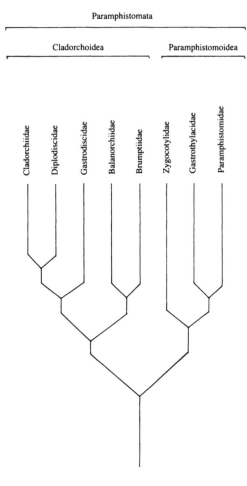

FIGURE 467.   Phylogenetic relationships of taxa of Paramphistomata.

are also inadequate on amphibians and reptiles. Some of the quoted species seem to be misidentified (it is very probable that neither *Paramphistomum cervi*[1027] nor *Calicophoron microbothrioides*[449] are found in India[843]) and nomenclaturally do not follow the present standard. There can be seen considerable differences not only in the types of chromosomes, but also in the chromosome number even in the same species which are well defined gross morphologically like *Gastrothylax crumenifer* or *Fischoederius elongatus*. These diverse results can be attributed to methods applied by various authors during examinations.

Bearing these remarks in mind, the author can confine himself to moderate conclusions only in his analyses.

The haploid number of chromosomes varies between 6 and 14. Heteromorphism was observed in some cases (*Megalodiscus*, No. 10,[378] *Calicophoron microbothrium*, No. 9),[639] and it was reported by Romanenko[787] that chromosomes of No. 1, 7, and 9 of *P. leydeni* of two different populations showed different structures.

Of the species studied the most frequent haploid number is 9 (54.8%), then 7 (22.5%), and 8 (19.3%), respectively. Hence, the modal number can be regarded to be 9, or, on the basis of the three most frequent chromosome numbers, it is 8. When various species of amphistomes possess same number of chromosomes, the chromosome architecture furnishes stable ground for differentiation (e.g., *Paramphistomum* spp., Table 10).

Karyotype analyses have revealed a newer scope of specific features which seem to be

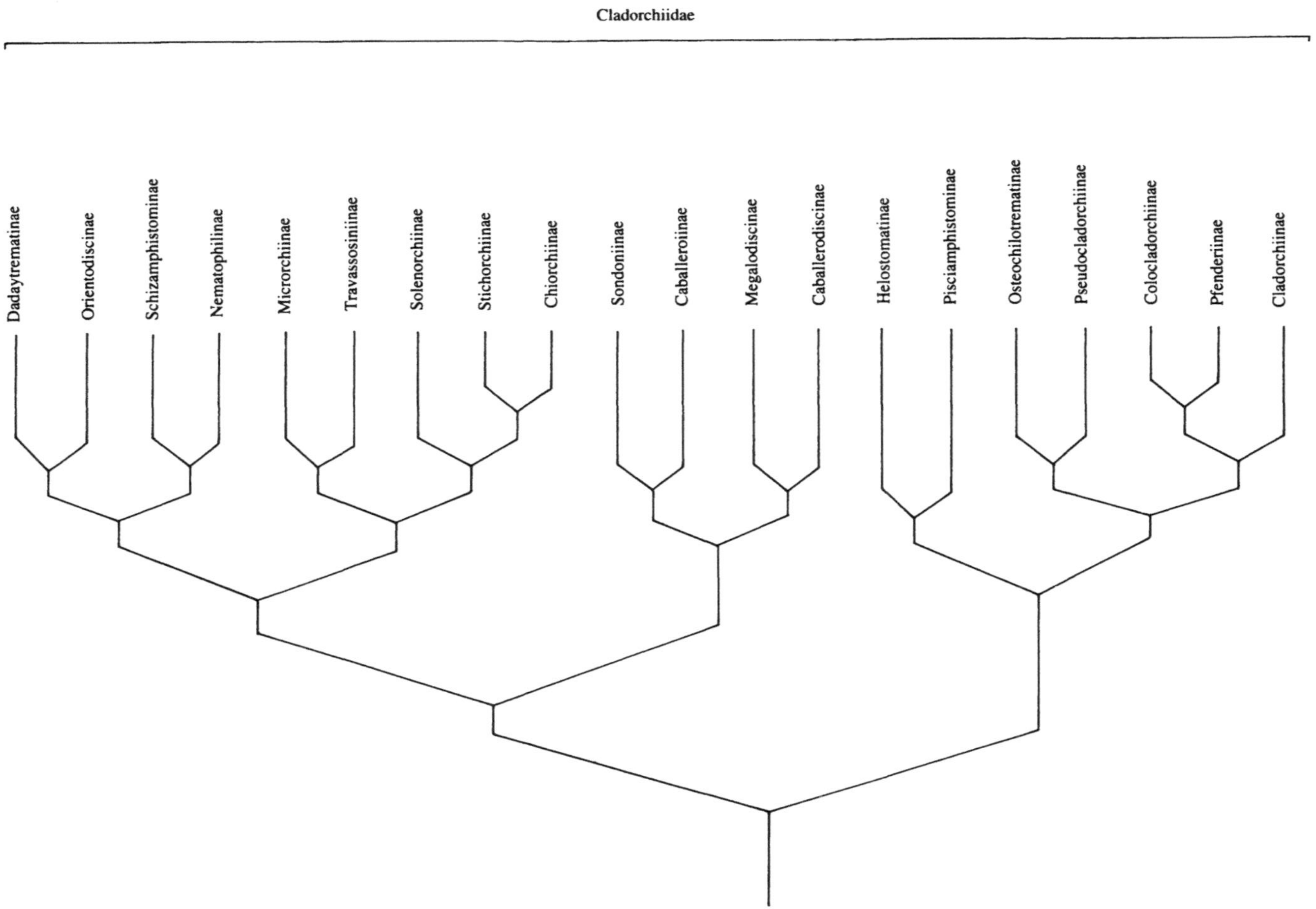

FIGURE 468. Phylogenetic relationships among subfamilies of Cladorchiidae.

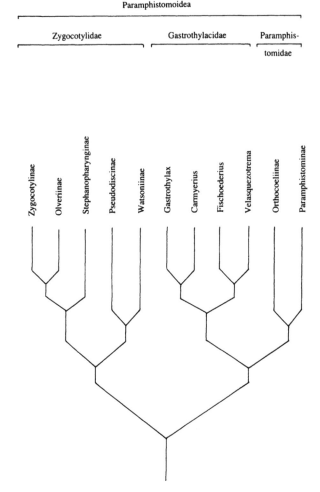

FIGURE 469. Phylogenetic relationships among taxa of Paramphisto-moidea.

more reliable and substantial than those of morphological characters upon which the present systematics of amphistomes have mainly been based. With the gradual increase of the data of karyotype structure of further amphistomes and reexamination of the literary information by uniform guiding principles, the karyosystematics will supply an important contribution to form the phylogenetic systematics of amphistomes.

Examining evolutionary relationships between species of various animals and their karyotypes within a phylogenetic group usually shows that the ancient forms are characterized by their possessing a higher number of acrocentric and telocentric chromosomes ("primitive karyotype"). Further evolution of chromosome number and form often was accompanied to form karyotypes with biarmed chromosomes. Such tendencies can be observed in amphistomes also from ancient forms to advanced species. In the amphistomes belonging to the ancient groups the ratio of the acrocentric/telocentric chromosomes is higher (*Diplodiscus amphichrus*, *Stichorchis subtriquetrus*; Table 10) in accordance with the general tendency. In species of the advanced groups the picture is much more variable; in certain species (*Orthocoelium* spp., *Calicophron microbothrium*, e.g.) the ratio of the biarmed chromosomes is higher than the acrocentric and telocentric ones together, but in other species of these groups this ratio shows irregular distribution. The latter results might be, however, in connection with methodological deficiencies.

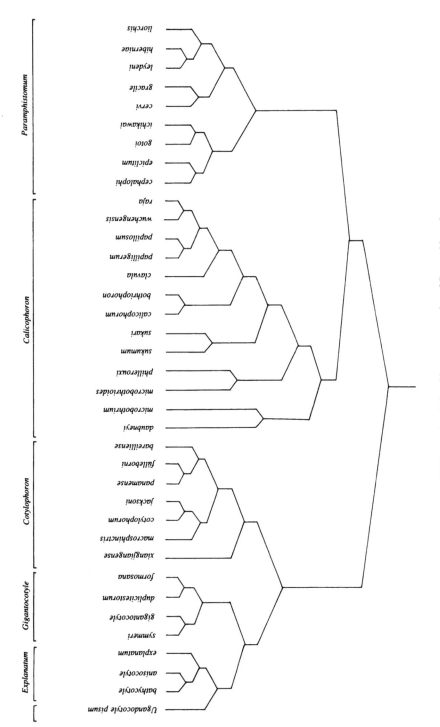

FIGURE 470. Phylogenetic relationships among species of Paramphistominae.

TABLE 10
Chromosome Number of Amphistomes

	2n	Metacentric	Submetacentric	Acrocentric	Telocentric	Ref.
Heronimus chelydrae (= *H. mollis*)	20	—	—	—	—	380
Megalodiscus temperatus	20	—	—	18	2	378
	16	—	—	—	—	159
	18	—	—	—		1009
Stunkardia dilymphosa	18	6	2	10	—	875
Stichorchis subtriquetrus	18	2	2	4	10	787
Diplodiscus amphichrus	18	2	6	6	4	818
	20					874
Zygocotyle lunata	14	4	—	10	—	1061
Gastrodiscoides hominis	14	—	—	—	—	449
Gastrothylax crumenifer	14	—	—	—	—	212
	18	4	6	2	6	929
	18	12	6	—	—	873
Fischoederius elongatus	16	6	6	—	4	873, 1029
	22	4	—	18		449
F. cobboldi	18	6	—	12	—	449
Orthocoelium dawesi	20	—	—	—	—	873
O. scoliocoelium	22	—	—	—	—	873
O. orthocoelium	14	—	—	—	—	873
O. dicranocoelium	18	4	10	4	—	1028
	18	4	10	—	4	929
Gigantocotyle (= *Explanatum*) *explanatum*	18	2	5	2	—	929, 1026
	18					774
Gigantocotyle (= *Explanatum*) *bathycotyle*	12	—	—	—	—	1064
Cotylophoron (= *Orthocoelium*) *elongatum*	16	—	—	—	—	213
C. indicum (= *Paramphistomum epiclitum*)	28	—	—	—	—	873
Calicophoron calicophorum	14	6	—	8	—	449
C. microbothrium (= *daubneyi*)	14	6	5	3	—	832
C. cauliorchis	18	—	—	—	—	873
C. crassum	14	—	—	—	—	910
	18	4	—	14	—	449
C. microbothrium	18	2	8	8	—	639
C. microbothrioides	16	4	—	12	—	449
Paramphistomum cervi	14	4	10	—	—	929, 1027
	18	2	3	—	4	775
P. hiberniae	16—18	—	—	—	—	1063
P. epiclitum	18	4	14	—	—	929
	18	—	—	—	—	873
P. ichikawai	18	2	4	12	—	787
Liorchis scotiae (= *Paramphistomum leydeni*)	18	2	6	10	—	786
	18	4	—	14	—	787
Paramphistomum scotiae (= *P. leydeni*)	12—16	—	—	—	—	1063
Cotylophoron cotylophorum	16	4	10	—	2	929

It is postulated that amphistomes, e.g., Digenea, originated from an ancient stock of Rhabdocoela. The species of rhabdocoeles possess a haploid number of chromosomes n = 2, 4, 8, 16, and 20[1044] in the majority of cases. There is evidence that both an increase and a decrease in chromosome number took place during speciation in Digenea by the usual cytogenetic mechanisms. The rhabdocoele ancestry with n = 4, 8, or 16 chromosomes seems to be the most probable and potential stock to develop into digenetic ancestor with n = 8 through all the possible mechanisms of aneuploidy.[449] Altogether, the karyotype analysis based on reliable results seems to be a useful means of studying phylogenetic relationships in groups of species having only neontological character states, as amphistomes have.

4. ZOOGEOGRAPHY OF THE AMPHISTOMES

4.1. GENERAL ACCOUNTS

There are three main theories available for zoogeographers of today in an analysis of animal distribution patterns. Their cardinal problems are to give answers to such questions as the origin of the center, role, and direction of dispersal.

Darlington[199] summarized the priniciples of the model, which are regarded as the vacuum or colonization theory, and he outlined clues to an analysis of historical zoogeography. Croizat[189] and his followers (e.g., Platnik and Nelson)[729] offered another alternative, the panbiogeographic model, which rejects essentially the theory of colonization. Brundin[129] has developed further the panbiogeographic model and formulated the paradigm of the phylogenetic biography. This theory emphasizes that (1) the population of the ancestral species gradually broke up due to local barriers, (2) the ancestral species gradually increased its distributional range, and (3) the processes, vicariance, and dispersal are corollaries of each other. According to this theory the origin of the area of the monophyletic group has always coincided with the area of the ancestral species. The role and direction of the dispersal is the peripheral apomorphy.

In the analysis of the distributional pattern of amphistomes the statements of phylogenetic biogeography have been used as general guidelines. Although the principles of these theories provide valuable clues for zoogeographical analysis even of the parasites, parasitologists have often been left to their own devices to explain distributional patterns of various parasites.

The distribution of the parasites, generally speaking, has also been governed by those factors which have been influenced in the dispersal of other animal groups. Specificities have to be stressed, however, originating from their parasitic way of life and the consequences that follow from this (viz., the environment of a parasite is double in nature: it is the host and the host's environment). It is evident from the dixenous type of the amphistomes' life history that the range of their distribution is limited by the range of the hosts required for completion of the life cycle and by environmental factors.

From the obligatory host-parasite relationships it has been well known among parasitologists for a long time that the recognition of a given parasite includes, at the same time, two other items of information: the type of host(s) and the distributional range. Of the helminthologists, Manter[587,588] studied more profoundly the regularities of host-parasite relationships and the distribution of parasites. He connected the distribution of the parasites with narrowness of the specificity of the host-parasite relationships and stated that (1) distribution of parasites is limited by the range of their hosts; (2) the narrower the specificity, the older the host-parasite relationship; (3) hosts have a greater variety of helminths in those areas where they have been living longer. It follows from the principles above that the distributional pattern of parasites can be concluded from the host-parasite relationship and the distribution of the hosts. The center of origin is thought to be the area having the highest diversity, endemicity, and specialization of parasites.

Amphistomes are parasitic mainly in freshwater and land vertebrates, but some taxa have adapted themselves to the marine way of life. The group Amphistomida, as has been formulated in this book, includes a total of 254 species. Their distribution has been concentrated in the tropical and subtropical areas. The rate of speciation has gradually decreased toward the temperate and boreal belts (Figures 471 and 472). The distributional patterns of the amphistomes of the submammalian hosts show intercontinental distribution (Figure 471) in accordance with the Croizat's generalized tracks, agree with the distribution of the intermediate and definitive hosts, and can be properly interpreted with the phenomenon of the main disjunctions caused by the continental drift. The distributional pattern of the mammalian

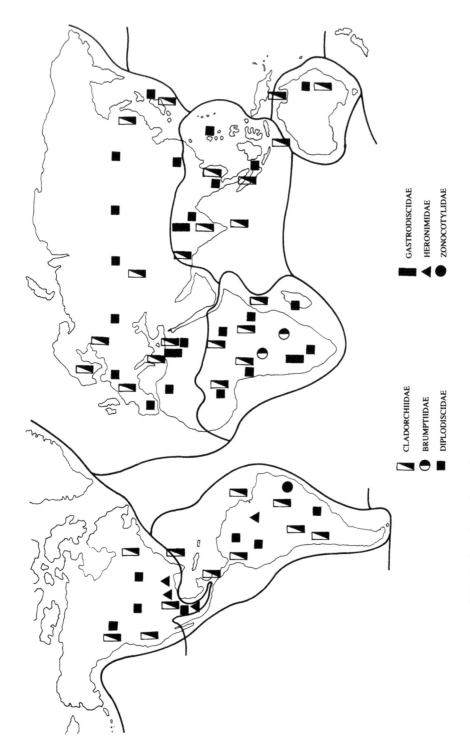

FIGURE 471.   Geographic distribution of members of amphistome families of various vertebrate groups.

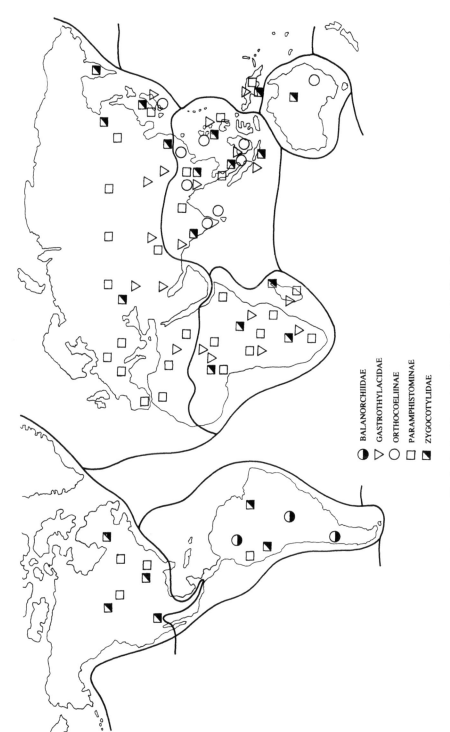

FIGURE 472.   Geographic distribution of members of amphistome families of various vertebrate groups.

BALANORCHIIDAE

GASTROTHYLACIDAE

ORTHOCOELIINAE

PARAMPHISTOMINAE

ZYGOCOTYLIDAE

amphistomes is much more variable; the role of the continental drift is recognizable to a lesser degree, similar to that of their definitive hosts (Figure 472). Along with the generalized tracks, several form-making areas represented by various taxonomic groups can be ascertained. As the evolution of the amphistomes has come into being simultaneously with that of their definitive hosts, understanding of the coevolutionary process can throw some light on the distributional pattern of amphistomes in space and time. In the following examinations species of the indigenous fauna only will be taken into account.

4.2. COEVOLUTIONARY RELATIONSHIPS

4.2.1. THE AMPHISTOMES OF FISHES

These are parasitic in species, with a few exceptions that belong to two major telostean fish groups: Ostariophysi and Acanthopterygii. The most preferred species are the representatives of the Cypriniformes, Siluriformes, and Perciformes (in cyprinids 25, in silurids 17, and in percids 13 amphistome species are parasitic); the other groups of fishes can be ignored in an analysis due to their low number (Table 6). The most important theories on the evolution and distributional pattern of freshwater fishes were outlined by Darlington,[198] Gery,[341] Golsin,[353] Cracraft,[186] Patterson,[706] Myers,[641] and Novacek and Marshall.[659] A short summary is given below.

The primary differentiation of freshwater fish took place on the supercontinent Pangaea (Laurasia and Gondwanaland). There were interchanges in both directions between the northern and southern fish faunas which are shown by the presence of the joint taxa, respectively. There are diverse opinions on the origin and dispersal of the species of ostaryophysid fishes which are the most important definitive hosts of the amphistomes parasitic in fishes. Darlington[198] supposed a South Asian; Gery,[341] an African; Novacek and Marshall,[659] a South American origin. On grounds of their dispersal routes it was supposed that it could happen by way of (1) marine radiation, (2) through land bridges, (3) holarctic continental dispersal, and (4) as a result of the continental drift. The latter theory, represented by Novacek and Marshall,[659] seems to be the most acceptable one because it is firmly in accordance with the connection of the continents in the past.

Accordingly, the ostaryophysid fishes were probably differentiated in South America, and they were divided into two sister groups (Cypriniformes and Siluriformes) at the beginning of the Cretaceous period. They were distributed both in South America and Africa, which was connected with South America by the first transgression of the Thethys (Paleocen), when they penetrated into Eurasia. The taxon Perciformes has several families with Gondwanaland elements and two families (Centrarchidae, Percidae) which contain Laurasian faunal elements. The former one is restricted to North America, the latter to the Laurasian distribution.

It is assumed that the amphistome parasitism began to evolve parallel with the adaptive radiation of their first vertebrate hosts (fish). The ancestor of the modern amphistomes originated presumably from a Rhabdocoele-like predecessor (Proto-amphistoma) whose population was generally distributed on the megacontinent Pangaea (primary cosmopolitanism). The breakup of Pangaea and formation of the continents were completed by the end of the Mesozoic era and the beginning of the Tertiary, which resulted in the great disjunctions of amphistome faunas of fish. In the presence of adequate hosts (e.g., ostariophysid fish), they were successfully colonized by the ancient forms of amphistomes. With the diverse development of the amphistomes of fishes on continents becoming gradually isolated the evolution of the modern forms of amphistomes characteristic of fishes today commenced.

Depicting the distributional pattern on a map showing the configuration of the ancient continents, the precretaceous expansion of the ancient amphistomes is evident (Figure 473).

The present distributional pattern that is ubiquitous in dispersal reflects the generalized

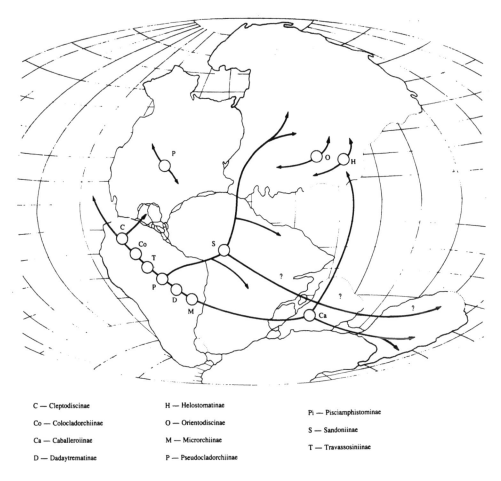

C — Cleptodiscinae	H — Helostomatinae	Pi — Pisciamphistominae
Co — Colocladorchiinae	O — Orientodiscinae	S — Sandoniinae
Ca — Caballeroiinae	M — Microrchiinae	T — Travassosiniinae
D — Dadaytrematinae	P — Pseudocladorchiinae	

FIGURE 473. Vicariant geographic distribution of subfamilies of fish amphistomes.

tracks which existed in the Mesozoic (Figure 474) period. Along the generalized tracks form-making areas are recognizable with specialized (South America: Microrchiinae, Travassosiniinae; North America: Pisciamphistominae; Asia: *Orientodiscus*, Helistominae), endemic (several places), and generalized groups of amphistomes (Dadaytrematinae, Pseudocladorchiinae).

4.2.2. THE AMPHISTOMES OF AMPHIBIANS

These are parasitic in the taxa Anura and Caudata. Their definitive hosts are mainly the species of the Ranidae, Bufonidae, Hylidae, Ambystomidae, and Salamandridae.

The distribution of the amphibians studied by Lynch,[568,569] Trueb,[994] Savage,[817] Kluge and Farris[490] and their results can be summarized as follows. The basic distributional pattern of amphibians derived from two primary sources: one of the two ancient evolutionary centers was on the norther landmass, Laurasia (Ambystomidae, Salamandridae, etc.) and the other on the southern one, Gondwanaland (Ranidae; Bufonidae, Hylidae, etc.). It is postulated that the Ranidae evolved in Africa, and they invaded the New World, South Asia, and Indomalayan-Australian areas. The most probable origin of the hylid frogs was the Gondwanaland and the South American and Central American stocks derived from this area. The North America species shows a relationship with the Asian forms. The taxa of the Bufonidae have a South American origin and reached North America, Eurasia, and Africa by subsequent invasions. The evolutionary history of the taxon Caudata commenced with the northern

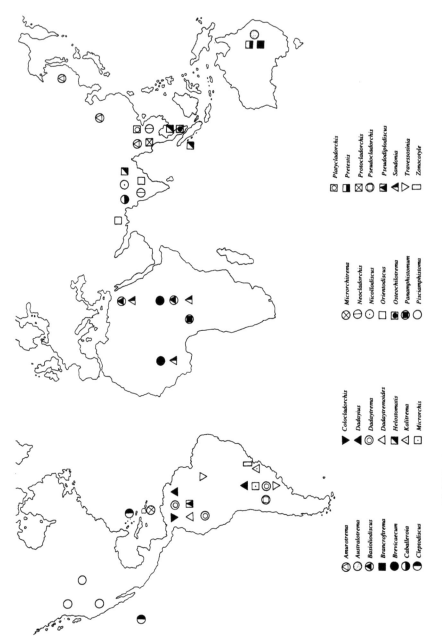

FIGURE 474. Geographic distribution of members of genera of fish amphistomes.

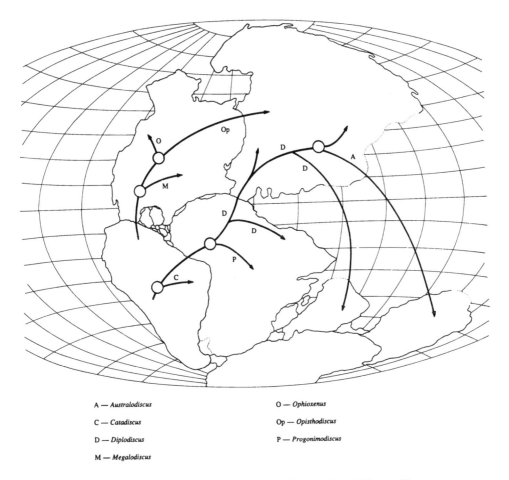

A — *Australodiscus*   O — *Ophioxenus*

C — *Catadiscus*   Op — *Opisthodiscus*

D — *Diplodiscus*   P — *Progonimodiscus*

M — *Megalodiscus*

FIGURE 475.   Vicariant geographic distribution of genera of amphibian amphistomes.

landmass. Ambystomidae is distributed in North America; only Salamandridae is present in both North America and Eurasia.

The acquisition of the amphistomes by amphibians revealed a diverse form from what took place in fishes. The amphistomes of amphibians presumably originated from generalized forms of amphistomes of fishes. One of the developmental lines of the amphistomes of amphibians (Diplodiscidae) is the sister group of a hypothetical pseudodiplodiscuslike stem species, and the other one (Megalodiscinae) is the sister group of the dadaytrematidlike stem species (Figure 475). The form-making areas of both of these two developmental lines are postulated to be in South America (*Catadiscus, Megalodiscus*). From these centers of origin they have been dispersed by way of their amphibian hosts. Their route has been accompanied by extensive diversification: on the one hand, *Megalodiscus, Ophioxenus, Opisthodiscus* and *Catadiscus, Progonimodiscus, Dilpodiscus, Australodiscus* on the other. A possible pathway on the ancient amphistomes is depicted in Figure 475, and the present dispersal is indicated by Figure 476 and Table 7.

4.2.3. THE AMPHISTOMES OF REPTILES

These are parasitic mainly in the taxa of Testudines and to a lesser extent of Squamata (snakes and rattlesnakes). The distribution of the recens turtles is in accordance with the generally accepted theory of the continental drift in the Mesozoic (Cracraft).[186] Chelydae is the single taxon which can be regarded to have Gondwanaland faunal elements with complete

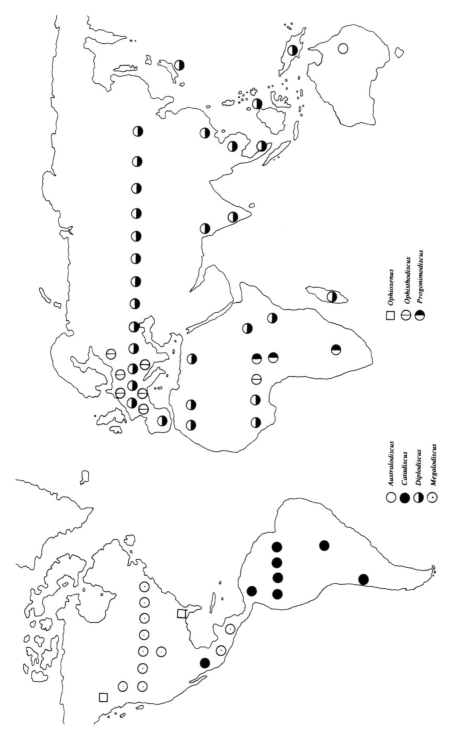

FIGURE 476. Geographic distribution of genera of amphibian amphistomes.

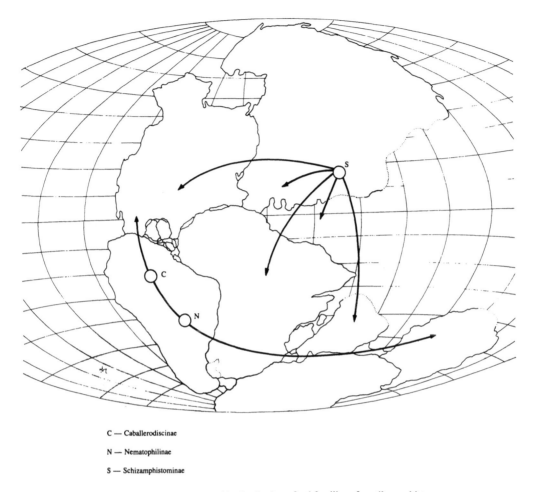

C — Caballerodiscinae

N — Nematophilinae

S — Schizamphistominae

FIGURE 477.   Vicariant geographic distribution of subfamilies of reptile amphistomes.

certainty. Its present distribution is confined to Australia-New Guinea-South America, and on the basis of fossil records this area had always been its dispersal range. The origin of the center of the major part of turtle-bearing amphistomes was either North American or Eurasian forms where they spread to the South, e.g., dermatemydids from Mexico to the northern part of Central America; chelydrids to South America; trionychids to Africa; testudinids now almost cosmopolitan, as they had been previously; chelonids cosmopolitan in the tropical seas. Emydids and testudinids probably originated from the tropical pelumedusids; now they are distributed in the Southern Hemisphere (Africa, Malagasy, South America), but in Cretaceous/Tertiary they were broadly distributed.

The amphistomes of the reptiles are confined mainly to the subfamilies Schizamphistominae, Nematophilinae, and Caballerodiscinae. The species of the latter two subfamilies are distributed in South and Central America (mainly endemic species) and in Australia (*Elseyatrema*). The taxa of Schizamphistominae have intercontinental dispersal with each endemic genera.

When the subcladograms of the three taxa in question are placed on a map indicating the outline of the Cretaceous continents, they show clear-cut intercontinental distribution, vicariance pattern, portraying the direction of the transformation of character states caused by geologic and geographic changes (Figure 477). Two general tracks can be observed: one of them is on Laurasia and the other on Gondwanaland, along with form-making areas. The *a posteriori* analysis of the general tracks shows that the radial center of the southern landmass was in South America. The Australian taxon *Elseyatrema*, which shows close similarity to the Central American taxon, *Caballerodiscus*, has moved along this pathway to its present area (it is in accordance with the phylogenetic ancestry of the definitive hosts and their Gondwanian origin and dispersal).

The center of radiation of the nothern generalized track was the Asian area (based on the high rate of endemicity), and they have reached North America and Africa (mainly by way of emydids). Some of the turtle amphistomes have adapted themselves to the condition of marine existence. The definitive affinity of the schizamphistomids species to emydids and to some of the groups having Laurasian origin (Trionychidae, Cheloniidae) can be explained with ecological factors and with the origin of these turtles. The presence of the sigle schizamphistomid species in Australia (*Lobatodiscus australiensis*) can be regarded as the consequence of a parallel evolution.

In the amphistome fauna of the Middle American turtles, Laurasian (*Pseudocleptodiscus*) and Gondwanian (*Nematophila*) elements can be observed alike. The southern forms supposedly reached the northern territories after formation of the Panama land bridge and vice versa (Figure 478 and Table 8).

4.2.4. THE AMPHISTOMES OF MAMMALS

These are parasitic in taxa of Metatheria and Eutheria, which could have been in connection with biotopes in which the genetically coded life cycle of amphistomes operated. The amphistomes have successfully colonized various groups of mammals (more than half of the species of amphistomes are parasitic in mammals). The high diversity of these amphistomes has been generated by two factors: (1) the adaptive radiation and speciation of the definitive hosts (mainly herbivorous) and (2) colonization of the anterior part (stomach) of the digestive tube. Adaptation of the new habitat has been accompanied by extensive speciation (78.2% of amphistomes of mammals are parasitic of forestomach) and with diversification (three separate families).

The history of the evolution of mammals can be traced back to the Cretaceous. As a result, further development of the taxa Eupantotheria, the Metatheria, and Eutheria and their lines has evolved. The major radiation of both groups took place in the upper Cretaceous, but the development of the larger modern forms of mammals originated from the beginning of the Tertiary; hence, the continental drift affected their distributional patterns to a lesser extent.

Of the mammals, a great number of species became adequate definitive hosts, but, of course, the preferred groups proved to be herbivorous ones (Table 9). Hence, the evolutionary history of these groups (Bovidae, Cervidae, and the omnivorous Suidae) presented shortly, but from the zoogeographic point of view amphistomes of other mammals may serve with confirming examples.

Bovids are a geologically younger group of even-toed ungulates. They originated on the northern landmass, and their evolutionary history began in the Pliocene and subsequent periods. In the Tertiary there was an ancient bovid fauna, ample in species, along the Himalayan territory, which invaded Africa through the Afro-Arab plate in the Miocene. The adaptive radiation of them in Africa was accompanied by extensive speciation and resulted in several new forms of amphistomes. Some of the bovids reached North America (e.g., *Bison*), but they never yielded so many species and groups as that of the Africans or the Asians.

The evolution of the cervids commenced in Asia, and several developmental lines evolved both in Eurasian and North America in the Miocene. After the emergence of the Panama land bridge, they penetrated into South America.

The suids both of the Old (Suidae) and the New World (Tayassidae) evolved in Eurasia

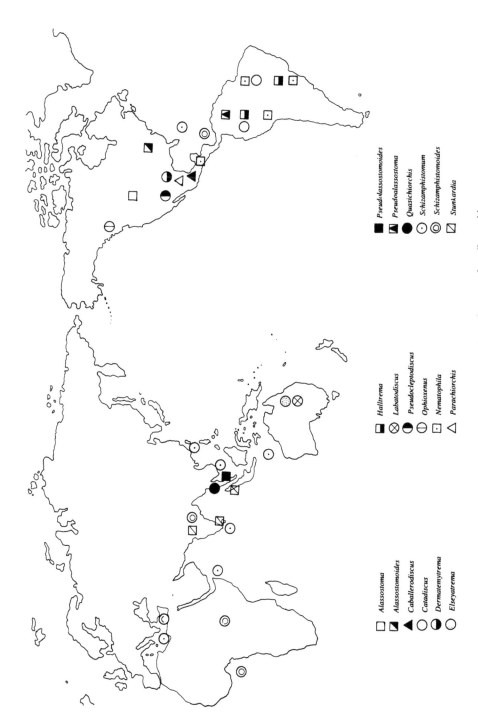

FIGURE 478. Geographic distribution of members of genera of reptile amphistomes.

in the early Oligocene, and their independent developmental lines diverted in the Oligocene. The tayassids reached South America in the Pleistocene.

Mammals have various groups of amphistomes which represent diverse developmental levels from the phylogenetic point of view. Some of them bear plesiomorphic (Balanorchiidae, Cladorchiidae), the others mixed plesiomorphic and apomorphic (*Zygocotyle, Gastrothylax*), and apomorphic (Paramphistominae) characters.

In the distributional patterns of the amphistomes of mammals, the formation of endemic areas, disjunct distributions caused by the continental drift, and dispersal brought about by the manifold moving of the definitive hosts can be recognized.

Of the eight developmental lines of the taxa Cladorchiidae (Figure 479), the species *Stichorchis subtriquetrus* and *Chiorchis fabaceus* show amphiatlantic dispersal while *Solenorchis* is distributed along the Indian and the eastern basin of the Pacific Ocean in general. The other taxa have limited distributional areas, and they show phyletic relationships with the local amphistome faunas (e.g., *Nematophila, Cladorchis, Taxorchis, Stichorchis* in South America or *Bancroftrema, Elseyatrema, Macropotrema, Gemellicotyle* developmental line in Australia).

The distribution of the Balanorchiidae-Brumptiidae taxa is confined to the southern continents (Figure 480). The monotypic *Balanorchis* is distributed in South America and can be considered an evolutionary relic of the formerly existing (Pliocene-Pleistocene) mammalian fauna rich of species. Species of Brumptiidae are endemic in Africa. The gastrodiscids have Afro-Asiatic distribution; they probably invaded Africa by dispersal of the equids, and they were also adapted to Africa suids (Figure 480).

Although the taxon Zygocotylidae is a uniform group from an anatomic point of view, their distributional patterns, however, show a varied picture with clearly seen homoplasy (Figure 481).

The form-making areas of the taxa of *Watsonius* and *Skrjabinocladorchis* were the African Continent, and some of them have been distributed with the definitive hosts (Primates) to Asia. The taxa *Gastrodiscoides, Homalogaster* have Asian form-making areas which colonized hosts that were phylogenetically related (*Homalogaster*: Bovidae, Cervidae) and others living under ecologically similar conditions (*Gastrodiscoides*: Suidae, Hominidae). The taxa *Pseudodiscus, Stephanopharynx*, and *Choerocotyloides* are African; *Olveria* is Indian, *Wardius* is North American, and *Macropotrema* is an Australian endemism. The *Zygocotyle* has Pan-American, African, and Asian distribution.

The taxa of the Gastrothylacidae have Afro-Asian distribution; apparently they never reached the New World (Figure 482). The genus *Fischoederius* is limited mainly to Asia. The form-making area of *Gastrothylax* was supposedly in Asia, and they were dispersed with the definitive hosts to Africa (Miocene). The species of carmyerids have Afro-Asian distribution with several endemic species in Africa (Figure 483). The presence of the joint species (*Carmyerius gregarius*) indicates an Afro-Asian migration of the definitive hosts.

The species of Orthocoeliinae have distribution along the countries of the Eastern Hemisphere, with many endemic forms in the African continent (Figure 484).

The *Orthocoelium scoliocoelium* is a joint species in Africa and Asia. The Australian orthocoelids (*Gemellicotyle*) can be regarded to be a result of the orthogenetic development of the *Bancroftrema-Elseyatrema-Gemellicotyle* developmental line.

The distribution of the taxa Paramphistominae concentrates on Africa and Eurasia. There are only a few species in the Americas (Figure 485). They probably reached the North American continent through the land connection (Beringia) with the definitive hosts, mainly with cervids (*C. microbothrioides* is a joint species of North America and Eastern Asia) (Figure 486). They penetrated (*Paramphistomum*) to South America after emergence of the Panama land bridge (Figure 487). The form-making area of *Cotylophoron* was probably *in* Africa from where they distributed to neighboring areas (Asia, India) and supposedly in the

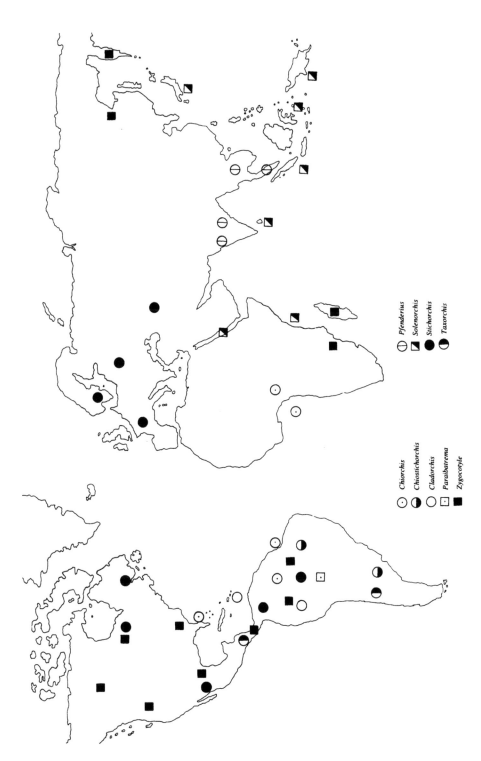

FIGURE 479. Geographic distribution of genera of bird and mammal amphistomes.

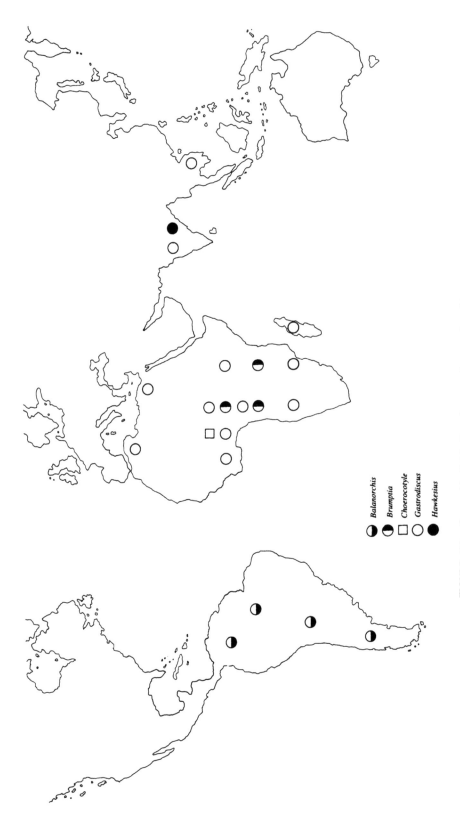

FIGURE 480.   Geographic distribution of genera of mammal amphistomes.

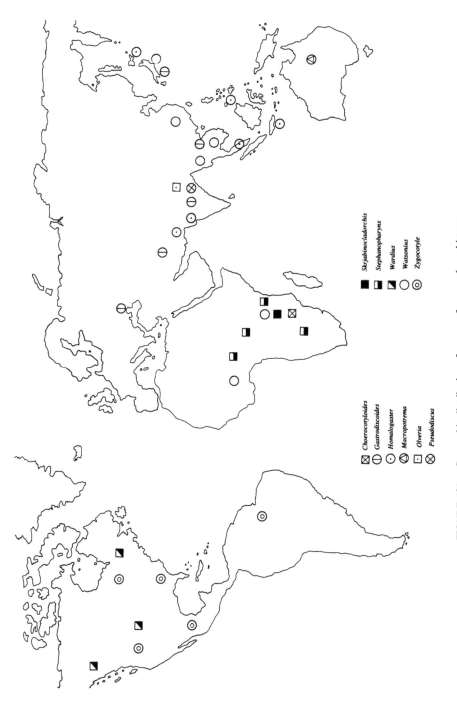

FIGURE 481. Geographic distribution of genera of mammal amphistomes.

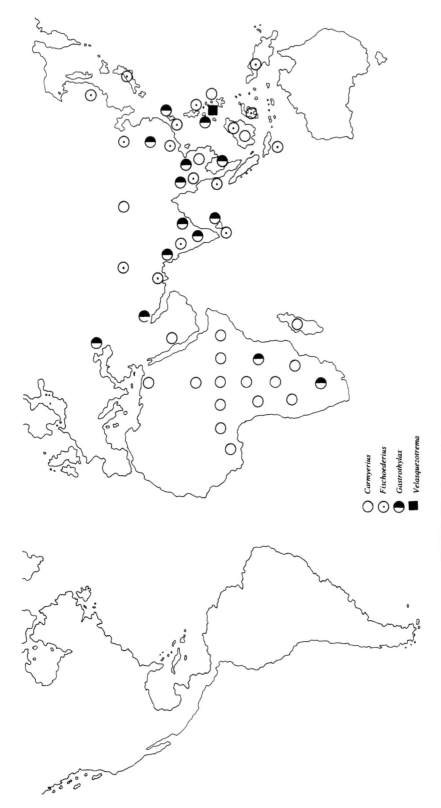

FIGURE 482. Geographic distribution of genera of Gastrothylacidae.

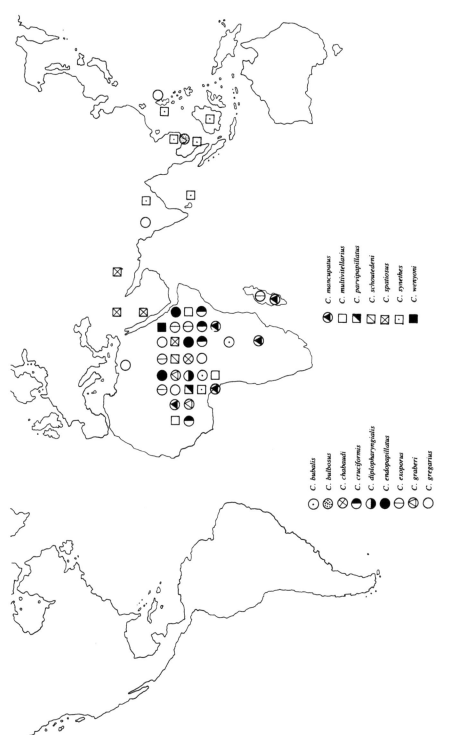

FIGURE 483. Geographic distribution of species of *Carmyerius*.

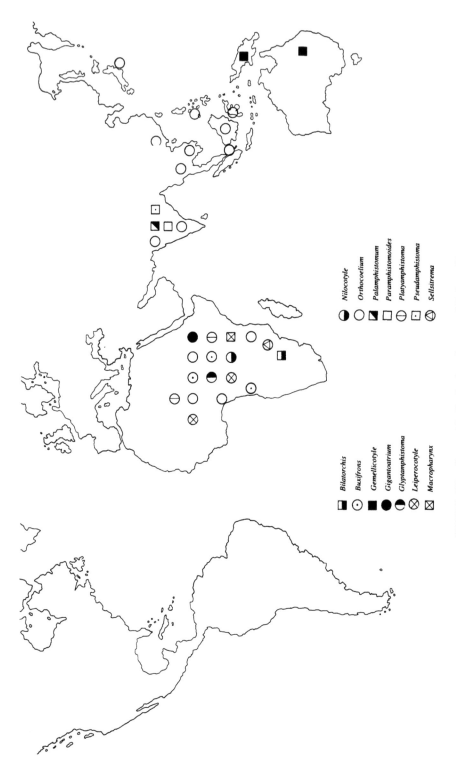

FIGURE 484.   Geographic distribution of genera of Orthocoeliinae.

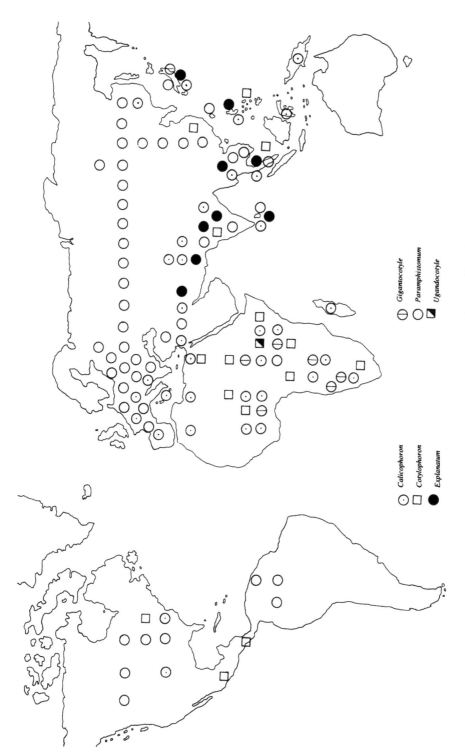

FIGURE 485. Geographic distribution of genera of Paramphistominae.

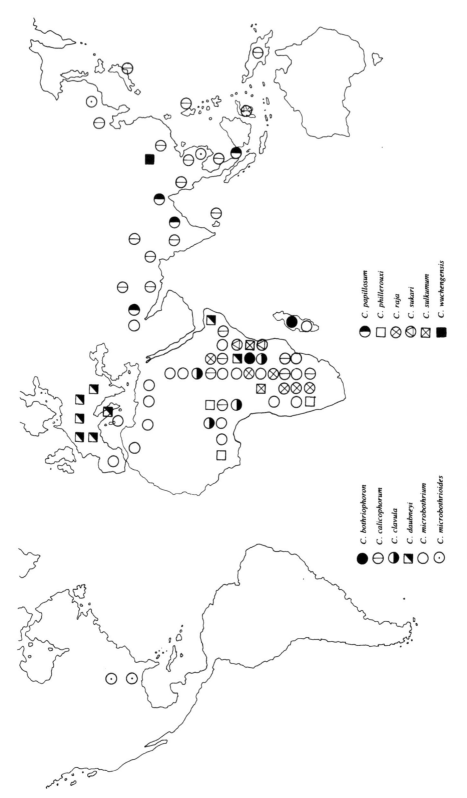

FIGURE 486.   Geographic distribution of species of *Calicophoron*.

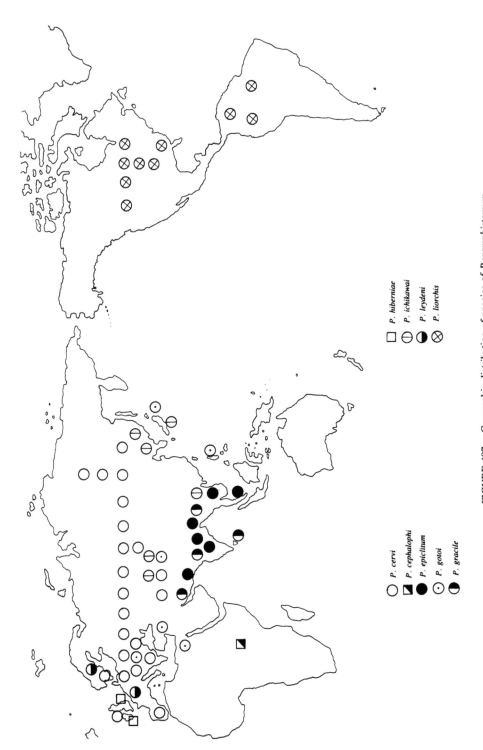

FIGURE 487. Geographic distribution of species of *Paramphistomum*.

far-removed North America. There are only two endemic groups (*Explanatum* in Asia and *Ugandocotyle* in Africa); the rest had an intercontinental distributional pattern (Figure 485). The distribution of the species of the two genera (*Calicophoron, Paramphistomum*), having the greatest diversity, is depicted in Figures 486 and 487. It seems that the center of origin of the *Calicophoron* is in Africa, and the species at the periphery (e.g., *C. daubneyi*) has apomorphic character states (terminal genitalium). On the other hand, the taxon *Paramphistomum* has a Eurasian origin of center, and some of them reached the northern and central part of Africa as well as North America.

4.3. DISTRIBUTION AND CHARACTERIZATION OF AMPHISTOMES BY ZOOGEOGRAPHIC REALMS

The areographic analysis of amphistomes seems to be reasonable to perform, due to the low number of species and their distributional data concerned, by zoogeographic realms. Such examinations present further details on the quality, similarity of faunas, and on the independent development of amphistomes of the given realms.

4.3.1. PALAEARCTIC REALM

The amphistome fauna of this realm includes 52 species (Table 11). The majority of them are parasitic in mammals (40 species 76.9%), and the rest are found in amphibians (6 species), fishes (3 species), reptiles (2 species), and birds (1 species) (Table 11). Of the species registered, 10 are endemic (19.2%), which are restricted mainly to mammals. The species of another group of amphistomes has probably a palaearctic origin (e.g., *Calicophoron daubneyi, Paramphistomum gotoi, P. ichikawai*, etc.) which, however, have also been distributed in other realms. The majority of amphistomes (26 species) of this realm belongs to the allochton species (*Sandonia sudanensis, Gastrodiscus aegyptiacus, Calicophoron microbothrium*, etc.) which have immigrated from adjacent areas. There are two amphiatlantic (*Stichorchis subtriquetrus, Paramphistomum cervi*) and one amphipacific (*Calicophoron microbothrioides*) faunal elements. The occurrence of some broadly distributed species (*Schizamphistomum scleroporum, Zygocotyle lunata*, etc.) has also been attached to this realm.

Due to the low number of endemic in this realm, its fauna shows the closest similarity to that of the Indomalayan (50.9), Africotropical (35.2), and Nearctic (21.4); (use of Simpson's[880] formula: $R_b = \dfrac{100 \times c}{b}$).

4.3.2. NEARCTIC REALM

This realm is composed of 29 amphistomes (Table 11). The majority of them (11 species) are parasitic in reptiles, and the rest are found in amphibians (8 species), mammals (6 species), fish (3 species), and birds (1 species) (Table 11). Of the species recovered, 16 are endemic (55.1%) which are restricted mainly to reptiles and amphibians. There is another group of amphistomes, including four species with probable nearctic origin (*Heronimus mollis, Calicophoron microbothrioides*, etc.) which, however, have a distributional range outside this realm. Two species (*Caballerodiscus tabascensis, C. panamense*) are allochton, two species amphiatlantic (*Chiorchis fabaceus, Stichorchis subtriquetrus*), and one species has amphipacific (*Calicophoron microbothrioides*) faunal elements. Besides these, broadly distributed species (*Schizamphistomum scleroporum, Zygocotyle lunata*, etc.) are added to the amphistomes of this realm.

Of the surrounding faunas the closest similarity is shown to that of the Neotropic (35.7) and Palaearctic ones (21.4).

TABLE 11
Distribution of Amphistomes in Zoogeographic Realms

Amphistome taxa	Palaearctic	Nearctic	Neotropical	Africotropical	Indomalayan	Australian	Oceanian
	1	2	3	4	5	6	7
Heronimidae							
Heronimus mollis	0	1	1	0	0	0	0
Zonocotylidae							
Zonocotyle bicaecata	0	0	1	0	0	0	0
Z. haroldtravassosi	0	0	1	0	0	0	0
Cladorchiidae							
Dadaytrema oxycephalum	0	0	1	0	0	0	0
Dadayius marenzelleri	0	0	1	0	0	0	0
Dadaytremoides grandistomis	0	0	1	0	0	0	0
Panamphistomum benoiti	0	0	0	1	0	0	0
Neocladorchis poonaensis	0	0	0	0	1	0	0
N. multilobularis	0	0	0	0	1	0	0
Cleptodiscus reticulatus	0	0	1	0	0	0	0
C. bulbosus	0	0	1	0	0	0	0
C. kyphosi	0	0	0	0	0	0	1
Macrorchitrema havanensis	0	0	1	0	0	0	0
Ophioxenus dienteros	0	1	0	0	0	0	0
O. singularis	0	1	0	0	0	0	0
O. microphagus	0	1	0	0	0	0	0
Travassosinia dilatata	0	0	1	0	0	0	0
Orientodiscus lobatus	0	0	0	0	1	0	0
O. jumnai	0	0	0	0	1	0	0
O. fossilis	0	0	0	0	1	0	0
O. mastacembeli	0	0	0	0	1	0	0
O. orchhaensis	0	0	0	0	1	0	0
Sandonia sudanensis	1	0	0	1	0	0	0
Basidiodiscus ectorchis	1	0	0	1	0	0	0
Pretestis australianus	0	0	0	0	0	1	0
Australotrema brisbanensis	0	0	0	0	0	1	0
Microrchis megacotyle	0	0	1	0	0	0	0
Microrchis ferrumequinum	0	0	1	0	0	0	0
Colocladorchis ventrastomis	0	0	1	0	0	0	0
Pseudocladorchis cylindricus	0	0	1	0	0	0	0
P. macrostomus	0	0	1	0	0	0	0
Brevicaecum niloticum	0	0	0	1	0	0	0
Kalitrema kalitrema	0	0	1	0	0	0	0
Nicollodiscus gangeticus	0	0	0	0	1	0	0
Osteochilotrema malayae	0	0	0	0	1	0	0
Caballeroia indica	0	0	0	0	1	0	0
C. bhavani	0	0	0	0	1	0	0
Brancroftrema neoceratodi	0	0	0	0	0	1	0
Platycladorchis microacetabularis	0	0	0	0	1	0	0
P. microacetabularis	0	0	0	0	1	0	0
Helostomatis helostomatis	0	0	0	0	1	0	0
H. sakrei	0	0	0	0	1	0	0
H. cirrhini	0	0	0	0	1	0	0
H. indica	0	0	0	0	1	0	0
H. mulleri	0	0	0	0	1	0	0
H. bundelkhandensis	0	0	0	0	1	0	0
H. fotedari	0	0	0	0	1	0	0

TABLE 11 (continued)
Distribution of Amphistomes in Zoogeographic Realms

Zoogeographic realms / Amphistome taxa	Palaearctic	Nearctic	Neotropical	Africotropical	Indomalayan	Australian	Oceanian
	1	2	3	4	5	6	7
H. cyprinorum	0	0	0	0	1	0	0
Protocladorchis pangasii	0	0	0	0	1	0	0
P. burmanicus	0	0	0	0	1	0	0
P. chinabutae	0	0	0	0	1	0	0
Autotrema dombrowskajae	1	0	0	0	1	0	0
Pisciamphistoma stunkardi	0	1	0	0	0	0	0
P. reynoldsi	0	1	0	0	0	0	0
Megalodiscus americanus	0	1	0	0	0	0	0
M. temperatus	0	1	1	0	0	0	0
M. intermedius	0	1	0	0	0	0	0
M. rankini	0	1	0	0	0	0	0
M. ferrissianus	0	1	0	0	0	0	0
Opisthodiscus diplodiscoides	1	0	0	0	0	0	0
Schizamphistomum scleroporum	1	1	1	0	0	1	0
S. erratum	0	0	1	0	1	1	0
S. taiwanense	0	0	0	0	1	0	0
Schizamphistomoides spinolosum	1	1	1	1	0	0	0
S. prescotti	0	0	0	0	1	0	0
S. constrictus	0	0	0	1	0	0	0
Stunkardia dilymphosa	0	0	0	0	1	0	0
S. burti	0	0	0	0	1	0	0
S. amboinensis	0	0	0	0	1	0	0
S. stunkardi	0	0	0	0	1	0	0
Alassostoma magnum	0	1	0	0	0	0	0
Alassostomoides parvus	0	1	0	0	0	0	0
A. chelydrae	0	1	0	0	1	0	0
A. louisianaensis	0	1	0	0	0	0	0
Pseudocleptodiscus margaritae	0	1	0	0	0	0	0
Pseudoalassostomoides rarus	0	0	0	0	1	0	0
Quasichiorchis purvisi	0	0	0	0	1	0	0
Lobatodiscus australiensis	0	0	0	0	0	1	0
Nematophila grandis	0	0	1	0	0	0	0
N. venezuelensis	0	0	1	0	0	0	0
Pseudoalassostoma heteroxenus	0	0	1	0	0	0	0
Halltrema avitellina	0	0	1	0	0	0	0
Parachiorchis parviacetabularis	0	1	0	0	0	0	0
Caballerodiscus tabascensis	0	1	1	0	0	0	0
C. resupinatus	0	0	1	0	0	0	0
Elseyatrema microacetabularis	0	0	0	0	0	1	0
Solenorchis travassosi	1	0	0	1	1	1	0
Chiorchis fabaceus	0	1	1	1	0	0	0
Chiostichorchis waltheri	0	0	1	0	0	0	0
Paraibatrema inesperata	0	0	1	0	0	0	0
Pfenderius papillatus	0	0	0	0	1	0	0
P. birmanicus	0	0	0	0	1	0	0
P. heterocaeca	0	0	0	0	1	0	0
Stichorchis giganteus	0	0	1	0	0	0	0
S. subtriquetrus	1	1	0	0	0	0	0
Cladorchis pyriformis	0	0	1	0	0	0	0

TABLE 11 (continued)
Distribution of Amphistomes in Zoogeographic Realms

Zoogeographic realms / Amphistome taxa	Palaearctic	Nearctic	Neotropical	Africotropical	Indomalayan	Australian	Oceanian
	1	2	3	4	5	6	7
C. asper	0	0	1	0	0	0	0
Taxorchis schistocotyle	0	0	1	0	0	0	0
T. ringueleti	0	0	1	0	0	0	0
Diplodiscidae							
Diplodiscus subclavatus	1	0	0	0	0	0	0
D. amphichrus	1	0	0	0	1	0	0
D. magnus	1	0	0	0	1	0	0
D. mehrai	1	0	0	0	1	0	0
D. sacculosus	0	0	0	0	1	0	0
D. pallascatus	0	0	0	1	0	0	0
D. lali	0	0	0	0	1	0	0
D. brevicoeca	0	0	0	1	0	0	0
D. chauhani	0	0	0	0	1	0	0
D. fischthalicus	0	0	0	1	0	0	0
D. nigromaculati	1	0	0	0	0	0	0
Australodiscus megalorchus	0	0	0	0	0	1	0
Ctadiscus dolichocotyle	0	0	1	0	0	0	0
C. cohni	0	0	1	0	0	0	0
C. pygmaeus	0	0	1	0	0	0	0
C. uruguayensis	0	0	1	0	0	0	0
C. marinholutzi	0	0	1	0	0	0	0
C. inopinatus	0	0	1	0	0	0	0
C. freitaslenti	0	0	1	0	0	0	0
C. mirandai	0	0	1	0	0	0	0
C. rodriguesi	0	0	1	0	0	0	0
C. propinguus	0	0	1	0	0	0	0
C. corderoi	0	0	1	0	0	0	0
C. eldoradiensis	0	0	1	0	0	0	0
C. longicaecalis	0	0	1	0	0	0	0
C. rochai	0	0	1	0	0	0	0
C. bathracorum	0	0	1	0	0	0	0
Dermatemytrema trifoliata	0	1	0	0	0	0	0
Progonimodiscus doyeri	0	0	0	1	0	0	0
Pseudodiplodiscus cornu	0	0	1	0	0	0	0
Gastrodiscidae							
Gastrodiscus aegyptiacus	1	0	0	1	1	0	0
G. secundus	0	0	0	0	1	0	0
Balanorchiidae							
Balanorchis anastrophus	0	0	1	0	0	0	0
Brumptiidae							
Brumptia bicuadata	0	0	0	1	0	0	0
Hawkesius hawkesi	0	0	0	0	1	0	0
Choerocotyle epuluensis	0	0	0	1	0	0	0
Zygocotylidae							
Zygocotyle lunata	1	1	1	1	1	0	0
Wardius zibethicus	0	1	0	0	0	0	0
Choerocotyloides onotragi	0	0	0	1	0	0	0
Oliveria indica	0	0	0	0	1	0	0
O. bosi	0	0	0	0	1	0	0

TABLE 11 (continued)
Distribution of Amphistomes in Zoogeographic Realms

Zoogeographic realms / Amphistome taxa	Palaearctic	Nearctic	Neotropical	Africotropical	Indomalayan	Australian	Oceanian
	1	2	3	4	5	6	7
Stephanopharynx compactus	0	0	0	1	0	0	0
Pseudodiscus collinsi	0	0	0	0	1	0	0
Macropotrema pertinax	0	0	0	0	0	1	1
Watsonius watsoni	0	0	0	1	1	0	0
W. noci	0	0	0	0	1	0	0
W. macaci	0	0	0	0	1	0	0
W. deschiensi	0	0	0	1	0	0	0
W. papillatus				1			
Homalogaster paloniae	1	0	0	1	1	0	0
Gastrodiscoides hominis	1	0	0	0	1	0	0
Skrjabinocladorchis jubilaricum	0	0	0	1	0	0	0
Gastrothylacidae							
Gastrothylax crumenifer	1	0	0	1	1	0	0
G. compressus	1	0	0	0	1	0	0
G. globoformis	1	0	0	0	0	0	0
Carmyerius gregarius	1	0	0	1	1	0	0
C. spatiosus	1	0	0	1	1	0	0
C. synethes	0	0	0	0	1	0	0
C. wenyoni	0	0	0	1	0	0	0
C. mancupatus	0	0	0	1	1	0	0
C. cruciformis	0	0	0	1	0	0	0
C. bubalis	0	0	0	1	0	0	0
C. exoporus	0	0	0	1	0	0	0
C. graberi	0	0	0	1	0	0	0
C. parvipapillatus	0	0	0	1	0	0	0
C. schoutedeni	0	0	0	1	0	0	0
C. endopapillatus	0	0	0	1	0	0	0
C. multivitellarius	0	0	0	1	0	0	0
C. diplopharyngialis	0	0	0	1	0	0	0
C. chabaudi	0	0	0	1	0	0	0
C. bulbosus	0	0	0	0	1	0	0
Fischoederius elongatus	1	0	0	0	1	0	0
F. cobboldi	1	0	0	0	1	0	0
F. japonicus	1	0	0	0	1	0	0
F. skrjabini	1	0	0	0	0	0	0
F. ovatus	1	0	0	0	0	0	0
F. compressus	1	0	0	0	0	0	0
Velasquezotrema brevisaccus	0	0	0	0	1	0	0
Paramphistomidae							
Orthocoelium orthocoelium	1	0	0	0	1	0	0
O. streptocoelium	1	0	0	1	1	0	0
O. dicranocoelium	0	0	0	0	1	0	0
O. scoliocoelium	1	0	0	1	1	0	0
O. parvipapillatum	0	0	0	0	1	0	0
O. dawesi	0	0	0	0	1	0	0
O. sinuocoelium	1	0	0	0	0	0	0
O. parastreptocoelium	1	0	0	0	0	0	0
O. gigantopharynx	0	0	0	0	1	0	0
O. brevicaeca	1	0	0	0	0	0	0
O. bovini	0	0	0	0	1	0	0

TABLE 11 (continued)
Distribution of Amphistomes in Zoogeographic Realms

Zoogeographic realms / Amphistome taxa	Palaearctic	Nearctic	Neotropical	Africotropical	Indomalayan	Australian	Oceanian
	1	2	3	4	5	6	7
O. narayanai	0	0	0	0	1	0	0
O. indonesiense	0	0	0	0	1	0	0
O. dinniki	1	0	0	0	1	0	0
O. serpenticaecum	0	0	0	0	1	0	0
Leiperocotyle okapi	0	0	0	1	0	0	0
L. congolense	0	0	0	1	0	0	0
L. gretillati	0	0	0	1	0	0	0
Paramphistomoides maplestonei	0	0	0	0	1	0	0
Platyamphistoma polycladiforme	0	0	0	1	0	0	0
Bilatorchis papillogenetalis	0	0	0	1	0	0	0
Buxifrons buxifrons	0	0	0	1	0	0	0
Gigantoatrium gigantoatrium	0	0	0	1	0	0	0
Nilocotyle pygmaea	0	0	0	1	0	0	0
N. minutum	0	0	0	1	0	0	0
N. wagandi	0	0	0	1	0	0	0
N. circularis	0	0	0	1	0	0	0
N. hippopotami	0	0	0	1	0	0	0
N. leiperi	0	0	0	1	0	0	0
N. microatrium	0	0	0	1	0	0	0
N. praesphinctris	0	0	0	1	0	0	0
N. hepaticae	0	0	0	1	0	0	0
N. duplicisphinctris	0	0	0	1	0	0	0
Sellsitrema sellsi	0	0	0	1	0	0	0
Pseudoparamphistoma cuonum	0	0	0	0	1	0	0
Macropharynx sudanensis	0	0	0	1	0	0	0
Palamphistomum dutti	0	0	0	0	1	0	0
P. lobatum	0	0	0	0	1	0	0
Gemellicotyle wallabicola	0	0	0	0	0	1	1
Glyptamphistoma paradoxum	0	0	0	1	0	0	0
Ugandocotyle pisum	0	0	0	1	0	0	0
Explanatum explanatum	1	0	0	1	1	0	0
E. bathycotyle	0	0	0	0	1	0	0
E. anisocotyle	0	0	0	0	1	0	0
Gigantocotyle gigantocotyle	0	0	0	1	0	0	0
G. formosana	1	0	0	1	1	0	0
G. symmeri	0	0	0	1	0	0	0
G. duplicitestorum	0	0	0	1	0	0	0
Cotylophoron cotylophorum	1	0	0	1	1	0	0
C. fülleborni	0	0	0	1	0	0	0
C. jacksoni	0	0	0	1	0	0	0
C. panamense	0	1	1	0	0	0	0
C. bareilliense	0	0	0	0	1	0	0
C. macrosphinctris	0	0	0	1	0	0	0
C. xiangjiangense	1	0	0	0	0	0	0
Calicophoron calicophorum	1	0	1	1	1	0	0
C. bothriophoron	0	0	0	1	0	0	0
C. microbothrium	1	0	0	1	0	0	0
C. papillosum	1	0	0	0	1	0	0
C. papilligerum	0	0	0	0	1	0	0
C. raja	0	0	0	1	0	0	0

TABLE 11 (continued)
Distribution of Amphistomes in Zoogeographic Realms

Zoogeographic realms / Amphistome taxa	Palaearctic	Nearctic	Neotropical	Africotropical	Indomalayan	Australian	Oceanian
C. clavula	1	0	0	1	0	0	0
C. microbothrioides	1	1	1	0	1	0	0
C. sukari	0	0	0	1	0	0	0
C. phillerouxi	0	0	0	1	0	0	0
C. daubneyi	1	0	0	1	0	0	0
C. sukumum	0	0	0	1	0	0	0
C. wuchengensis	1	0	0	0	0	0	0
Paramphistomum cervi	1	1	0	0	0	0	0
P. liorchis	0	1	1	0	0	0	0
P. gracile	1	0	0	0	1	0	0
P. epiclitum	0	0	0	0	1	0	0
P. gotoi	1	0	0	0	1	0	0
P. ichikawai	1	0	0	0	1	0	0
P. leydeni	1	0	1	0	0	0	0
P. hiberniae	1	0	0	0	0	0	0
P. cephalophi	0	0	0	1	0	0	0

4.3.3. NEOTROPICAL REALM

The amphistome fauna of this realm includes 57 species (Table 11). They occur closely similar in frequency in various vertebrate grades except for birds. Sixteen species are parasitic in fishes, 16 in amphibians, 14 in mammals, 10 in reptiles, and 1 in birds (Table 11). The amphistome fauna is characterized by the high percentage of endemicity in the lower vertebrates. Of the species described, 44 are endemic (77.1%) which are parasitic in fish (16 species), in amphibians (14 species), in mammals (8 species), and in reptiles (6 species). Beside the endemic species there are three species of allochton (*Heronimus mollis, Megalodiscus temperatus, Paramphistomum liorchis*) and one species is of amphiatlantic (*Chiorchis fabaceus*) faunal elements. Besides these, some broadly distributed species (*Schizamphistomum scleroporum, Zygocotyle lunata*, etc.) are also found among the amphistomes of this realm.

Except for the San Francisco River, fishes of the larger river systems of this realm harbored amphistomes. The Magdalenean has two monoipic genera (*Dadaytrematoides, Colocladorchis*) whose distribution is confined only to this area. There is one species in the Orinoco-Venezuelan region (*Dadaytrema oxycephalum*) having broad distribution in the realm. The Guinean-Amazonian and the Paranean regions have the greater number of amphistomes of fish. The Amazonas system has one endemic species (*Pseudodiplodiscus cornu*), and the rest (*Dadayius, Dadaytrema*, and *Travassosinia*) is common with that of the Paranean region. The Paraná river system has four endemic genera (*Microrchis, Pseudocladorchis, Zonocotyle*, and *Kalitrema*) while the other amphistomes of fishes are common with the amphistomes of the Guinean-Amazonian region. The common species of these regions indicate the relationship of these river systems in ancient times. Of the amphistomes of mammals, *Balanorchis anastrophus* is regarded to be an evolutionary relic which is the only representative of a former (Plio-Pleistocene) period which was much more abundant with mammalian fauna (Figure 488).

The high percentage of endemicity of the amphistomes of this realm assures its inde-

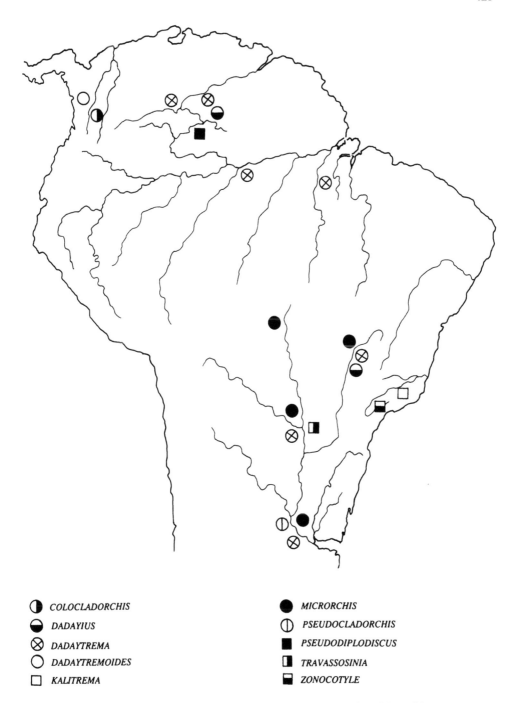

FIGURE 488. Geographic distribution of genera of South American fish amphistomes.

◖ COLOCLADORCHIS		● MICRORCHIS	
◓ DADAYIUS		◐ PSEUDOCLADORCHIS	
⊗ DADAYTREMA		■ PSEUDODIPLODISCUS	
○ DADAYTREMOIDES		▮ TRAVASSOSINIA	
□ KALITREMA		◒ ZONOCOTYLE	

pendence and, hence, shows low similarity index with the neighboring Nearctic realm (Rb = 35.7).

4.3.4. AFRICOTROPICAL REALM

This realm includes the second largest amphistome fauna (79 species) (Table 11). They are parasitic mostly in mammals (68 species, 86.0%) and to a lesser extent in fish (4 species), in amphibians (4 species), in reptiles (2 species), and birds (1 species) (Table 11). Of the total sum 57 species are endemic which are parasitic mainly in mammals (50 species, 72.1%). Furthermore, there are 11 species with probable Africotropical origin, 7 of allochton species (*Homalogaster paloniae, Orthocoelium streptocoelium, Calicophoron daubneyi,* etc.), 1 of amphiatlantic (*Chiorchis fabaceus*) and 3 of broadly distributed species of birds and marine turtles.

Due to the high percentage of endemicity of amphistomes of this realm, it shows low similarity to the Palaearctic (35.2) and Indomalayan (17.5) realms.

4.3.5. INDOMALAYAN REALM

The amphistome fauna of this realm is characterized by the highest number of species (98), high diversity (33 genera), and endemicity (65 species, 66.2%) (Table 11). The species of this realm are parasitic in mammals (55 species), in fish (26 species), in reptiles (10 species), in amphibians (6 species), and in birds (1 species). The endemic forms are concentrated in fishes (24 species), mammals (30 species), and amphibians (3 species) (Table 11). There are 18 species of probable Indomalayan origin (*Amurotrema dombrowskajae, Gastrodiscoides hominis, Paramphistomum gracile,* etc.), and 11 species of allochton (*Homalogaster paloniae, Calicophoron calicophorum, P. ichikawai,* etc.) faunal elements. Four species, parasitic in birds and in chelonians, having broad distribution are also included in this realm.

The fauna of this realm shows the closest similarity to the Palaearctic (50.9) and Africotropical (17.5) realms.

4.3.6. AUSTRALIAN AND OCEANIC REALMS

The amphistome fauna of both of these realms is characterized by the low number of species. This can be explained by the historical events of the Australian continent and of the oceanic islands as well as by the lacking of the actual and the potential definitive hosts in large number.

The Australian realm has 11 species of amphistomes (Table 11) from which 6 are autochton, 2 have (*Macropotrema pertinax, Gemellicotyle wallabicola*) probably Australian origin, and 3 of broadly distributed species being parasitic of chelonians and the sea cow of the coastal water. Of the endemic species, 3 are parasitic in fishes, 2 in turtles, and 1 in amphibians.

The Oceanic realm has an autochton species (*Cleptodiscus kyphosi,* parasitic in marine fish) and two species, parasitic in marsupials (Papua New Guinea) which are common faunal elements with the previous realm (Table 11).

The present distributional picture given in the above pages can be altered with the help of the enlargement of the distributional range of the definitive hosts of by human activity. Information referring to the latter case will be given in the following chapter.

5. DYNAMIC ZOOGEOGRAPHY OF THE AMPHISTOMES

It is probable that the first larger anthropochore dispersal of various animals and chiefly domestic ones was the great migration in prehistoric times. Later, with discovery of the continents and the oceanic islands and with the advance of the conveyances, the first colonists carried with them their domestic animals. By the middle of the 20th Century, the introduction of animals spread several times into many parts of the world. It is now difficult to ascertain whether a species is native to an area or is an adventive one. During recent decades a great variety of food (fish and mammals) and game, pets, and familiar wild animals, etc., have been introduced under government-controlled measures. In spite of these measurements, several parasites have spread with these animals from one continent to the other.

Of the vertebrate hosts there are several submammalian groups bearing amphistome parasites, which have been introduced (e.g., fish) from their native areas, but seemingly none of the amphistomes has been transported with them. We have much more information of the parasites introduced which were carried with domestic animals and game.

The prerequisites of the successful establishment of a given amphistome are the simultaneous presence of the worm, the susceptible intermediate, and the adequate definitive hosts. As the amphistomes show narrow specificity to their intermediate host, hence, the fate of the amphistomes carried with the definitive hosts depends either on the degree of the specificity of the flukes or on the presence of the same or closely related intermediate hosts (e.g., helminth parasites of the muskrat in the introduced areas).

There are numerous countries, islands, or other territories where domestic animals, mainly domestic ruminants, are not indigenous (Australia, New Zealand, Papua New Guinea, Cuba, Pacific Oceanic Islands, etc.), and their amphistomes, undoubtedly, have been introduced by these definitive hosts, and they successfully established their populations. In this respect the following species have involved.

Gastrodicus aegyptiacus
  Australia, Guadeloupe, Guyana
Fischoederius elongatus
  Borneo, Celebes, Mariana Islands
F. cobboldi
  Borneo
Carmyerius synethes
  Borneo
C. mancupatus
  Malagasy
Orthocoelium streptocoelium
  Australia, Borneo, Cuba
O. scoliocoelium
  Celebes, Borneo
O. gigantopharynx
  Borneo
Cotylophoron cotylophorum
  Cuba
C. panamense
  Trinidad, Dominica, Cuba, North America
Explanatum explanatum
  Cuba, Celebes
Calicophoron calicophorum
  Australia, Celebes, Pacific Oceanic Islands, New Caledonia, Cuba

C. clavula
  Cuba
C. raja
  Cuba
C. bothriophoron
  Malagasy
C. microbothrium
  Malagasy, Cuba
C. microbothrioides
C. sukumum
  Cuba
C. phillerouxi
  Mauritius
Paramphistomum cervi
  Costa Rica, Celebes, probably North America, Cuba
P. ichikawai
  Australia

The species *G. aegyptiacus* was reported from Australia, Guadeloupe, and Guyana outside the indigenous areas, but seemingly this amphistome could not establish population in these territories.

There are several other countries (e.g., Japan, Philippines, Brazil, etc.) where the number of amphistomes of ruminants, besides the endemic species, was considerably increased by the introduction of domestic ruminants.

This unwanted experience suggests that in certain species, even of the intermediate host, specificity may be potentially broader than it could be judged from the relevant data of the indigenous areas. The dispersal data arising from human activities also show that without efficient sanitary measures, the spread of amphistome parasites is likely.

6. EVOLUTION OF AMPHISTOMES

Amphistomida is one of the natural groups of the digenetic trematodes whose representatives have successfully colonized each grade of vertebrates. If we visualize the evolution of the amphistome parasitism as it is accepted in general, then it becomes evident that the evolution of the amphistomes has been accomplished parallel with the development of their definitive hosts. It means in a time perspective that the sequence of the vertebrate hosts represented newer and newer areas for their colonization. This process included a series of adaptations, thereafter an adaptive radiation with specialization and diversification of parasites to their colonized hosts. Hence, each vertebrate grade has its own special group of amphistomes, so it can be said that the hosts or host groups are as characteristic of parasites as of the parasites which are their hosts. Nevertheless, there is no direct connection in every respect between the taxonomic level of the hosts and their amphistome parasites. The tempo of the higher level differentiation of amphistomes seems to be slower than that of their definitive hosts. Namely, the amphistomes of the lower vertebrates and that of some of the mammals have fundamentally the same general plan in their structure, indicating that amphistomes have their own phylogeny (species of the family Cladorchiidae). Hence, the groups of hosts and their sequence reveal rather the evolutionary trends of the amphistomes than those of their developmental levels. The evolution of the ruminants and their adaptive radiation accompanied the colonization of the anterior part of the digestive tube (rumen, reticulum, liver), and this has given rise to a new developmental line with several new taxa (Gastrothylacidae, Paramphistomidae).

The amphistome type of parasites, similarly to the other flukes, have supposedly developed from Rhabdocoele-like ancestors which, in the prevertebral geologic time, existed in the form of mollusk-turbellarian relationships. The evolution of the modern amphistomes commenced parallel with the evolution of the vertebrate groups. The ancient parasitic population, representing the developmental line which led toward the amphistomes, has been named Proto-amphistoma,[859] a hypothetical form of the primitive amphistomes. They had, besides the generalized character states, a broad geographic distribution on the Pangean megacontinent. They were furnished with the preadaptive ability to evolve coaccommodation with ancient groups of vertebrates (fishes), and they had developed their obligatory parasites. The life-cycle pattern of the amphistomes became fixed at this early time, and its similarity in principles has been preserved in every species. The speciation of the definitive hosts has been accompanied by speciation of the amphistomes in different combinations, and the diversity of the latter seemed to be in accordance with the number of the adequate hosts found on certain continents. This conception can explain the distribution of the amphistomes on every continent and their extreme diversification. Accordingly, the Neotropical and Indomalayan realms, which were furnished with the greatest freshwater fish faunas, have the greatest number of fish amphistomes, involving both species of plesiomorphic and apomorphic traits. Due to the specificity which has come into being in the increasing time of coaccommodation, the scope of the amphistomes, which is characteristic of fish, has evolved. It should be supposed, however, that besides the specific forms there have existed species with generalized traits which were able to colonize the amphibians, as the following grade in the time sequence of the vertebrate's evolution. The fish amphistomes can be traced from the ancient cosmopolitan proto-amphistome stock, distributed along the whole Pangea. Figure 473 shows the hypothetical form-making areas of the fish-amphistomes superimposed on the map having the Mesozoic configurations of the continents. The form-making area of the amphibian-amphistomes is easily recognizable. The two developmental lines of the amphibian-amphistomes (Megalodiscinae, Diplodiscinae) can be designated in the Neotropical realm. The generalized forms (Megalodiscinae) have developed from a dadaytrematid-like ancestor; the other line (Diplodiscinae) including the specialized forms has derived from a pseudodiplodiscid-like predecessor (Figure 475).

The two primary form-making areas of the endemic species are the Neotropical (Nematophilininae, Caballereodiscinae) and Indomalayan (Schizamphitominae) realms. The Neotropical forms can be contacted with the dadaytrematid-like ancestor of the fish, and a single species has become adapted to the reptiles (*Dermatemytrema trifoliata*) from the diplodiscid stock. The amphistomes of the Indomalayan radial center show phyletic relationships with the fish-amphistomes (e.g., *Orientodiscus*) of these areas (Figure 477).

The single species of the bird amphistomes (*Zygocotyle lunata*) is the sister species of the Nearctic mammalian endemic amphistome *Wardius*) which shows similarity to the local reptilian amphistome fauna (Figure 481).

The amphistome parasites of the submammalian vertebrates bear ancient morphological and ecological characters, and their intercontinental distributional pattern of today is a reflection of the conditions that existed before the breakup of the Pangaea. At the end of the Cretaceous era, with the formation of the continents, the great disjunctions have isolated the amphistome-faunas, and their development has continued independently. The exchange of faunas was possible through the subsequently disappearing Beringia and Afro-Arabic plate and the newly formed Panama land bridge. With the adaptive radiation of the mammals in the Tertiary era, changes in several directions have been ensured for the amphistomes which represented a uniform organizational level at that time. Looked at from an anatomic point of view, the cirrus sac disappeared from the terminal genitalium, the position of the pores of the excretory and Laurer's canals relative to each other exchanged, and the pharyngeal appendages gradually regressed. From the ecological point of view, the colonization of the anterior part of the digestive tract (stomach, liver) and the adaptive radiation accompanying this process have been evolutionary events with considerable consequences.

The formation of the mammalian-amphistomes has led through five combinations of the morphological character states: (1) plesiomorphic habitat and morphological states, except for the terminal genitalium (Zygocotylidae); (2) plesiomorphic morphological traits, apomorphic habitat (*Balanorchis, Brumptia*); (3) plesiomorphic state of the position of the pores of the excretory and Laurer's canals, apomorphic habitat, and terminal genitalium (*Stephanopharynx*); (4) plesiomorphic trait of position of the pores of the excretory and Laurer's canals apomorphic habitat and morphological states (Gastrothylaciidae); (5) apomorphic characters in both morphological and ecological traits (Paramphistominae). The combination of numbers 1 to 3 did not prove to be successful developmental lines, as the taxa involved are mainly monotypic ones. The greatest diversity, in both morphological and ecological characters, is found in the combinations of numbers 4 and 5, which are parasitic mainly in ruminants. The role of the ruminant's stomach in the generation of the apomorphic characters and speciations is obvious. There seems to be correlation in the herbivorous mammals, between the numbers of the cavities of the stomach and the numbers of amphistomes, e.g., the Indian elephant, having a simple stomach, has four species of amphistomes; the hippopotamus, having a stomach consisting of three parts, harbors 26 amphistomes; and in cattle, having ruminant's stomach, 55 species of amphistomes are parasitic.

Besides the above-mentioned relationships, a correlation can be found between the habitat of the amphistomes and their hosts as well as between the morphological strucuture of the adults and the larval forms.

The habitat of the lower digestive tube is regarded to be a plesiomorphic ecological trait. The amphistomes, living in this part of the digestive tube, are characterized by plesiomorphic states (presence of cirrus sac, pharyngeal appendages and the position of the pores of the excretory and Laurer's canal in relation to each other). Structurally, cercariae belonging to the species of these groups of amphistomes are characterized among others by their having either simple excretory trunks (without cross-connecting tubes) or with branched main excretory tubes (Cercaria diplocotylea, Cercaria intermedia). These amphistomes, organized by the ancient plan of structure, constitute the major part of the species.

Parallel with the adaptive radiation of the mammals in the Tertiary several morphological changes have taken place in the structure of the amphistomes known then. From the anatomic point of view, the cirrus sac disappeared from the end part of the male genital system, the position of the pores of the excretory and Laurer's canals were modified, the pharyngeal appendages have gradually regressed; and from the ecological point of view, the colonization of the anterior part of the digestive tube (stomach, rumen) was one of the most important evolutionary event. This adaptive radiation has been accompanied by an extensive diversity of species.

The highest diversity can be perceived in those taxa whose character states are in greater part (Gastrothylacidae, Orthocoelinae) or completely (Parampihstominae) apomorphic. The colonization of the forestomach and the consequence on the adaptive radiation have yielded to the evolution of the derived groups of amphistomes comprising three families and a total of 42.5% of the known species of the Amphistomida.

REFERENCES

1. **Ábrahám, A. A.**, Az *Opisthodiscus diplodiscoides nigrivasis* Méhelÿ in degrendszere (in Hungarian), *Allattani Tanulmányok*, 1, 136, 1929.
2. **Achan, P. D.**, A survey of the helminth fauna parasitic in the freshwater fishes of the Madras State, *Fish. Stat. Rep. Yearb.*, 1954-1955, 385, 1956.
3. **Adiwinata, R. T.**, Tjatjing jang berparasit pada hewan menjusui dan unggas di Indonesia (in Indonesian), *Hemera Zoa*, 62, 229, 1955.
4. **Afzal, M., Shafique, M., Hussain, A., and Saeed, M.**, A study of helminths of cattle and buffaloes in Lahore, *Pak. J. Sci.*, 33, 14, 1981.
5. **Agarwal, G. P. and Agrawal, S. C.**, On *Orientodiscus* sp. (Digenea: Paramphistomidae) Fischoeder, 1901 from the large intestine of freshwater fish, *Mastacembelus armatus* (Lac.), *Ann. Number. Proc. Natl. Acad. Sci. India*, 110, 1979.
6. **Agarwal, G. P. and Agrawal, S. C.**, Studies on the trematode parasites of Bundelkhand region. II. *Helistomatis bundelkhandensis* n. sp. from the intestine of freshwater eel, *Mastacembelus armatus* (Loc.), *Riv. Parasitol.*, 41, 11, 1980a.
7. **Agarwal, G. P. and Agrawal, S. C.**, On *Orientodiscus* sp. (Digenea: Paramphistomidae) from the large intestine of freshwater fish, *Mastacembelus armatus*, *Indian J. Parasitol.*, 3, 9, 1980b.
8. **Agrawal, V.**, *Schizamphistomoides prescotti* n. sp. (Trematoda: Paramphistomidae), from *Hardella thurgi* (Gray), *Ann. Parasitol. (Paris)*, 42, 427, 1967.
9. **Agrawal, N.**, Further observation on *Diplodiscus lali* Pandey et Chakrabarti, 1968, *Indian J. Zootomy*, 15, 127, 1976.
10. **Agrawal, R. D.**, Infective stages of two helminth parasites from intermediate-paratenic hosts with notes in their partial life-cycle in experimental mammals and on two amphistome cercariae from two of the common snails their encystment and development in experimental mammals,, *Agra Univ. J. Res.*, 20, 137, 1971a.
11. **Agrawal, R. D.**, An amphistome cercaria from *Gyraulus euphraticus* (Mousson), *Curr. Sci.*, 40, 378, 1971b.
12. **Agrawal, R. D. and Pande, B. P.**, Cercaria of *Fischoederius elongatus* its encystment and development in experimental animals, *Indian J. Agric. Sci.*, 41, 1055, 1971.
13. **Agrawal, R. D. and Ahluwalia, S. S.**, A note on the occurrence of *Gastrothylax crumenifer* in sambar *(Cervus unicolor)*, *Indian Vet. J.*, 57, 436, 1980.
14. **Ahluwalia, S. S.**, *Gastrodiscoides hominis* (Lewis et McConnel) Leiper, 1913 — (The amphistome parasite of man and pig), *Indian J. Med. Res.*, 48, 315, 1960.
15. **Akhmerov, A. H.**, A new trematode genus from fish (in Russian), *Tr. Gelmintol. Lab. Akad. Nauk SSSR*, 19, 45, 1959.
16. **Albaret, J. L., Bayssade-Dufour, Ch., Guilhon, J., Kulo, S. D., and Picot, H.**, Cycle biologique de *Paramphistomum togolense* n.sp. (Trematoda: Paramphistomidae), *Ann. Parasitol. (Paris)*, 53, 495, 1978.
17. **Albaret, J. L., Bayssade-Dufour, Ch., Diaw, O. T., Vassiliades, G., Sey, O., and Gruner, L.**, Disposition des organites argyrophiles superficiels du miracidium et de la cercaire de *Paramphistomum philleroui* Dinnik, 1961 (Trematoda: Paramphistomidae), *Ann. Parasitol. Hum. Comp.*, 56, 147, 1981.
18. **Aleksandrova, I. V.**, Instances of intensive invasion of elks with *Paramphistomum cervi* in Kirov region (in Russian), *Zool. Zh.*, 41, 780, 1962.
19. **Aleksandrowska, I., Borowski, E., Iwanowski, J., Smieckowicz, R., Szczuka, E., Walkowiak, M., and Kawelicz, M.**, Straty poubojowe spowodowane przez pasożyty występujące u owiec (in Polish), *Med. Weter.*, 26, 417, 1970.
20. **Alho, C. J. R.**, Contribuiçâo ao conhecimento da fauna helmintologica de quelônios do Estadado Porá, Brasil, *Bol. Mus. Para. Emilio Goeldi Nova Ser. Zool.*, 58, 56, 1956.
21. **Aliff, J. V., Smith, D., and Lucas, H.**, Some metazoan parasites from fishes of middle Georgia, *Trans. Am. Microsc. Soc.*, 96, 145, 1977.
22. **Allen, J. A.**, Parasites of fur-bearing animals, *Proc. 5th Pac. Sci. Congr.*, 1933, 5, 1981, 1934.
23. **Altaif, K. J., Al-Abassy, S. N., Al-Saqur, I. M., and Jawad, A. K.**, Experimental studies on the suitability of aquatic snails as intermediate hosts for *Paramphistomum cervi* in Iraq, *Ann. Trop. Med. Parasitol.*, 72, 151, 1978.
24. **Alwar, V. S.**, Amphistomiasis (a review of the literature), *Indian Vet. J.*, 25, 417, 1948.
25. **Amato, J. F. R. and Gutierras, V. C.**, *Balanorchis* Fischoeder, 1901 (Trematoda: Balanorchiidae), em rúmen de *Bos taurus* L., no Rio Grande do Sul, Brasil, *Arq. Fac. Vet. UFRGS, Porto Alegre*, 2, 7, 1974.
26. **Amato, J. F. R., Velazquez-Maldonado, J. J., and Amato, S. B.**, Intecções conjuntans de bovinos por *Balanorchis anastrophus* Fischoeder, 1901 e *Paramphistomum* sp. (Trematoda: Paramphistomatidae) no Rio Grande do sul Brasil, *Rev. Bras. Biol.*, 42, 371, 1982.
27. **Anantaraman, M.**, Various aspects of protozoans, helminths and arthropod parasitism have been investigated, *Madras Vet. Coll. A.*, 23, 78, 1965.

28. **Anderson, D. R. and Beaudoin, R. L.,** Host habitat and age as factors in the prevalence of intestinal parasites of the muskrat, *Bull. Wildl. Dis. Assoc.,* 2, 70, 1966.

29. **Anderson, I. G.,** The prevalence of helminths in impala, *Aepyceros melampus* (Lichteinstein, 1812) under game ranching conditions, *S. Afr. J. Wildl. Res.,* 13, 55, 1983.

30. **André, E.,** Recherches parasitologique sur les Amphibiens de la Suisse, *Rev. Suisse Zool.,* 20, 471, 1912.

31. **Angel, L. M.,** *Bancroftrema neoceratodi,* gen. et sp. n. a paramphistomatid trematode from the Australian lungfish, *J. Parasitol.,* 52, 1058, 1966.

32. **Angel, L. M. and Manter, H. W.,** *Pretestis australianus* gen. et sp. nov. (Digenea: Paramphistomidae) from Australian fish, and a closely related cercaria, *Cercaria acetabulapapillosa* sp. nov., with notes on the life history, *An. Inst. Biol. Univ. Nac. Auton. Mex.,* 41, 1, 1970.

33. **Applewhaise, L. M. and Ruiz, A.,** *Gastrodiscus aegyptiacus* in horses Guyana, *Br. Vet. J.,* 139, 265, 1983.

34. **Arfaa, F.,** A study of *Paramphistomum microbothrium* in Khurzistan S. W. Iran, *Ann. Parasitol. Hum. Comp.,* 37, 549, 1962.

35. **Arora, R. G. and Kalra, D. S.,** Some observations on the pathology of ovine liver infected with *Gigantocotyle explanatum* (Näsmark, 1937), *Haryana Agric. Univ. J. Res.,* 1, 90, 1971.

36. **Arru, E. and Deiana, S.,** Gli ospiti intermedi di *Paramphistomum* (*P. cervi*) in Sardegna, *Atti Soc. Ital. Sci. Vet.,* 23, 909, 1969.

37. **Arru, E. and Papadopoulos, I.,** Gli ospiti intermedi di alcuni fra i piu comuni trematodi della sardegna con particolare riguardo a *Fasciola hepatica* — sistemi di lotta, *Vet. Ital.,* 22, 151', 1971.

38. **Artigas, P. and Pacheco, G.,** *Stichorchis myopotami* n. sp. (Trematoda), *C. R. Seances Soc. Biol. Paris,* 112, 404, 1933a.

39. **Artigas, P. and Pacheco, G.,** *Chiostichorchis myopotami* (Artigas e Pacheco, 1932), parasito do *Myocaster coypus Chiostichorchis* n. gn. (Trematoda: Paramphistomidae), *Rev. Biol. Hig.* (1932), 3, 103, 1933b.

40. **Artigas, P. T. and Pérez, M. D.,** *Catadiscus eldoradiensis* n. sp., Trematoda: Paramphistomata de *Leptodactylus ocellatus, Mem. Inst. Butantan Sao Paulo,* 31, 5, 1964.

41. **Asadov, S. M.,** Helminth parasites of the coypu, acclimatized in Azerbaïdzhan (in Russian), *Izv. Akad. Nauk Az. SSR,* 9, 41, 1951.

42. **Asadov, S. M.,** *Helminth Parasites of Ruminants in the U.S.S.R. and Their Eco-Geographical Analysis* (in Russian), Izv. Akad. Nauk Az. SSR, Baku, 1960a, 388.

43. **Asadov, S. M.,** Helminths of domestic and wild ruminants in Azerbaïdzhan (in Russian), *Tr. Inst. Zool. Akad. Nauk Az. SSR,* 21, 97, 1960b.

44. **Asadov, S. M., Kolesnichenko, M. L., Melikov, Yu. F., Ismailov, G. D., and Gadzhieva, S. A.,** Contribution to the study of distribution of paramphistomes of ruminants in Azerbaidzhan (in Russian), in *Examinations of Helminth Parasites in Azerbaïdzhan,* Izv., "Elm", Baku, 1975, 24.

45. **Asadov, S. M., Malikov, Yu. F., and Gadzhieva, S. A.,** Ram's horn — intermediate host of two species of paramphistomids in Azerbïdzhan (in Russian), *Dokl. Akad. Nauk. Az. SSR,* 26, 65, 1972.

46. **Ashizawa, H., Nosaka, D., Yamaguchi, H., and Snagaki, K.,** Pathological findings of the forestomach of cattle infected with Paramphistomidae (in Japanese), *Bull. Fac. Agric. Miyazaki Univ.,* 16, 101, 1969.

47. **Ashizawa, H., Nosaka, D., and Tateyama, S.,** Infestation of the forestomach of goats and sheep with Paramphistomidae (in Japanese), *Bull. Fac. Agric. Miyazaki Univ.,* 17, 84, 1970.

48. **Audi, S.,** Paramfistomoza buraga ovce (in Serbo-Croat), *Vet. Arch.,* 16, 102, 1946.

49. **Aziz, J. A., Behlert, O., Behlert, C., and Johara, M. Y.,** Amphistomosis in cattle in Malaysia — case report, *Kajian Vet. Malays.,* 15, 69, 1983.

50. **Azzie, M. A. J.,** Pathological infection of thoroughbred horses with *Gastrodiscus aegyptiacus, J. S. Afr. Vet. Assoc.,* 46, 77, 1975.

51. **Babero, B. B.,** Some helminth parasites of Alaskan beavers, *J. Parasitol.,* 39, 674, 1953.

52. **Babić, P. B.,** Helminti digestivnich organa goveda Srbije sa naračitim osvrtom na njikovo pojavljivanje i nedovoljno opisane vrste (in Serbo-Croat), *Acta Vet. Belgrade,* 16, 291, 1966.

53. **Badanin, N. V.,** The occurrence of *Gastrodiscoides hominis* — parasitic in man and boar in Kazahstan (in Russian), *Russ. Zh. Trop. Med.,* 7, 514, 1929.

54. **Baer, J. G.,** Résultats Zoologiques du voyage du Dr. P. A. Chappuis an Nil supérieur. III. Helminthes, *Rev. Suisse Zool.,* 30, 337, 1923.

55. **Baer, J. G.,** Un trématode parasite de l'Okapi (*Cotylophoron congolense* n. sp.), *Rev. Zool. Bot. Afr.,* 28, 310, 1936.

56. **Baer, J. G.,** Etude critique des helminthes parasites de l'Okapi, *Acta Trop.,* 1, 164, 1950.

57. **Baer, J. G.,** Helminthes parasites, *Explor. Parcs Natl. Congo Belge,* 1, 3, 1959.

58. **Baer, J. G. and Joyeux, C.,** Classe des trématodes (Trematode Rudolphi), in *Traité de Zoologie, Anatomie, Systematique, Biologie,* Vol. 4, Grassé, P. P., Ed., Masson, Paris, 1961, 561.

59. **Bailenger, J. and Chanseau, J.,** Etudee des vers parasites des Amphibiens Anoures de la region de Bordeaux, *Ann. Parasitol. Hum. Comp.,* 29, 546, 1954.

60. **Bain, G. A. and Threlfall, W.**, Helminth parasites of hooded mergansers, *Lophodytes cucullatus* (L.) from Ontario, *Proc. Helminthol. Soc. Wash.*, 44, 219, 1977.

61. **Bakke, T. A.**, Tormparasitter hos bever fra Norge (in Norwegian), *Fauna (Oslo)*, 31, 128, 1978.

62. **Balasingam, E.**, Some helminth parasites of Malayan reptiles, *Bull. Natl. Mus. Singapore*, 32, 103, 1963.

63. **Balasubramaniam, G., Anandan, R., Ganeskale, D., and Alwar, V. S.**, Outbreak of amphistomiasis in sheep in Tamil Nadu, *Cheiron*, 2, 55, 1973.

64. **Bali, H. S.**, Incidence of helminth parasites in sheep in Bihar, *J. Anim. Health Prud.*, 1, 35, 1973.

65. **Bali, H. S. and Fotedar, D. N.**, On the morphology of a paramphistome in sheep, *Ceylonocotyle scolicoelium* from Jammu and Kashmir, *J. Res.*, 9, 199, 1972.

66. **Bali, H. S. and Fotedar, D. N.**, On a new amphistome, *Oliveria thapari* n. sp. from the rumen of crossbred sheep from Kashmir, *Indian J. Helminthol.*, 24, 36, 1974.

67. **Bali, H. S., Chhabra, R. C., and Srivastava, G. C.**, A check list of the helminth parasites of domestic ruminants in the Punjab, *J. Res. Punjab Agric. Univ.*, 22, 401, 1985.

68. **Ball, L.**, Notes on the helminth parasites of muskrat from Western Colorado, *J. Parasitol.*, 38, 83, 1952.

69. **Bangham, R. V.**, Parasites from fish of Buckeye Lake, Ohio, *Ohio J. Sci.*, 41, 441, 1941.

70. **Barker, F. D.**, Parasites of the American muskrat *(Fiber zibethicus)*, *J. Parasitol.*, 1, 184, 1915.

71. **Barker, F. D. and Parsons, S.**, A new species of monostome from the painted terrapin, *Chrysemys marginata* Agassiz, *Zool. Anz.*, 45, 193, 1914.

72. **Baruš, V., Groschaft, J. and Otčenášek, M.**, Helmintofauna ocasatých obojživelniků z územi Československa (in Czech.), *Czech. Parasitol.*, 10, 43, 1963.

73. **Barrois, T.**, Sur un Paramphistome nouveau *(Chiorchis noci,* nov. sp.) parasite du caecum du *Macacus cynomolgus*, *C. R. Seances Soc. Biol. Paris*, 64, 791, 1908.

74. **Bayssade-Dufour, Ch., Albaret, J. L., Grabda-Kazubska, B., and Kulo, S. D.**, Étude comparée des systémes excrétuer et sensoriel de deux cercaires de *Diplodiscus* (Trematoda, Diplodiscidae) parasites d'amphibiens européens et africains, *Ann. Parasitol. Hum. Comp.*, 53, 595, 1978.

75. **Beaver, P. C.**, Studies of the development of *Alassostoma parvum* Stunkard, *J. Parasitol.*, 16, 13, 1929.

76. **Beck, C. and Forrester, D. J.**, Helminths of the Florida manatee, *Trichechus manatus latirostris*, with a discussion and summary of the parasites of sirenians, *J. Parasitol*, 74, 628, 1988.

77. **Becklung, W. M. A.**, Revised check list of internal and external parasites of domestic animals in the United States and in Canada, *Vet. Res.*, 25, 1380, 1964.

78. **Benmokhtar-Betkonche, Z.**, Helminthes d' Algérie. Trematodes digenees de *Rana ridibunda perezi* de l'Ouest Algerien, *Bull. Soc. Hist. Nat. Afr. Nord*, 65, 207, 1974.

79. **Bennett, H. J.**, The life history of *Cotylophoron cotylophorum*, a trematode from ruminants, *Ill. Biol. Monogr.*, 14, 1, 1936.

80. **Bennett, H. J.**, A partial checklist of the trematodes of Louisiana vertebrates, *Proc. La. Acad. Sci.*, 4, 178, 1938.

81. **Bennett, H. J. and Humes, A. G.**, An unusual amphistome miracidium, *Proc. La. Acad. Sci.*, 4, 249, 1939.

82. **Bennett, H. J. and Allison, R.**, Observations on the life-cycle of the treamatode, *Stichorchis subtriquetrus*, *Proc. La. Acad. Sci.*, 20, 10, 1958.

83. **Bergeon, F.**, Rapport an gouvernement du Cambodge an l'enseignement de la parasitologie des animaux domestiques et la production animale an Cambodge, *FAO Rep. (2047), Expand. Tech. Assist. Program*, 1, 1965.

84. **Beverley-Burton, M.**, Some digenetic trematodes from amphibians and reptiles in Southern Rhodesia including two new species and a new genus: *Sakimitrema hytatorchis* n. gen., n. sp. (Plagiorchiidae) and *Halipegus rhodesiensis* n. sp. (Halipegidae), *Proc. Helminthol. Soc. Wash.*, 30, 49, 1963.

85. **Beverley-Burton, M.**, *Ophioxenus microphagus* (Ingles, 1936) comb. n. (Digenea: Paramphistomidae) from Ectotherms in Western North America with comments on host-parasite relationships, *Proc. Helminthol. Soc. Wash.*, 54, 197, 1987.

86. **Beverley-Burton, M. and Margolis, L.**, *Ophioxenus lampetrae* sp. nov. (Digenea: Paramphistomidae) from ammocoetes of the western brook lamprey *(Lampetra richardsoni* Vladykov et Follett) in British Columbia, *Can. J. Zool.*, 60, 2514, 1982.

87. **Bhalerao, G. D.**, Two new trematodes from reptiles, *Paryphostomum indicum* n. sp. and *Stunkardia dilymphosa* n.g., n.sp., *Parasitology*, 23, 99, 1931.

88. **Bhalerao, G. D.**, The trematode parasites of the Indian elephant, *Elephas indicus, Indian J. Vet. Sci.*, 3, 103, 1933.

89. **Bhalerao, G. D.**, Helminth parasites of the Indian elephant from the Andamans and Burma, *Indian J. Vet. Sci.*, 5, 35, 1935.

90. **Bhalerao, G. D.**, Studies of the helminths of India. Trematoda. IV, *J. Helminthol.*, 15, 97, 1937.

91. **Bhalerao, G. D.**, The cercarial fauna of the irrigated tract of the Nizam's Dominions with suggestions regarding the relationship to the trematode parasites in man and domesticated and other animals, *Indian J. Vet. Sci.*, 13, 294, 1943.

92. **Bhatnagar, A. K., Gupta, A. N., and Mehta, S.,** Proposed synonymy of Zygocotyle and Stunkardia (Digenetic trematodes), Indian J. Helminthol., 32, 117, 1980.

93. **Bhutta, M. S. and Khan, D.,** Digenetic trematodes of vertebrates from Pakistan, Bull. Dep. Zool. Univ. Punjab, 8, 44, 1975.

94. **Bilqees, F. M.,** Freshwater fish trematodes of Pakistan. Decription of a metacercarial form Tetracotyle muscularis Chakrabarti, 1970 and Orientodiscus fossilis n. sp., Phylodistomum kalrai n.sp., in Helminth Parasites of Some Vertebrates Chiefly from Fishes of West Pakistan, Bilqees, F. M., Saeed, R., Rehana, R., Khatoon, A., and Kaikabad, S. H., Eds., Karachi, Pakistan Agric. Res. Council, Goon. Pakistan, 1972, 41.

95. **Bilqees, F. M.,** Two species of trematodes of Chelonia mydas from Karachi coast, Pakistan, Acta Parasitol. Pol., 22, 295, 1974.

96. **Blair, D.,** Indosolenorchis hirudinaceus Crusz, 1951 (Playtyhelminthes: Digenea) from the dugong, Dugong dugong (Müller) (Mammalia: Sirenea), Ann. Parasitol. (Paris), 55, 511, 1980.

97. **Blair, D.,** Paramphistomes (Digenea: Paramphistomidae) parasitic in marine turtles, Austr. J. Zool., 31, 851, 1983.

98. **Blair, D., Beveridge, I. and Speare, R.,** Macropotrema pertinax gen. et sp. nov. (Digenea: Paramphistomidae) from a wallaby, Macropus agilis, in northern Australia, and associated pathology, Ann. Parasitol. (Paris), 54, 585, 1979.

99. **Bobkova, A. F.,** Contribution to a study of helminth fauna of cattle in the Bielorussian Poles'ya (in Russian), in Rab. Gel'mint. K 80-let. Akad. K. J. Skrjabina, Izdv. Hozya., Moscow, 1959, 17.

100. **Boch, J., Schmid, K., Rückrich, H. U., Erich, E., Keller, B., Weiland, G., and Göbel, E.,** Stomach fluke infection (Paramphistomum cervi) of ruminants, Berl. Münch. Tieraerztl. Wochenschr., 96, 338, 1983.

101. **Boer, E.,** Frequentia van voor komen van einige bekende parasieten bij de karbouw (in Indonesian), Hemera Zoa Buitenzorg., 57, 310, 1950.

102. **Boero, J. J. and Boehringer, J. K. de.,** El parasitismo de nuestra fauna autóctona. II. Los parasitol del carpincho Hydrochoerus hydrochoerus y del guiya (Myocastor coypus), Rev. Fac. Cienc. Vet. La Plata, 21, 161, 1967.

103. **Bogitsh, B. J.,** Cytochemical and ultrastructural observations on the tegument of the trematode Megalodiscus temperatus, Trans. Am. Micros. Soc., 87, 477, 1968.

104. **Bogitsh, B. J. and Cheng, T. C.,** Pisciamphistoma reynoldsi (Paramphistomatidae), a new parasite of Lepomis spp. in Virginia, J. Tenn. Acad. Sci., 34, 159, 1959.

105. **Boray, J. C.,** Studies on intestinal amphistomosis in cattle, Aust. Vet. J., 35, 282, 1959.

106. **Boray, J. C.,** Studies on intestinal paramphistomosis in sheep due to Paramphistomum ichikawai Fukui, 1922, Vet. Med. Rev., 4, 290, 1969.

107. **Boray, J. C.,** Molluscan host and trematodes in the Pacific Basin, in Biology and Control of Endoparasites, Symons, L. E. A., Donald, A. D., and Dinee, J. K., Eds., Academic Press, Sydney, 1982, 81.

108. **Bortnovskiĭ, P. F.,** Contribution to the study of eggs and miracidia of Paramphistomum cervi (Zeder, 1790) (Paramphistomatidae: Trematoda) (in Ukrainian), in Tvarinii Svit Zahidnik Raioniv Ukraini, Vidav. "Naukova Dumka," Kiev, 1964, 12.

109. **Bouchard, J. L.,** The Platyhelmintes parasitizing some northern Maine amphibia, Trans. Am. Microsc. Soc., 70, 245, 1951.

110. **Bourgat, R.,** Trématodes d'Amphibiens du Togo, Bull. Mus. Natl. Hist. Nat. Paris, 1, 597, 1979.

111. **Bourgat, R. and Kulo, S.-D.,** Recherches sur le cycle biologique d'un Paramphistomidae (Trematoda) d'amphibiens en Afrique, Ann. Parasitol. Hum. Comp., 52, 7, 1977.

112. **Bozhkov, D.,** Experimental study of the transfer of mature helminths from Rana ridibunde to R. temporaria (in Bulgarian), Khelmintologiya, 6, 7, 1978.

113. **Brace, E. C., Casey, B. R., Cossham, R. B., McEwan, J. M., McFarlane, B. G., Macken, J., Monro, P. A., Moreland, J. M., Northen, J. B., Street, R. J., and Yaldwyn, J. C.,** The frog, Hyla aurea as a source of animal parasites, Tautara, 5, 13, 1953.

114. **Bracegirdle, J. R.,** A case of Gastrodiscus aegyptiacus infection in a horse in Ethiopia, Vet. Rec., 93, 561, 1973.

115. **Brandes, G.,** Die Gattung Gastrothylax, Abh. Naturforsch. Ges. Halle, 21, 195, 1898.

116. **Brandt, B. B.,** Parasites of certain North Carolina Salientia, Ecol. Monogr., 6, 491, 1936.

117. **Bravo-Hollis, M.,** Revision de los generos Diplodiscus Diesing, 1836 y Megalodiscus Chandler, 1923 (Trematoda: Paramphistomoidea). I, II, An. Inst. Biol. Univ. Nac. Auton. Mex., 12, 127, 1941.

118. **Bravo-Hollis, M.,** Estudio systematico de los trematodos de los "ajolotes" de Mexico, An. Inst. Biol. Univ. Nac. Auton. Mex., 14, 141, 1943.

119. **Bravo-Hollis, M. and Caballero Deloya, J.,** Catalogue of the helminthological collection of the Institute of Biology, Publ. Estec. Inst. Biol. Univ. Nat. Auton. Mex., 2, 138, 1973.

120. **Brenes, R. R.,** Catalogo de los helmintos parasitos de costa Rica, Rev. Biol. Trop., 9, 67, 1961.

121. **Brenes, R. R ., Jiménez-Quirós, O., Sancho, G. A., and Flores, E. D.,** Helmintos de la República de Costa Rica. XIII. Algunos tremátodos de *Rana pipiens.* Decriptión de *Glypthelmins facioci* n.sp., *Rev. Biol. Trop.*, 7, 191, 1959.

122. **Brooks, D. R.,** A review of the genus *Allassostomoides* Stunkard, 1924 (Trematoda: Paramphistomidae) with a redescription of *A. chelydrae* (MacCallum, 1919) Yamaguti, 1958, *J. Parasitol.*, 61, 882, 1975.

123. **Brooks, D. R.,** Parasites of amphibians of the Great Plains. II. Platyhelminthes of amphibians in Nebraska, *Bull. Univ. Nebr. State Mus.*, 10, 65, 1976.

124. **Brooks, D. R.,** New records for amphibian and reptile trematodes, *Proc. Helminthol. Soc. Wash.*, 42, 286, 1979.

125. **Brooks, D. R. and Mayers, M. A.,** Plathyhelminths of Nebraska turtles with description of two new species of spirorchiids (Trematoda: Spirorchiida), *J. Parasitol.*, 61, 403, 1975.

126. **Brooks, D. R. and Buckner, R. L.,** Some platyhelminth parasites of sirens (Amphibia: Sirenidae) from North America, *J. Parasitol.*, 62, 906, 1976.

127. **Brooks, D. R. and Fusco, A. C.,** Some digenetic trematodes from caudate amphibians in the southeastern United States, *J. Miss. Acad. Sci.*, 23, 95, 1978.

128. **Brumpt, E.,** Contribution a l'etude de l'evolution des paramphistomids *P. cervi* et cercaire de *Planorbis exustus. Ann. Parasitol. Hum. Comp.*, 14, 552, 1936.

129. **Brundin, L. Z.,** Croizat's panbiography versus phylogenetic biogeography, in *Vicariance Biogeography: a Critic*, Nelson, G. and Rosen, D. E., Eds., Columbia University Press, New York, 1981, 94.

130. **Bryden, M. M.,** The anthelmintic efficiency of niclosamide and menichlopholan in the treatment of intestinal paramphistomosis in sheep, *Aust. Vet. J.*, 45, 133, 1969.

131. **B'tchvarov, G. K.,** The helminth fauna of tailed amphibians (Amphibia-Caudata) in Bulgaria (in Bulgarian), *Nauchn. Tr. Plovdiv. Univ. P. Khilendarski*, 10, 163, 1972.

132. **B'tchvarov, G .,** Study of helminth fauna of ecaudata amphibians (Amphibia-Ecaudata) in the north east of Bulgaria (in Bulgarian), *Nauchn. Tr. Plovdiv. Univ. P. Khilendarska*, 11, 70, 1973.

133. **B'tchvarov, G.,** Study of helminth fauna of ecaudata amphibians (Amphibia-Ecaudata) in the north-east of Bulgaria (in Bulgarian), *Nauchn. Tr. Plovdiv. Univ. P. Khilendarski*, 11, 70, 1978.

134. **Buckley, J. J. C.,** On a new amphistome cercaria (Diplocotylea) from *Planorbis exustus, J. Helminthol.*, 17, 25, 1939.

135. **Buckley, J. J. C.,** A helminthological survey in Northern Rhodesia, *J. Helminthol.*, 21, 111, 1946.

136. **Burgu, A.,** Belisli bazi laboratuvar hayvalarinin *Paramphistomum cervi* metaserkerleri ile deneysel enfeksiyonu (in Turkish), *Ankara Univ. Vet. Fak. Derg.*, 28, 93, 1981.

137. **Burgu, A.,** Behavior of *Paramphistomum cervi* cercariae towards different coloures during encystation, *Ankara Univ. Vet. Fak. Derg.*, 29, 143, 1982.

138. **Bush, A. O. and Samuel, W. M.,** A review of helminth communities in beaver (*Castor* spp.) with a survey of *Castor canadensis* in Alberta (Canada), in *Proc. First Worldwide Furbearer Conf.*, Chapman, J. A. and Pursley, D., Eds., 1981, 678.

139. **Butler, R. W. and Yeoman, G. H.,** Acute intestinal paramphistomiasis in zebu cattle in Tanganyika, *Vet. Rec.*, 74, 222, 1962.

140. **Bwangamoi, O.,** Helminth parasites of domestic and wild animals in Uganda, *Bull. Epizoot. Dis. Afr.*, 16, 429, 1968.

141. **Byrd, E. E. and Reiber, R. J.,** Larval flukes from Tennessee. III. Studies on cercariae from *Helisoma trivolvis* Say, with description of new species, *Rec. Rellfoot Lake Biol. Stat.*, 4, 132, 1940.

142. **Caballero, C. E.,** Revision de las especies que actualmente formal el género *Heronimus* MacCallum, 1902, *An. Inst. Biol. Univ. Nac. Auton. Mex.*, 11, 225, 1940a.

143. **Caballero, C. E.,** Trematádos de las tortugas de Mexico. I, *An. Inst. Biol. Univ. Nac. Auton. Mex.* 11, 1259, 1940b.

144. **Caballero, C. E.,** *Zygocotyle lunata* (Diesing, 1835) en el gonado vanuco de Mexico, *An. Inst. Biol. Univ. Nac. Auton. Mex.*, 11, 209, 1940c.

145. **Caballero, C. E.,** Trematodos de los Tortugas de Mexico. III. Decripcion de un nuevo género de la subfamilia Cladorchinae Fischoeder, 1901, y redescripcion de *Dermatemytrema trifoliata* Price, 1937, *An. Inst. Biol. Univ. Nac. Auton. Mex.*, 14, 161, 1943.

146. **Caballero, C. E.,** *Stichorchis subtriquetrus* (Rudolphi, 1814) en los castrores del estado Nuevo Leon, *Mexico, An. Inst. Biol. Univ. Nac. Auton. Mex.*, 18, 165, 1947.

147. **Caballero, C. E.,** Helmintos de la Republica de Panamá. XIII. Una nueva especie de *Catadiscus* Cohn, 1904 (Trematoda: Digenea), *Rev. Iber. Parasitol. Libro-Homenaja Lopez-Neyra*, 23, 1955.

148. **Caballero, C. E.,** Trematodos de los Tortugas de Mexico. VIII. Descripcion de un nuevo género que parasita a torugas de agua dulce, *Ciencia*, 3, 61, 1961.

149. **Caballero, C. E. and Sokoloff, D.,** Un nuevo trematodo amfistoma parasito del intestino de una tortuga de agua dulce *Dermatemys mawii* Gray, *Schizamphistomoides tabascensis* n.sp., *An. Inst. Biol. Univ. Nac. Auton. Mex.*, 5, 41, 1934.

150. **Caballero, C. E., Zerecero, M. C., and Grocott, R. G.**, Helmintos de le Republica de Panamá. XV. Trematodos de *Chelone mydas* (L.), totuga marina comestible del Oceano Pacifico del morte 2a parte, *An. Inst. Biol. Univ. Nac. Auton. Mex.*, 26, 149, 1955.

151. **Caballero, C. E., Zerecero, D. M., and Grocott, R. G.**, Helmintos de la Republica de Panamá. XXI. Algunos trematodos de quelonios de agua dulce (2a parte), *An. Inst. Biol. Univ. Nac. Auton. Mex.*, 29, 181,, 1958.

152. **Caballero, C. E. and Diaz-Ungria, C.**, Intento de un catalogo de los trematodos digeneos registrados en territorio Venezolano, *Mem. Soc. Cienc. Nat. La Salle*, 18, 18, 1958.

153. **Cable, R. M.**, *Opistholebes diodontis* n. sp., its development in the final host, the affinities of some amphistomous trematodes from marine fishes and the Allocreadidoid problem, *Parasitology*, 46, 1, 1956.

154. **Caeiro, V. M. P.**, Acerca de alguns "Paramphistominae" não assinalados em territórios Portugueses, *Rev. Cienc. Vet.*, 56, 68, 1961

155. **Calzada-Verella, C.**, *Balanorchis anastrophus* Fischoeder trematodes, parásito del estómago de dos bovinos del Uruguay, *Bol. Mens. Dir. Ganad. Urug.*, 24, 287, 1940.

156. **Cameron, T. W. M.**, Studies on the endoparasitic fauna of Trinidad mammals, *Can. J. Res.*, 14, 165, 1936.

157. **Canyngham, H. F.**, A new trematode of man *(Amphistoma watsoni)*, *Bull. Med. J.*, 2, 663, 1904.

158. **Căpuşe, I. and Dancău, D.**, Contribuţii la studial helmintofaunei anurelor din R. P. R. (in Rumanian), *An. Univ. C. I. Parhon, Bucuresti*, 16, 141, 1957.

159. **Cary, L. R.**, The life history of *Diplodiscus temperatus* Stafford with especial reference to the development of parthenogenetic eggs, *Zool. J. Anat.*, 28, 595, 1909.

160. **Catalano, P. A. and White, A. M.**, Helminths of the salamanders *Gyrinophilus porphyriticus*, *Pseudotriton ruber*, and *Pseudotriton montanus* (Caudata: Plethodontidae) from Ohio, *Ohio J. Sci.*, 82, 120, 1982.

161. **Cawston, F. G.**, Bilharziosis in South Africa, *J. Am. Med. Assoc.*, 70, 439, 1918.

162. **Chandler, A. C.**, Three new treamtodes from *Amphiuma means*, *Proc. U. S. Natl. Mus.*, 63, 55, 1923.

163. **Chatterji, R. C.**, Helminth parasites in fresh-water turtles, *Rec. Ind. Mus.*, 38, 81, 1936.

164. **Chatterji, R. C.**, On a new genus of amphistomes (Trematode) from a silurid fish of Rangoon, *Rec. Ind. Mus.*, 40, 337, 1938.

165. **Chellappa, D. J.**, Observations on the metazoan parasitic infestations of sheep in Tamil Nadu (India), *Abst. Int. Congr. Parasitol. Baroda, Indai*, 45, 1977.

166. **Chen, H. T.**, Animal parasites of hogs, goats and buffaloes from Hong Kong, *Hong Kong Nat.*, 6, 102, 1935.

167. **Chen, H. T.**, Parasites in slaughter houses in Canton. III. Trematodes and Cestodes parasitic in the alimentary tract of buffaloes, *Lingnan Sci. J.*, 16, 157, 1937.

168. **Chen, H. T. et al.**, Paramphistomata (Szidat, 1936) Skrjabin et Schulz, 1937, in *Fauna sinica, Platyhelminthes, Trematoda, Digenea. I.*, Science Press, Beijing, 1985, 230.

169. **Chinone, S.**, Paramphistomes of domestic animals (in Japanese), *J. Jpn. Vet. Med. Assoc.*, 32, 661, 1979.

170. **Chinone, S. and Itagaki, H.**, A preliminary report on the development of *Homalogaster paloniae* Poirier, 1883, a bovine intestinal paramphistome, *Jpn. J. Vet. Sci.*, 39, 665, 1977.

171. **Chinone, S., Itagaki, H., Miyazawa, M., Asai, M., Mochizuki, A., Ito, K., Usui, K., and Mochizuki, T.**, *Homalogaster paloniae* Poirrier, 1883, a caecal paramphistome, obtained Shizuoka city in Japan (in Japanese), *Bull. Azabu Vet. Coll.*, 2, 317, 1977.

172. **Chiriac, E. and Malcoci, E.**, Contribuţii la cunoaşterea helmintilor amfibienilor din regiunea Jaşi (in Rumanian), *An. Stiint. Univ. Al. I. Cuza Iasi*, 14, 301, 1968.

173. **Chiriac, E. and Udrescu, M.**, *Fauna Republicii Socialiste Romănia, Plathelminthes, Trematoda*, (in Rumanian), Acad. Rep. Soc. Rom., 1973, 218.

174. **Christian, F. A. and White, L. L.**, The genus *Allassostomoides* Stunkard, 1924, with description of *Allassostomoides louisianaensis* n. sp. (Trematoda: Paramphistomidae) from the pig frog, *Rana grylio*, in Lousiana, *Am. Midl. Nat.*, 90, 218, 1973.

175. **Chroust, K.**, Efficiency of the anthelmintic Terenol (Resorantel) against the fluke *Paramphistomum (Liorchis) scotiae* in cattle and against the tapeworms *Monezia* spp. in lambs, *Acta Vet. (Brno)*, 42, 281, 1973.

176. **Chu, J. K.**, Studies on amphistomes in Korean cattle, *Korean J. Parasitol.*, 10, 34, 1972.

177. **Cobbold, T. S.**, On the destruction of elephants by parasites: with remarks on two new species of Entozoa on the so-called earth-eating habits of elephants and horses in India, *Veterinarian*, 733, 1875.

178. **Cohn, L.**, Zur Kenntnis einiger Trematoden, *Zentralbl. Bakteriol. Parasitenkde Infektionskr.*, 34, 35, 1903.

179. **Cohn, L.**, Helmintologische Mittheilungen II., *Arch. Naturgesch.* 70, 229, 1904.

180. **Colvin, H. J.**, The longevity of miracidia of *Paramphistomum microbothrioides* in various media, *Proc. La. Acad. Sci.*, 25, 109, 1962.

181. **Combes, C., Léger, N., and Vidal, D.**, Inventaire des helminthes de *Rana esculenta* L. (amphibien, anoure) dans l'ile de Corse, *Ann. Parasitol. Hum. Comp.*, 58, 761, 1974.

182. **Cordero, E. H. and Vogelsang, E. G.**, Nuevos trematodos. II. Cuatro "Paramphistomidae" de los quelonios sudamericanos, *Rev. Med. Vet. Parasitol. Caracas*, 2, 3, 1940.

183. **Correa, A. A. S. and Artigas, P. T.**, *Catadiscus rochai* n. sp. (Trematoda: Paramphistomidae), parasito de *Dromicus typhlus* (L.) (Ophidia: Colubridae), *Mem. Inst. Butantan Sao Paolo*, 42, 145, 1978.

184. **Cort, W. W.**, Larval trematodes from North American fresh water snails. Preliminary report, *J. Parasitol.*, 1, 65, 1914.

185. **Cort, W. W.**, Some North American larval trematodes, *Ill. Biol. Monogr.*, 1, 447, 1915.

186. **Cracraft, J.**, Phylogenetic models and classification, *Syst. Zool.*, 23, 71, 1974.

187. **Craig, J. P. and Davies, G. O.**, *Paramphistomum cervi* in sheep, *Vet. Rec.*, 49, 1116, 1937.

188. **Crandall, R. B.**, The life-history and affinities of the turtle lung-fluke *Heronimus chelydrae* MacCallum, 1902, *J. Parasitol.*, 46, 289, 1960.

189. **Croizat, L.**, *Space, Time and Form: the Biological Synthesis*, published by the author, Caracas, 1964, 76.

190. **Crusz, H.**, A new amphistome fluke, *Indosolenorchis hirudinaceus* gen. et sp. nov. from the caecum of a dugong from the Indian Ocean, *Ceylon J. Sci. Biol. Sci.*, 24, 135, 1951.

191. **Crusz, H.**, The nature, incidence and geographical distribution of amphistome infestations in neat cattle, buffaloes and goats in Ceylon, *Ceylon J. Sci.*, 25, 60, 1952.

192. **Crusz, H.**, Parasites of endemic and relict vertebrates: a biogeographical review, in *Ecology and Biography in Sri Lanka*, Fernando, C. H., Ed., Junk, The Hague, 1984, 321.

193. **Crusz, H. and Fernand, V. S. V.**, The trematode parasites of the dugong with descriptions of two monostome and histopathological changes in the host, *J. Parasitol.*, 40, 499, 1954.

194. **Crusz, H. and Nagaliyadde, L.**, Parasites of the relict fauna of Ceylon. VII. General considerations and first host-parasite checklist, *C. R. Seances Soc. Biogeogr.* 477, 85, 1978; *Addenda and corrigenda, C. R. Seances Soc. Biogeogr.*, 489, 87, 1980.

195. **Cruz e Silva, J. A.**, Contribuičao para o estudo dos helmintoses dos espéecies pecuárias do sul do Save, *Vet. Mosambb. Louruco Morques,*, 4, 33, 1971.

196. **Cvetković, Lj.** Acute intestinal paramphistomiasis of sheep. First outbreak reported from Yugoslavia, *Vet. Glas.* 22, 41, 1968.

197. **Daday, E.**, In Südamerikanischen Fischen lebende Trematoden-Arten, *Zool. Jahrb.*, 24, 470, 1907.

198. **Darlington, P. J.**, The geographical distribution of cold-blooded vertebrates, *Q. Rev. Biol.* 23, 1, 1948.

199. **Darlington, P. J.**, *Zoogeography: the Geographical Distribution of Animals*, John Wiley & Sons, New York,, 1957, 675 pp.

200. **Davydova, I. V.**, To the demonstration of the calicophoroniasis in cattle and in sheep in the U.S.S.R. (in Russian), *Helminthologia*, 1, 31, 1959.

201. **Davydova, I. V.**, A new trematode-*Ceylonocotyle petrowi* sp. nov. (Paramphistomatidae) from a deer (*Cervus nippon*) in the Far-East of the U.S.S.R. (Littoral region) (in Russian), *Helminthologia*, 3, 63, 1961.

202. **Davydova, I. V.**, *Cotylophoron vigisi* n. sp. from cattle in the Primorsk region (in Russian), *Tr. Vses. Inst. Gelminthol.*, 10, 3, 1963.

203. **Davydova-Velitchko, I. V.**, On the geographical distribution of various paramphistomid infections in ruminants in the U.S.S.R. (in Russian), *Tezisy Dokl. Nauchn. Konf. Vses. Ova. Gelminthol.*, 1, 49, 1962.

204. **Davydova-Velitchko, I. V.**, Occurrence of *Paramphistomum ichikawai* Fukui, 1922 in cattle in the U.S.S.R. (in Russian), *Tr. Vses. Inst. Gelminthol.*, 10, 3, 0964.

205. **Dawes, B.**, On a collection of Paramphistomatidae from Malaya, with revision of the genera *Paramphistomum* Fischoeder, 1901, *Parasitology*, 28, 330, 1936.

206. **Daynes, P.**, Note sur les helminthoses des animaux domestiques recornues a Madagascar, *Rev. Elev. Med. Vet. Pays Trop.*, 17, 477, 1964.

207. **Daynes, P. and Graber, M.**, Principales helminthoses des animaux domestiques en Ethiopic, *Rev. Elev. Med. Vet. Pays Trop.*, 27, 301, 1974.

208. **Deiana, S.**, L'infestione del *Bulinus contortus* da cercariae di *Schistosoma bovis* (Sonsino, 1876) e del *Paramphistomum cervi* (Schrank, 1970) in diverse stagioni dell'anno, *Boll. Soc. Ital. Sper.*, 29, 1939, 1953.

209. **De Leon, D. D. and Juplo, R. J.**, Gastrointestinal helminth parasites of Philippine carabaos (*Bubalus bubalis* L.), *J. Parasitol.*, 52, 1214, 1966.

210. **Deschiens, R.**, Relation de quatte cos l'infestation par *Watsonius watsoni* chez le papion, *Bull. Soc. Pathol. Exot.*, 33, 369, 1940.

211. **Deveral, M.**, Redescription and early development of *Isoparorchis hypselobagri* (Billet, 1898) (Paramphistomidae; Caballeroinae n. subf.) together with an account of a cryptogonimid metacercariae from the fishes of the Bhavanisgar Reservoir, *Indian J. Anim. Sci.*, 42, 51, 1972.

212. **Dhingra, O. P.**, Spermatogenesis of a digenetic trematode, *Gastrothylax crumenifer*, *Res. Bull. Panjab Univ. Sci.*, 65, 11, 1955a.

213. **Dhingra, O. P.**, Spermatogenesis of a digenetic trematode *Cotylophoron elongatum*, *Res. Bull. Panjab Univ. Sci.*, 64, 1, 1955b.

214. **D'Huart, J. P.**, Note sur la pathologie et le parasitisme chez le phacochere (*Phacochoerus aethiopicus* Pallas), *Acta Zool. Pathol. Antverp.*, 58, 41, 1974.

215. **Diaw, O. T., Samnaliev, P., Pino, L. A., Bayssade-Dulfour, Ch., Albaret, J. L., and Vassiliades, G.,** Structures argyrophiles des formes larvaires de deux souches de *Paramphistomum microbothrium*: l'une parasite d' *Isidora guernei* et *Ovis aries*, l'autse parasite d' *Isidora truncata* et *Bos taurus*, *Ann. Parasitol. Hum. Comp.*, 58, 455, 1983.

216. **Diaw, O. T., Bayssade-Dufour, Ch., Pino de Morales, L. A., Albaret, J.-L., and Vassiliades, G.,** Chétotaxie de la cercaire de *Gastrodiscus aegyptiacus* (Trematoda: *Paramphistomoidae*) parasite du cheval, *Ann. Parasitol. Hum. Comp.*, 61, 289, 1986.

217. **Diaw, O. T., Seye, M., and Sarr, Y.,** Epidemiologie des trematodoses du betail dans la region de Kolda, Casamance (Senegal), *Rev. Elev. Med. Vet. Pays Trop.*, 41, 257, 1988.

218. **Diaz-Ungria, C.,** Identificatión de una colleción de parásitos de vertebrados venezolanos, *Bol. Soc. Venez. Cienc. Nat.*, 27, 113, 1968.

219. **Diaz-Ungria, C.,** Estudio de una colleción de parásitos en el Estado Zulia, *Acta Cient. Venez.*, 20, 126, 1969.

220. **Diaz-Ungria, C.,** Algunas especies de helmintos nuevos para Venezuela, *Rev. Iber. Parasitol.*, 39, 313, 1979.

221. **Diesing, C. M.,** Monographie der Gattungen *Amphistoma* und *Diplodiscus*, *Ann. Wien Mus. Naturgesch.*, 1, 235, 1836.

222. **Diesing, C. M.,** Abbildungen neuer Gattungen brasilianischer Binnerwürmer (Entozoa), *Ber. Vers. Dtsch. Naturforsch. Aerzte Prag*, 189, 1838.

223. **Diesing, C. M.,** Neue Gattungen von Binnenwürmer nebst einem Nachtrage zur Monografic d er Amphistomen, *Ann. Wien Mus. Naturgesch.*, 2, 219, 1839.

224. **Dinnik, J. A.,** On *Ceylonocotyle scoliocoelium* (Fischoeder, 1904) and its intermediate host in Kenya, East Africa, *J. Helminthol.*, 30, 149, 1956.

225. **Dinnik, J. A.,** *Paramphistomum phillerouxi* sp. nov. (Tremata: Paramphistomatidae) and its development in *Bulinus forskalii*, *J. Helminthol.*, 35, 69, 1961.

226. **Dinnik, J. A.,** *Paramphistomum daubneyi* sp. nov. from cattle and its snail host in the Kenya Highland, *Parasitology*, 52, 143, 1962.

227. **Dinnik, J. A.,** *Paramphistomum sukumum* sp. nov. and other stomach-flukes from cattle in the Sukumaland area of the lake Region, Tanganyika, *Parasitology*, 54, 201, 1964.

228. **Dinnik, J. A.,** The snail hosts of certain Paramphistomatidae and Gastrothylacidae (Trematode) discovered by the late Dr. P. L. LeRoux in Africa, *J. Helminthol.*, 39, 141, 1965.

229. **Dinnik, J. A. and Dinnik, N. N.,** The life-cycle of *Paramphistomum microbothrium* Fischoeder, 1901 (Trematoda, Paramphistomidae), *Parasitology*, 44, 285, 1954.

230. **Dinnik, J. A. and Dinnik, N. N.,** Stomach flukes (Trematoda: Paramphistomidae) found in cattle, sheep and goats in the Highlands of Kenya, *Annu. Rep. East Afr. Vet. Res. Org.*, 1954/55, 84, 1955.

231. **Dinnik, J. A. and Dinnik, N. N.,** Developoment of *Paramphistomum sukari* Dinnik, 1954 (Treematoda: Paramphistomidae) in a snail host, *Parasitology*, 47, 209, 1957.

232. **Dinnik, J. A. and Dinnik, N. N.,** Development of *Carmyerius exoporus* Maplestone (Trematoda: Gastrothylacidae) in a snail host, *Parasitology*, 50, 469, 1960.

233. **Dinnik, J. A. and Dinnik, N. N.,** The growth of *Paramphistomum microbothrium* Fischoeder to maturity and its longevity in cattle, *Bull. Epizoot. Dis. Afr.*, 10, 27, 1962.

234. **Dinnik, J. A., Walker, J. B., Barnett, S. F., and Brocklesby, D. W.,** Some parasites obtained from game animals in western Uganda, *Bull. Epizoot. Dis. Afr.*, 11, 37, 1963.

235. **Dinnik, J. A. and Hammond, J. A.,** Division of helminth diseases, *Annu. Rep. East Afric. Vet. Res. Org. 1967*, 47, 1968.

236. **Dobbin, J. Ee.,** Fauna helminthologica de batraquios de Pernambuco, Brasil. I. Trematoda, *An. Soc. Biol. Pernambuco*, 15, 23, 1957.

237. **Dodbiba, A.,** Mbi helmintofaunen gastrointestinale te dheve (*Ovis aries*) (in Albanian), *Bul. Shken. Bujq.*, 8, 130, 1969.

238. **Dollfus, R. Ph.,** Mission Saharienne Augiéras-Draper 1927—28. Trematodes de Mammiferas, Oiseaux et Poissons, *Bull. Mus. Natl. Hist. Nat. Paris*, 4, 555, 1932.

239. **Dollfus, R. Ph.,** Trematodes Récolté an Congo Belge par le Professeur Paul Brien (mai—aout 1937), *Ann. Mus. R. Congo Belge Ser. 5*, 1, 1, 1950.

240. **Dollfus, R. Ph.,** Miscellanea helminthologica Maroccana. I. Quelgues Trematodes, Cestodes et Acanthocephales, *Arch. Inst. Pasteur Maroc*, 4, 104, 1951.

241. **Dollfus, R. Ph.,** Station experimentale de parasitologie de Richelien (Indre-ettorie). Contribution a la faune parasitaise regionale. II, *Ann. Parasitol. (Paris)*, 36, 266, 1961.

242. **Dollfus, R. Ph.,** Variations intraspecifiques chez un *Carmyerius* (Trematoda: Gastrothylacidae) parasite de buffle du Congo belge, *Ann. Parasitol. Hum. Comp.,,* 37, 108, 1962.

243. **Dollfus, R. Ph.,** Un trématodee peu conno parasite du caecum de primate catorrhinien cynomorphe du Vietnam, *Bull. Soc. Pathol. Exot.*, 56, 442, 1963a.

244. **Dollfus, R. Ph.,** Hotes et lieux de récolte de quelques Trématodes digénétiques de Vertébrés de la Collection du Musée royal de l'Afrique centrale, *Rev. Zool. Bot. Afr.*, 48, 323, 1963b.

245. **Drożdż, J. and Malczewski, A.,** Endoparasites et maladies parasitaires des animaux domestiques au Vietnam, *Lab. Parasitol. Acad. Pol. Sci. Warsaw,* 1967, 43.

246. **Duarte, M. J. F.,** Helmintos parasitos dos animais domesticos no Estado do Rio de Janeiro, *Arq. Esc. Vet. Belo Horizonte,* 33, 67, 1981.

247. **Dubinin, V. B.,** Parasitic fauna of the boars in the Volga delta (in Russian), *Ref. rabot uchrezd. otd. nauk. ANUSSR, 1940, Izdv. ANUSSR,* Moscow - Leningrad, 1941, 193.

248. **Dubinina, N. M.,** Ecological examination of parasitic fauna of marsh-frog (*Rana ridibunda* Pall.) in the Volga estuary (in Russian), *Parazitol. Sb. Akad. Nauk SSSR,* 12, 300, 1950.

249. **Dubinina, M. N.,** Dynamics of parasitic fauna of the grass snakes along the coastal part of the delta of the river Volga (in Russian), *Tr. Zool. Inst. Akad. Nauk SSSR.,* 13, 171, 1953.

250. **Dubois, G.,** Trématoda (Matérioux de la mission scieintifique Suisse an Angola), *Bull Soc. Neuchatel. Sci. Nat.,* 55, 73, 1931.

251. **Dunn, T. S., Hanna, R. E. B., and Nizami, W. A.,** Ultrastructural and histochemical observations on the epidermis, presumptive tegument and glands of the miracidium of *Gigantocotyle explanatum* (Trematoda: Paramphistomidae), *Int. J. Parasitol.,* 17, 885, 1987.

252. **Durie, P. H.,** The paramphistomes (Trematoda) of Australian ruminants. I. Systematics, *Proc. Linn. Soc. N. S. W.,* 76, 41, 1951.

253. **Durie, P. H.,** The paramphistomes (Trematoda) of Australian ruminants. II. The life history of *Ceylonocotyle streptocoelium* (Fischoedere) Näsmark and of *Paramphistomum ichikawai* Fukui, *Aust. J. Zool.,* 1, 193, 1953.

254. **Durie, P. H.,** The paramphistomes (Trematoda) of Australian ruminants. III. The life-history of *Calicophoron calicophorum* (Fischoeder) Näsmark, *Aust. J. Zool.,* 4, 152, 1956.

255. **Dutt, S. C.,** Paramphistomes of bovins with description of *Gastrothylax indicus* n. sp., *Indian J. Parasitol.,* 2, 39, 1978.

256. **Dutt, S. C. and Srivastava, H. D.,** The intermediate host and the cercaria of *Gastrodiscoides hominis* (Trematoda: Gastrodiscidae). Preliminary report, *J. Helminthol.,* 40, 45, 1966.

257. **Dutt, S. C. and Srivastava, H. D.,** The life history of *Gastrodiscoides hominis* (Lewis and McConnel, 1876) Leiper, 1913 — the amphistome parasite of man and pig, *J. Helminthol.,* 46, 35, 1972.

258. **Dvoryadkhin, V. A. and Besprozvannykh, V. V.,** The biology of *Paramphistomum ichikawai* and *Calicophoron ijimai* (Paramphistomatidae) — the agents of paramphistomiasis of domestic ruminants in the Primor'e and Priamur'e (U.S.S.R.), (in Russian), in *Biologiya i sistematika gel' mintov zhivotnykh Dal' nevo Vostoka, Vladivostok,* Izdv. AN SSR, 1981, 3.

259. **Dvoryadkhin, V. A., Ermolenko, A. V., and Besprozvannykh, V. V.,** On the biology of *Paramphistomum petrowi* — a parasite of sika dear in the Primor'e region (in Russian), *Parazitologiya,* 17, 311, 1983.

260. **Dwivedi, M. P.,** *Kachugotrema amboinensis* n.g., n.sp. (Paramphistomidae), *Indian J. Helminthol.,* 19, 45, 1967.

261. **Dwivedi, M. P. and Chauhan, B. S.,** On some digenetic trematodes. II, *J. Zool. Soc. India,* 22, 87, 1970.

262. **Edelényi, B.,** A Szeged környéki békák belső élősködő férgei (in Hungarian), *Állattani Közl.,* 39, 1, 1942.

263. **Edgar, G.,** Paramphistomiasis of young cattle, *Aust. Vet. J.,* 14, 27, 1938.

264. **Eduardo, S. L.,** *Carmyerius synethes* (Fischoeder, 1901) Stiles and Goldberger, 1910 (Gastrothylacidae) a pauched amphistome of Philippine carabao, *Philipp. J. Vet. Med.,* 14, 117, 1975.

265. **Eduardo, S. L.,** On two species of the genus *Calicophoron* Näsmark, 1937 (Digenea: Paramphistomidaee) from Philippine carabao, *Philipp. J. Vet. Med.,* 15, 96, 1976.

266. **Eduardo, S. L.,** Taxonomic value of tegumental structures in the identification of some species of the family paramphistomidae Fischoeder, 1901 occurring in mammals, *Proc. Third EMOP, Cambridge,* 1980a, 196.

267. **Eduardo, S. L.,** *Orthocoelium indonesiense,* a new species of amphistome from ruminants of Indonesia, *Syst. Parasitol.,* 1, 203, 1980b.

268. **Eduardo, S. L.,** A new genus, *Leiperocotyle* for *Cotylophoron okapi* Leiper, 1935 and *C. congolense* Baer, 1936 and redescription of *C. okapi, Syst. Parasitol.,* 1, 255, 1980c.

269. **Eduardo, S. L.,** *Bilatorchis papillogenitalis* n.g., n.sp. (Paramphistomidae: Orthocoeliinae), a parasite of the red leche (*Kobus leche* Gray, 1850) from Zambia, *Syst. Parasitol.,* 1, 14, 1980d.

270. **Eduardo, S. L.,** On the specific identity of *Amphistoma tuberculata* Cobbold, 1875, *Syst. Parasitol.,* 2, 117, 1980e.

271. **Eduardo, S. L.,** The taxonomy of the family Paramphistomidae Fischoeder, 1901, with special reference to the morphology of species occurring in ruminants. I. General considerations, *Syst. Parasitol.,* 4, 7, 1982a.

272. **Eduardo, S. L.,** The taxonomy of the family Paramphistomidae Fischoeder, 1901 with special reference to the morphology of species occurring in ruminants. II. Revision of the genus *Paramphistomum* Fischoeder, 1901, *Syst. Parasitol.,* 4, 189, 1982b.

273. **Eduardo, S. L.,** The taxonomy of the family Paramphistomidae Fischoeder 1901, with special reference to the morphology of species occurring in ruminants. III. Revision of the genus *Calicophoron* Näsmark, 1937, *Syst. Parasitol.,* 5, 25, 1983.

274. **Eduardo, S. L.,** The taxonomy of the family Paramphistomidae Fischoeder, 1901 with special reference to the morphology of species occurring in ruminants. IV. Revision of the genus *Gigantocotyle* Näsmark, 1937 and the elevation of the subgenus *Explanatum* Fukui, 1929 to full generic status, *Syst. Parasitol.,* 6, 3, 1984.

275. **Eduardo, S. L.,** The taxonomy of the family Paramphistomidae Fischoedere, 1901 with special reference to the morphology of species occurring in ruminants. VI. Revision of the genus *Orthocoelium* (Stiles et Goldberger, 1910) Price et McIntosh, 1953, *Syst. Parasitol.,* 7, 125, 1985a.

276. **Eduardo, S. L.,** The taxonomy of the family Paramphistomidae Fischoeder, 1901, with special reference to the morphology of species occurring in ruminants. VII. Redescription of *Leiperocotyle congolense* (Baer, 1936) Eduardo, 1980 and a new name, *Leiperocotyle gretillati* for *Ceylonocotyle scoliocoelium var. benoiti.* Gretillat, 1966, *Syst. Parasitol.,* 7, 231, 1985b.

277. **Eduardo, S. L.,** The taxonomy of the family Paramphistomidae Fischoeder, 1901, with special reference to the morphology of species occurring in ruminants. V. Revision of the genus *Cotylophoron* Stiles et Goldberger, 1910, *Syst. Parasitol.,* 7, 3, 1985c.

278. **Eduardo, S. L.,** The taxonomy of the family Paramphistomidae Fischoeder, 1901 with special reference to the morphology of species occurring in ruminants. VIII. The genera *Stephanopharynx* Fischoeder, 1901, and *Balanorchis* Fischoeder, 1901, *Syst. Parasitol.,* 8, 57, 1986.

279. **Eduardo, S. L. and Manuel, M. F.,** Amphistomes (Trematoda: Digenea) of cattle and carabaos slaughtered in the Great Manila Area, *Philipp. J. Vet. Med.,* 14, 33, 1975.

280. **Eduardo, S. L. and Peralta, R. C.,** *Orthocoelium serpenticaecum* (Paramphistomidae: Orthocoeliinae), a new species from the swamp buffalo (*Bubalus bubalis* Linnaeus) in the Philippines, *Syst. Parasitol.,* 9, 221, 1987.

281. **Eduardo, S. L. and Javellana, C. R. H.,** A new genus, *Velasquezotrema,* for *Fischoederius brevisaccus* Eduardo, 1981, *Philipp. J. Vet. Med.,* 24, 29, 1987.

282. **Edwards, S. R. and Nahhas, F. M.,** Some endoparasites of fishes from the Sacramento-San Joaquin delta, California, *Calif. Fish Game,* 54, 247, 1968.

283. **Efford, J. E. and Tsumura, K.,** Observations on the biology of the trematode *Megalodiscus microphagus* in amphibians from Marion Lake, British Columbia, *Am. Midl. Nat.,* 82, 197, 1969.

284. **Eisa, A. M.,** Normal worm burden of cattle in upper Nile province, *Sudan J. Vet. Sci. Anim. Husb.,* 4, 63, 1963.

285. **El-Moukdad, A. R.,** Zur Helminthen fauna syrisher Rinder, *Angew. Parasitol.,* 20, 11, 1979.

286. **Erbolatov, K. M.,** The species composition and distribution of paramphistomes in ruminants in Kazakhstan (in Russian), *Byull. Vses. Inst. Gelmintol.,* 13, 55, 1974.

287. **Erhardová, B.,** K systematikému postavenií motilic z podčelcdi Paramphistominae ve stredni a jihovychodni Europě na základě morfologických znaků (in Czech.), *Čslká Parasit.,* 11, 145, 1964.

288. **Erhardová, B. and Kotrlý, A.,** Cizopasni červi zažívacího ústrojí našich volně žijících přežvykavců, (in Czech.) *Cslka Parasitol.,* 2, 41, 1955.

289. **Erich, K.,** Histologische Untersuchungen an Sporozysten, Redien und Zerkarien von *Paramphistomum cervi,* Zeder, 1790, in der Zwischen wirts schnecke *Planorbis planorbis,* Inau-Dissert., Tierarztlichen Fakultat, Ludwig-Maximilians-Universitat, München, 1983, 1.

290. **Erickson, A. B.,** Parasites of beavers, with a note on *Paramphistomum castori* Kofoid et Park, 1937 a synonym of *Stichorchis subtriquetrus, Am. Midl. Nat.,* 31, 625, 1944.

291. **Eslami, A. and Faizy, A.,** Gastrointestinal helminths of goats in Iran, *J. Vet. Univ. Teheran,* 31, 74, 1976.

292. **Euzeby, J.,** Les helminthes du bétail et du porc dans la Federation de Malaya, *Rev. Elev. Med. Vet. Pays Trop.,* 10, 15, 1957.

293. **Euzeby, J.,** Helminthic diseases of domestic ruminants and horses in France, in *Helminth Diseases of Cattle, Sheep and Horses in Europe,* Urquhart, G. M. and Amour, J., Eds., Glasgow, 1973, 148.

294. **Evranova, V. T.,** On the helminth parasites of moose (*Alces alces* L.) (in Russian), *Uchen. Zap. Kazan. Gos. Vet. Inst. Imeni N.E. Baumana,* 61, 151, 1954.

295. **Ezzat, M. A. E.,** Helminth Parasites of Some Ungulates from the Giza Zoological Gardens, Egypt, with an Appendix on Some Nematodes from the African Rhinoceros, Bull. 241, Vet. Sect., Tech. Sci. Serv., Ministry of Agriculture, Egypt (1943), 1, 1945.

296. **Fagbemi, B. O.,** The effects of environmental factors on the development, behaviour and survival of *Paramphistomum microbothrium* miracidia, *Vet. Parasitol.,* 16, 71, 1984.

297. **Fahmy, M. A. M., Mandour, A. M., Arafa, M. S., and Omran, L. A. M.,** On the larval trematodes recovered from *Cleopatra bulimoides* in Assiut governorate, *Assiut Vet. Med.,* 3, 4, 93, 1977.

298. **Fain, A.,** Contribution a l'étude des formes larvaires des Trematodes au Congo belge et spécialement de la larve de *Schistosoma mansoni, Mem. Acad. R. Sci. colon, Cl. Sci. Nat. Med.,* 22, 1, 1953.

299. **Farooq, M.,** *Schizamphistomoides lissemysi* a new trematode from a fresh water turtle *Lissemys punctata punctata, J. Sci. Phys. Sec.,* 1, 156, 1972.

300. **Farooq, M., Khanum, Z. and Ansar, R.,** Helminth parasites of fresh water tortoise and poisonous snake from Sind, Pakistan, *Pak. J. Zool.,* 15, 31, 1983.

301. **Faust, E. C.,** A biological survey of described cercariae in the United States, *Am. Midl. Nat.,* 53, 85, 1919.

302. **Faust, E. C.,** Notes on trematodes from the Philippines, *Philipp. J. Sci.,* 17, 627, 1920.

303. **Faust, E. C.,** Notes on larval flukes from China, *Parasitology,* 14, 248, 1922.

304. **Faust, E. C.,** Further observation South African larval trematodes, *Parasitology,* 18, 101, 1926.

305. **Faust, E. C.,** Notes on helminths from Panama. I. *Taxorchis schistocotyle* (Fischoeder, 1901) from the Panamien capybara, *Hydrocoerus isthmius* Goldman, 1912, *J. Parasitol.,* 21, 323, 1935.

306. **Fedorchenko, N. G.,** Bithional — an effective anthelmintics against apramphistomiasis (in Russian) *Veterinariya,* 43, 39, 1966.

307. **Fernandez, J. C. and Travassos, T. E.,** Lista dos helmintos parasitos des animais domésticos de Pernambuco, *An. Univ. Fed. Rural Pernambuco Cienc. Biol. Recife,* 3, 221, 1976.

308. **Fernando, A.,** Report on the helminth parasites of an Asian elephant which died in Singapure, *Ceylon Vet. J.,* 9, 99, 1961.

309. **Fishoeder, F.,** Die Paramphistomiden der Säugetiere, *Zool., Anz.,* 24, 367, 1901.

310. **Fischoeder, F.,** Die Paramphistomiden der Säugethiere, Inaug.-Dissert., Königsberg im Preussen, 1902, 1.

311. **Fischoeder, F.,** Die Paramphistomiden der Säugethiere, *Zool. Jahrb.,* 17, 485, 1903.

312. **Fischoeder, F.,** Beschreibung dreier Paramphistomiden-Arten aus Säugethieren, *Zool. Jahrb.,* 20, 453, 1904.

313. **Fischthal, J. H. and Kuntz, R. E.,** Trematode parasites of fishes from Egypt. I. *Basidiodiscus ectorchis,* n. gen. n. sp. and *Sandonia sudanensis* McClelland, 1957 (Paramphistomidae), *Proc. Helminthol. Soc. Wash.,* 26, 32, 1959.

314. **Fischthal, J. H. and Thomas, J. D.,** Digenetic trematodes of amphibians and reptiles from Ghana, *Proc. Helminthol. Soc. Wash.,* 35, 1, 1968.

315. **Fischthal, J. H. and Thomas, J. D.,** Digenetic trematodes of fish from the Volta river drainage system in Ghana prior to the constructuon of the Volta dam at Akosombo in May, 1964, *J. Helminthol.,* 46, 991, 1972.

316. **Fischthal, J. H. and Kuntz, R. E.,** Some trematodes of amphibians and reptiles from Taiwan, *Proc. Helminthol. Soc. Wash.,* 42, 1, 1975.

317. **Fischthal, J. H. and Kuntz, R. E.,** Additional records of trematodes of amphibians and reptiles from Taiwan, *Zool. Scr.,* 10, 161, 1981.

318. **Fitzsimmons, W. M.,** A host check of helminth parasites from domestic animals in Nyasaland, *Br. Vet. J.,* 120, 186, 1964.

319, **Fitzsimmons, W. M.,** Report on the helminthiasis research project, 1967—1970, Maseru, Lesotho, *Agric. Inf. Serv.,* 1, 1971.

320. **Fox, J. G. and Hall, W.,** Fluke (*Gastrodiscoides hominis*) infection in a rhesus monkey with related intussusception of the colon, *J. Am. Vet. Med. Assoc.,* 157, 714, 1970.

321. **Freitas, J. F. T.,** Novo trematodeo paramfistomideo parasito de rã — *Catadiscus inopinatus* n. sp., *Rev. Bras. Biol.,* 1, 121, 1941.

322. **Freitas, J. F. T.,** *Catadiscus mirandai* n. sp., *parasito de Hemipipa carvalhoi* Mir., *Rev. Bras. Biol.,* 3, 411, 1943.

323. **Freitas, J. F. T.,** Rapidas informações sôbre hospedadones es distribução geografica de alguns trematodeos parasitos de batraguios, *Atas Soc. Biol. Rio de Janeiro,* 4, 29, 1960.

324. **Freitas, J. F. T. and Lent., H.,** Revisão do gênero *Catadiscus* Cohn, 1904 (Trematoda: Paramphistomidea), *Bol. Inst. Biol. Bahia,* 4, 305, 1939.

325. **Freitas, J. F. T. and Lent, H.,** A propósito de *Halltrema avitellina* Lent et Freitas, 1939, *Rev. Bras. Biol.,* 2, 115, 1942.

326. **Freitas, J. F. T. and Dobin, J. E.,** Novo parasito de rã: *Catadiscus propinguis* sp. n. (Trematoda, Paramphistomidae), *Rev. Bras. Biol.,* 16, 439, 1956.

327. **Freitas, M. G. and Costa, H. M. A.,** Lista de helmintos parasitos dos animals domésticos do Brasil, *Arq. Esc. Sup. Vet.,* 12, 443, 1959.

328. **Freitas, M. G. and Costa, H. M. A.,** Nota sobre ocorrência de helmintos em animéésticos do Brasil, *Arq. Esc. Vet. Belo Hirozonte,* 21, 176, 1969.

329. **Fried, B.,** Infectivity, growth, development, excystation and transplantation of *Zygocotyle lunata* (Trematoda) in the chick, *J. Parasitol.,* 56, 44, 1970.

330. **Fried, B. and Nelson, P. D.,** Host-parasite relationship of *Zygocotyle lunata* (Trematoda) in the domestic chick, *Parasitology,* 77, 49, 1978.

331. **Fuhrmann, O.,** Zweite Klasse des Cladus Plathelminthes Trematodes, *Handbuch der Zoologie,* Vol. 2, Waltere der Gruyter, Berlin, 1928, 9.

332. **Fukui, T.,** A new amphistome parasite found in the stomach of *Bos taurus, Dobutsugaku Zasshi,* 34, 588, 1922.

333. **Fukui, T.,** Amphistomes of elephant (in Japanese), *Dobutsugaku Zasshi,* 38, 79, 1926.

334. **Fukui, T.,** Studies on Japanese amphistomatous parasites, with revision of the group, *Jpn. J. Zool.,* 2, 219, 1929.

335. **Gagarin, V. G.,** Helminth parasites of wild ruminants in Kirgiziya (in Russian), *Akad. Nauk Kirgiz. USSR Biol. Nauk,* 1, 133, 1959.

336. **Gassmann, M.,** Etude des Trematodes et Acanthocéphale d'Amphibiens di Jura (Note préliminaire), *Rev. Suisse Zool.,* 79, 980, 1972.

337. **Gassmann, M.,** Contribution á l' étude des trématodes d'amphibiens du Cameroon, *Ann. Parasitol. Hum. Comp.,* 50, 559, 1975.

338. **Gentile, R., Sciscioli, V., and Sobrero, L.,** Trematodi della famiglia Paramphistomidae della Somalia. I. *Paramphistomum bothriophoron* (Braun, 1892) Fischoedet, 1901, *Riv. Parassitol.,* 41, 67, 1980.

339. **Georgiev, B.,** Contribution to a study of helminth parasites of sheep in Bulgaria (in Bulgarian), *Izv. Inst. Eksp. Vet. Med. Bulg. Akad. Nauk.,* 5, 155, 1956.

340. **Georgiev, B.,** The dynamics of the development of *Paramphistomum microbothrium* in the intermediate host *Galba truncatula,* in a natural environment (in Bulgaria), *Khelmintologiya,* 7, 28, 1979.

341. **Gery, J.,** The fresh water fishes of South America, in *Biogeography and Ecology in South America,* Fittkau, E. J., Illies, J., Klinge, H., Schwabe, G. H., and Sioli, H., Eds., Junk, The Hague, 1969, 828.

342. **Ghafoor, M. A.,** Histopathological studies on *Gigantocotyle bathyocotyle* (Fischoeder, 1901) Näsmark, 1937: infection of the liver of buffalo (*Bos bubalis*) in Maharashtra, *Proc. Indian Sci. Congr. Assoc.,* 57, 498, 1970.

343. **Ginechinskaya, T. A.,** On the study of cercariae of the Rybinskiĭ water reservoir. II. (in Russian), *Vestn. Leningr. Gos. Univ.,* 21, 62, 1959.

344. **Gluzman, I. Ya.,** On the study of the method to propagate paramphistomid cercariae for experimental purposes, *Veterinariya (Kiev),* 11, 44, 1967.

345. **Gluzman, I. Ya.,** Biology of *Planorbis planorbis* and seasonal variation in their infection with *Liorchis scotiae* in the Polesie area (in Russian), *Veterinariya (Kiev),* 17, 60, 1968a.

346. **Gluzman, I. Ya.,** The viability of the adolescariae of *Liorchis scotiae* (Willmott, 1950) Velitchko, 1966 in laboratory and natural conditions (in Russian), *Tr. Vses. Inst. Gelmintol.,* 14, 176, 1968b.

347. **Gluzman, I. Ya.,** Morphology and biology of *Liorchis scotiae* (Willmott, 1950) Velitchko, 1966) of eggs and miracidia (in Russian), *Vet. Resp. Mezhved. Temat. Nauchn. Sb.,* 23, 28, 1969.

348. **Gluzman, I. Ya.,** Experimental study of the life-cycle of *Paramphistomum ichikawai* (in Ukrainian), *Veterinariya (Kiev),* 33, 90, 1972a.

349. **Gluzman, I. Ya.,** Winter survival rate of the infective stages of *Liorchis scotiae* (Trematoda: Paramphistomatidae) in the Ukrainian Polesie area (in Ukrainian), *Probl. Parazitol.,* 1, 205, 1972b.

350. **Gluzman, I. Ya. and Artemenko, Yu. G.,** Experimental infection of calf with *Liorchs scotiae* (Willmott, 1950) Velitchko, 1966 (in Russian), *Tr. Gelmintol. Lab. Akad. Nauk SSSR,* 15, 87, 1969.

351. **Goeze, J. A. E.,** Versuch einer Naturgeschichte der Eeingeweidewürmer thiersischer Körper, Blakerburg, 178, 1782.

352. **Gohar, N.,** Liste des trématodes parasites et leurss hôtes vertebreés signalés dans la vallée du Nil, *Ann. Parasitol. Hum. Comp.,* 12, 322, 1934.

353. **Golsin, W. A.,** Considerations regarding the phylogeny of Cypriniformes fishes with special reference to structures associated with feeding, *Copeia,* 4, 761, 1973.

354. **Gottschalk, C.,** Zur Verbreitung des Lebegesels, des Lanzettegls und des Pansenegels in den thüringischen Bezirken Erfurt und Gera, *Angew. Parasitol.,* 3, 101, 1962.

355. **Grabda-Kazubska, B.,** Observations on the life-cycle of *Diplodiscus subclavatus* (Pallas, 1760) (Trematoda, Diplodiscidae), *Acta parasitol. Pol.,* 27, 261, 1980.

356. **Grabda-Kazubska, B.,** *Opisthodiscus diplodiscoides* Cohn, 1904 (Trematoda, Diplodiscidae) in *Rana esculenta* s.l. in Poland, with remarks on its synonymy and geographical distribution, *Acta Parasitol. Pol.,* 27, 141, 1980.

357. **Graber, M.,** Helminthes parasites de certains animoux domestiques et sauvage due Tchad, *Bull. Epizoot. Dis. Afr.,* 17, 403, 1969.

358. **Graber, M.,** Helminthes et helminthiases des équidés (anas et chevaux) de la réepublique du Tchad, *Rev. Elev. Med. Vet. Pays Trop.,* 23, 207, 1970.

359. **Graber, M.,** Helminthes et helminthiases des animaux domestiques et sauvages d'Éthiopie, *Inst. Elev. Med. Vet. Pays Trop. Maisons-Afort,* 1, 1973.

360. **Graber, M.,** Les trématodoses hépatiques et gastriguees des zébus d'Afrique Centrale, *Rev. Elev. Med. Vet. Pays Trop.,* 28, 311, 1975.

361. **Graber, M.,** Endoparasites in domestic and wild animals of the Central African Republic, *Bull. Anim. Health Prod. Afr.,* 29, 25, 1981.

362. **Graber, M., Doutre, M., Finelle, P., Keravic, I., Ducroz, G., and Mototaingar, P.,** Les helminthes de quelques ortiodactyles sauvage apparttenant aux familles des bovidés et des suidés. Ces mammféres, en République du Tchad et en R.C.A., sont-ils des réservoirs de parasites pour les animaux domeestiques vivant á leur contact, *Rev. Elev. Med. Vet. Pays Trop.,* 17, 377, 1964.

363. **Graber, M., Fernagut, R., and Oumatie, O.,** Helminthes des zébus adultes de la region de Maroua (Nord-Cameroon), *Rev. Elev. Med. Vet. Pays Trop.,* 19, 149, 1966.

364. **Graber, M. and Daynees, P.,** Mollusques vecteurs de trématodeses humanies et animals en Ethiopie, *Rev. Elev. Vet. Pays Trop.,* 27, 307, 1974

365. **Graber, M. and Thal, J.,** Paramphistomatoidea (Trematoda) de divers Ruminants sauvages du Tchad, du Nord Cameroon et de la Répub005luie Centrafricaine, *Bull. Inst. Fr. Afr. Noire,* 42, 261, 1980.

366. **Graber, M., Blanc, P., and Delavenay, R.,** Helminthes des anumaux sauvages d'Ethiopie. I. Mammiféras, *Rev. Elev. Meed. Vet. Pays Trop.,* 33, 143, 1980.

367. **Graubmann, H. D., Gräfner, G., and Odening, K.,** Zur Paramphistomose des Rot und Rehwildes, *Mh. Vet. Med.,* 33, 892, 1978.

368. **Gretillat, S.,** Maintien du genre *Bothriophoron* Stiles et Goldberger, 1910 et valeur de l'espece *Paramphistomum bothriophoron* (Braun, 1892) Fischoeder, 11901 (Trematoda: Paramphistomatidae), parasite du reticulum du zébu malgache, *Ann. Parasitol. Hum. Comp.,* 33, 240, 1958.

369. **Gretillat, S.,** Structure anatomique du diverticule pharyngien dans l'espece *Stephanopharynx compactus,* *C. R. Acad. Sci.,* 250, 4064, 1960a.

370. **Gretillat, S.,** Cycle évolutif de *Carmyerius dollfusi* Golvan, Chabaud et Gretillat, 1957. Premieres recherches. Formes larvaires et hotes intermediaires, épidémiologie de la Gastrothylose bovine a Madagascar, *Ann. Parasitol. Hum. Comp.,* 35, 45, 1960b.

371. **Gretillat, S.,** Amphistomes (trématodes) des ruminanats domestiques de la République du Tchad. Description d'un Gastrothylacidae nouvean, *Carmyerius graberi* n. sp., *Ann. Parasitol. Hum. Comp.,* 35, 509, 1960c.

372. **Gretillat, S.,** *Carmyerius papillatus* n. sp. et *Carmyerius parvipapillatus* n. sp. (Trematoda: Gastrothylacidae), parasite des réservoirs gastriques de l'antilope, *Kobus defassa* Rupp., *Ann. Parasitol. Hum. Comp.,* 37, 121, 1962.

373. **Gretillat, S.,** Sur quelques Paramphistomidae (Trematoda) d'une collection du Musée Royal de l'Afrique Centrale, *Rev. Zool. Bot. Afr.,* 69, 351, 1964.

374. **Gretillat, S.,** Notes et remarques au sujet d'une collection de trematodes du Muséee Royal de l' Afrique Centrale, *Rev. Zool. Bot. Afr.,* 73, 166, 1966.

375. **Grieder, H.,** Seltene Nutriaparasiten, *Schweiz. Arch. Tierheilkd.,* 73, 520, 1973.

376. **Grobbelaar, C. S.,** On South African Paramphistomidae (Fisch.). II. Some trematodes in South African Anura and the relationships and distribution of their hosts, *Trans. R. Soc. S. Afr.,* 10, 181, 1922.

377. **Groschaft, J., Otero, A. C., and Tenora, F.,** Trematodes (Trematoda) from Cuban turtles, *Chelonia mydas mydas* (L.) and *Eretmochelys imbricata imbricata* (L.) (Testudinata: Cheloniidae), *Acta Univ. Agric. Brno,* 25, 155, 1977.

378. **Grossman, A. I. and Cain, G. D.,** Karyotypes and chromosome morphologies of *Megalodiscus temperatus* and *Philophthalmus gralli,* *J. Helminthol.,* 55, 71, 1981.

379. **Gubanov, N. M.,** *Helminth fauna of economically important mammals in the Yakut ASSR* (in Russian), Izdv. "Nauka", Moscow, 1964, 148.

380. **Guilford, H. G.,** Gametogenesis in *Heronimus chelydra* MacCallum, *Trans. Am. Microsc. Soc.,* 74, 182, 1955.

381. **Guilford, H. G.,** Some helminth parasites found in turtles from northwestern Wisconsin, *Trans. Wis. Acad. Sci.,* 48, 121, 1959.

382. **Guilhon, J., and Priouzeau, M.,** La paramphistomose bovine en France, *Recl. Meed. Vet. Ec. Alfort,* 181, 225, 1945.

383. **Gupta, N. K.,** Anatomy of *Paramphistomum (Cauliorchis) crassum, Res. Bull. East Panjub Univ.,* 8, 91, 1950.

384. **Gupta, N. K.,** On the morphology of *Paramphistomum bathycotyle* Fischoeder, 1901 a common amphistome in the bile ducts of Indian bovins. *Res. Bull. East Panjab Univ.,* 15, 33, 1951.

385. **Gupta, N. K.,** On a new species, *Ceylonocotyle dawesi* from *Bos indicus* Linn. in Madras (South India), *Res. Bull. Panjab Univ.,* 140, 67, 1958.

386. **Gupta, N. K.,** On *Paramphistomum epiclitum* Fischoeder, 1904, a parasite of farm animals in the Punjab, *Res. Bull. Panjab Univ.,* 14, 307, 1963.

387. **Gupta, N. K.,** On two trematode parasites of the genus *Ceylonocotyle* Näsmark, 1937, collected from the rumen of cattle in Uttar Pradesh, *Res. Bull. Panjab Univ. Sci.,* 16, 277, 1965a.

388. **Gupta, N. K.,** On two amphistomid parasites of the genus *Calicophoron* from ungulates of economic importance in India, *Res. Bull. Panjab Univ. Sci.,* 16, 283, 1965b.

389. **Gupta, N. K.,** On *Gigantocotyle explanatum* (Creplin, 1847) an amphistomid parasite of cattle, *Res. Bull. Panjab Univ. Sci.,* 17, 77, 1966.

390. **Gupta, N. K. and Dutta, T.,** On *Fischoederius cobboldi* a pouched amphistome from cattle in India, *Res. Bull. Panjab Univ. Sci.,* 18, 41, 1967.

391. **Gupta, N. K. and Walia, S.,** *Pseudodiscus collinsi* (Cobbold, 1875) Stiles et Goldberger, 1910, an amphistomid parasite of equines in India, *Res. Bull. Panjab Univ. Sci.,* 20, 49, 1969.

392. **Gupta, N. K. and Gupta, P.,** *Cochinocotyle bovini* n. gen., n. sp. (Family: Paramphistomidae: subfamily Paramphistominae) from cattle at Ernakulam (South India), *Res. Bull. Panjab Univ. Sci.,* 21, 323, 1970.

393. **Gupta, N. K. and Kumari Adarsh,** One new and one already known amphistome belonging to genus *Helostomatis* Travassos, 11934 (Trematoda: Paramphistomatidae) from fresh water fish *Labeo dero* and *Cirrhina mrigala* from Nangal and Ropar, *J. Parasitol.,* 54, 126, 1970.

394. **Gupta, N. K. and Gupta, P.,** New species of the genus *Ceylonocotyle* Näsmark, 1937 from cattle at Ernakulam (South India), *Res. Bull. Panjab Univ. Sci.,* 23, 31, 1972a.

395. **Gupta, N. K. and Gupta, P.,** *Cotylophoron chauhani* n. sp. from sheep at Ernakulum (South India), *Res. Bull. Panjab Univ. Sci.,* 23, 37, 1972b.

396. **Gupta, N. K. and Nakhasi, U.,** On some amphistomid parasites from India. II, *Rev. Iber. Parasitol.,* 37, 251, 1977.

397. **Gupta, S. P. and Verma, S. L.,** On three trematode parasites of fishes from Lucknow, *Proc. Indian Sci. Congr. Assoc.,* 57, 461, 1970.

398. **Gupta, S. P. and Tandon, V. L.,** On some digenetic trematodes from marine fishes of Puri, Orissa, *Indian J. Helminthol.,* 35, 112, 1983.

399. **Gupta, P. P., Singh, B., Mandal, P. C., Gill, B. S., and Grewal, G. S.,** A post-mortem study of mortality pattern in adult buffaloes in Punjab, India, *Indian J. Anim. Sci.,* 48, 669, 1978.

400. **Gupta, V. and Ahmad, J.,** Digenetic trematodes of marine fishes. On five new digenetic trematodes from marine fishes of Bengal, Puri, Osissa, *Helminthologia,* 16, 161, 1979.

401. **Gupta, R. P., Yadav, C. L., and Ruprah, N. S.,** Studies on the bionomics of some aquatic snails and their cercarial fauna of Haryana State, *Indian Vet. Med. J.,* 11, 77, 1987.

402. **Güralp, N. and Oğuz, T.,** Yurdumuz tiftik keçilerinde görülen parazit türlei ve bunlarin yayiliş orani (in Turkish), *Ankara Univ. Vet. Fak. Derg.,* 14, 55, 1967.

403. **Gvozdev, E. V., Agapova, A. I., and Belyakova, Yu.V.,** Life-cycle of the trematode *Amurotrema dombrowskajae* (Trematoda, Diplodiscidae) (in Russian), *Parazitologiya,* 20, 288, 1986.

404. **Haderlie, E. C.,** Parasites of the fresh-water fishes of northern California, *Univ. Calif. Berkeley Publ. Zool.,* 57, 303, 1953.

405. **Halvorsen, O. and Wissler, K.,** *Elephastsongylus* sp. (Nematoda, Protostrongylidae) and other animals in faeces of moose (*Alces alces* L.) in north Norway, *Fauna Norvegica,* 4, 37, 1983.

406. **Hanson, M. L.,** Some digenetic trematodes of plectognath fishes of Hawaii, *Proc. Helminthol. Soc. Wash.,* 22, 75, 1955.

407. **Harmos, C. E.,** Checklist of parasites from catfishes of northeastern Kansas, *Trans. Kans. Acad. Sci.,* 62, 262, 1959.

408. **Harwood, P. D.,** The helminth parasitic in the Amphibia and Reptilia of Houston, Texas and vicinity, *Proc. U.S. Natl. Mus.,* 81, 1, 1932.

409. **Haseeb, M. and Khan, D.,** Studies on larval trematodes infecting freshwater snails in Pakistan. IV. Monostome and amphistome cercariae, *Pak. J. Zool.,* 14, 21, 1982.

410. **Helle, O.,** A survey of helminthiases in cattle, sheep and horses in Norway, in *Helminth Diseases of Cattle, Sheep and Horses in Europe,* Urquhart, G. M. and Armour, J., Eds., University of Glasgow, 1973, 160.

411. **Hennig, W.,** *Phylogenetic Systematics,* University of Illinois, Urbana, 1966.

412. **Henriksen, S. A. and Nansen, P.,** *Paramphistomum* spp. — vomikter (in Danish), *Dan. Vet. Tidsskr.,* 59, 44, 1976.

413. **Herber, E. C.,** On the mother redia of *Diplodiscus temperatus* Stafford, 1905, *J. Parasitol.,* 24, 549, 1938.

414. **Herber, E. C.,** Studies on the biology of the frog amphistome *Diplodiscus temperatus* Stafford, *J. Parasitol.,* 25, 189, 1939.

415. **Herd, R. P. and Hull, B. L.,** *Paramphistomum microbothrioides* in American bison and domestic beef cattle, *J. Am. Vet. Med. Assoc.,* 179, 1019, 1981.

416. **Heyneman, D., Brenes, M. R. R., and Diaz-Ungria, C.,** Trématodos de Venezuela. II. Algunos tremátodos de pescees, reptiles y aves con descripcio de una nueva especies de género Lubens, *Mem. Soc. Cienc. Nat. La Salle,* 20, 138, 1960.

417. **Hilmy, I. S.,** New paramphistomes from Red Sea dugong, *Helicore helicore* with description of *Solenorchis* gen. n. and Solenorchinae subfamily n., *Proc. Egypt. Acad. Sci.,* 4, 1, 1949.

418. **Holl, F. J.,** A paramphistome from fishes, *J. Parasitol.,* 16, 35, 1929.

419. **Horak, I. G.,** Studies on paramphistomiasis. IV. Modified critical and controlled anthelmintic test on the conical fluke *Paramphistomum microbothrium, J. S. Afr. Vet. Med. Assoc.,* 33, 203, 1962.

420. **Horak, I. G.,** Studies on paramphistomiasis. VI. The anthelmintic efficacy of Lintex and Freon against *Paramphistomum* spp. in sheep and cattle, *J. S. Afr. Vet. Med. Assoc.,* 35, 161, 1964.

447

421. **Horak, I. G.,** The anthelmintic efficacy of Bithional against *Paramphistomum microbothrium, Fasciola* spp. and *Schistosoma mattheei, J. S. Afr. Vet. Med. Assoc.,* 36, 561, 1965.
422. **Horak, I. G.,** Host-parasite relationships of *Paramphistomum microbothrium* Fischoeder, 1901, in experimentally infested ruminants, with particular reference to sheep, *Onderstepoort J. Vet. Res.,* 34, 451, 1967.
423. **Horak, I. G.,** Paramphistomiasis of domestic ruminants, *Adv. Parasitol.,* 9, 33, 1971.
424. **Horak, I. G. and Clark, R.,** Studies on paramphistoomiasis. V. The pathological physiology of the acute disease in sheep, *Onderestepoort J. Vet. Res.,* 30, 145, 1963.
425. **Hovorka, J. Pačenovský, J. and Mitterpák, J.** Druhové zastúpenie trematódov podradu Paramphistomata na Kube (in Slovak), *Vet. Med. (Prage)* 19, 265, 1974.
426. **Hsu, Y. C.,** Helminths of cows in Soochow, *Lingnan Sci. J.,* 14, 605, 1935.
427. **Hunter, G. W.,** *Diplodiscus intermedius,* sp. nov. from *Rana catesbeiana* Show, *J. Parasitol.,* 17, 74, 1930.
428. **Huq, M. M. and Rahman, M. S.,** Morphology and incidence of *Homalogaster paloniae* in cattle in Mymensingh district, East Pakistan, *Pak. J. Vet. Sci.,* 2, 84, 1968.
429. **Ibrović, M. and Levi, I.,** Paramphistomosa goveda-klinička slika: terapija (in Serbo-Croat), *Veterinaria (Sarajevo),* 25, 147, 1976.
430. **Impand, P., Thirachandra, S., and Bunnag, T.,** Helminth faunas of rats and domestic animals and their zoonotic potential role in north and northeast Thailand, *J. Parasitol. Top. Med. Assoc. Thailand,* 6, 105, 1983.
431. **Ingles, L. G.,** Worm parasites of California amphibia, *Trans. Am. Microsc. Soc.,* 55, 72, 1936.
432. **Innes, J. A.,** *Gastrothylax bubalis* n. sp. with a few notes on the genus *Gastrothylax* (Poirier), *Parasitology,* 5, 127, 1912.
433. **Iqbal, M. K.,** A survey of helminth parasites of livestock of Karachi, *Agric. Pak.,* 14, 29, 1963.
434. **Issa, G. I. and Ebaid, N. M.,** Parasitic trematodes of Nile fishes. IV. The parasitic species of *Synodontis membranaceus, J. Egypt. Vet. Med. Assoc.,* 29, 169, 1969.
435. **Ito, J.,** Studies on Cercariae in Japan, in commemoration of 25th Anniv. at Prof. Shizuoka University, 1980, 463.
436. **Ito, J., Parasarathorn, T., and Tongkoon, B.,** Studies on cercariae from freshwater snails in Thailand, *Jpn. J. Med. Sci. Biol.,* 15, 249, 1962.
437. **Ito, J. and Blas, B. L.,** Studies on the freshwater cercariae in Leyte Island, Philippines. V. Cercariae from Planorbidae, *Jpn. J. Med. Sci. Biol.,* 47, 445, 1977.
438. **Ivashkin, V. M.,** *Helminth Parasites of Sheep* (Olvis aries), Goat (Capra hircus), Cattle (Bos taurus) and of Yak (Poephagos gruniens) (in Russian), Izdv. Akad. Nauk SSSR, Moscow, 1955, 203.
439. **Jain, S. P.,** Studies on the biology and comparative morphology of some amphistomatous parasites of domestic animals, *Agra Univ. J. Res.,* 28, 75, 1969.
440. **Jain, S. P.,** *Cercaria onkari* n. sp. an amphistome cercaria from India, *Zool. Anz.,* 188, 261, 1972a.
441. **Jain, S. P.,** On the validity of characters used in the classification of amphistome cercariae, *Zool. Anz.,* 188, 267, 1972b.
442. **Jain, S. P.,** Studies on amphistomes. I. Amphistome cercariae and their life histories, *Agra Univ. J. Res.,* 22, 63, 1973.
443. **Jain, S. P.,** Studies on amphistome. II. A survey of the incidence and nature of amphistome in aquatic snails, *Agra Univ. J. Res.,* 25, 81, 1976.
444. **Jain, S. P. and Srivastava, H. D.,** The life history of *Ceylonocotyle scoliocoelium* (Fischoeder, 1904) Näsmark, 1937 — a common amphistome parasite of ruminants in India, *Agra Univ. J. Res.,* 18, 1, 1970.
445. **Jain, P. C. and Kamalapur, S. K.,** On the occurrence of *Calicophoron cauliorchis* (Stiles et Goldberger, 1910) Näsmark, 1937 (Paramphistomidae: Trematoda) in sheep in India, *Indian Vet. J.,* 47, 307, 1970.
446. **Jain, S. P., Gupta, A. N., and Sharma, P. N.,** Redescription of *Cercariae indicae XXXII* Sewell, 1922, an amphistome cercaria from the snail *Bulinus pulchellus* (Benson), *Acta Parasitol. Pol.,* 19, 251, 1971.
447. **Janchev, J.,** Helminths from Indian elephant *(Elephas maximus)* (in Bulgarian), *Izv. Zool. Inst. Sofia,* 18, 189, 1965.
448. **Jesus, Z. and Waramontri, J.,** Parasites of domesticated animals in Thailand. I. Worm parasites of the pig, *J. Natl. Res. Counc. Thailand,* 2, 11, 1961.
449. **Jha, A. G.,** Cytogenetics, evolution and systematics of Digenea (Trematoda: Platyhelminthes), *Egypt. J. Genet. Cytol.,* 4, 201, 1975.
450. **Johnson, C. A.,** *Sternotherus minor peltifer* (Chelonia), a new host record for *Heronimus chelydrae* MacCallum, 1902 (Trematoda: Digenea), *J. Parasitol.,* 53, 617, 1967.
451. **Johnston, S. L.,** On some trematode parasites of Australian frogs, *Proc. Linn. Soc. N. S. W.,* 37, 285, 1912.
452. **Jonathan, S. R.,** The life history of *Calicophoron ijimai* (stomach fluke of cattle) in New Zealand, *Trans. R. Soc. N. Z.,* 79, 518, 1952.

453. **Jones, A.,** A comparative morphological study of the subfamily Gastrodiscinae Monticelli, 1892 (Paramphistomidae) with comments on some members of related subfamilies, *J. Nat. Hist.,* 20, 863, 1986.

454. **Jones, A.,** A new species of *Protocladorchis* (Paramphistomidae, Dadaytrematinae) from *Pangasius nasutus* (Bleeker, 1863) in Thailand, *J. Nat. Hist.,* 21, 1021, 1987.

455. **Jones, A. and Seng, L. T.,** Amphistomes from Malaysian fishes, including *Osteochilotrema malayae* gen. nov., sp. nov. (Paramphistomida: Osteochilotrematinae subfam. nov.), *J. Nat. Hist.,* 20, 117, 1986.

456. **Joszt, L.,** The helminth parasites of the European beaver, *Castor fiber* L. in Poland, *Acta Parasitol. Pol.,* 12, 85, 1964.

457. **Joyeux, C. E. and Baer, J. G.,** Trématodes, in *Recherches sur les Helminthes d'Afrique Occidentale Francaise,* Joyeux, C. E., Gendre, E., and Baer, J. G., Eds., Paris, 1928a, 9.

458. **Kadenatsii, A. N.,** The helminth fauna of domestic and wild ruminants in the Khabarovsk region and a study of its epizootiology (in Russian), *Tezisy Dokl. Nauchn. Konf. Vses. Ova. Gelmintol.,* 1, 66, 1962.

459. **Kadenatsii, A. N.,** Contribution to the study of amphistomes of ruminants in the Habarovsk region (in Russian), *Tr. Vses. Inst. Gelmintol.,* 13, 12, 1963.

460. **Kadhim, J. K., Altaif, K. I., and Hawa, N. J.,** The occurrence of paramphistomes in ruminants in Iraq, with a description of *Gigantocotyle explanatum* in cattle and buffaloes, *Bull. Endem. Dis.,* 12, 109, 1970.

461. **Kalabekov, A. L.,** Helminth parasites of the frog of Asia Minor (*Rana macrocnemis* Boul.) on the northern slopes of central Caucasus (in Russian), *Sb. Zool. Rab. Ordzhonikidze,* 19, 1973.

462. **Kamburov, P., Vassilev, I., and Samnalie, P.,** On the species composition of Paramphistomidae Fischoeder, 1901, in Bulgaria (in Bulgarian), *Khelmintologiya,* 1, 19, 1976.

463. **Kamburov, P., Vassilev, I., Samnaliev, P., and Kanev, I.,** Occurrence of *Paramphistomum microbothrioides* Price et McIntosh, 1944 in Bulgaria (in Bulgarian), *Khelmintologiya,* 4, 22, 1977.

464. **Kamburov, P. and Osikovski, E.,** Establishing of *Paramphistomum microbothrium* Fischoeder, 1901 in the buffalo (*Bos bubalus*) in Bulgaria (in Bulgarian), *Khelmintologiya,* 2, 61, 1976.

465. **Karavaev, D. K. and Amangaliev, M.,** Paramphistomosis of sheep in Gur'ev region (in Russian), *Vestn. Skh. Nauki (Alma-Ata),* 11, 65, 1964.

466. **Katiyar, R. D. and Varshney, T. R.,** Amphistomiasis in sheep and goats in Uttar Pradesh, *Indian J. Vet. Sci.,* 33, 94, 1963.

467. **Katiyar, R. D. and Garg, R. K.,** Comparative efficacy of various chemotherapeutic agents in amphistomiasis, *Indian Vet. J.,* 42, 761, 1965.

468. **Katkov, M. V.,** The morphology of miracidia of *Liorchis hiberniae* (Willmott, 1950) Velitchko, 1966 (Trematoda: Paramphistomatata) (in Russia), *Tr. Vses. Inst. Gelmintol.,* 15, 141, 1969.

469. **Katkov, M. V.,** Morphology of the miracidium of *Paramphistomum ichikawai* Fukui, 1922 (Trematoda: Paramphistomidae (in Russian), *Tr. Vses. Inst. Gelmintol.,* 18, 111, 1971.

470. **Katkov, M. V.,** The intermediate hosts of paramphistomides in cattle in the U.S.S.R. (in Russian), *Tr. Vses. Inst. Gelmintol.,* 20, 91, 1973.

471. **Katkov, M. V.,** Experimental study on the biology of *Liorchis scotiae* (Willmott, 1950) — Velitchko, 1966 (Trematoda: Paramphistomidae) — a parasite of *Planorbis planorbis* (Mollusca: Pulmonata) (in Russian), *Tr. Vses. Inst. Gelmintol.,* 25, 42, 1980.

472. **Katkov, M. V., Velitchko, I. V., and Gluzman, I. Ya.,** About the list of species of the genus *Liorchis* Velitchko, 1966 (Trematoda: Paramphistomidae) in the U.S.S.R. (in Russia), *Tr. Vses. Inst. Gelmintol.,* 18, 115, 1971.

473. **Kaw, B. L.,** Studies in helminthology; helminth parasites of Kashmir. I. Trematoda, *Indian J. Helminthol.,* 2, 67, 1950.

474. **Kawaoe, U., Cordeiro, N. S., and Artigas, P. L.,** *Taxorchis caviae* sp. n. (Trematoda: Paramphistomidae), parasito intestinal de *Cavia aperea aperea* Erxleben, 1777 (Rodentia: Caviidae), *Mem. Inst. Oswaldo Cruz,* 76, 1, 1981.

475. **Keith, R. K. and Keith, K.,** *Ceylonocotyle streptocoelium* in feral ruminants in the Northern Territory of Australia, *Aust. Vet. J.,* 45, 594, 1969.

476. **Kelly, B. J. G.,** *Paramphistomum cervi* in cattle in Ireland., *Ir. Vet. J.,* 2, 241, 1948.

477. **Kelly, J. D. and Henderson, A. W. K.,** *Calicophoron calicophorum* (Trematoda: Paramphistomatidae) and paramphistomiasis in domestic cattle in the East Kimberley district of western Australia, *Trop. Anim. Health Prod.,* 5, 192, 1973.

478. **Khaidarov, U. K.,** Identification of the intermediate host of *Calicophoron* in southern Uzbekistan, *Tr. Uzb. Nauchno Issled. Inst. Vet.,* 22, 90, 1974.

479. **Khalil, M.,** A description of *Gastrodiscoides hoominis* from the napu mouse deer, *Proc. R. Soc. Med.,* 16, 8, 1923.

480. **Khalil., L. F.,** On a redescription of *Brevicaecum niloticum* McClelland, 1957 (Trematoda: Paramphistomidae), and the erection of a new subfamily, *J. Helminthol.,* 97, 215, 1963.

481. **Khalil, L. F.,** Studies on the helminth parasites of fresh-water fishes of Sudan, *J. Zool. London,* 158, 143, 1969.

482. **Khalil, L. F.,** *Australotrema brisbanensis* n.g., n.sp. (Paramphistomidae: Dadaytrematinae) from the Australian freshwater mullett *Trachystoma petardi* (Costleran), *Syst. Parasitol.*, 3, 65, 1981.

483. **Khan, M. I.,** A survey of helminth parasites of livestock of Karachi, *Agric. Pak.*, 14, 290, 1963.

484. **Khuan Shen-i,** The first case of the finding of the trematode *Zygocotyle lunatum* (Diesing, 1835) in the territory of the U.S.S.R. (in Russia), *Tr. Vses. Inst. Gelmintol.*, 11, 319,, 1961.

485. **Kim, J. H.,** The list of parasites arranged by each domestic animal recorded in Korea, *Bull. Off. Int. Epizoot.*, 49, 618, 1958.

486. **Kinsella, J. M. and Forrester, D. J.,** Helminth of the Florida duck, *Anas platyrhynchos fulnigula, Proc. Helminthol. Soc. Wash.*, 39, 173, 1972.

487. **Kisilene, V. and Mitskus, A.,** The helminth fauna of *Ondatra zibethica* in Lithuania (in Russian), *Acta Parasitol. Litu.*, 14, 43, 1976.

488. **Kisilev, N. P.,** On the biology of *Paramphistomum ichikawai* Fukui (in Russian), *Veterinariya (Moscow)*, 44, 51, 1967.

489. **Klesov, M. D. and Mereminskiĭ, A. I.,** Fasciolosis and paramphistomosis of ruminants and the protection against them in the Ukrainian Pol'esya (in Russian), *Prob. Ova. Prikl. Gelmintol.*, 1973, 289.

490. **Kluge, A. G. and Farris, J. S.,** Quantitative phyletics and the evolution of anurans, *Syst. Zool.*, 18, 1,, 1969.

491. **Knight, I. M.,** Diseases and parasites of the muskrat (*Ondatra zibethica*) in British Columbia, *Can. J. Zool.*, 29, 188, 1951.

492. **Kobayashi, H.** Une nouvelle espéce de *Watsonius* chez in singe, *Dobutsugaku Zasshi*, 27, 421, 1915.

493. **Kobulej, T. and Udvarhelyi, J.,** Efficacy of Terenol in the treatment of cattle with natural ruminal paramphistome infestations, *Acta. Vet. Hung.*, 22, 219, 1972.

494. **Kofoid, C. A. and Park, J. T.,** A new trematode, *Paramphistomum castori* sp. nov. from *Castor canadensis baileyi* Nelson, from Mary's Rives Nevada, *Univ. Calif. Berkeley Publ. Zool.*, 41, 419, 1937.

495. **Kohn, A. and Fróes, O. M.,** *Saccocoelium godoyi* n.sp. (Aplolopoidae) and other trematode parasites of fishes of Guaiba estuary, RS, Brazil, *Mem. Inst. Oswaldo Cruz*, 81, 67, 1986.

496. **Kohn, A., Ferenandez, B. M. M., Macedo, B., and Abramson, B.,** Helminth parasite of freshwater fishes from Pirassununga, SP, Brazil, *Mem. Inst. Oswaldo Cruz*, 80, 327, 1985.

497. **Komiya, Y.,** Metacercariae in Japan and adjacent territories, *Meguro Parasitol. Mus.*, Tokyo, 1965, 303.

498. **Kosaroff, G. and Mihailova, P.,** The intermediate host and some aspects of the life history of *Paramphistomum* from the district of Razlog (in Bulgarian), *God. Sofii. Univ.*, 51, 133, 1959.

499. **Kotrlá, B. and Prokopič, J.,** Paramphistomiasis of cattle in Cuba, *Acta Vet. Brno*, 42, 35, 1973.

500. **Kotrlá, B., Prokopič, J., and Vishnyakov, Yu.,** Contribution to the morphology and distribution of *Paramphistomum microbothrium* Fischoeder, 1901 in cattle and sheep in the Balkans, *Folia parasit. (Praha)*, 21, 215, 1974.

501. **Kotrlá, B., Blažek, K., and Amin, A.,** Trematodes of domestic ruminants of Afghanistan and their role in pathology, *Folia Parasitol. (Prague)*, 23, 217, 1976.

502. **Kotrlá, B. and Chroust, K.,** *Paramphistomum ichikawai* in cattle in southern Moravia, *Acta Vet. Brno*, 47, 97, 1978.

503. **Kotrlá, B. and Kotrlý, A.,** Vyskyt motolice rodu *Paramphistomum* v CSSR (in Czech.), *Vet. Med. (Prague)* 27, 483, 1982.

504. **Kozák, A.,** Trematoden der Frösche in Theissebene, *Helminthologia*, 10, 277, 1969.

505. **Kraneburg, W.,** Beiträge zur Biologie und Pathogenitat des einheimischen Pansenegels *Paramphistomum cervi.* I. Entwicklungsstadien in der Aussenwelt und im zwischenwist, *Berl. Muench. Tierarztl. Wochenschr.*, 90, 316 1977

506. **Kraneburg, W. and Hasslinger, M.-A.,** Unteresuchungen zu Biologie und vorkommen einheimischer Pansenegel, *J. Parasitenkd.* 50, 215, 1976.

507. **Kraneburg, W. and Boch, J.,** Beiträge zur Biologie und Pathogenitat des einheimischen Pansenegels *Paramphistomum cervi.* III. Entwicklung in Rind, Schaf und Réh, *Berl. Muench. Tierarztl. Wochenschr.*, 91, 71, 1978.

508. **Kraneveld, F. C. and Douwes, J. B.,** Aanwullenda lijst von voor nederlandsch indië meriwe parasitaire wormen bij zoogdieren en vogels, *Ned. - Indische Blad. Diergeneeskd.*, 25, 178, 1940.

509. **Krull, W. H.,** The snail *Pseudosuccinea columella* and *Galba bulimoides techella* new host for *Paramphistomum cervi* (Schrank, 1790) Fischoeder, 1901, *J. Parasitol.*, 20, 108, 1933.

510. **Krull, W. H.,** Life history studies on *Cotylophoron cotylophorum* (Fischoeder, 1901) Stiles et Goldberger, 1910, *J. Parasitol.*, 19, 173, 1934.

511. **Krull, W. H. and Price, H. F.,** Studies on the life history of *Diplodiscus temperatus* Stafford from the frog, *Occas. Pap. Mus. Zool. Univ. Mich.*, 237, 1, 1932.

512. **Kryukova, K. A.,** Biology of *Paramphistomum cervi* (in Russian), *Tezisy Dokl. Nauchn. Konf. Vses. Ova. Gelmintol.*, 2, 162, 1957.

513. **Kulasiri, C. and Seneviratne, R. D.,** *Gigantocotyle explanatum* (Creplin) (Trematoda: Paramphistomidae) infection of the liver of the buffalo in Ceylon, *J. Comp. Pathol.*, 66, 83, 1956.

514. **Kumar, V., Dutt., S. C., and Jain, S. P.,** *Cercaria helicorbis* n. sp. an amphistome cercaria from the snail *Helicorbis coenosus* (Benson), *Indian J. Helminthol.*, 20, 40, 1968.

515. **Kurtpinar, H. and Latif, B. M.,** Paramphistomiasis of cattle and buffaloes in Iraq, *Vet. Rec.*, 87, 668, 1970.

516. **Kutzer, E. and Hinaidy, H. K.,** Die Parasiten der Wildlebenden Wieederkaufer Österreeichs, *Z. Parasitenkd.*, 32, 354, 1969.

517. **Kuzmovich, L. G.,** On the study of the development of the larval stages of *Liorchis scotiae* (Willmott, 1950) Velitchko, 1966 in permanent and seasonal biotopes (in Russian), *Sb. Probl. Parazitol.*, 1, 288, 1975.

518. **Lahille, F. and Joan, T.,** Nota preliminar sobre in nuevo génoro de trematodees, *Physis (Buenos Aires)*, 3, 216, 1917.

519. **Lal, M.,** Occurrence of a pigment layer in *Gastrothylax crumenifer* (Croplin, 1847), *Experientia*, 15, 176, 1959.

520. **Lancaster, W. E.,** A check list of helminths of domestic livestock in Malaya, *J. Malay. Vet. Med. Assoc.*, 1, 151, 1957.

521. **Lang, A.,** Über die Cercariae von *Amphistomum subclavatum*, *Ber. Naturforsch. Ges.* Freiburg *im Breisgau*, 6, 81, 1892.

522. **Lankester, M. W., Snider, J. B., and Jerrard, R. E.,** Annual maturation of *Paramphistomum cervi* (Trematoda: Paramphitomidae) in moose, *Alces alces* L., *Can. J. Zool.*, 57, 2355, 1979.

523. **Larson, O. R.,** Larval trematodes of freshwater snails of Lake Itaska, Minnesota, *Proc. Minn. Acad. Sci.*, 29, 252, 1961.

524. **Lämler, G., Sahai, B. N., and Herzog, H.,** Anthelminthic efficacy of 2.6-dihydroxybenzoic acid-4′-bromanilide (Hoe 296V) against mature and immature *Paramphistomum microbothrium* in goats, *Acta Vet. Hung.*, 19, 447, 1969.

525. **Lee, S. K.,** Paramphistomiasis in cattle and buffaloes slaughtered in Kuala Lumpur, *Kajian Vet.*, 1, 73, 1967.

526. **Lee, S. K.,** A light microscopic study of the nervous system of three species of amphistomes (Trematoda: Paramphistomidae), *Zool. Anz.*, 187, 1, 1971.

527. **Lee, S. K. and Lowe, C. Y.,** Comparative histological and anatomical studies on amphistomes (Trematoda) from Malayan-Thai buffaloes and Malayan cattle, *Zool. Anz.*, 187, 25, 1971.

528. **Lee, C. C., Sheikh-Omar, A. R., and Jones, A.,** *Calicophoron microbothrioides* (Price et McIntosh, 1944) (Paramphistomidae: Paramphistominae) in Malaysian sambar deer *(Cervus unicolor)*, *N. Z. Vet. J.*, 35, 190, 1987.

529. **Leidy, J.,** A synopsis of Entozoa and some of their ecto-congeners, observed by the author, *Proc. Acad. Natl. Sci. Philipp.*,, 8, 42, 1856.

530. **Leiper, R. T.,** An account of Some Helminths Contained in Dr. C. M. Wenyon's Collection from the Sudan, 3rd Rep. Wellcome Res. Lab., Khartoum, 1908, 187.

531. **Leiper, R. T.,** The entoza of the hippopotamus, *Proc. Zool. Soc. London*, 1, 233, 1910.

532. **Leiper, R. T.,** Observations on certain helminths of man, *Trans. R. Soc. Trop. Med. Hyg.*, 6, 265, 1913.

533. **Leiper, R. T.,** Helminth parasites obtained from the okapi at post morten, *Proc. Zool. Soc. Lond.*, Abstr. No. 389, 949, 1935.

534. **Leiper, J. W. G.,** Report to the Government of Iraq on Animal Parasites and Their Control, FAO Rep. No. 610, Food and Agriculture Organization, Rome, 1957, 1.

535. **Leitão, J. L. D. S.,** Parasites dos animals domesticos em Portugal Metropolitano, *Inst. Alta Cultura (Lisbon)*, 13, 1963.

536. **Lengy, J.,** Study on *Paramphistomum microbothrium* Fischoeder, 1901. A rumen parasite of cattle in Israel, *Bull. Res. Counc. Isr. Sect. B: Zool.*, 9B, 71, 1960.

537. **Lengy, J.,** Some observations on the biochemistry and haematology of *Paramphistomum microbothrium* and *Schistosoma bovis* infections in lambs, *Refu. Vet.*, 19, 115, 1962.

538. **Lent, H. and Freitas, J. F. T.,** Pesqusas helminthológicos realizados no Estado do Pará. VII. Trematoda, Paramphistomoidea, *Bol. Inst. Biol. Bahia*, 4, 82, 1939.

539. **Lent, H., Freitas, J. F. T., and Proenca, M. C.,** Alguns helmintos de batráquios colecoandos no Paraguai, *Mem. Inst. Oswaldo Cruz*, 44, 195, 1946.

540. **Leon, D. D. and Juplo, R. J.,** Gastrointestinal parasites of Philippine caraboas *Bubalus bubalis* L., *J. Parasitol.*, 52, 1214, 1966.

541. **Leong, T. S., Khoo, K. H., Soon, F. L., Eddy, S. P. T., and Wong, S. Y.,** Parasites of fishes from Tasik Temengor in Perak, Malaysia, *Malay. Nat. J.*, 41, 75, 1987.

542. **Leonov, V. A., Ryzhikov, K. M., Tsimbaliuk, A. M., and Belogurov, O. I.,** Trematodes of anserine birds of Kamchatka (in Russia), *Tr. Vses. Inst. Geellmintol.*, 13, 196, 1963.

543. **Lepojev, O. and Cvetković, Lj.,** Prilog poznavanju epizootologije paramfistomoze u SR (in Serbo Croate), *Acta Parasitol. Jug.*, 7, 15, 1976.

544. **Le Roux, P. L.,** Helminthiasis of domestic stock in the Union of the South Africa, *J. S. Afr. Vet. Med. Assoc.,* 1, 43, 1930.

545. **Le Roux, P. L.,** List of Helminths Collected from Mammals and Birds in the Mazabuka Area, Northern Rhodesia, Annu. Rep. (1931), Department of Animal Health, Northern Rhodesia, Appendix B., 1932, 31.

546. **Le Roux, P. L.,** Report of the Assistant Veterinary Research Officer, Annu. Rep. (1933), Department of Animal Health, Northern Rhodesia, 28, 1934.

547. **Le Roux, P. L.,** Annu. Rep. (1938) Veterinary Department Northern Rhodesia, Appendix C, 58, 1939.

548. **Le Roux, P. L.,** Game conservation and severe helminthiasis of the stock in the National Game Reserves, *Trans. R. Soc. Trop. Med. Hyg.,* 49, 293, 1955.

549. **Le Roux, P. L.,** Life cycle of *Gastrodiscus aegyptiacus* (Cobbold, 1876), *Trans. R. Soc. Trop. Med. Hyg.,* 52, 14, 1958.

550. **Lewis, T. R. and McConnell, J. F. P.,** *Amphistoma hominis* n. sp., a new parasite affecting man, *Proc. Asiat. Soc. Bengal,* 182, 1876.

551. **Li, L. Y.,** Some trematode parasites of frogs with a description of *Diplodiscus sinicus* sp. nov., *Lignan Sci. J.,* 16, 61, 1937.

552. **Lim, L. H. S. and Furtado, J.,** Two new trematode species from freshwater fishes of Peninsular Malaysia, *Parasitol. Hung.,* 17, 37, 1984.

553. **Linton, E.,** Helminth fauna of the dry Tortugas. II. Trematodes, *Pap. Tortugas Lab. Carn. Inst. Wash.,* 4, 15, 1910.

554. **Lluch, J., Roca, V., and Navarro, P.,** Contribución al conocimiento de la helmintofauna de los herpetos ibéricos. III. Digenea Paramphistomatidae, Hemiuridae, Gorgoderidae, Plagiorchiidae, Haematolechidae y Cephalogonimidae de *Rana perezi* Seoane, 1985 (Amphibia: Ranidae), *Rev. Iber. Parasitol.,* 46, 387, 1986.

555. **Lombardero, O. J. and Moriena, R. A.,** Nuevos helmintos del carpincho (*Hidrochoerus hidrochoeris*) para la Argentine, *Rev. Med. Vet. (Buenos Aires),* 54, 265, 1973.

556. **Lombardero, O. J. and Moriena, R. A.,** Nuevos helmintos para la Argenina en *Phrynops hilarii* (Dumérily Bribon), *Rev. Med. Vet. (Buenos Aires),* 58, 964, 1977.

557. **Looss, A.,** Über *Amphistomum subclavatum* und seine Entwicklung, *Fortschr. 70. Geburst. R. Leukart,* Leipzig, 1892, 147.

558. **Looss, A.,** Recherches sur les fauna parasitaire de l'Egypt. I. *Mem. Inst. Egypt,* 3, 1, 1896.

559. **Looss, A.,** Notizen zur Helminthologie Egyptens. IV. Über Trematoden aus Seeschildkröten der egyptischen Küsten, *Zentralbl. Bakteriol. Parasitenkd. Infektionskr.,* 30, 618, 1901.

560. **Looss, A.,** Über neue und bekannte Trematoden aus Seeschildkröten, *Zool. Jahrb.,* 16, 21, 1902.

561. **Looss, A.,** On some parasites in the museum of the School of Tropical Medicine,, Liverpool, *Ann. Trop. Med. Parasitol.,* 1, 123, 1907.

562. **Looss, A.,** Über den Bau eineger anscheinend seltner Trematoden Arten, *Zool. Jahrb.,* 15, 325, 1912.

563. **López-Neyra, C. R.,** *Helminthos de los Vertebrados Ibéricos,* Imprenta Urania, Granada, 3, 1947, 134.

564. **Lowe, C. Y.,** Comparative studies of lymphatic system of four species of amphistomes, *Z. Parasitenkd.,* 27, 169, 1966.

565. **Lutta, A. S.,** Die Fauna der parthenogenetischen Trematoden Gernerationen in den Süsswassermollusken Peterhofs, *Tr. Leningr. Obva. Estest. Voispyt.,* 63, 261, 1934.

566. **Lutz, A.,** *Estudios de Zoologia y Parazitologia Venezolanas,* Rio de Janeiro, 1928, 1.

567. **Lynch, J. E.,** The miracidium of *Heronimus chelydrae* McCallum, *Q. J. Microsc. Sci.,* 76, 13, 1933.

568. **Lynch, J. D.,** Evolutionary relationships, osteology and zoogeography of the leptodactyloid frogs, *Univ. Kans. Mus. Nat. Hist. Mus. Publ.,* 53, 1, 1971.

569. **Lynch, J. D.,** The transition from archaic to advenced frogs, in *Evolutionary Biology of the Anurans,* Vial, J. L., Ed., University of Missouri Press, Columbia, 1973, 133.

570. **MacCallum, G. A.,** *Heronymus chelydrae* nov. gen., nov. sp. A new monostome parasite of the American snappping turtle, *Zentralbl. Bakteriol. Parasitenkd. Infektionskr.,* 32, 632, 1902.

571. **MacCallum, G. A.,** On two new amphistome parasites of Sumatra fishes, *Zool. Jahrb.,* 22, 667, 1905.

572. **MacCallum, G. A.,** Some new species of parasitic trematodes of marine species, *Zoopathology,* 1, 31, 1916.

573. **MacCallum, G. A.,** A new species of trematode (*Cladorchis gigas*) parasitic in elephants, *Bull. Amer. Mus. Nat. Hist.,* 37, 865, 1917.

574. **MacCallum, G. A.,** Notes on the genus *Telorchis* and other trematodes, *Zoopathology,* 1, 41, 1918.

575. **MacCauley, J. E.,** The paramphistome *Megalodiscus microphagus* Ingles, 1936 from the giant salamander from Oregon, *J. Parasitol.,* 45, 614, 1959.

576. **MacKinnon, B. M. and Burt, M. D. B.,** Platyhelminth parasites of muskrats (*Ondatra zibethica*) in New Brunswick, *Can. J. Zool.,* 56, 350, 1978.

577. **Macnae, W., Rock, L., and Makowski, M.,** Platyhelminths from the South African clowed toad, or platana (*Xenopus laevis*), *J. Helminthol.,* 47, 199, 1973.

578. **Macy, R. W.,** On the life-cycle of *Megalodiscus microphagus* Ingles (Trematoda: Paramphistomatidae), *J. Parasitol.,* 46, 662, 1960.

579. **Maeder, A.-M., Combes, C., and Knoepffler, L.-Ph.**, Parasites d'Amphibians du Gabon: Plagiorchiidae et Paramphistomatidae (Digenea), *Biol. Gabonica*, 5, 283, 1969.

580. **Malek, E. A.**, Trematode infections in some domesticated animals in the Sudan, *J. Parasitol.*, 45, 21, 1959.

581. **Malek, E. A.**, *Bulinus forskalii* Ehrenberg 1831: intermediate host of *Gastrodiscus aegyptiacus* (Cobbold, 1876) Looss, 1896, *J. Parasitol.*, 45 (5 Suppl. 2), 16, 1960.

582. **Malek, E. A.**, The life cycle of *Gastrodiscus aegyptiacus* (Cobbold, 1876) Looss, 1896 (Trematoda: Paramphistomatidae: Gastrodiscinae), *J. Parasitol.*, 57, 975, 1971.

583. **Malik, B. S., Rai, P., and Ahluwalia, S. S.**, A note on helminths of the Indian elephant. I. Amphistomatous parasites, *Indian J. Vet. Sci.*, 29, 11, 1959.

584. **Mañé-Garzón, F.**, Un nouveau trématode des batraciens de l'Uruguay: *Catadiscus corderoi* n.sp., *Comun. Zool. Mus. Hist. Nat. Montevideo*, 4, 1, 1958.

585. **Mañé-Garzón, F. and Gortari, A. M.**, Sobre algunos tremátodos de ofidios del Uruguay, *Commun. Zool. Mus. Hist. Nat. Montevideo*, 8, 1, 1965.

586. **Manter, H. W.**, A collection of trematodes from Florida Amphibia, *Trans. Am. Microsc. Soc.*, 57, 26, 1938.

587. **Manter, H. W.**, The zoogeography of trematodes of marine fishes, *Exp. Parasitol.*, 4, 62, 1955.

588. **Manter, H. W.**, Some aspects of the geographical distribution of parasites, *J. Parasitol.*, 53, 1, 1967.

589. **Manter, H. W. and Pritchard, M. H.**, Mission de Zoologie Médicale an Maniema (Congo, Léopoldville) (P.L.G. Benoit, 1959). V. Vermes-Trematoda, *Ann. Mus. R. Congo Belge Ser. 8vo*, 132, 75, 1964.

590. **Manuel, M. F. and Madriaga, C. L.**, A study on the helminth fauna of Philippine goat, *Philipp. J. Vet. Med.*, 5, 79, 1966.

591. **Maplestone, P. A.**, A revision of the Amphistomata of mammals, *Ann. Trop. Med. Parasitol.*, 17, 113, 1923.

592. **Markusheva, U. M. and Sagalovich, V. M.**, The helminth fauna of amphistomes on the territory of Bielorussia (in Russian), *Tr. Gelmintol. Lab. Akad. Nauk SSSR*, 26, 159, 1974.

593. **Martinez, F. A.**, Helmintofauna de los mammiferos silvestres, Trematodes, *Vet. Argent.*, 3, 544, 1986.

594. **Martinez-Fernandez, A. R., Simon-Vicente, F., and Cordereo del Campillo, M.**, On the morphology of *Opisthodiscus nigrivasis* (V. Méhely, 1929) Odeneing, 1959 (Trematoda: Paramphistomidae) of *Rana ridibunda*, *Rev. Iber. Parasitol.*, 48,, 9, 1988.

595. **Masi Pallarés, R., Benitez Usher, C. A., and Vergara, G.**, Helminthes en peces y reptiles del Paraguay, *Rev. Parag. Microbiol.*, 8, 61, 1973.

596. **Masi Pallarés, R., Benitez Usher, C., and Maciel, S.**, Lista de helmintos del Paraguay, *Rev. Parag. Microbiol.*, 11, 43, 1976.

597. **Maxwell, J. P.**, Intestinal parasitism in South Futzien, *Chin. Med. J.*, 35, 377, 1921.

598. **McClelland, W. F. J.**, Two new genera of amphistomes from Sudanense freshwater fishes, *J. Helminthol.*, 31, 247, 1957.

599. **McCoy, O. R.**, Notes on cercariae from Missouri, *J. Parasitol.*, 15, 199, 1929.

600. **McCulley, R. M., Nickerr, J. W., and Kruger, S. P.**, Observation on the pathology of birharziasis and other parasite infestations of *Hippopotamus amphibius*, *Oderstepoort J. Vet. Res.*, 34, 563, 1976.

601. **McKenzie, C. E. and Welch, H. E.**, Parasite fauna of the muskrat, *Ondatra zibethica* (Linnaeus, 1766) in Manitoba, Canada, *Can. J. Zool.*, 57, 640, 1979.

602. **Méhely, L.**, *Opisthodiscus diplodiscoides nigrivasis* n. subsp. Ein neuer Saugwurm der Ungarischen fauna, *Stud. Zool.*, 2, 84, 1929.

603. **Melo, H. J. and Ribeiro, H. S.**, Helmintos parasitas dos animais domésticos no Estado de Mato Grosso, *Arq. Esc. Sup. Vet. Est. Minas Gerais*, 29, 161, 1977.

604. **Merdivenci, A.**, Ehli koyum (*Ovis aries*) larimizda buldugumuz *Paramphistomum cervi* (Zeder, 1790, Schrank, 1790) (Fam. Paramphistomatidae) (in Turkish), *Turk. Vet. Hekim. Dern. Derg.*, 29, 12, 1959.

605. **Mereminskiĭ, A. I.**, Paramphistomiasis of ruminants in forest areas of the Ukraine (in Ukrainian), *Visn. Silskogospod. Nauki*, 12, 93, 1973.

606. **Mereminskiĭ, A. I., Gluzman, I. Ya., and Artemenko, Yu. G.**, Paramphistomiasis of cattle in Poles'e (in Russian), *Veterinariya*, 12, 51, 1968.

607. **Mereminskiĭ, A. I., Gluzman, I. Ya., and Artemenko, Yu. G.**, Paramphistomosis of cattle in Ukraine (in Russian), in *Sb. Rab. Gel'mint. Posv. 90-let. Rozhd. K. I. Skrjabina*, Izdv. "Kolos", Moscow, 1971, 217.

608. **Merotra, V. and Gupta, N. K.**, On a reptilian amphistome, *Stunkardia dilymphosa* Bhalerao, 1931, with a discussion on the synonymy of an allied genus and some species, *Res. Bull. Panjab Univ. Sci.*, 30, 13, 1981.

609. **Meskal, F. H.**, Trematodes of anurans from Ethiopia, *Arbok Univ. Bergen, Mat. Naturvitensk. Ser.*, 1, 1, 1970.

610. **Mettam, R. W. M.**, Identification list of helminths from Departmental Collection 1920—1931, Annu. Rep., (1931) Veterinary Department, Uganda, Appendix IB, 1932, 20.

611. **Mettam, R. W. M.,** Annu. Rep., (1934), Veterinary Department, Uganda, Apendix II, 1935, 32.

612. **Mettrich, D. F.,** A re-description of *Zygocotyle lunata* (Diesing, 1836), Stunkard, from *Anas platyrhyncha* in Southern Rhodesia, *Rhod. Agric. J.,* 56, 197, 1959.

613. **Mettrich, D. F.,** Some trematodes and cestodes from mammals of Central Africa, *Rev. Biol. (Lisbon),* 3, 149, 1963.

614. **Milogradova, G. P. and Spasskii, A. A.,** Helminth fauna of anurans of the Eastern Siberia (in Russian), *Tezisy Dokl. Nauchn. Konf. Vses. Org. Gelmintol.,* 1, 200, 1957.

615. **Mishra, N. and Tandon, V.,** Nervous system in the bovine pouched paramphistome, *Fischoederius cobboldi* (Trematoda: Digenea), *Indian J. Parasitol.,* 8, 33, 1984.

616. **Mishra, N. and Tandon, V.,** Nervous system in *Olveria indica,* a rumen paramphistome (Digenea) of bovines, as revealed by non-specific esterase staining, *J. Helminthol.,* 60, 193, 1986.

617. **Mitskevich, V. Y.,** *Cotylophoron skrjabini* n. sp. from *Rangifer tarandus* (in Russian), in *Rab. Gelmint. 80-let K. I. Skrjabina,* Izdv. Minist. Sel'sk. Hozya. SSR, Moscow, 1959, 231.

618. **Mohandas, A.,** Studies on the freshwater cercariae of Kerala. V. Paramphistomid and Opisthorchioid cercariae, *Vestn. Cesk. Spol. Zool.,* 40, 196, 1976.

619. **Moravec, F.,** Some digenetic trematodes from Egyptian fishes, *Vestn. Cesk. Spol. Zool.,* 41, 52, 1977.

620. **Moravec, F.,** Some helminth parasites from amphibians of Vancouver Island, B.C., Western Canada, *Věstn. Cesk. Spol. Zool.,* 48, 107, 1984.

621. **Morel, P. C.,** Les helminthes des animaux domestiques de Afrique occidentale, *Rev. Elev. Med. Vet. Pays Trop.,* 12, 153, 1959.

622. **Moskvin, S. N.,** Helminth parasites of domestic animals in the People's Republic of Albania (in Russian), *Tr. Mosk Vet. Akad.,* 27, 172, 1958.

623. **Moskvin, A. S.,** Experimental infection of the snail, *Lymnaea truncatula* with miracidia of paramphistomids (in Russian), *Byull. Vses. Inst. Gelmintol.,* 37, 55, 1984.

624. **Mottl., S. and Pav, J.,** Muflon jako hostitel motolice jeleni *Paramphistomum cervi* (Schrank, 1790), (in Czech), *Cslka Parasitol.,* 5, 153, 1958.

625. **Muchlis, A.,** A short report on amphistomosis in cattle due to *Homalogaster paloniae* Poiriere, 1883, *Commun. Vet. Bogor,* 8, 16, 1964.

626. **Mukherjee, R. P.,** Studies on the life history of *Ceylonocotyle scolicoelium* (Fischoeder, 1904) Näsmark, 1937, an amphistome parasite of sheep and goats, *Proc. 47th Indian Sci. Congr.,* 3, 438, 1960.

627. **Mukherjee, R. P.,** Studies on some amphistome trematodes of domesticated animals, *Agra Univ. J. Res.,* 11, 131, 1962.

628. **Mukherjee, R. P.,** On two new species of amphistomes from Indian sheep and goat. *Indian J. Helminthol.,* 15, 70, 1963.

629. **Mukherjee, R. P.,** A study on the development and morphology of the miracidium of *Stunkardia dilymphosa* Bhalerao, 1931 an amphistome parasite of fresh water turtles, *Proc. 2nd All Indian Congr. Zool.,* 2, 370, 1966a.

630. **Mukherjee, R. P.,** Studies on the life history of *Fischoederius elongatus* (Poirier, 1883) Stiles et Goldberger, 1910, an amphistome parasite of cow and buffalo in Indian, *Indian J. Helminthol.,* 18, 5, 1966b.

631. **Mukherjee, R. P.,** On some amphistomes of India, *Indian J. Helminthol.,* 18, 94, 1966c.

632. **Mukherjee, R. P.,** Seasonal variations of cercarial infections in snails, *J. Zool. Soc. India,* 18, 39, 1966d.

633. **Mukherjee, R. P.,** Studies on the life history of *Cotylophoron indicum* Stiles et Goldberger, 1910, an amphistomous parasite of ruminants in India, *J. Zool. Soc. India,* 20, 105, 1968.

634. **Mukherjee, R. P.,** On a new amphistome cercaria from *Bulinus pulchellus, Sci. Cult.,* 38, 418, 1972.

635. **Mukherjee, R. P.,** Studies on the life history of *Ceylonocotyle scoliocoelium* (Fischoeder, 1904) Näsmark, 1937, an amphistome parasite of ruminants, in B.S. Chauhan Commemoration Volume, Tiwari, K. K. and Srivastava, C. B., Eds., Zoology Society of India, Bhubaneswar, 1975, 251.

636. **Mukherjee, R. P. and Srivastava, H. D.,** Studies on the life history of *Gigantocotyle explanatum* (Creplin, 1847) Näsmark, 1937, a common amphistome parasite in the bile duct and gall bladder of buffaloes, *Proc. 47th Indian Sci. Congr.,* 3, 440, 1960.

637. **Murrell, K. D.,** *Helisoma antrosum,* first intermediate host for *Wardius zibethicus* Barker et East, 1915 (Trematoda: Paramphistomidae), *J. Parasitol.,* 49, 116, 1963.

638. **Murrell, K. D.,** Stages in the life cycle of *Wardius zibethicus* Barker, 1915, *J. Parasitol.,* 51, 600, 1965.

639. **Mutafova, T.,** Study of the karyotype of *Paramphistomum microbothrium* Fischoeder, 1901 (in Bulgarian), *Khelminologiya,* 16, 37, 1983.

640. **Myers, B. J., Wolfgang, R. W., and Kuntz, R. E.,** Helminth parasites from vertebrates taken in the Sudan (East Africa), *Can. J. Zool.,* 38, 833, 1960.

641. **Myers, G. S.,** Fresh-water fishes and West Indian zoogeography, *Smithson. Rep.,* 339, 1938.

642. **Nagaty, H. F.,** On some parasites collected in Egypt from food mammals, *J. Egypt. Med. Assoc.,* 25, 110, 1942.

643. **Nagaty, H. F.,** A revised list of the helminth parasite of man and food mammals in Egypt, *J. R. Egypt Med. Assoc.,* 32, 423, 1949.

644. **Nama,, H. S.,** On the occurrence of *Ceylonocotyle cuonum* (Bhalerao, 1937) (Trematoda: Paramphistomidae), *Indian Vet. J.,* 53, 263, 1976.

645. **Nasimov, H.,** Occurrence of *Liorchis scotiae* (Willmott, 1950) Velitchko, 1966 in ruminants in Uzbekistan (in Russian), *Mater. Nauchn. Konf. Vses. Obva. Gelmintol. Uzbek. Tashkent,* 90, 1968.

646. **Nazarova, N. S.,** Helminth infections of deer (*Cervus nippon*) on state farms in the Primorsk region of the U.S.S.R. (in Russian), *Byull. Vses. Inst. Gelmintol.,* 10, 73, 1973.

647. **Näsmark, K. E.,** A revision of the family Paramphistomidae, *Zool. Bidr. Uppsala,* 16, 301, 1937.

648. **Nekrasov, A. V. and Smirnov, M. N.,** Contribution to the study of the helminth and ectoparasite fauna of Artiodactyla in the Buryat ASR (in Russian), *Tr. Buryat. Inst. Estest. Nauk Buryat Fil. Sib. Otd. Akad. Nauk SSSR,* 18, 40, 1977.

649. **Nickel, S.,** Über einen Elefantentrematoden der Gattunk *Hawkesius, Dtsch. Tieraerztl. Wochenschr.,* 56, 184, 1949.

650. **Nikitin, V. F.,** Biology of *Gastrothylax crumenifer* (Creplin, 1847) (Trematoda: Paramphistomatata) (in Russian), *Mater. Nauchn. Konf. Vses. Obva. Gelmintol.,* Part 4, 178, 1965.

651. **Nikitin, V. F.,** The biology of *Gastrothylax crumenifer* (Creplin, 1847) -caused gastrothylaxosis of large and small ruminants (in Russian), *Tr. Uzb. Nauchno. Issled. Inst. Vet.,* 18, 198, 1967a.

652. **Nikitin, V. F.,** Contribution to the study of paramphistomids in the lower Volga reaches (in Russian), *Mater. Zool. Konf. Ped. Inst. RSFSR. Volvograd,* 3, 215, 1967b.

653. **Nikitin, V. F.,** Contribution to a study of *Liorchis hiberniae* (Paramphistomatata) (in Russian), *Byull. Vses. Inst. Gelmintol.,* 1, 80, 1967c.

654. **Nikitin, V. F.,** The study on the biology of some species of the Paramphistomatata in ruminants in the U.S.S.R. (in Russian), *Tr. Vses. Inst. Gelmintol.,* 14, 251, 1968.

655. **Nikitin, V. F.,** Paramphistome infection in cattle (in Russian), *Veterinariya (Moscow),* 49, 79, 1972.

656. **Nilsson, O.,** The interrelationship of endoparasites of wild cervids (*Capreolus capreolus*) L. and *Alces alces* L.) and domestic ruminants in Sweden, *Acta Vet. Scand.,* 12, 36, 1971.

657. **Nitzsch, C. L.,** Artikele Amphistoma, in *Ersch und Gruber, Allgemeine Encyclopedie wiss. Künste,* Leipzig, 3, 398, 1819.

658. **Nollen, P. M. and Nadakvukaren, M. J.,** *Megalodiscus temperatus:* scanning electron microscopy of the tegumental surface, *Exp. Parasitol.,* 36, 123, 1974.

659. **Novacek, M. J. and Marshall, L. G.,** Early biogeographic history of Ostariophysan fishes, *Copeia,* 1, 1, 1976.

660. **Nöller, W. and Schmid, F.,** Zur Kenentnis der Entwicklung von *Paramphistomum cervi* (Schrank) s. *Amphistotmum conicum* (Zeder), *Sitzungsber. Ges. Naturforsch. Freunde Berlin,* 1924, 8—10, 148, 1927.

661. **Núñez, M. O.,** Fauna de agua dulce de la Republica Argentina. IX. sobre representantes de la familia Paramphistomatidae (Trematoda), *Physis (Buenos Aires),* 38, 55, 1979.

662. **Odening, K.,** Über die Parasitenfauna des Wasser Frösche (*Rana esculenta* Linné) in einigene mitteldeutschen Biotopen, *Wiss. Z. Friedrich Schiller Univ. Jena Math. Naturwiss. Reihe,* 487, 1955.

663. **Odening, K.,** Über die Diplodiscidae der einheimischen Frösche (Trematoda: Paramphistomatata), *Z. Parasitenkd.,* 19, 54, 1960.

664. **Odening, K.,** Historische und moderne Gesichtspunkte beim Aufbau eines natürlichen Systems der digenetischen Trematoden, *Biol. Beitr.,* 1, 73, 1961.

665. **Odening, K.,** Some freshwater cercariae from North Vietnam, in *Jub. Vol. 60th Birthday H. D. Srivastava,* 1968a, 455.

666. **Odening, K.,** Einige Trematoden aus Fröschen und Schildkröten in Vietnam and Kuba, *Zool. Anz.,* 181, 289, 1968b.

667. **Odening, K.,** Einige Süswasser cercariaen aus Nordvietnam, *Parasitol. Schriftenr.,* 21, 187, 1971.

668. **Odening, K.,** Verwandtschaft, System und zykloontogenetische Besonderheiten der Trematoden, *Zool. Jahrb.,* 101, 345, 1974.

669. **Odening, K., Bockhardt, I., and Gräfner, G.,** Zur Frage der Pansenegelarten in der DDR (Trematoda, Paramphistomidae) und ihre Zwischenwirtschenecken, *Mh. Vet. Med.,* 33, 179, 1978.

670. **Odening, K., Bockhardt, I., and Gräfner, G.,** Zwischenwirtsspezifitat, Cerecarien und Einmerkmale der drei einheimischen *Paramphistomum* — Arten (Trematoda), *Zool. Jahrb.,* 106, 214, 1979.

671. **Odening, K. and Samnaliev, P.,** A new amphistome cercaria from *Lymnaea truncatula* in Europe, *Ann. Parasitol. Hum. Comp.,* 62, 117, 1987.

672. **Odhner, T.,** Zum natürlichen System der digenen Trematoden, I., *Zool. Anz.,* 37, 181, 1911.

673. **Oliveira Sobrinko, S. A.,** Um caso de esticorocose vesificado no matadouro de Guarulhos, *Rev. Ind. Anim. (Sao Paulo),* 2, 158 1939.

674. **Olteanu, G. and Lungu, V.,** Helminţi şi helmintoze la animale domestic in R.P.R.I. Helmintofauna la ovine (in Romanian), *Lucr. Stiint. Inst. Patol. Ig. Enim.,* 12, 401, 1963.

675. **Orlov, l. V.,** Studies of the castor's trematode, *Stichorchis subtriquetrus* (Rudolphi, 1814) (in Russian), *Parasites and Diseases of Wild Animals,* Moscow, 1948, 134.

676. **O'Roke, E. C.,** Larval trematodes from Kansas freshwater snails, *Kans. Univ. Sci. Bull.,* 10, 161, 1917.

677. **Ortlepp, R. J.,** On a collection of helminths from a South African farm (in the northern district of Natal, South Africa), *J. Helminthol.*, 4, 127, 1926.

678. **Ortlepp, R. J.,** N'oorsig van Suid-Afrikaanse helminte vesal met verwysing na die wat in ons wild herkouers voorkom, *Tydskr. Natuurwet.*, 1, 203, 1961.

679. **Ortlepp, R. J.,** Some helminths recorded from red and yellow billed hornbill from the Kruger National Park, *Onderstepoort J. Vet. Res.*, 31, 39, 1964.

680. **Oshmarin, P. G. and Oparin, P. G.,** Helminth fauna of livestock in the Primorsk region with separate notes on the epizootiology of helminthiases (in Russian), in *Parasitic Worms of Mammals and Birds in the Primorsk Region*, Oshmarin, P. G., Ed., Izdv. Akad. Nauk, Moscow, 1963, 280.

681. **Oshmarin, P. G. and Demshin, N. I.,** On the helminths of domestic and of some wild animals in the Vietnam Democratic Republic (in Russian), in *Examinations of Fauna, Systematics and Biochemistry of Helminths of the Far East*, Izd., Akad. Nauk, Vladivostok, 1972, 5.

682. **Otto, R.,** Beiträge zur Anatomie und Histologie der Amphistomeen, Inaug.-Dissertation, Leipzig, 1896, 1—78.

683. **Ovcharenko, D. A.,** Changes in the helminth fauna with the age of *Cervus nippon hortulorum* kept in parks in Far-Eastern Soviet Union (in Russian), *Vestn. Leningr. Univ. Ser. Biol.*, 18, 5, 1963.

684. **Overstreet, R. M.,** Digenetic trematodes of marine teleost fishes from Biscayne Bay, Florida, *Tulane Stud. Zool. Bot.*, 15, 110, 1969.

685. **Owen, I. L.,** personal communication, 1985.

686. **Ozaki, Y.,** Studies on the trematode families Gyliauchenidae and Opistholebetidae with special reference to lymph system, *J. Sci. Hiroshima Univ.*, 1, 167, 1937.

687. **Ozaki, Y.,** Studies on the miracidium of trematodes. I. The miracidia of *Paramphistomum explanatum* (Creplin), *P. orthocoelium* Fischoeder and *Gastrothylax cobboldi* Fischoeder, *J. Sci. Hiroshima Univ.*, 12, 99, 1951.

688. **Pačenovský, J., Hovorka, J., and Krupicer, L.,** Arten bestimmung der paramphistomatose der Rinder und schafe in der Mongolischen Volksrepublik, *Folia Vet.*, 19, 191, 1975.

689. **Pačenovský, J., Hovorka, J., and Krupicer, I.,** Studies on the ventral sucker (acetabulum) in *Liorchis* (Velitchko, 1966), in *Abstr. 3rd Int. Symp. Helminth. Inst., Kosice*, 1976, 1.

690. **Pačenovský, J., Záhoř, Z., and Krupicer, I.,** First report of *Paramphistomum daubneyi* in cattle in Algeria, *Vet. Med. (Prague)*, 32, 379, 1987.

691. **Padilha, T. N.,** Caracterização de familia Zonocotylidae com redecrição de *Zonocotyle bicaecata* Travassos, 1948 e descrição de um novo gênero (Trematoda, Digenea), *Rev. Bras. Biol.*, 38, 415, 1978.

692. **Pagenstecher, H. A.,** Trematodenlarven und Trematoden, *Helminth. Beitr. (Heidelberg)*, 1, 1, 1857.

693. **Pallas, I. S.,** De infestis viventibus intra viventia, *Lugduni Batavor*, 1, 1760.

694. **Palmieri, J. R. and Sullivan, J. T.,** *Stunkardia minuta* sp. n. (Trematoda: Paramphistomidae) from the malayan box-tortoise *Cuora amboinensis*, *J. Helminthol.*, 51, 121, 1977.

695. **Pande, P. G.,** Acute amphistomosis of cattle in Assam (a preliminary report), *Indian J. Vet. Sci.*, 5, 364, 1935.

696. **Pande, B. P.,** On some digenetic trematodes from *Rana cyanophlyctis* of Kumaon Hills, *Proc. Indian Acad. Sci.*, 6, 109, 1937.

697. **Pandey, K. S.,** On a new trematode *Diplodiscus chauhani* n.sp. from the common Indian frog, *Rana cyanophlyctis* Schneider, *Proc. Indian Acad. Sci.*, 69, 203, 1969.

698. **Pandey, K. C.,** Studies on some known and unknown trematode parasites, *Indian J. Zootomy*, 14, 197, 1973.

699. **Pandey, K. C. and Chakrabarti, K. K.,** On a new trematode *Diplodiscus lali* n.sp. from the common Indian frog, *Rana tigrina Daud*, *Ceylon J. Sci. Biol. Sci.*, 8, 38, 1968.

700. **Pandey, K. C. and Jain P.,** A new amphistome cercariae from an Indian aquatic snail *Indoplanorbis exustus* (Deshayes), *Proc. Zool. Soc. (Calcutta)*, 24, 29, 1971.

701. **Pandey, K. C. and Agrawal, N.,** On the miracidium of *Diplodiscus lali* Pandey et Chakrabarti, 1968, *Folia Morphol.*, 27, 57, 1979.

702. **Parker, M. W.,** The trematode parasites from a collection of amphibians and reptiles, *J. Tern. Acad. Sci.*, 16, 27, 1941.

703. **Patnaik, M. M.,** A note on helminth parasites of the black buck *(Antilope cervicapra)*, *Curr. Sci.*, 33, 120, 1964.

704. **Patnaik, M. M.,** Common disease conditions of domestic ruminants in Orissa, *Orissa Vet. J.*, 4, 33, 1969.

705. **Patnaik, M. A. and Achrjyo, L. N.,** Notes on the helminth parasites of vertebrates in Barang Zoo (Orissa), *Indian Vet. J.*, 47, 723, 1971.

706. **Patterson, C.,** The distribution of Mesozoic freshwater fishes, *Mem. Mus. Nat. Hist.*, 88, 156, 1975.

707. **Paudi, B. C., Mohanty, A. K., and Misra, S. C.,** A note on the occurrence of amphistomes in the rumen of spotted deer *(Axis axis)* at the Nandakanan Biological Park, Barang, Orissa, *Indian Vet. J.*, 64, 893, 1987.

708. **Pav, J., Sobotka, A., Zajiček, D., and Zima, L.,** K výskyta *Paramphistomum cervi* (Schrank, 1790) u naši spárkaté zvěře (in Czech), *Sb. Cesk. Akad. Zemed. Ved Vet. Med.*, 35, 161, 1962.

709. **Pavlović, D.**, Prilog ponavanju ekstenzitete infekcije nekih endo-parazita nadenih kod zaklanik bivola različitik starosnik grupa sa teritorije sap Kosovo (in Serbo-Croate), *Vet. Glas.*, 29, 285, 1975.

710. **Penso, G.**, Su due paramfistomi parassiti dei bovini dell' Africa Orientale Italiana, *Inst. Sanit. Publica*, 111, 363, 1940.

711. **Pereira, R. C. S.**, Redescrição de *"Chiostichorchis waltheri"* (Sprehn, 1932) Travassos, 1934, *Atas Soc. Biol. Rio de Janeiro*, 12, 161, 1968.

712. **Pester, F. R. N. and Keymer, I. F.**, *Gastrodiscoides hominis* from an orangutan *Pongo pygmaeus*, S.E. Asia, *Trans. R. Soc. Trop. Med. Hyg.*, 62, 10, 1968.

713. **Peter, C. T.**, Studies on cercarial fauna in Madras. IV. The amphistome and gymnocephalus group of cercariae, *Indian J. Vet. Sci.*, 26, 27, 1956.

714. **Peter, C. T.**, Studies on the life history of *Gastrodiscus secundus* Looss, 1907, an amphistomatous parasite of equines in India, *Indian J. Helminthol.*, 12, 18, 1960.

715. **Peter, C. T. and Mudaliar, S. V.**, On a new cercaria determined to be the larva of *Gastrodiscus secundus* Looss, 1907, *Curr. Sci.*, 17, 303, 1948.

716. **Peter, C. T. and Srivastava, H. D.**, Studies on the life history of *Pseudodiscus collinsi* (Cobbold, 1875) Sonsino, 1895, an amphistomatous parasite of equines in India, in *Proc. 41st Ind. Sci. Congr.*, 41, 221, 1954.

717. **Peter, C. T. and Srivastava, H. D.**, On five new species of amphistome cercariae from India, *Proc. 42nd Indian Sci. Congr.*, Pt 3, 221, 1955.

718. **Peter, C. T. and Srivastava, H. D.**, On amphistome cercariae in India with a description of some new species, *Indian J. Helminthol.*, 12, 51, 1960a.

719. **Peter, C. T. and Srivastava, H. D.**, Studies on the life history of *Pseudodiscus collinsi* (Cobbold) of equines and elephants in India, *Indian J. Helminthol.*, 12, 1, 1960b.

720. **Peter, C. T. and Srivastava, H. D.**, On *Cercaria chungathi* Peter and Srivastava, 1955 and its relationship to *Gastrothylax crumenifer* (Creplin), *Parasitology*, 54, 111, 1961.

721. **Petriashvili, L. J.**, Helminth fauna of the marsh-frog (*Rana ridibunda* Pall.) in Lake Bazaletsky (in Russian), *Soobshch. Akad. Nauk. Gruz. GSSR*, 36, 457, 1964.

722. **Petrochenko, V. I. and Romanovskiĭ, A. A.**, On the role of rodents in dispersal of coenurus of sheep and helminth parasites of rodents in Kizlyarsh steppe (in Russian), *Mat. Nachn. Konf. Vses. Ova. Gelmintol.*, 1, 222, 1966.

723. **Pick, F.**, Sur on nouveau trematode du genre *Watsonius* chez *Papio sphinx*, *Bull. Soc. Pathol. Exot.*, 44, 59, 1951.

724. **Pick, F.**, Information nouvelles sur la distomatose a *Watsonius watsoni*, *Bull. Soc. Pathol. Exot.*, 57, 502, 1964.

725. **Pick, F. and Deschiens, R. E. A.**, La distomatose a *Watsonius watsoni* (Conyngham, 1904), *Bull. Soc. Pathol. Exot.*, 40, 202, 1947.

726. **Pike, A. W.**, Helminth parasites of the amphibians *Dicroglossus occipitalis* (Günther) and *Bufo regularis* Reuss, in Khartoum, *J. Nat. Hist.*, 13, 337, 1979.

727. **Pike, A. W. and Condy, J. B.**, *Fasciola tragelaphi* sp. nov. from the sitatunga, *Tragelaphus spekei* Rothschild, with a note on the prepharyngeal pouch in the genus *Fasciola* L., *Parasitology*, 56, 511, 1966.

728. **Pinto, C. and Almeida, J. L.**, III. Helmintologia, in *Zooparasots de Interesse Médico e Veterinario*, Pinto, C., Ed., Editora Cientifica, Rio de Janeiro, 1945, 364.

729. **Platnik, N. I. and Nelson, G. J.**, A model of analysis for historical biogeography, *Syst. Zool.*, 27, 1, 1978.

730. **Podlesnii, G. V.**, On the epizootiology of *Paramphistomum cervi* of calves and the use of prophylactic measure against this desease in the collective forms of Volhynia Oblast (in Ukrainian), *Matter. Ses. Viddil. Tvar. Ukr. Akad. Sils. Nauk*, 109, 1962.

731. **Pogorelii, A. I. and Mereminskiĭ, A. I.**, Period of life-cycle of *Paramphistomum cervi* (Schrank, 1790) in the intermediate host (in Russian), *Mater. Nauchn. Konf. Obva. Gelmintol.*, 2, 40, 1963.

732. **Poirier, J.**, Trématodes parasites de l'éléphant d'Afrique, *Assoc. Fr. Avance, Sci. C. R.* (1908), 37, 580, 1909.

733. **Porter, A.**, The history of some trematodes occurring in South Africa, *S. Afr. J. Sci.*, 18, 156, 1921.

734. **Porter, A.**, The larval trematode found in certain South African mollusca, with special reference to schistosomiasis (Bilharziasis), *S. Afr. Inst. Med. Res.*, 62, 1, 1938.

735. **Poumarau, E. M. C.**, *Catadiscus longicoecalis* nueva espeie parasita de ofidios (Trematoda: Paramphistomidae) con una lista de espedies del genero *Catadiscus* Cohn, 1904, *Physis (Buenos Aires)*, 25, 277, 1965.

736. **Pratt, I. and McCauley, J. E.**, Trematodes of the Pacific Northwest, an annoted catalog, *Oreg. State Monogr. Stud. Zool.*, 11, 118, 1961.

737. **Premvati, G. and Agarwal, M.**, On *Stunkardia dilymphosa* Bhalerao, 1931 (Trematoda: Paramphistomatidae), and synonymy of allied flukes f rom fresh-water chelonians of Malaya and India, *Acta Parasitol. Pol.*, 28, 33, 1981.

738. **Presidente, P. J. A.,** Ectoparasites, endoparasites and some diseases reported from sambar deer throughout its native range and in Australia and New Zealand, *Proc. 49th Deer Refresh. Course,* 72, 543, 1979.

739. **Prestwood, A. K., Smith, J. F., and Mohan, W. E.,** Geographical distribution of *Gongylonema pulchrum, Gongylonema verrucosum* and *Paramphistomum liorchis* in white-tailed deer of the southeastern United States, *J. Parasitol.,* 56, 123, 1970.

740. **Prestwood, A. K., Forest, E., and Kellogg, F. E.,** Helminth parasitisms among intermingling insular populations of white-tailed deer, feral cattle and feral swine, *J. Am. Vet. Med. Assoc.,* 166, 787, 1975.

741. **Price, E. W.,** The host relationship of the trematode genus *Zygocotyle, J. Agric. Res.,* 36, 911, 1928.

742. **Price, E. W.,** The trematode parasites of marine mammals, *Proc. U.S. Natl. Mus.,* 81, 1, 1932.

743. **Price, E. W.,** Two new trematodes from African reptiles, *Proc. Helminthol. Soc. Wash.,* 3, 67, 1936.

744. **Price, E. W.,** Three new genera and species of trematodes from cold-blooded vertebrates, in K. J. Skrjabin *Jubilee Volume,* 1937, 483.

745. **Price, E. W.,** The fluke situation in American ruminants, *J. Parasitol.,* 39, 119, 1953.

746. **Price, R. P. and Buttner, J. K.,** Gastrointestinal helminths of the Central newt, *Notophthalmus viridescens lousianensis* Wolterstorff, from Southern Illinois, *Proc. Helminthol. Soc. Wash.,* 49, 285, 1982.

747. **Price, E. W. and McIntosh, A.,** Paramphistomes of North American domestic ruminants, *J. Parasitol. Suppl.,* 30, 9, 1944.

748. **Price, E. W. and McIntosh, A.,** Two new trematodes of the genus *Cotylophoron* Stiles et Goldberger, 1910, from American sheep, in *G. S. Thaper Commemoration Volume,* 1953, 227.

749. **Pritchard, M. H.,** Notes on four helminths from the clawed toad, *Xenopus laevis* (Daudin) in South Africa, *Proc. Helminthol. Soc. Wash.,* 31, 121, 1964.

750. **Prod'hon, J., Richard, J., Brygoo, E. R., and Daynes, P.,** Presence de *Paramphistomum microbothrium* Fischoeder, 1901 a Madagascar, *Arch. Inst. Pasteur Madagascar,* 37, 27, 1968.

751. **Prokopič, J.,** K gelmintofaunee ámfibij Albanii (in Czech), *Cslka. Parasitol.,* 7, 151, 1960.

752. **Prokopič, J. and Kotrlá, B.,** Difusion de los trematodos de la familia Paramphistomatidae en los bovinos y ovejas en Cuba, *Rev. Cubana Med. Trop.,* 26, 149, 1974.

753. **Prokopič, J. and Křivanec, K.,** Helminths of amphibians, their interaction and host-parasite relationships, *Acta Sci. Natl. Acad. Sci. Bohem. Brno,* 9, 1, 1975.

754. **Prudhoe, S.,** Some trematodes from Ceylon, *Ann. Mag. Nat. Hist.,* 11, 1, 1944.

755. **Prudhoe, S.,** Some roundworms and flatworms from the West Indies and Surinam. III. Trematodes, *J. Linn. Soc. London,* 41, 415, 1949.

756. **Prudhoe, S.,** Trematoda-Exploration du Parc National de l'Upemba. *Mission G. F. de Witte (1946-49),* Brussels, 48, 1, 1957.

757. **Prudhoe, S.,** A new genus of paramphistome trematode from a Wallaby, in *B. S. Chauhan Commemoration Volume,* Tiwari, K. K. and Srivastava, C. B., Eds., Zoology Society of India, Bhubaneswar, 1975, 63.

758. **Prudhoe, S., Yeh, L.-S. and Khalil, L. F.,** A new amphistome trematode from ruminants in Northern Rhodesia, *J. Helminthol.,* 38, 57, 1964.

759. **Quiroz, R. H., Garcia, R., and Davalos, E.,** Identificacion de *Cotylophoron cotylophorum* (Fischoeder, 1901) en un Ovino de Mexico, *Tec. Pecu. Mex.,* 21, 61, 1972.

760. **Quiroz, R. H. and Ochoa, R.,** Presencia de *Paramphistomum cervi* (Schrank, 1790), en un ovino de raza Tabasco o peliguey en Mexico, *Tec. Pecu. Mex.,* 21, 59, 1973.

761. **Rahman, M. H.,** A survey of helminthiasis in East Pakistan, *Indian Vet. J.,* 35, 539, 1958.

762. **Rai, P.,** A redescription of *Pseudodiscus collinsi* (Cobbold, 1875) Stiles et Goldberger, 1910 — a common parasite of equines, *Natl. Acad. Sci. India,* 29, 201, 1959.

763. **Rai, P. and Srivastava, J. S.,** Preliminary list of helminth parasites of donkey *Equus asinus),* *Curr. Sci. Bangalose,* 27, 456, 1958.

764. **Railliet, A.,** Les helminthes de animaux domestiques et de l'homme en Indochine, *Bull. Soc. Zool. Fr.,* 49, 594, 1924.

765. **Raina, M. K., Sey, O., and Khan, M. D.,** Paramphistomes (Trematoda: Amphistomida) of domestic ruminants in Kashmir, India, *Miscnea, Zool. Hung.,* 4, 5, 1987.

766. **Rankin, J. S.,** An ecological study of parasites of some North Carolina salamanders, *Ecol. Mongr.,* 7, 169, 1937.

767. **Rao, M. A. N. and Ayyar, L. S. P.,** A preliminary report on two amphistome cercariae and their adults, *Indian J. Vet. Sci.,* 2, 402, 1932.

768. **Rao, A. T. and Achrjyo, L. N.,** Pathological lesions in livers of two Indian sambars (*Cervus unicolor niger*) infected with *Paramphistomum explanatum* (Creplin, 1847) Näsmark, 1937 *Gigantocotyle explanatum, Indian Vet. J.,* 46, 916, 1969.

769. **Rapić, D.,** *Lymnaea truncatula* posrednik u razvoju metilja *Paramphistomum microbothrium* (in Serbo-Croat), *Acta Parasitol. Jug.,* 10, 51, 1979.

770. **Rausch, R.,** Parasites of Ohio muskrats, *J. Wildl. Manage.,* 10, 79, 1946.

771. **Rausch, R. W.,** Observations on some helminths parasitic in Ohio turtles, *Am. Midl. Nat.,* 38, 434, 1947.

772. **Reinhardt, S.,** Zur Entwicklung, anatomie und Histologie von *Paramphistomum microbothrium* (Trematoda, Digenea), Inaug. Diss. Bonn, 1969, 1.

773. **Reisinger, E.**, Utersuchungen über Bau und Function des Exkretionsapparates digenetische Trematoden, I. Die Emunktorien des Miracidium von *Schistosoma haematobium* Bilharz einigen Beiträge zur dessen anatomie und Histologie, *Zool. Anz.*, 57, 1, 1923.

774. **Rhee, J. K., Kang, C. W., and Lee, H. I.**, The karyotype of *Paramphistomum explanatum* (Creplin, 1849) obtained form Korean cattle (in Korean), *Korean J. Parasitol.*, 24, 42, 1986.

775. **Rhee, J. K., Kim, Y. H., and Pork, B. K.**, The karyotype of *Paramphistomum cervi* (Zeder, 11790) from Korean cattle (in Korean), *Korean J. Pparasitol.*, 25, 154, 1987.

776. **Rice, E. W. and Heck, O. B.**, A survey of the gastrointestinal helminths of the muskrat, *Ondatra zibethicus* collected from two localities in Ohio, *Ohio J. Sci.*, 75, 263, 1975.

777. **Richard, J. and Daynes, P.**, *Zygocotyle lunata* (Diesing, 1836) (Trematoda) chez un canard sauvage a Madagascar, *Bull. Mus. Natl. d'Hist. Nat. Paris*, 38, 949, 1966.

778. **Richard, J., Chabaud, A. G., and Brygoo E. R.**, Notes sur la morphologie et la biologie des trematodes digenes parasites des grenoulles du jardin de l'Institut Pasteur a Tananarive, *Arch. Inst. Pasteur Madagascar*, 37, 31, 1968.

779. **Rinses, J.**, *Paramphistomum cervi*, *Tijdschr. Diergeneeskd.*, 73, 81, 1948.

780. **Roach, R. W. and Lopes, V.**, Mortality in adult ewes resulting from intestinal infestation with immature paramphistomes complicated by severe fascioliasis, *Bull. Epizoot. Dis. Afr.*, 14, 317, 1966.

781. **Rockett, N., Freire, J. J., and Fi Primio, R. A.**, Nota prévia sobre a presença de *Paramphistomum cervi* (Schrank, 1790) no Estado do Rio Grande do Sul, 1965 apud Freise, J. J., Fauna parasitária riograndense. I. Introdução, boi, ovelha e cabra, *Rev. Med. Vet. Sao Paulo*, 3, 40, 1967.

782. **Rohde, K.**, Über einigee Malayische Trematodoen, *J. Parasitenkd.*, 22, 268, 1963.

783. **Rohde, K.**, Two amphistomes, *Lobatodiscus australiensis* n.g., n.sp. and *Elseyatrema microacetabularis* ng., n.sp. from the Australian turtle *Elseya dentata* (Gray), *Syst. Parasitol.*, 6, 219, 1984.

784. **Rolfe, P. F. and Boray, J. C.**, Chemotherapy of paramphistomosis in sheep, *Aust. Vet. J.*, 64, 1, 1987a.

785. **Rolfe, P. F. and Boray, J. C.**, Chemotherapy of paramphistomosis in cattle, *Aust. Vet. J.*, 64, 328, 1987b.

786. **Romanenko, L. N.**, Karyotype of *Liorchis scotiae* (Willmott, 1950) (Paramphistomatidae) (in Russian), *Tr. Vses. Inst. Gelmintol.*, 19, 153, 1972.

787. **Romanenko, L. N.**, Study of the chromosome number of some trematodes of the suborder, Paramphistomata (in Russian), *Tr. Gelmintol. Lab.*, 26, 226, 1974.

788. **Romashov, V. A.**, Contribution to the study of the helminth fauna of red deer in the Voronezh rezervation (in Russian), *Mater. Nauchn. Konf. Vses. Obva. Gelmintol.*, 2, 64, 1963.

789. **Romashov, V. A.**, Helminth fauna of European beaver in its aboriginal colonies of Europe, *Acta Parasitol. Pol.*, 17, 55, 1969.

790. **Rosen, R. and Manis, R.**, Trematodes of Arkansas amphibians, *J. Parasitol.*, 62, 833, 1976.

791. **Roth, H. H. and Dalchow, W.**, Untersuchungen über den Wurmbefall von Antilopen Rhodesien, *Z. Angew. Zool.*, 54, 203, 1967.

792. **Roveda, R. J. and Ringuelet, R.**, Lista de los parásitos de los animales domesticos en la Argentina, *Gac. Vet.*, 9, 67, 1947.

793. **Roy, T. K.**, Cytochemical studies of esterases in the bovine *Ceylonocotyle scoliocoelium*, *Indian J. Exp. Biol.*, 18, 872, 1980.

794. **Rozman, M.**, *Diplodiscus subclavatus* in *Rana esculenta* and its developmental forms in snail *Planorbis planorbis*, *Veterinaria (Sarajevo)*, 4, 233, 1970.

795. **Rudolphi, C. A.**, *Entozoorum sive vermium intestinalium historia naturalis*. II. Amstelaedami: sumtibus tabernae Librariae et Artium, 1809, 340.

796. **Rudolphi, C. A.**, Erster Nachtung zu meiner Naturgeschichte der Eingeweidewürmer, *Ges. Naturforsch. Freide Berlin Mag. neuest. Entdeck.* 6, 83, 1814.

797. **Ruiz, J.**, *Catadiscus freitaslenti* sp. n. (Trematoda: Paramphistomidae), parasito de ofídeo neotrópico, observação, sòbre a presença de dois canais eferentes no gêenero *Catadiscus* Chon, 1904, *Mem. Inst. Butantan*, 17, 29, 1943.

798. **Ruziev, Sh. M.**, Trematodiasis of the proventriculus in cattle and sheep in Karakalpak ASSR (in Russian), Samarkand, Izdv. FAN, 1972, 147.

799. **Rybaltovskiĭ, O. V.**, *Gastrothylax* infestation in cattle (in Russian), *Dokl. Vses. Akad. Skh. Nauk*, 22, 38, 1957.

800. **Ryšavý, B., Moravec, F., Baruš, V., and Yousif, F.**, Some helminths of *Bulinus truncatus* and *Biomphalaria alexandrina* from the irrigation system near Cairo, *Folia Parasitol. (Prague)*, 21, 97, 1974.

801. **Ryzhikov, K. M., Sharpilo, V. P., and Shevchenko, N. N.**, *Helminth parasites of amphibians in the USSR* (in Russian), Izd. "Nauka", Moscow, 1980, 59.

802. **Sachs, R. and Sachs, C.**, A survey of parasitic infections of wild herbivores in the Serengeti region northern Tanzania and the Lake Rukwa region in southern Tanzania, *Bull. Epizoot. Dis. Afr.*, 16, 455, 1968.

803. **Sadana, J. R., Gupta, R. K. P., Mahajan, S. K., and Kuchroo, V.**, Cholelithiasis associated with *Gigantocotyle explanatum* infection in a bullock, *Haryana Vet.*, 16, 84, 1977.

804. **Sadykhov, I. A.**, On the helminth parasites of the water vole in Azerbadzhan (in Russian), *Izv. Akad. Nauk Az. SSR*, 3, 77, 1960.

805. **Sakamoto, T., Tsumura, I., and Izawa, S.,** *Homalogaster paloniae* from a cow in Tottori, Japan (in Japanese), *Jpn. J. Parasitol.*, 13, 501, 1964.

806. **Salami-Cadoux, M.-L. and Grégorio, R. de,** Présence de *Diplodiscus subclavatus* an Togo. Considérations sur le genre *Diplodiscus* (Digenea: Paramphistomidae) en Afrique et a Madagascar, *Bull. Inst. Fr. Afr. Noire*, 38, 785, 1976.

807. **Salimov, B., Ruziev, Sh., and Khaidarov, U.,** On the paramphistomids of large and small ruminants in Izbekistan (in Russian), in *Diseases of Farm Animals, Vol.* 19, Izdv., "FAN" Tashkent, 1971, 97.

808. **Samnaliev, P., Sey, O., and Dimitrov, V.,** Nomenclature of the argentophile structures of miracidia. Argentophile structure of miracidium *Paramphistomum ichikawai* Fukui, 1922, EMOP, 4, Izmir,, 1964, 143.

809. **Samnaliev, P. and Vassilev, L.,** Development of *Paramphistomum microbothrioides* Price and McIntosh, 1944 in *Lymnaea (Galba) truncatula,*, *Khelmintologiya*, 11, 62, 1981.

810. **Samnaliev, P., Bayssade-Dufour, Ch., Albaret, J.-L., Albaret, Dimitrov, V., Cassone, J., and Kamburov, P.,** Structures argyrophiles tegumentaires du miracidium, de la rédie et de la cercaire de *Paramphistomum daubneyi* Dinnik, 1962 (Trematoda, Paramphistomidae), *Ann. Parasitol. (Paris)*, 56, 155, 1981.

811. **Samnaliev, P. and Poljakova-Krusteva, O.,** On the possibility of a penetration of *Bulinus truncatus* by *Paramphistomum cf. daubneyi* Dinnik, 1962 miracidia, and of *Lymnaea (Galba) truncatula* by *Paramphistomum microbothrium* Fischoeder, 1901) miracidia (in Bulgarian), *Khelmintologiya*, 13, 67, 1982.

812. **Samnaliev, P., Pino, L. A., Bayssade-Dufour, Ch., and Albaret, J.-L.,** Structures argyrophiles superficielles du miracidium et de la cercaire de *Paramphistomum leydeni* Näsmark, 1937, *Ann. Parasitol. Hum. Comp.*, 59, 151, 1984.

813. **Samnaliev, P., Albaret, J.-L., Bayssade-Dufour, Ch., Dimitrov, V., and Cassonee, J.,** Structures argyrophiles superficielles du miracidium et de la cercaire de *Paramphistomum microbothrioides* Price et McIntosh, 1944 (Trematoda, Paramphistomidae), *Ann. Parasitol. Hum. Comp.*, 61, 625, 1986.

814. **Samuel, W. M., Barrett, M. W., and Lynts, G. A.,** Helminths in moose of Alberta, *Can. J. Zool.*, 54, 307, 1976.

815. **Sandground, J. H.,** *Gastrodiscus hominis* as a parasite of rats in Java, *J. Parasitol.*, 26 (Suppl.), 34, 1940.

816. **Saoud, M. F. A. and Wannas, M. Q. A.,** A qualitative and quantitative survey on the helminth parasites of fishes from the Aswan high dam lake in Egypt, *Qatar Univ. Sci. Bull.*, 4, 129, 1984.

817. **Savage, J. M.,** The geographical distribution of frogs patterns and prediction, in *Evolutionary Biology of the Anurans* Vial, J. L., Ed., University of Missouri Press, Columbia, 1973, 351.

818. **Saxena, J. N.,** On the eggs and miracidia of *Diplodiscus amphichrus magnus* Srivastava, 1934, *All Ind. Congr. Zool.*, 2, 375, 1962.

819. **Schad, G. A., Kuntz, R. E., Anteson, R. K., and Webster, G. F.,** Amphistomes (Trematoda) from domestic ruminants in North Borneo (Malaysia), *Can. J. Zool.*, 42, 1964.

820. **Schiffo, H. P. and Lombardero, O. J.,** Mortandad de vacunos producida por *Balanorchis anastrophus*, *Gac. Vet.*, 36, 139, 1974.

821. **Schillhorn, Van Veen T. W., Shonekan, R. A. O., and Febiyi, J. P.,** Host-parasite cheklist of helminth parasites of domestic animals in Northern Nigeria, *Bull. Anim. Health Prod. Afr.*, 23, 269, 1975.

822. **Schmid, K., Rückrich, H. U., and Boch, J.,** The development of *Paramphistomum cervi* from miracidium to metacercaria, *Berl. Muench. Tieraerztl. Wochenschr.*, 94, 463, 1981.

823. **Schoon, O.,** *Paramphistomum cervi* (in Dutch), *Tijdschr. Diergeneesk*, 72, 831, 1947.

824. **Schwartz, B.,** Internal metazoan parasites collected from ruminants in the Philippines, *Philipp. J. Sci.*, 26, 521, 1925.

825. **Seddon, H. R.,** Diseases of domestic animals in Australia. I. Helminth infections, *Serv. Publ. Dep. Hlth. Aust. Vet. Hyg.*, 8, 1, 1950.

826. **Segal, D. B., Humprey, J. M., Edwards, S. J., and Kirby, M. D.,** Parasites of man and domestic animals in Vietnam, Thailand, Laos and Cambodia. Host list and bibliography, *Exp. Parasitol.*, 23, 412, 1968.

827. **Self, J. T. and Bouchard, J. L.,** Parasites of the wild turkey, *Meleagris gallopavo intermedia* Sinnet from the Wichita Mountains Wildlife Refuge, *J. Parasitol.*, 36, 502, 1950.

828. **Seneviratna, P.,** A check list of helminths in the department of veterinary pathology, University of Ceylon, Peradenija, *Ceylon Vet. J.*, 3, 32, 1955.

829. **Serrano, F. M. H.,** Fauna helmintoloógica dos animais domésticos de Angola, *Pecu. Loanda*, 20, 51, 1962.

830. **Sewell, R. B. S.,** Cercariae indicae, *Indian J. Med. Res.*, 10 (Spec. Suppl.), 1, 1922.

831. **Sey, O.,** Tanulmányok a magyarországi parazita féregfaunáról. I (in Hungarian), *Pécsi Tanárképz. Föisk. Tud. Kozl.*, 413, 1964.

832. **Sey, O.,** Gametogenesis in *Paramphistomum microbothrium* Fischoeder, 1901, *Acta Vet. Acad. Sci. Hung.*, 21, 93, 1971.

833. **Sey, O.,** Investigation on the eggs, the process of embryo formation and of the structure of miracidium of *Paramphistomum daubneyi* Dinnik, 1962, *Parasitol. Hung.*, 5, 17, 1972.

834. **Sey, O.,** Histological characterization of the pharynx, genital atrium and the acetabulum of *Nematophila grande* (Diesing, 1839) (Trematoda: Paramphistomata), *Parasitol. Hung.*, 6, 63, 1973a.

835. **Sey, O.,** A *Paramphistomum daubneyi* Dinnik, 192 petéinek kikelési mechanizmusa (in Hungarian), *Allattani Kozl.*, 60, 95, 1973b.

836. **Sey, O.,** On the species of *Paramphistomum* of cattle and sheep in Hungary, *Acta Vet. Acad. Sci. Hung.*, 24, 19, 1974.

837. **Sey, O.,** Histological examination on the muscular organs of some amphistomes (Trematoda: Paramphistomata), *Parasitol. Hung.*, 8, 55, 1975.

838. **Sey, O.,** Studies on the stomach flukes of buffalo in Egypt (Trematoda: Paramphistomata), *Folia Parasitol. (Prague)*, 23, 237, 1976.

839. **Sey, O.,** Examination of helminth parasites of marine turtles caught along the Egyptian coast, *Acta Zool. Hung.*, 23, 387, 1977a.

840. **Sey, O.,** Examination of amphistomes (Trematoda: Paramphistomata) parasitizing in Egyptian ruminant, *Parasitol. Hung.*, 10, 47, 1977b.

841. **Sey, O.,** Examination of rumen flukes (Trematoda: Paramphistomata) of cattle in Rumania, *Paraditol. Hung.*, 11, 23, 1978.

842. **Sey, O.,** Life-cycle and geographical distribution of *Paramphistomum daubneyi* Dinnik, 1962 (Trematoda: Paramphistomata), *Acta Vet. Acad. Sci. Hung.*, 27, 115, 1979a.

843. **Sey, O.,** Examination of validity and systematic position of some paramphistomids of Indian ruminants, *Parasitol. Hung.*, 12, 31, 1979b.

844. **Sey, O.,** Amphistome parasites of the dugong and a revision of the subfamily Solenorchiinae (Trematoda: Paramphistomatidae), *Acta Zool. Acad. Sci. Hung.*, 26, 223, 1980a.

845. **Sey, O.,** Revision of the amphistomes of European ruminants. *Parasitol. Hung.*, 13, 13, 1980b.

846. **Sey, O.,** Re-examincaton of an amphistome (Trematoda) collection deposited in the Geneve Museum with a description of *Orthocoelium saccocoelium* sp. n., *Rev. Suisse Zool.*, 87, 431, 1980c.

847. **Sey, O.,** Revision of the genus *Cotylophoron* Stiles et Goldberger, 1910 (Trematoda: Paramphistomata), *Helminthologia*, 19, 11, 1982a.

848. **Sey, O.,** The morphology, life-cycle and geographical distribution of *Paramphistomum cervi* (Zeder, 1790) (Trematoda: Paramphistomata), *Miscnea Zool. Hung*, 1, 11, 1982b.

849. **Sey, O.,** Reconstruciton of the systematics of the family Diplodiscidae Skrjabin, 1949 (Trematoda: Paramphistomata), *Parasitol. Hung.*, 16, 63, 1983a.

850. **Sey, O.,** Revision of the family Gastrothylacidae Stiles et Goldberger, 1910 (Trematoda, Paramphistomata), *Acta Zool. Acad. Sci. Hung.*, 29, 223, 1983b.

851. **Sey, O.,** Scanning electron microscopic examination of the tegumental surface of some amphistomes (Trematode: Amphistomida), *Parasitol. Hung.*, 17, 45, 1984a.

852. **Sey, O.,** Description of *Watsonius papillatus* sp. n. and the revision of the subfamily Watsoniinae Näsmark, 1937 (Trematoda: Amphistomida), *Acta Zool. Hung.*, 30, 493, 1984b.

853. **Sey, O.,** Review of pouched amphistomes of Vietnamese ruminants, with a description of *Carmyerius bulbosus* sp. n. (Trematoda: Amphistomata), *Miscnea Zool. Hung.*, 3, 31, 1985a.

854. **Sey, O.,** Amphistomes of Vietnamese vertebrates (Trematoda: Amphistomida), *Parasitol. Hung.*, 18, 17, 1985b.

855. **Sey, O.,** Description some new taxa of amphistomes (Trematoda, Amphistomida) from Vietnamese freshwater fishes, *Acta Zool. Hung.*, 32, 161, 1986.

856. **Sey, O.,** Typifying and classifying of the muscular organs of amphistomes (Trematoda: Amphistomida), *Parasitol. Hung.*, 20, 45, 1987.

857. **Sey, O.,** Chemotherapy of paramphistomosis of domestic ruminants, *Programme and Abstracts, ICOPA* 5, Budapest, 69, 1988a.

858. **Sey, O.,** Rumen flukes of wild ruminants in Hungary, in 18th cong. Hung. Biol. Soc. (Abstr. 1988b., 78.

859. **Sey, O.,** Scope of and proposal for systematics of the Amphistomida (Lühe, 1909), Odening, 1974 (Trematoda), *Parasitol. Hung.*, 21, 17, 1988c.

860. **Sey, O. and Abdel-Rahman, M. S.,** Studies on *Paramphistomum* species of cattle and sheep in Egypt, *Assuit Vet. Med. J.*, 2, 145, 1975.

861. **Sey, O. and Sayed, R. I.,** Examination of the pre-parasitic stages of two species of fish amphistomes (Trematode), *Acta Zool. Acad. Sci. Hung.*, 22, 165, 1976.

862. **Sey, O. and Visnyakov, Yu.,** Examination of paramphistomid species of Bulgarian domestic ruminants, *Parasitol. Hung.*, 9, 25, 1976.

863. **Sey, O. and Arru, E.,** A revision of species of *Paramphistomum* Fischoeder, 1901 occurring in Sardinian domestic ruminants, *Riv. Parassitol.*, 38, 295, 1977.

864. **Sey, O. and Graber, M.,** Examinations of amphistomes (Trematoda: Paramphistomata) of some African mammals, *Rev. Elev. Med. Vet. Pays Trop.*, 32, 161, 1979.

461

865. **Sey, O. and Graber, M.**, *Nilocotyle duplicisphinctris* sp. n. (Trematoda: Paramphistomidae) from *Hippopotamus amphibius*, *J. Helminthol.*, 54, 123, 1980.

866. **Sey, O. and Eslami, A.**, Review of amphistomes (Trematoda: Paramphistomata) of Iranian domestic ruminants, *Parasitol. Hung.*, 14, 61, 1982.

867. **Sey, O. and Böröczky, K.**, Bendömételykór vizsgálata gemenci gimszarvasokban és özekben, *Nimrod*, 10, 22, 1983.

868. **Sey, O. and Moravec, F.**, An interesting case of hyperparasitism of the nematode *Spironoura babei* Ha Ky, 1971 (Nematoda: Kathlaniidae), *Helminthologia*, 23, 173, 1986.

869. **Shagraev, M. A. and Zhaltsanova, D. S. D.**, A study of the helminth fauna of domestic deer in the Buryat ASSR (in Russian), in *Fauna and Resorsy of the Vertebrates of Lake Baykal Basin*, Izdv. Akad. Nauk, U.S.S.R., Sibirs. Otdel. Buryats. Filial, 1980, 128.

870. **Shakhurina, E. A. and Tyhmanyants, A. A.**, Biological characteristics of species causing paramphistomosis of cattle (in Russian), *Parasites of Animals and Man in the Lower Reaches of the Amu Darya*, Izdv. FAN, Tashkent, 1969, 82.

871. **Shanta, C. S.**, A revised check list of helminths of domestic animals in West Malaysia, *Malay. Vet. J.*, 7, 180, 1982.

872. **Sharma Deorani, V. P. and Jain, S. P.**, Reasons for involvement of duodenum in parasitic part of life cycle of rumen amphistomes (Paramphistomatidae: Trematoda), *Indian J. Helminthol.*, 21, 177, 1969.

873. **Sharma, G. P., Mittal, O. P., and Madhu Bala**, Chromosome studies in the family Paramphistomidae (Digenea: Trematoda), *Proc. 55th Ind. Sci. Congr.*, 3, 472, 1968.

874. **Sharma, G. P., Mittal, O. P., and Nakhasi, H.**, Karyological studies on two species of digenetic trematodes (Abstr.), *Proc. Ind. Sci. Congr. Assoc.*, 61, 1974.

875. **Sharma, G. P. and Nakhasi, V.**, Studies on the chromosomes of three species of the Indian digenetic trematodes, *ICOPA, Munich*, 3, 23, 1974.

876. **Sharma, P. N. and Ratnu, L. S.**, Morphology, histochemistry and the biological significance of the lymph system of the trematode *Orthocoelium scoliocoelium*, *J. Helminthol.*, 56, 59, 1982.

877. **Shipley, A. E.**, *Cladorchis watsoni* (Conyngham) a human parasite from Africa, *Thompson yates and Johnston Lab. Rep.*, 8, 129, 1905.

878. **Siddiqi, A. H.**, Three new species of *Orientodiscus* (Trematoda: Paramphistomata) from freshwater turtles, *J. Helminthol.*, 39, 377, 1965.

879. **Simon- Vicente, F., Martinez-Fernandez, A., and Cordero del Campillo, M.**, Some observations on the redia, cercaria and metacercaria of *Opisthodiscus nigrivasis* (V. Mehlij (sic), 1929) Odening, 1959 (Trematoda: Paramphistomidae), *J. Helminthol.*, 48, 187, 1974.

880. **Simpson, G. G.**, Mammals and the nature of continents, *Am. J. Sci.*, 24, 1, 1943.

881. **Singh, K. S.**, Some trematodes collected in India, *Trans. Am. Microsc. Soc.*, 73, 202, 1954.

882. **Singh, K. S.**, On a new amphistome cercaria, *C. lewerti*, from India, *Trans. Am. Microsc. Soc.*, 76, 366, 1957.

883. **Singh, K. S.**, A redescription and life-history of *Gigantocotyle explanatum* (Creplin, 1847) Näsmark, 1937 (Trematoda: Paramphistomidae) from India, *J. Parasitol.*, 44, 210, 1958.

884. **Singh, K. S.**, On *Srivastavia indica* n.g., n.sp. (Paramphistomatidae) a parasite of ruminants and its life-history, in *H.D. Srivastava Commemorational Volume*, Izatnagar, 1970, 117.

885. **Singh, C. D. N. and Lakra, P.**, Pathologic changes in naturally occurring *Cotylophoron cotylophorum* infection in cattle, *Am. J. Vet. Res.*, 32, 659, 1971.

886. **Singh, K. S. and Malaki,, A.**, Parasitological survey of Kumaum region. XVIII. One known and two new cercariae from fresh-water snails, *Indian J. Helminthol.*, 15, 54, 1963.

887. **Singh, P. N. and Kuppuswami, P. B.**, Pathology of caprine liver infested with *Gigantocotyle explanatum* (Creplin, 1847), *Indian Vet. J.*, 46, 209, 1969.

888. **Sivtseva, M. Z.**, Breadth of distribution of Paramphistomata of the rumen of cattle in the European section of RSFSR, (in Russian), *Mater. Dokl. Nauchn. Konf. Posv. 90-let. Kazans. Vet. Inst.*, 176, 1963.

889. **Skrjabin, K. I.**, Parasitic trematodes and nematodes collected by the expedition of Prof. V. Dogiel and I. Sokolov in *British East Africa*, *Rev. Zool. Exped. to British East Africa by Prof. Dogiel and Sokolov, 1914*, 1, 736, 1916.

890. **Skrjabin, K. I.**, *Trematodes of Animals and Man Vol. 3* (in Russian), Izdv. Akad. Nauk Moscow, 1949, 623 pp.

891. **Skrjabin, K. I.**, Reconstruction of systematics of the trematode order Paramphistomata Skrjabin et Schulz, 1937 (in Russian), *Dokl. Akad. Nauk SSSR*, 45, 919, 1949.

892. **Smith, R. J.**, The life-history of *Megalodiscus ferissianus* n. sp. (Trematoda: Paramphistomatidae), thesis, University of Michigan, Ann Arbor, 1953.

893. **Smith, R. J.**, The miracidium of *Wardius zibethicus* (Trematoda: Paramphistomidae), *J. Parasitol.*, 44, 195, 1958.

894. **Smith, R. J.**, Ancylid snail: first intermediate host to certain trematodes as a new host for *Magalodiscus* and *Haematoloechus*, *Trans. Am. Microsc. Soc.*, 78, 228, 1959.

895. **Smith, R. J.,** Ancylid snails as intermediate hosts of *Megalodiscus temperatus* and other digenetic trematodes, *J. Parasitol.* 53, 287, 1967.

896. **Snider, J . B. and Lankester, M. W.,** Biology of rumen flukes (*Paramphistomum liorchis*) in moose (*Alces alces*) of Northwestern Ontario, Canada, *ICOPA*, 5, 469, 1982.

897. **Sobrero, R.,** Ricostruzione del ciclo di vita di *Paramphistomum clavula* (Näsmark, 1937), parassita dei ruminanti in Somalia, *Parassitologia*, 4, 165, 1962.

898. **Sogandares-Bernal, F.,** *Cleptodiscus kyphosi,* a new trematode (Paramphistomatidae) in *Kyphosus sectatrix* (Linn.) from Bimini, B.W.I., *J. Parasitol.*, 45, 148, 1959.

899. **Sokoloff, D. and Caballero, E.,** Una nueva especies de trematodo del intestino del manati. *Schizamphistoma manati* sp. n., *An. Inst. Biol. Univ. Mex.*, 3, 163, 1932.

900. **Sokoloff, D. and Caballero, E.,** Primera contribucion al conocimiento de los parásitos de *Rana montezumae* (Trematoda), *An. Inst. Biol. Univ. Mex.*, 4, 15, 1933.

901. **Sonsino, P.,** Studi sui parassitidi molluschi di acqua dolce nei dintorni di Cairo in Egitto, *Festschr. 70 Geburtst. R. Leukart's*, 134, 1892.

902. **Southwell, T. and Kirshner, A.,** A description of a new species of amphistome, *Chiorchis purvisi*, with notes on the classification of the genera within the group, *Ann. Top. Med. Parasitol.*, 31, 215, 1937.

903. **Speare, R.,** Comparative morphology of the eggs of the paramphistomid trematodes of the agile wallaby, *Macropus agilis* (Gould, 1842), *J. Wildl. Dis.*, 19, 368, 1983.

904. **Speare, R., Beveridge, I., and Johnson, P. M.,** Parasites of the agile wallaby, *Marcropus agilis* (Marsupialia), *Aust. Wildl. Res.*, 10, 89, 1983.

905. **Sprehn, C. E. W.,** *Lehrbuch der Helminthologie*, Gebrüder Borntraeger, Berlin, 1932, 184.

906. **Srivastava, H. D.,** On the trematodes of frogs and fishes of the United Provinces, *Proc. Acad. Sci. United Provinces, India*, 4, 113, 1934.

907. **Srivastava, H. D.,** Studies on the amphistomatous parasites of Indian food-fishes. I. Two new genera of amphistomes from an Indian fresh-water fish, *Silundia gangetica* Cuv. and Val., *Indian J. Vet. Sci.*, 8, 367, 1938a.

908. **Srivastava, H. D.,** A study of the life history and pathogenecity of *Cotylophoron cotylophorum* (Fischoeder, 1901) Stiles and Goldberger, 1910, of Indian ruminants and a biological control to check the infestation, *Indian J. Vet. Sci.*, 8, 381, 1938b.

909. **Srivastava, H. D. and Tripathi, H. N.,** Amphistomes of Indian ruminants. I. Two species of amphistomes referable to *Palamphistomum* Srivastava et Tripathi (1980) from sheep, goats and buffaloes, *Indian J. Agric. Sci.*, 57, 936, 1987.

910. **Srivastava, M. D. L. and Jha, A. G.,** Structure and behaviour of the chromosomes of *Paramphistomum crassum* Stiles et Goldberger (Trematoda: Digenea), *Proc. Natl. Acad. Sci. India*, 126, 1964.

911. **Srivastava, C. B. and Gosh, R. K.,** Occurrence of *Pseudodiscus hawkesi* (Cobbold, 1875) Sonsino, 1895 (Trematoda: Paramphistomatidae) from Indian rhino, *Newsl. Zool. Surv. India*, 3, 131, 1977.

912. **Stafford, J.,** Trematodes from canadian vertebrates, *Zool. Anz.*, 28, 681, 1905.

913. **Starobogatov, Ya, I.,** Systematics and phylogeney of Planorbidae (Gastropoda: Pulmonata) (in Russian), *Bull. Mosk. Obsch. Ispyt Otd. Biol.*, 63, 37, 1958.

914. **Stepanov, N. A.,** On the study of the embryonic development of *Paramphistomum gotoi* Fukui, 1922 in laboratory conditions (in Russian), *Uchen. Zap. Gorkov. Gos. Ped. Inst.*, 99, 21, 1969.

915. **Stiles, C. W. and Goldberger, J.,** A study of the anatomy of *Watsonius* (n.g.) *watsoni* of man and of nineteen allied species of mammalian trematode worms of the superfamily Paramphistomoidea, *Bull. Gyg. Lab. Public Health Marine Hospt. Serv. U.S.*, 60, 1, 1910.

916. **Stock, M. and Barrett, M. W.,** Helminth parasites of the gastrointestinal tracts and lungs of moose (*Alces alces*) and wapiti (*Cervus elaphus*) from Cypress Hills, Alberta, Canada, *Proc. Helminthol. Soc. Wash.*, 50, 246, 1983.

917. **Stoĭkova, R.,** Prinos k'm helminto faunata na *Rana temporaria* L. v Bulgarija (in Bulgarian), *Izdv. Zool. Inst. S. Muz.*, 33, 195, 1971.

918. **Strong, R. P. and Shattock, G. C.,** XXI. Animal parasitic infections, in: *The African Republic of Liberia and the Belgian Congo Based on the Observations Made and Material Collected during the Harvard African Expedition 1926—1927*, Strong, R. P., Ed., Contr. Dept. Trop. Med. Inst. Trop. Biol. Med. No. 5, Vol. 1, 412, 1930.

919. **Strong, P. A. and Bogitsh, B. J.,** Ultrastructure of the lymph system of the trematode *Megalodiscus temperatus*, *Trans. Am. Microsc. Soc.*, 92, 570, 1973.

920. **Strydonck, Van D.,** Contribution a l'étude de l'anatomie de la morphologie et de la systématique des Paramphistomidae Africains, *Ann. Mus. R. Afr. Cent. Ser. 4*, 183, 1, 1970.

921. **Stunkard, H. W.,** Studies in North American Polystomidae, Aspidogastridae and Paramphistomidae, *Ill. Biol. Monogr.*, 3, 287, 1917.

922. **Stunkard, H. W.,** On the specific identity of *Heronimus chelydrae* MacCallum and *Aorchis extensus* Baker et Parsons, *J. Parasitol*, 6, 11, 1919.

923. **Stunkard, H. W.**, The present status of the amphistome problem, *Parasitology,* 17, 137, 1925.

924. **Stunkard, H. W.**, On the specific identity of *Amphistomum bicaudatum* Poirier and *Cladorchis gigas* MacCallum, *Anat. Rec.,* 34, 165, 1926.

925. **Stunkard, H. W.**, The parasitic worms collected by the American Museum of Natural History expedition to the Belgian Congo, 1909—1914, Trematoda, *Bull. Am. Mus. Nat. Hist.,* 58, 233, 1929.

926. **Stunkard, H. W.**, Morphology and relationships of the trematode *Opisthoporus aspidonectes* (MacCallum, 1917) Fukui, 1929, *Trans. Am. Microsc. Soc.,* 49, 210, 1930.

927. **Stunkard, H. W.**, On the specific identity of the digenetic trematode *Monostomum molle* Leidy, 1856 and *Heronimus chelydrae* MacCallum, 1902, *J. Parasitol.,* 50, 99, 1964.

928. **Subbarao, T., Venkataratnam, A., and Satyanarayanacharyulu, N.**, A note on the occurrence of *Ceylonocotyle scoliocoelium* (Fischoeder, 1904). *Indian Vet. J.,* 46, 633, 1969.

929. **Subramanyam, M. D. L. and Venkat Reddy, P.**, The role of chromosomes in the taxonomy of some digenetic trematodes, *Nucleus (Calcutta),* 20, 128, 1977.

930. **Sultanov, M. A., Sarymsakov, F. S., Muminov, P., and Davlyatov, N.**, Helminth parasites of animals of Kara-Kalpakskoi ASSR (in Russian), in *Parasites of Animals and Man in the Lower Reaches of the Amu Darya,* Izdv. FAN, Tashkent 1969, 3.

931. **Sumwalt, M.**, Trematode infestation of the snakes of San Juan Island, Puget Sound, *Wash. Univ. Stud. Sci.,* 2, 73, 1926.

932. **Sutton, C. A.**, Contribution al conocimiento de la fauna parasitoligica Argentina. II, *Neotropica (Buenos Aires),* 21, 72, 1975.

933. **Swales, W. E.**, A review of Canadian helmintology. I. The present status of knowledge of the helminth parasites of domesticated and semidomesticated mammals and economically important birds in Canada as determined from work published prior to 1933, *Can. J. Res.,* 8, 468, 1933.

934. **Swart, P. J.**, The identity of so-called *Paramphistomum cervi* and *P. explanatum,* two common species of ruminant trematodes in South Africa, *J. S. Afr. Vet. Med. Assoc.,* 26, 463, 1954.

935. **Swart, P. J.**, *Nilocotyle hepaticae* n. sp. from *Hippopotamus amphibius, Onderstepoort, J. Vet. Res.,* 28, 55, 1961.

936. **Swart, P. J.**, A redescription of *Nilocotyle (Nilocotyle) praesphinctris* Näsmark, 1937 (Trematoda: Paramphistomidae) from *Hippopotamus amphibius, Onderstepoort, J. Vet. Res.,* 33, 73, 1966.

937. **Swart, P. J.**, A study of the epidermal structures of the miracidia of *Calicophoron calicophorum* (Fischoeder, 1901) Näsmark, 1937 and *Paramphistomum microbothrium* Fischoeder, 1901, *Onderstepoort J. Vet. Res.,* 34, 129, 1967.

938. **Swart, P. J. and Reinecke, R. K.**, Studies on paramphistomiasis. I. The propagation of *Bulinus tropicus* Krauss, 1848, *Onderstepoort J. Vet. Res.,* 29, 183, 1962.

939. **Szidat, L.**, Parasiten aus Liberia und Französisch-Guinea. II. Trematoden, *Z. Parasitenkd.,* 4, 506, 1932.

940. **Szidat, L.**, Über die Entwicklungsgeschichte und den ersten Zwischenwirt von *Paramphistomum cervi* Zeder, 1790 aus dem Magen Wiederkauern, *Z. Parasitenkd.,* 9, 1, 1937.

941. **Szidat, L.**, Beiträge ezum Aufbau eines natürlichen System der Trematoden. I. Die Entwicklung von *Echinocercaria choanocephala* U. Szidat zu *Cathaemasia hians* und die Ableitung der Fasciolidae von den Echinostomidae, *Z. Parasitenkd.,* 11, 239, 1939.

942. **Szidat, L. and Núñez, O.**, Un trematode del estomago de rumiantes sudamericanos. *Balanorchis anastrophus* como cazador y Predator (Paramphistomidae, Balanorchinae), *Neotropica (Buenos Aires),* 8, 93, 1962.

943. **Szulc, W.**, Pryzwry (Trematoda) plazów Wyżyny lódzkiej (in Polish), *Fragm. Faun. (Warsaw),* 10, 99, 1962.

944. **Takahashi, So.**, The intermediate host of *Diplodiscus subclavatus* in Japan (in Japanese), *Fukuoka Ika Daigaku Zasshi,* 20, 712, 1927a.

945. **Tkahashi, So.**, The life cycle of *Paramphistomum cervi* (Zed.), particularly its intermediate host (in Japanese), *Fukuoka Ika Daigaku Zasshi,* 20, 617, 1927b.

946. **Tandon, R. S.**, On a new amphistome, *Olveria bosi* n.sp. from the rumen of buffalo, *Bos bubalis* from Lucknow, *Indian, J. Helminthol.,* 3, 93, 1951.

947. **Tandon, R. S.**, On a new amphistome *Paramphistomum spinicephalus* n.sp. from the rumen of buffalo, *Bos bubalis,* from Lucknow, *Indian J. Helminthol.,* 7, 35, 1955.

948. **Tandon, R. S.**, Development of the miracidium and its morphology in *Oliveria indica* Thapar et Singha 1945, an amphistome (Trematoda), parasite of cattle and buffalo in India, *Trans. Am. Microsc. Soc.,* 71, 353, 1957a.

949. **Tandon, R. S.**, Life history of *Gastrothylax crumenifer* (Creplin, 1847), *Z. Wiss. Zool.,* 160, 39, 1957b.

950. **Tandon, R. S.**, Studies on the excretory system of amphistomes of ruminants. I. *Carmyerius spatiosus* (Stiles et Goldberger, 1910), *Proc. Natl. Acad. Sci. (India),* 28, 340, 1958a.

951. **Tandon, R. S.**, Development and morphology of the cercaria of an amphistome *Fischoederius elongatus* (Stiles et Goldberger, 1910), recovered from a naturally infected *Lymnaea luteola* at Lucknow, *Zool. Anz.,* 161, 200, 1958b.

952. **Tandon, R. S.,** Studies on the lymphatic system of amphistomes of ruminants. II. The genera *Gastrothylax* and *Fischoeder, Zool. Anz.,* 164, 217, 1960.

953. **Tandon, R. S.,** The amphibian genus of amphistomes, *Pseudochiorchis* Yamaguti, 1958. Recovered from chelonians at Lucknow, and its synonymity with the genus *Kachugotrema* Dwivedi, 1967, *Folia Parasitol. (Prague),* 17, 293, 1970.

954. **Tandon, R. S.,** Studies on "crowding effect" on *Gastrothylax crumenifer* and *Fischoederius elongatus,* the common amphistome parasites of ruminants, observed under natural conditions, *Res. Bull. Meguro Parasitol. Mus.,* 7, 12, 1973.

955. **Tandon, V. and Maitra, S. C.,** Stereoscan observations on the surface topography of *Gastrothylax crumenifer* (Creplin, 1847) Poirier, 1883 and *Paramphistomum epiclitum* Fischoeder, 1904 (Trematoda), *J. Helminthol.,* 55, 231, 1981.

956. **Tandon, V. and Maitra, S. C.,** Scanning electron microscopic observations on the tegumental surfaces of two rumen flukes (Trematoda: Paramphistomata), *J. Helminthol.,* 56, 95, 1982.

957. **Tandon, V. and Maitra, S. C.,** Surface morphology of *Gastrodiscoides hominis* (Lewis et McConnel, 1876) Leiper, 1913 (Trematoda: Digenea) as revealed by scanning electron microscope, *J. Helminthol.,* 57, 339, 1983.

958. **Tandon, V. and Maitra, S. C.,** Scanning electron microscopy of the tegumental surface of some *Orthocoelium* species (Trematoda: Paramphistomata), *J. Helminthol.,* 24, 171, 1987.

959. **Tandon, V. and Sharma, V.,** Amphistome fauna of ruminants in Himachal Pradesh, *Indian J. Parasitol.,* 5, 241, 1981.

960. **Taranko-Tulecka, H.,** Helmintofauna traszki zwyczajnej-*Triturus ulgaris* L. okolic Lublina (in Polish), *Acta Parasitol. Pol.,* 7, 423, 1959.

961. **Tchertkova, A. A.,** A new trematode from the intestine of a chimpanzee, *Skrjabinocladorchis jubilaricum* n.g., n. sp. (in Russian), in *Rab. Gel'mint. 80-let. K.I. Skrjabina,* Izdv. Min. Sel'sk. Hozy. U.S.S.R., Moscow, 1959, 188.

962. **Tendeiro, J.,** Subsidios para o conhecimento da fauna parasitológica da Guiné, *Bol. Cult. Guine Port.,* 3, 638, 1948.

963. **Tenora, F., Kotrlá, B., and Blažek, K.,** Finding of the trematode *Gigantocotyle siamense* (Stiles et Goldberger, 1910) in buffalo in Afghanistan, *Acta Vet. Brno,* 43, 111, 1974.

964. **Thal, J. A.,** Les maladies a la peste bovine, etudes et lutte, Ndélé République Centrafricaine. Euquuěte sur le peste bovine et les maladies similaires, *AGA:SF/CAF 13. Rapp. Techn. 1,* Rome, 1, 1972.

965. **Thapar, G. S.,** Systematic survey of helminth parasites of domesticated animals in India, *Indian J. Vet. Sci.,* 24, 211, 1956.

966. **Thapar, G. S.,** A new genus of amphistomatous parasites from the intestine of a fish, *Cirrhina fulungel,* from India, in *Libro Home, Caballero, E. C.,* 1960, 315.

967. **Thapar, G. S.,** The life history of *Olveria indica,* an amphistome parasite from the rumen of Indian cattle, *J. Helminthol.,* (R. T. Leiper Suppl.), 179, 1961.

968. **Thapar, G. S. and Singha, B. B.,** On the morphology of a new genus of amphistomes from the rumen of cattle in the United Province, *Indian J. Vet. Sci.,* 15, 219, 1945.

969. **Thatcher, V. E.,** Some helminths parasitic in *Clemmys marmorata, J. Parasitol.,* 40, 481, 1954.

970. **Thatcher, V. E.,** Trematodes of turtles from Tabasco, Mexico, with a description of a new species of *Dadaytrema* (Trematoda: Paramphistomidae), *Am. Midl. Nat.,* 70, 347, 1963.

971. **Thatcher, V. E.,** Estudios sobre los trematodos de Tabasco, Mexico: lista de huespedes y sus parasitos, *An. Esc. Nac. Cienc. Biol. Mex.,* 13, 91, 1964.

972. **Thatcher, V. E.,** Paramphistomidae (Trematoda: Digeenea) de peixes de água doce: dois novos genesos de Colômbia e uma redescrição de *Dadaytrema oxycephala* (Diesing, 1836) Travassos, 1934, da Amazonia, *Acta Amazonica,* 9, 203, 1979.

973. **Threlfall, W.,** Parasites of moose (*Alces alces*) in Newfoundland, *J. Mammal.,* 48, 668, 1967.

974. **Threlfall, W.,** A preliminary check list of the helminth parasites of the common snipe, *Capella gallinago* (Linnaeus), *Am. Midl. Nat.,* 84, 13, 1970.

975. **Thurston, J. P.,** Studies on some Protozoa and helminth parasites of *Xenopus,* the African clawed toad, *Rev. Zool. Bot. Afr.,* 82, 349, 1970.

976. **Tidswell, F.,** Parasites, Rep., Bureau of Microbiology, New South Wales, 1910, 79.

977. **Toktouchikova, M. G.,** Helminth parasites of cattle in Kirgizia (in Russian), in *Helminthological Investigations in Kirgiziva,* "JLIM", Frunze, 1975, 63.

978. **Toshchev, A. P.,** Helminth fauna of domestic animals of eastern Siberia (in Russian), *Tr. Irkuts K. Nauchno. Issled. Vet. Opyt. Stants.,* 1, 134, 1949.

979. **Travassos, L.,** Contribuçio para a systematica dos Paramphistomoidea com uma nota sobre o emprego do phenol em helminthologia, *Bras. Med.,* 25, 357, 1921.

980. **Travassos, L.,** Notas helmintológicas, *Bras. Med.,* 36, 256, 1922.

981. **Travassos, L.,** *Catadiscus cohni* nova especie, nova tramatódeo de batrachio, *Sci. Med.,* 4, 278, 1926.

982. **Travassos, L.,** Notas helminthologicas, *Bol. Biol.,* 19, 148, 1931.

983. **Travassos, L.,** Sur un nouveau de poissons de la vallee du flueuve Parahyba, *C. R. Seances Soc. Biol. Paris,* 114, 839, 1933a.

984. **Travassos, L.,** Observations sur *Zygocotyle lunatum* (Dies, 1835) (Trematoda: Paramphistomidae), *C. R. Seances Soc. Biol. Paris,* 114, 598, 1933b.

985. **Travassos, L.,** *Atractis trematophila* n. sp., nematodes parasito do ceco de um trematodeo Paramphistomidea, *Mem. Inst. Oswaldo Cruz,* 28, 267, 1934a.

986. **Travassos, L.,** Synopse dos Paramphistomoidea, *Mem. Inst. Oswaldo Cruz,* 29, 19, 1934b.

987. **Travassos, L.,** Contribuição ao conhecimento dos helmintos dos peixas d'água doce do Brasil I. (Trematoda, Aspidogastridae), *Mem. Inst. Oswaldo Cruz,* 45, 513, 1948.

988. **Travassos, L., Pinto, C., and Muniz, J.,** Excursão scientifica no Estado de Mato Grosso na zona do Pantanal (Margens dos rios S. Lourenço e Cuyabá) realizada em 1922, *Mem. Inst. Oswaldo Cruz,* 20, 249, 1927.

989. **Travassos, L., Artigas, P., and Pereira, C.,** Fauna helmintológica dos peixes de água doce do Brasil, *Arch. Inst. Biol.,* 1, 5, 1928.

990. **Tripathi, H. N. and Srivastava, H. D.,** Three new amphistome cercariae from *Indoplanorbis exustus, Indian J. Parasitol.,* 3, 1980a.

991. **Tripathi, H. N. and Srivastava, H. D.,** Life history of *Palamphistomum lobatum,* an amphistome of Indian sheep, goats and buffaloes, Abstr. D60, 3rd Natl. Congr. Parasitol., Hissar, India, 1980b, 93.

992. **Tripathi, H. N. and Srivastava, H. D.,** Life history of *Palamphistomum dutti,* an amphistome of Indian sheep, goats and buffaloes, Abstr. D61, 3rd Natl. Congr. Parasitol., Hissar, India, 1980c, 94.

993. **Troncy, P. M., Graber, M., and Thal, J.,** Enquêtesur la pathologie de la faune sauvage en Afrique centrale le parasitisme des Suides sauvages Premiers resultats d'enquête, *Rev. Elev. Med. Vet. Pays Trop.,* 25, 205, 1972.

994. **Trueb, L.,** Bones, frogs and evolution, in *Evolutionary Biology of Anurans,* Vial, J. L., Ed., University of Missouri Press, Columbia, 1973, 5.

995. **Tscherner, W.,** Helminthofaunische untersuchungen an *Rana esculenta* L. und *Rana ridibunda* Pall, mit besondere berücksichtigung der Europaischen *Prosotocus*-arten (Trematoda: Lecithodendriidae), *Mitt. Zool. Mus. Berlin,* 42, 59, 1966.

996. **Tubangui, M. A.,** Metazoan parasites of Philippines domesticated animals, *Phillip. J. Sci.,* 28, 11, 1925.

997. **Tubangui, M. A.,** Trematode parasites of Philippine vertebrates. VI. Description of new species and classification, *Philipp. J. Sci.,* 52, 167, 1933.

998. **Tubangui, M. A.,** A summary of the parasitic worms reported from Philippines, *Philipp. J. Sci.,* 76, 225, 1947.

999. **Tubangui, M. A. and Masilungan, V.,** The cercaria of *Diplodiscus amphichrus* Tubangui, 1933 (Trematoda), *Bull. Natl. Res. Counc.,* 23, 178, 1939.

1000. **Tudor,, G. and Anton, E.,** Cercetări privind Parampfistomoza remegătoorelor în judetul Tulcea (in Rumanian), *Rev. Zootech. Med. Vet. Buc.,* 18,, 72, 1968.

1001. **Ueta, M. T., Deberaldini, E. G., Cordeiro, N. S., and Artigas, P. T.,** Ciclo biologico de *Paraibatrema inesperata* n.g., n.sp. (Trematoda: Paramphistomidae), a portir de metacercarias desenvolvidas em *Biomphalaria tenagophila* (D'Orbigny, 1935) (Mollusca, Planorbidae), *Mem. Inst. Oswaldo Cruz,* 76, 15, 1981.

1002. **Ukoli, F. M. A.,** On two amphitome trematodes, *Brevicaecum nilocitum* McClelland, 1957 and *Sandonia sudanensi* McClelland, 1957 of fishes from the river Niger, *Niger. J. Sci.,* 6, 3, 1972.

1003. **Ulmer, M. J.,** *Physa sayii,* a new intermediate host for the turtle lung fluke, *Heronimus chelydrae* (Trematoda: Heronimidae), *J. Parasitol.,* 46, 813, 1960.

1004. **Ulmer, M. J.,** Studies on the helminth fauna of Iowa. I. Trematodes of amphibians, *Am. Midl. Nat.,* 83, 38, 1970.

1005. **Ulmer, M. J. and Sommer, S. C.,** Development of sporocyst of the turtle lung fluke, *Heronimus chelydrae* MacCallum (Trematoda: Heronimidae), *Proc. Iowa Acad. Sci.,* 64, 601, 1957.

1006. **Upadhay, D. S., Swarup, D., and Bhatia, B. B.,** A note on biliary amphistomiasis in goat, *Indian J. Vet. Med.,* 6, 134, 1986.

1007. **Vagin, V. L., Lyubarskaya, O. L., and Sololina, F. M.,** Contribution to the study of parasites of invertebrates in Tatar SSR (in Russian), *Prob. Parazitol., Tr. Nauchn. Konf. Parazitol. U.S.S.R.,* 1, 117, 1972.

1008. **Vaidyanathan, S. N.,** Experimental infestation with *Fischoederius elongatus* in a calf at Madras, *Indian J. Vet. Sci.,* 11, 243, 1941.

1009. **Van der Wooude, A.,** Germ cell cycle of *Megalodiscus temperatus* (Stafford, 1905) Harwood, 1932 (Paramphistomidae: Trematoda), *Am. Midl. Nat.,* 45, 1, 1954.

1010. **Varma, A. K.,** Human and swine *Gastrodiscoides, Indian J. Med. Res.,* 42, 475, 1954.

1011. **Varma, A. K.,** On a collection of paramphistomes from domesticated animals in Bihar, *Indian J . Vet. Sci.,* 27, 67, 1957.

1012. **Varma, A. K.,** Observations on the biology and pathogenecity of *Cotylophoron cotylophorum* (Fischoeder, 1901), *J. Helminthol.,* 35, 161, 1961.

1013. **Vassiliades, G.,** Les affections parasitaires dues á des helminthes chez les bovins du Sénégal, *Rev. Elev. Med. Vet. Pays Trop.,* 31, 157, 1978.

1014. **Vasil'ev, A. A., Velitchko, I. V., Gluzman, I. Yu., Mereminskiĭ, A. I., Nikitin, V. F., and Ovechkin, N. A.,** Efficacy of bithional in acute paramphistomid infections of cattle (in Russian), *Byull. Vses. Inst. Gelmintol.,* 4, 23, 1970.

1015. **Vassilev, I. P. and Samnaliev, P.,** Development of *Paramphistomum microbothrium* Fischoeder, 1901, in intermediate host, *Galba truncatula* (in Bulgarian), *Khelmintologiya,* 6, 13, 1978.

1016. **Vaz, Z.,** *Contribuição ao conhecimento dos trematóides de peixes fluviaes do Brasil,* São Paulo, 1932, 47 pp.

1017. **Vaz, Z.,** Adaptacao ao porco domestico do *Stichorchis giganteus,* parasita de porcos selvagens, Redescricao, *Arch. Inst. Biol. Sao Paulo,* 6, 45, 1935.

1018. **Vegli, F.,** I vermi parassiti negli animali del Sud-Africa, *Ann. R. Acad. Angrar. Torino,* 62, 16, 1919.

1019. **Velázquez-Moldano, J. J.,** Estudo taxônomico dos trematodeos paramphistomi-formes do rúmen de bovinos do Estado do Rio Grande do Sul, Brasil, *Fundação Cargill, Sgo Paulo,* 1976, 1.

1020. **Velitchko, I. V.,** A new trematode — *Paramphistomum petrowi* Davydova, 1961 nov. comb. (Paramphistomatidae) of the sika deer (*Cervus nippon*) from the Primorsk territory, U.S.S.R. (in Russian), *Mater. Nauchn. Konf. Vses. Ova. Gelmintol.,* 3, 60, 1966.

1021. **Velitchko, I. V.,** *Liorchis* n.g. (Trematoda: Paramphistomatidae) (in Russian), *Mater. Nauchn. Konf.,* 5, 70, 1967.

1022. **Velitchko, I. V.,** The distribution of Paramphistomatoidea in ruminants on the territory of the U.S.S.R. (in Russian), *Byull. Vses. Inst. Gelmintol.,* 2, 34, 1968.

1023. **Velitchko, I. V.,** On the paramphistomids of ruminants in the U.S.S.R. (in Russian), in *Sb. Rab. Gelmintol. Posv. 90-let.* K-I, Skrjabina, Izdv. "KOLOS", Moscow, 1971, 61.

1024. **Velitchko, I. V.,** On the taxonomy of the genus *Calicophoron* Näsmark, 1937 (Paramphistomatidae) (in Russian), *Tr. Vses. Inst. Gelmintol,* 20, 55, 1973.

1025. **Velitchko, I. V.,** Species of the Paramphistomata in game of the U.S.S.R. (in Russian), *Byull. Vses. Inst. Gelmintol.,* 16, 24, 1975.

1024. **Venkat Reddy, P. and Subramanyam, S.,** Chromosome number of the paramphistome, *Gigantocotyle explanatum* Näsmark, 1937, *Curr. Sci.,* 44, 400, 1975.

1027. **Venkat Reddy, P. and Subramanyam, S.,** Chromosome studies in *Paramphistomum cervi* Zeder, 1790 (Trematoda - Digenea - Paramphistomatidae), *Caryologia,* 28, 181, 1975b.

1028. **Venkat Reddy, P. and Subramanyam, S.,** Chromosomee studies in the digenetic trematode *Ceylonocotyle dicranocoelium* (Fischoeder, 1901) Näsmark, 1937, *Rev. Roum. Biol. Biol. Anim. (Bucarest),* 21, 49, 1976.

1029. **Venkat Reddy, P. and Subramanyam, S.,** Karyological studies on the pouched amphistome, *Fischoederius elongatus, Cienc. Cult. (Sao Pa lo),* 35, 70, 1983.

1030. **Vercammen-Grandjean, P. H.,** Les trematodes du lac Kivu sud, *Ann. Mus. R. Congo Belge Ser.,* 8vo, 5, 7, 1960.

1031. **Vercruysee, R.,** Un nouveau parasite de l'éléphant d'Afrique *Loxodonta africana: Cotylophoron cotylophorum* (Fischoeder, 1901) Stiles et Goldberger, 1910, *Ann. Soc. R. Zool. Belg.,* 81, 21, 1950.

1032. **Viana, L.,** Tentative de catalogação dos especies brasileiras de trematodeos, *Mem. Inst. Oswaldo Cruz,* 18, 95, 1924.

1033. **Vicente, J. J., Santos, E. D., and Souza, S. V.,** Helmintos de peixes de rios amazonicus da colecao helmintologica do Instituto Oswaldo Cruz. I. Trematoda, *Atas Soc. Biol. Rio de Janeiro,* 19, 9, 1978.

1034. **Vigueras, I. P.,** Notas sôbre algunas especies nuevas de trematodes y sobre otras poco conocidas, *Rev. Univ. Habana,* 28-29, 3, 1940.

1035. **Vigueras, I. P.,** Contribucion al conocimiento de la fauna helmintologica cubana, *Mem. Soc. Cub. Hist. Nat.,* 22, 21, 1955.

1036. **Vogelsang, E. G.,** Parasitos del suino, *Rev. Ganadera (Caracas),* 11, 20, 1935.

1037. **Vogelesang, E. G.,** Contributión al estudio de la parasitología animal en Venezuela. XVI. Ecto y endoparásitos en animals domésticos y salvajes de la Guayana Venezolana, *Rev. Med. Vet. Parasitol.,* 7, 145, 1948.

1038. **Vogelsang, E. G. and Rodriguez, C.,** Ecto y endoparasitos de animales en cautiverio del jardin Zoológico de Maracay, *Rev. Med. Vet. Parasitol. Caracas,* 11, 311, 1952.

1039. **Vojtkova, L.,** Závislost trematodofauny žab na jejich životním prostředi (in Czech), *Scr. Fac. Sci. Nat. Univ. Purkynianae Brun.,* 2, 13, 1972.

1040. **Vojtkova, L.,** Motolice obojzivelnikii CSSR. I. Dospele motolice (in Czech), *Folia Fac. Sci. Nat. Univ. Purkynianae Brun.,* 15, Biol. 45, 1, 1974.

1041. **Volna-Nábělkova, L.,** Příspěvek k poznání helmintofauny žab, (in Czech), *Publ. Fac. Sci. Univ. Purkyne Brno,* 57, 1964.

1042. **Vujić, B.,** Paramphistomosa prezivara i principi odredivanja pripadnosti ovik parazita (in Serbo-Croat), *Veterinaria (Sarajevo),* 14, 471, 1965.

467

1043. **Vujić, B., Petrović, K., Palić, D., and Perić, Ž.**, Jedan slučaj akutne paramfistomoze goveda (in Serbo-Croat), *Vet. Glas.*, 25, 265, 1971.

1044. **Walton, A. C.**, Some parasites and their chromosomes, *J. Parasitol.*, 45, 1, 1959.

1045. **Wandera, J. G.**, Reflection on sheep diseases in Kenya, *Bull. Epizoot. Dis. Afr.*, 17, 121, 1969.

1046. **Wang, P. Q.**, Studies on amphistomes (Trematoda: Paramphistomata (Szidat, 1936)) of Fujian Province (in Chinese), *Acta Univ. Fujian (Zool.)*, 1, 237, 1959.

1047. **Wang, X. Y.**, Examinations of flukes in the People's Republic of China. I. Description of new species (in Chinese), *Acta Parasitol. Sinica*, 3, 205, 1966.

1048. **Wang, P. Q.**, Some digenetic trematodes of vertebrates mainly from Fujian province, China (in Chinese), *Wuyi Sci. J.*, 3, 40, 1983.

1049. **Wang, P. Q.**, On two new flukes and on the examinations of life histories of Chinese flukes (in Chinese), *Acta Univ. Fujian*, 2, 62, 1977.

1050. **Wang, X. Y.**, Systematic studies on amphistome trematodes from China. II. Paramphistomidae: Paramphistominae and Gastrothylacidae, with notes on some new species (in Chinese), *Acta Zootax. Sinica*, 4, 327, 1979.

1051. **Ward, H. B.**, On the structure and classification of North American parasitic worms, *J. Parasitol.*, 4, 1, 1917.

1052. **Westhuysen, O. P.**, A monograph on the helminth parasites of the elephant, *Onderstepoort, J. Vet. Sci.*, 10, 49, 1938.

1053. **Whitten, L. K.**, Paramphistomiasis in sheep, *N. J. Vet. J.*, 3, 144, 1955.

1054. **Wiedmann, F. D.**, A contribution to the anatomy and embryology of *Cladorchis (Stichorchis) subtriquetrus* Rud. 1814 (Fischoeder, 1901), *Parasitology*, 10, 267, 1917.

1055. **Wieczorowski, S.**, Paramphstomoza bydla. I. Ekstensywność inwazji na terenie woj. bialostockiego wedlug danych opracowanych przez Z H W w Bialymstoku (in Polish), *Med. Wet.*, 27,, 146, 1971.

1056. **Wikerhauser, T., Brgleez, J., and Kutičic, V.**, O anthelmintickom djelovanju Terenol na *Paramphistomum microbothrium* u goveda (in Serbo-Croat), *Acta Parasitol. Jug.*, 6, 25, 1975.

1057. **Willey, C. H.**, Studies on the morphology and systematic position of the trematode *Protocladorchis pangasii* n.g. *(Cladorchis pangasii* MacCallum, 1905), *Trans. Am. Microsc. Soc.*, 54, 8, 1935.

1058. **Willey, C. H.**, The morphology of the amphistome cercaria *C. poconensis* Willey, 1930, from the snail, *Helisoma antrosa*, *J. Parasitol.*, 22, 68, 1936.

1059. **Willey, C. H.**, The life history and bionomics of the trematode, *Zygocotyle lunata* (Paramphistomidae), *Zool. Sci. Contrib. N. Y. Zool. Soc.*, 26, 65, 1941.

1060. **Willey, C. H.**, The relation of lymph and excretory systems in *Zygocotyle lunatum*, *Anat. Rec.*, 120, 68, 1954.

1061. **Willey, C. H. and Godman, G. C.**, Gametogenesis, fertilization and cleavage in trematode, *Zygocotyle lunata* (Paramphistomidae), *J. Parasitol.*, 37, 283, 1951.

1062. **Williams, R. W.**, Helminths of snapping turtle, *Chelydra serpentina* from Oklahoma, including the first report and description of the male *Capillaria serpentina* Harwood, 1932, *Trans. Am. Microsc. Soc.*, 72, 175, 1953.

1063. **Willmott, S.**, On the species of *Paramphistomum* Fischoeder, 1901, occurring in Britain and Ireland with notes on some material from the Netherlands and France, *J. Helminthol.*, 24, 155, 1950a.

1064. **Willmott, S.**, Gametogenesis and early development in *Gigantocotyle bathycotyle* (Fischoeder, 1901) Näsmark, 1937, *J. Helminthol.*, 24, 1, 1950b.

1065. **Willmott, S.**, The development and morphology of the miracidium of *Paramphistomum hiberniae* Willmott, 1950, *J. Helminthol.*, 26, 123, 1952.

1066. **Willmott, S.**, The morphology of *Brumptia bicaudata* (Poirier, 1908) Odhner, 1926 (Trematoda: Paramphistomoidea), *Proc. Zool. Soc. London*, 134, 623, 1960.

1067. **Willmott, S. and Pester, F. R. N.**, The discovery of *Paramphistomum hiberniae* Willmott, 1950 and its intermediate host in the Channel Islands, *J. Helminthol.*, 29, 1, 1955.

1068. **Willmott, S. and Pester, F. R. N.**, A study of the Gastrodiscinae Monticelli, 1892, (Gastrodiscidae: Paramphistomoidea), *J. Helminthol.*, (R. T. Leiper Suppl.), 205, 1961.

1069. **Wotton, R. M. and Sogandares-Bernal, F.**, A report on the occurrence of microvillus-like structures in the caeca of certain trematodes (Paramphistomatidae), *Parasitology*, 53, 157, 1963.

1070. **Wu, S. C., Yen, W. C., and Shen, S. S.**, A survey of the helminths of the domestic animals in Southwestern China, *Acta Zool. Sinica*, 17, 373, 1965.

1071. **Yamaguti, S.**, Research notes — *Fischoederius elongatus* (Poirrier, 1883) and *F. siamensis* Stiles et Goldberger, 1910, *J. Parasitol.*, 21, 416, 1935.

1072. **Yamaguti, S.**, Studies on the helminth fauna of Japan. XXVII. Trematodes of mammals. II, *Jpn. J. Med. Sci.*, 1, 131, 1939.

1073. **Yamaguti, S.**, Zur Entwicklungsgeschichte von *Diplodiscus amphichrus japonicus* Yamaguti, 1936, *Z. Parasitenkd.*, 11, 652, 1940.

1074. **Yamaguti, S.**, *Ceylonocotyle scoliocoelium* (Fischoeder, 1904) Näsmark, 1937 (Trematoda: Paramphistomidae) from Japanese goat, in G. S. Thapar Commemorational Volume, 1953, 301.

1075. **Yamaguti, S.**, Parasitic worms mainly from Celebes. V. Trematodes of mammals, *Acta Med. Okayama*, 8, 341, 1954.

1076. **Yamaguti, S.**, *Systema Helminthum. Vol. 1. Digenetic Trematodes of Vertebrates*, (Parts 1 and 2) Interscience, New York, 1958, 353.

1077. **Yamaguti, S.**, *Synopsis of Digenetic Trematodes of Vertebrates*, Vols. 1 and 2, Keigaku, Tokyo, 1971, 81.

1078. **Yamaguti, S.**, *A Synoptical Review of Life Histories of Digenetic Trematodes of Vertebrates with Special Reference to the Morphology of their Larval Forms*, Keigaku, Tokyo, 1975, 85.

1079. **Yeh, S. L.**, On a new paramphistome trematode *Gigantocotyle leuroxi* sp. n. from the stomach of the red leche, *Onotragus leche* from Northern Rhodesia, *Parasitology*, 47, 432, 1957.

1080. **Yuen, P. H.**, Three trematodes from Malayan amphibians including two new species, *J. Parasitol.*, 48, 532, 1962.

1081. **Yusuf, I. A. and Chaudhry, M. A.**, Species of paramphistomids (Trematoda) from buffaloes in Peshawar region of West Pakistan, *Trop. Anim. Health Prod.*, 2, 235, 1970.

1082. **Zablotskii, V. I.**, Early development of *Gastrodiscoides hominis* Lewis et McConnel, 1876 (in Russian), *Sb. Parasitol. Rab. Tr. Astrakh. Zapovedn.*, 9, 119, 1964.

1083. **Zablotskii, V. I.**, Biology of the trematode *Gastrodiscoides hominis* (Paramphistomatata, Gastrodiscidae) (in Russian), *All Union Konf. (Alma Ata)*, 1969, 28.

1084. **Zablotskii, V. I.**, The helminth fauna of the racoon-dog and of muskrat acclimatized in the Volga Delta (in Russian), *Tr. Astrakh. Zapovedn.*, 13, 364, 1970.

1085. **Zadura, J.**, *Paramphistomum cervi* (Schrank, 1790) as the cause of a serious disease in stag (*Cervus elaphus* L.), *Acta Parasitol. Polon.*, 8, 345, 1960.

1086. **Zadura, I. and Niec, L.**, Robaczyca Zoladki u krowy wywolana przez przywry *Paramphistomum cervi* (in Polish), *Med. Wet.*, 8, 370, 1952.

1087. **Zahedi, M., Vellayan, S., Krishnasamy, M., and Jeffery, J.**, *Paramphistomum epiclitum* Fischoeder, 1904 (Paramphistomidae: Paramphistominae) in a Malaysian *Bos gaurus* Hubbachi: a new host record, *Malay. Vet. J.*, 7, 224, 1983.

1088. **Zajiček, D.**, Development of miracidia of *Cotylophoron cotylophorum* (Fischoeder, 1901) under condition of Cuba, *Acta Vet. Brno*, 43, 251, 1974.

1089. **Zajiček, D. and Valenta, Z.**, Příspěvek k poznáni střevnich helmintu dovážených opic rodu *Macacus* jejich pategonity a vztahu k helmintum člověcka (in Czech), *Vet. Med. Prague*, 31, 397, 1958.

1090. **Zdun, V. I.**, On the infection of molluscs (family: Planorbidae) with larval forms of *Paramphistomum cervi* Zeder, 1790 and with other digeneans from the water bodies of Ukraine (in Russian) in *Rab. Gel'mint. k 80-let. Akad. K.I. Skrjabina, Izdv. Min. Sel's. Hozya. SSR.*, Moscow, 1958, 135.

1091. **Zdun, V. I.**, *Larval trematodes of freshwater mulluscs in Ukraina* (in Ukrainian), Markevich, O. P., Ed., Izd. Akad. Nauk. Ukrains SSR., Kiev, 1961, 13.

1092. **Zdzitowiecki, K., Dŕózdż, J., Malczewski, E., and Zarnowski, E.**, Trematodes from the genus *Paramphistomum* Fischoeder, 1901 in ruminants of Poland, *Bull. Acad. Pol. Sci. Cl. II. Ser. Sci. Biol.*, 25, 537, 1977.

1093. **Zgardin, E. S. and Frukhtman, E. A.**, Helminth fauna of cattle and sheep in Moldavia (in Russian), *Nat. Nauchn. Konf. Vses. Ova. Gelmintol.*, 1, 95, 1965.

1094. **Zhaltsanova, D. D.**, On the species of paramphistomids in Buryatia (in Russian), *Mater. Nauchn. Konf. Vses. Ova. Gelmintol.*, 1, 88, 1969a.

1095. **Zhaltsanova, D. D.**, The influence of winter conditions on the viability of paramphistomid eggs in Buryatia (in Russian), *Byull. Vses. Inst. Gelmintol.*, 3, 33, 1969b.

1096. **Zhaltsanova, D. D.**, Occurrence of *Fischoederius elongatus* in the Buryat ASSR. (in Russian), *Mater. Nauchn. Issled. Chlen. Vses. Ova. Gelmintol.*, 24, 49, 1972.

1097. **Zhaltsanova, D. D.**, Experimental study of the life-cycle of *Liorchis scotiae* (Willmott, 1950) Velitchko, 1966 (in Russian), *Tr. Buryat. Inst. Estest. Nauk Buryat. Fil. Sib. Otd. Akad. Nauk SSR*, 15, 11, 1977.

1098. **Zharikov, I. S.**, Biological basis of devastation of trematodosis of ruminants (in Russian), *Izdv. UROZHAI, Minsk*, 1973.

1099. **Zhu, H., Tong, Y. Y., and Chen, S. M.**, Notes on the digenetic trematodes in some wild animals from Guangdong province, China, with description of a new species (in Chinese), *Acta Zootax. Sinica*, 7, 15, 1982.

1100. **Zwicker, G. M. and Carlton, W. W.**, Fluke *(Gastrodiscoides hominis)* infection in a rhesus monkey, *J. Am. Vet. Med. Assoc.*, 161, 702, 1972.

INDEX

Milton Keynes UK
Ingram Content Group UK Ltd.
UKHW051930141024
449569UK00027B/1423